W9-BQW-092

Competition and Succession in Pastures

Contents

Contributors   vii

Preface   ix

1   Competition and Succession in Pastures – Some Concepts and Questions   1
     P.G. Tow and A. Lazenby

2   Measurement of Competition and Competition Effects in Pastures   15
     N.R. Sackville Hamilton

3   Genotype and Environmental Adaptation as Regulators of Competitiveness   43
     IG.M.O. Nurjaya and P.G. Tow

4   Competition between Grasses and Legumes in Established Pastures   63
     A. Davies

5   Plant Competition in Pastures – Implications for Management   85
     D.R. Kemp and W.McG. King

6   Diversity and Stability in Humid Temperate Pastures   103
     E.A. Clark

7   The Population Dynamics of Pastures, with Particular Reference to Southern Australia   119
     E.C. Wolfe and B.S. Dear

8   Formulation of Pasture Seed Mixtures with Reference to Competition and Succession in Pastures   149
     W. Harris

9   Effects of Large Herbivores on Competition and Succession in Natural Savannah Rangelands   175
     C. Skarpe

10   Competition and Environmental Stress in Temperate Grasslands   193
     D.A. Peltzer and S.D. Wilson

11   Interaction of Competition and Management in Regulating Composition and Sustainability of Native Pasture   213
     D.L. Garden and T.P. Bolger

12   Global Climate Change Effects on Competition and Succession in Pastures   233
     B.D. Campbell and D.Y. Hunt

13   Competition and Succession in Re-created Botanically Diverse Grassland Communities   261
     R. Chapman

14   Implications of Competition between Plant Species for the Sustainability and Profitability of a Virtual Farm Using a Pasture–Wheat Rotation   283
     B.R. Trenbath

15   Some Concluding Comments   305
     A. Lazenby and P.G. Tow

Index   315
  Botanical Composition   315
  Competition   316
  Grass–Legume Relationships   317
  Plant Species   318
  Succession   320

Contributors

T.P. **Bolger**, CSIRO Plant Industry, GPO Box 1600, Canberra, ACT 2601, Australia

B.D. **Campbell**, AgResearch, Grasslands Research Centre, Private Bag 11008, Palmerston North, New Zealand

R. **Chapman**, c/o Orchard House, Carlton Scroop, Grantham, Lincolnshire NG32 3AS, UK

E.A. **Clark**, Department of Plant Agriculture, University of Guelph, Guelph, Ontario N1G 2W1, Canada

A. **Davies**, Institute of Grassland and Environmental Research, Plas Gogerddan, Aberystwyth, Ceredigion SY23 3EB, UK

B.S. **Dear**, NSW Agriculture, Wagga Wagga, NSW 2650, Australia

D.L. **Garden**, NSW Agriculture, GPO Box 1600, Canberra, ACT 2601, Australia

W. **Harris**, Lincoln Botanical, 27A Edward St, Lincoln, Canterbury, New Zealand

D.Y. **Hunt**, AgResearch, Grasslands Research Centre, Private Bag 11008, Palmerston North, New Zealand

D.R. **Kemp**, CRC for Weed Management Systems and Pasture Development Group, University of Sydney, Faculty of Rural Management, Orange, NSW 2800, Australia

W.McG. **King**, CRC for Weed Management Systems and Pasture Development Group, NSW Agriculture, Orange Agricultural Institute, Orange, NSW 2800, Australia

A. **Lazenby**, 63 Kitchener Street, Hughes, ACT 2605, Australia

IG.M.O. **Nurjaya**, Department of Biology, Faculty of Mathematical and Natural Sciences, Udayana University, Kampus Bukit Jimbaran, Denpasar, Bali, Indonesia

D.A. **Peltzer**, Landcare Research, PO Box 69, Lincoln 8152, New Zealand

N.R. **Sackville Hamilton**, Institute of Grassland and Environmental Research, Plas Gogerddan, Aberystwyth, Ceredigion SY23 3EB, UK

C. **Skarpe**, Norwegian Institute for Nature Research, Tungasletta 2, N-7485 Trondheim, Norway

P.G. **Tow**, Department of Agronomy and Farming Systems, University of Adelaide, Roseworthy Campus, Roseworthy, SA 5371, Australia

B.R. **Trenbath**, Centre for Legumes in Mediterranean Agriculture, University of Western Australia, Nedlands, WA 6907, Australia

S.D. **Wilson**, Department of Biology, University of Regina, Regina, Saskatchewan S4S 0A2, Canada

E.C. **Wolfe**, School of Agriculture, Charles Sturt University, Locked Bag 588, Wagga Wagga, NSW 2678, Australia

Preface

———————————

This book has its origins in the desire to reassess research and understanding on competition and succession in pastures. It comes in response to a need for better utilization, conservation and, in some cases, rehabilitation of the world's grasslands. Our goals for the book are:

1. To review current and past research and its applications.
2. To provide for agronomists, ecologists and others an understanding of how competition and succession operate in natural and sown pastures.
3. To show the outcomes and practical significance of competition and succession in situations of agricultural, pastoral and ecological importance.
4. To show how competition is influenced by and interacts with environmental and biological factors in grasslands and pastures.
5. To show how competition and succession may be regulated to optimize botanical composition, productivity and persistence.
6. To examine critically the ways in which competition and succession are analysed, evaluated and measured.

The topics covered are complementary and relevant to the theme of the book, while authors have been selected for their expert knowledge of these topics, gained in different parts of the world. Some contributors focus on more fundamental aspects of ecology and some on practical agronomic objectives, while others combine something of each.

The definition and principles of plant competition receive attention throughout the book. Many authors consider them in relation to the aims of their chapters, thus adding to the overall understanding. While competition may be seen as a concept as simple as one plant gaining potentially limiting resources for growth at the expense of another, plants in a community may affect each other, beneficially as well as adversely, in many complex and diverse ways. Furthermore, the use to which such resources are put in the plant may also affect its competitiveness. What is observed or measured as the outcome of competition is often the net outcome of the integration of all relevant influences and effects. In some chapters of this book, competition is discussed in terms of this broad definition. Other chapters provide more understanding of the nature, assessment and measurement of competition.

A number of authors focus on the role of competition in succession – changes with time in the botanical structure of pasture under the influence of a range of biotic and abiotic factors. In considering these changes, it is often difficult to distinguish between the effects of competition and those of other influences, such as environmental factors and management. Nevertheless, the provision of this information, within a framework of clear ecological objectives, facilitates the formulation and testing of hypotheses for understanding the processes and principles of competition and succession in pastures. It is the aim of this book to increase such understanding and thus lead to further progress in both research and management.

There are 15 chapters in this book. The first covers a number of concepts and questions basic to under-standing competition and succession in pastures and using such knowledge to develop optimum systems of management. Chapter 2 seeks to clarify the concept of competition, highlights some of the pitfalls in its study and describes and analyses the use of a range of experiments for answering specific questions which bear on competition. The roles of genotype and environmental adaptation in regulating competition are considered at some length in Chapter 3, together with the association of a number of plant characteristics with competitiveness. Chapter 4 is focused on grass–legume relationships in established temperate pas-tures, and indicates how a number of environmental and management factors can influence the balance of grasses and clovers present. It is one of several contributions (the others are Chapters 5, 7, 9, 11 and 13) where the effect of the grazing animal is considered in depth.

In Chapter 5, the implications of competition and succession for the management of pastures in south-eastern Australia are evaluated, and a method is proposed for monitoring changes in pasture composition. A strong case for associating diversity and stability in humid, temperate pastures is presented in Chapter 6, together with an analysis of deficiencies in the present practice of compiling seeds mixtures and evaluating cultivars. Chapter 7 covers the population dynamics of southern Australian pastures, relating competition and succession to agricultural development since European settlement, and considers why some plants are successful and others fail in temperate Australia. Principles for formulating seed mixtures are presented in Chapter 8, together with a critique of the relevance of competition and succession in selecting appropriate species and cultivars to provide yield, quality and stability in pastures. Chapter 9 examines how large herbi-vores influence competition between plants in African savanna rangelands, driving or changing succession and influencing the structure and function of the ecosystem. An analysis of competition in natural pastures of north America, the focus of Chapter 10, includes a consideration of the association of traits of stress-tol-erant species and competitive ability in stressful environments.

Chapter 11 covers the interaction of competition and management in regulating the composition of native pastures in south-eastern Australia and indicates the importance of perennial grasses in grassland sustainability. Potential effects of global climate change on competition and succession in pastures are examined in Chapter 12, with short-term responses in plant development shown to be better understood than long-term effects on resources and disturbance. Re-establishing botanically diverse grasslands is advo-cated in Chapter 13, with an outline of difficulties of achieving this objective in high-fertility conditions and on land disturbed by mining or engineering. Chapter 14 describes a whole-farm model used to explore the biological and economic effects of various methods of managing competition between annual ryegrass and subterranean clover in the pasture phase. A summation and critical commentary on some important findings presented in the preceding 14 contributions are attempted in Chapter 15, indicating progress in the better understanding of competition and succession in pastures, and using this knowledge for optimal management of grassland.

It is a pleasure to record grateful thanks for the help given in a number of ways in the preparation of this book, by various people. These include the Head, Professor David Coventry, and staff of the Department of Agronomy and Farming Systems, the University of Adelaide, Mrs Ann Lazenby and Mrs Margaret Tow.

Philip G. Tow
Alec Lazenby
January 2001

1 Competition and Succession in Pastures – Some Concepts and Questions

Philip G. Tow[1] and Alec Lazenby[2]

[1]Department of Agronomy and Farming Systems, University of Adelaide, Roseworthy Campus, Roseworthy, Australia; [2]63 Kitchener Street, Hughes, Australia

Background

Extensive observations made during the 19th and early 20th centuries on the behaviour of plants growing together in communities led to the development of a number of principles on competition and successional changes (Clements *et al.*, 1929). Sustained interest in interplant relations over the past half-century has resulted in a large volume of research and many theories on the subject. Publications such as Society for Experimental Biology (1961), Harper (1977), Wilson (1978), Grace and Tilman (1990), Begon *et al.* (1996) and Radosevich *et al.* (1997) demonstrate both the progress made in understanding competition and succession and the inadequacies of such understanding, which is required for the optimum management and preservation of the world's plant communities.

This book is focused on competition and succession among plants in pastures; the term 'pasture' is defined as vegetation used for grazing by domestic or wild animals (Fig. 1.1a, b) and cutting by humans for fodder conservation (Fig. 1.2). Grasses are a universal component of such vegetation, which is thus often termed 'grassland'. Legumes and other herbs are other common pasture components (Fig. 1.3a, b, c). Shrubs and trees may also coexist with grasses provided they are spaced widely enough to prevent crowding out of smaller plants (Figs 1.1b and 1.4).

Naturally occurring grasslands (Fig. 1.5a, b), once occupying vast areas of the world, have now been drastically reduced, largely because of cultivation for cropping. The areas remaining generally owe their continuing existence to climatic, edaphic or topographic limitations to cropping or to their being set aside as conservation areas (Fig. 1.6a, b, c). Species originally characteristic of natural grasslands were well adapted to their environment. However, overgrazing and deterioration in soil chemical and physical attributes have resulted in the degradation of many such grasslands, and inferior plants, often regarded as weeds, have replaced some of the original components. Some degraded natural pastures are now being restored to conserve biodiversity and increase the attractiveness of the landscape (see Chapman, Chapter 13, this volume). Effective rehabilitation and appropriate management of such grasslands are each dependent on a proper understanding of the principles involved in plant competition and succession.

Where climate, soil, landscape and financial incentives are favourable, large areas of improved pastures have been sown in many countries. Cultivars selected and bred for high levels of productivity, persistence and feed value are sown in the most intensive farming systems as single species (Fig. 1.7) or as mixtures of grasses or of grasses and legumes (Fig. 1.3a, b). The principles of competition and succession are directly relevant to the challenge of maintaining desirable pasture composition, free of weeds, while achieving high levels of both productivity and utilization. These objectives become even more difficult to achieve when the

(a)

(b)

Fig. 1.1. (a) Sheep grazing in winter on a pasture of annual medic (*Medicago truncatula*) in the cereal–pasture zone of South Australia, Australia. (b) A mix of grasses, shrubs and trees grazed by native animals in the Masai Mara National Reserve in Kenya.

Fig. 1.2. A mixture of oats (*Avena sativa*) and common vetch (*Vicia sativa*) grown for silage at the Roseworthy Campus of the University of Adelaide, South Australia, Australia. The oats are a strong competitor against vetch, but the legume can compensate by climbing upwards with long, twining stems and tendrils.

origin of the sown species is outside the country of use – a fact which may add complexity to management, to compensate for imperfect climatic and edaphic adaptation. Conversely, similar complexity arises from climatic change (see Campbell and Hunt, Chapter 12, this volume) or from the effects of change in land use on soil conditions (see Chapman, Chapter 13, this volume). Because of the many environmental and technological changes currently affecting grassland and other plant communities, an understanding of interplant and plant–environment relationships is important for interpreting, predicting and managing change in species composition.

The term 'composition' may be used in various ways. For example, the term 'botanical or species composition' can refer to: the presence of particular species; a list of species present; or the proportion (%) of various species in terms of plant numbers, tiller or stem numbers, dry matter (DM) yield, leaf area or ground cover. Botanical composition is influenced by competitive relationships and may be an indicator of a stage in succession.

The functioning and effects of competition and succession in pastures differ from those in other natural and man-made plant communities, such as forests and annual crops. The most important of such differences are caused by the grazing animal. Grazing and trampling reduce the height and sometimes the density of the canopy. This reduces competition for light between shoots and may indirectly reduce the intensity of root competition if part of the root system reacts to defoliation by growing more slowly or dying. Defoliation by grazing and cutting reduces the competitive advantage gained by plants which emerge earlier, have a larger embryo to begin growth and have favourable attributes such as higher initial relative growth rates, tillering rates, leaf expansion, root spread and stature (Milthorpe, 1961).

Grazing of grasses encourages renewed growth of existing tillers and development of new ones. In mixtures, the timing and intensity of grazing affect competitive relationships and resulting proportions of the community components (Milthorpe, 1961). Severe defoliation and trampling may kill existing plants and leave gaps for invasion by others or result in soil erosion, perhaps on a large scale (see Davies, Chapter 4; Kemp and King, Chapter 5; Wolfe and Dear, Chapter 7; Skarpe, Chapter 9; Peltzer and Wilson, Chapter 10; Garden and Bolger, Chapter 11; and Chapman, Chapter 13, this volume).

(a)

(b)

(c)

Fig. 1.3. (a) A well-balanced mixture of perennial ryegrass (*Lolium perenne*) and white clover (*Trifolium repens*) in its third year, near Lutterworth, Leicestershire, England. (b) A 4-year-old unirrigated pasture, composed principally of the perennial grass cocksfoot (*Dactylis glomerata*) and the annual legume subterranean clover (*Trifolium subterraneum*), near Mt Barker in the Adelaide Hills region of South Australia. Each year, following opening autumn or winter rain, the legume must re-establish from seed in competition with the established grass. (c) A balanced, irrigated pasture mixture of Grasslands Puna chicory (*Cichorium intybus*), WL516 lucerne (*Medicago sativa*) and Maru phalaris (*Phalaris aquatica*), in late winter, at Dalby Agricultural College on the Darling Downs of southern Queensland, Australia. Weeds have been excluded by the dense, diverse mixture of sown species.

Fig. 1.4. Vigorous spring growth of native grasses following winter burning, in open *Eucalyptus* woodland in the Undara Volcanic National Park in far north Queensland, Australia.

Competition – Definitions and Concepts

Definitions of plant competition and succession are based largely on observation and experience or measurements of effects on plants, rather than on an understanding of mechanisms (Tilman, 1990). The following definition of competition proposed by Clements *et al.* (1929) is still accepted by many:

> Competition is a purely physical process. With few exceptions, such as the crowding up of tuberous plants when grown too close, an actual struggle between competing plants never occurs. Competition arises from the reaction of one plant upon the physical factors about it and the effect of these modified factors upon its competitors. When the immediate supply of a single, necessary factor falls below the combined demands of the plants, competition begins.

In Milthorpe's (1961) words, the term competition describes 'those events leading to the retardation in growth of a plant which arise from association with other plants. It results from the modification by adjacent individuals of the *local* environment of each particular individual'. This general definition could include such effects as allelopathy (the adverse effects on one plant of a toxic substance derived from another).

The following 'working definition', proposed by Begon *et al.* (1996), is applicable to pastures: 'Competition is an interaction between individuals, brought about by a shared requirement for a resource in limited supply and leading to a reduction

(a)

(a)

(b)

(b)

Fig. 1.5. (a) A dense, vigorous stand of native
bluegrass (*Dichanthium sericeum*) in central
Queensland, Australia. The originally vast areas of
such native grasses have been greatly reduced by
overgrazing and/or cultivation for cropping. (b) A
remnant of original prairie preserved in Texas, USA.

(c)

in the survivorship, growth and/or reproduction of
at least some of the competing individuals concerned.' However, the statement that competition is
an interaction between individuals is not a definition
in itself; competition is defined by the cause (limited
resources) and the net effect (yield reduction) of
competition. This illustrates the difficulty of providing a precise definition of competition without a
clear understanding of the mechanisms involved (see
Sackville Hamilton, Chapter 2, this volume).

Goldberg (1990) has proposed a simple, mechanistic framework for studying interactions
between plants, based on her observation that
'most interactions between individual plants actually occur through some intermediary'. In the case
of competition for resources (e.g. plant nutrients
and water) one or more of the competitors will
have an effect on the abundance of the resources
(intermediaries) and they will also respond to

Fig. 1.6. (a) Arid rangeland in Wadi Rum in the
south of the Kingdom of Jordan. Such areas are
grazed periodically by the migratory sheep, goats
and camels of the Bedouin people. (b) Rangeland
on uncultivated slopes near Digne, Alpes-de-Haute-
Provence, France. (c) Grassland in the Masai Mara
National Reserve, Kenya, grazed periodically by
migratory herds of wild animals and domesticated
native cattle.

changes in abundance of the resources. Plants can
be good competitors either by rapidly pre-empting
and depleting a resource (by uptake) or by being
able to continue growth at depleted resource levels.

Fig. 1.7. A pure stand of the annual legume barrel medic (*Medicago truncatula*) at the Roseworthy Campus of the University of Adelaide, South Australia. Such stands are kept as free of weeds as possible, to maximize nitrogen input for livestock feed and following crops, as well as for annual seed production.

The latter case (classed as low net competitive response to competitors) may occur through continued uptake at low levels of resource, decreased resource loss from plant parts or increased efficiency of conversion of internal stores of the resource to new growth (see also Peltzer and Wilson, Chapter 10, this volume).

The relative importance of above- and below-ground competition has often been questioned. Milthorpe (1961) concluded from various experiments and observations on crops and pastures that 'competition between roots usually commences long before the shoots are sufficiently developed to cause mutual shading'. Further, following an analysis of 23 competition studies, Wilson (1988) concluded that root competition is usually more important than shoot competition in determining competitive balance between species. However, species vary in their response to root competition, as found by Bolger (1998). He conducted an experiment to compare the capacity of seedlings of a number of southern Australian pasture plants to 'invade' an established sward of phalaris, a perennial grass. Experimental treatments comprised varying degrees of shoot and root competition from phalaris and varying levels of plant nutrient supply. The species differed greatly both in the ability of their recruiting (intersown) seedlings to compete with established phalaris root systems and in their relative response to root and shoot competition (see also Nurjaya and Tow, Chapter 3, this volume).

The depletion of light resources has sometimes been measured in mixed-plant canopies (Stern and Donald, 1962; Rhodes and Stern, 1978), but depletion of soil water and nutrients by components of mixtures is much more difficult to quantify. Yet Tilman (1982, 1988, 1994) has based his definition of competitive ability on the theory that, over a number of years, the winning competitor is the species (among those initially present) which is able to reduce the concentration of the limiting soil resources (e.g. available N) to the lowest level and still maintain its population, i.e. it is the one with the lowest resource requirement or R^*. This mechanism of competition has been called the resource reduction model. The R^* values for a group of species, if known, would predict the final (equilibrium) outcome of competition among these species for a limiting resource; it should be independent of the timing of establishment of competing species, their starting proportions and the initial sizes of individual plants. However, the experimental work supporting this definition has been done in small, ungrazed plots. The resource reduction model may therefore be more applicable to lightly grazed, low-input grasslands, rather than to intensively grazed pastures receiving regular, high inputs of nutrients (see also Kemp and King, Chapter 5; and Peltzer and Wilson, Chapter 10, this volume).

An alternative view of competition to that of Tilman is that a plant will be competitively superior if it has the capacity to capture (pre-empt) resources faster than others. This can be related to particular plant traits, such as high potential relative growth rate (Grime, 1979). Plants that can tolerate low levels of resource (e.g. plant nutrient) availability are classed as stress tolerators. It has been hypothesized that differences among competing species in resource acquisition rates, once established, are maintained and magnified during competition because of positive feedback between growth and resource capture (Harper, 1977; Grime, 1979; Keddy, 1990; Begon *et al.*, 1996). This proposed mechanism of resource competition has been called resource pre-emption or asymmetric competition. It occurs, for example, when large plants intercept a disproportionate share of light, while small plants have very little effect on the light reaching the larger plants. In comparing the models of Tilman and Grime, Wedin and Tilman (1993) explain that, while both of the above mechanisms of competition allow for an initial pre-emption of resources by one species, Grime and Tilman differ on which mechanism determines the long-term outcome of competition.

Goldberg (1990) suggests that the two models agree over a successional sequence which progresses from fast-growing species with rapid resource uptake rates to slower-growing species that are tolerant of low resource levels. Tilman's R* value for species with the lowest resource requirement would refer to dominant, highly stress-tolerant species in equilibrium (non-successional) communities.

Asymmetric competition may occur between plants of the same species (as part of intraspecific competition) or of different species (interspecific competition). It accounts for self-thinning, particularly in newly established pasture. Populations experiencing the greatest degree of crowding (intensity of competition) have the greatest size inequality, i.e. competition exaggerates underlying size inequalities (Begon et al., 1996). Thus self-thinning occurs in response to plant density, but the level of thinning is also modified by the availability of resources, such as moisture and light.

Plants which establish earliest not only have a large adverse effect on later-appearing plants, but are themselves little affected by the latter. Thus the earliest-established plants tend to persist, while attempts to invade their environs continue to fail, at least where the initial density of the earliest plants is high (see Fig. 1.3). This principle is used where possible in pasture management to exclude weeds. However, it also means that the introduction of desirable species into existing swards is unlikely to succeed unless adequately sized gaps are created by the use of cultivation, herbicides or heavy grazing. Once gaps are created, the way is open for rapidly establishing 'opportunists' (ruderals) to fill them. Gaps are also created and weeds allowed to enter when desirable species die due to extreme climate conditions or attack by insect pests and diseases. For example, the grassy weeds *Vulpia* spp. invaded large areas of southern Australian lucerne-based pastures when Hunter River lucerne (*Medicago sativa* cv. Hunter River) was decimated by the spotted alfalfa aphid in 1978/79. *Vulpia* then spread to other pasture areas as opportunity arose. A survey of farms in south-eastern South Australia and western Victoria showed that, by 1998, *Vulpia fasciculata* (silvergrass) and other *Vulpia* species were at a serious level of infestation on 1.8 million ha (Silvergrass Task Force, 1998). The average loss of gross farm income due to silvergrass was about 22% compared with a silvergrass-free environment.

Dominance of one species by another may also occur in an established pasture as a result of a differential response to seasonal climate variation or to selective grazing or simply as a result of differences in growth habit. One of the tasks of grazing management is to prevent excessive or prolonged dominance of one desirable component of a pasture over another, thus preventing excessive lowering of the 'presence' of the latter (see Kemp and King, Chapter 5; and Harris, Chapter 8, this volume).

Competition – Quantification of Effects

In any practical consideration of competition in plant communities, it is of value to be able to quantify the effect of competitive interactions on the components of the community and on the course of competition over time. In pasture communities, it is useful if the results can be related to environmental factors or to management treatments that have been applied. This should also lead to the defining of appropriate management for regulating interplant relations. Keddy (1989, 1990) has defined competition intensity as 'the combined (negative) effects of all neighbours on the performance of an individual or population'. It is measured by comparing the performance of components in a mixture with those in monoculture, or comparing the performance of 'target' plants surrounded by neighbours with that of the plants in plots cleared of neighbours. Grace (1995) has argued the inadequacy of using absolute differences between yields in monoculture and mixture as a measure of competition intensity. This is because the magnitude of the difference would depend not only on the relative competitive abilities, but also on the relative magnitude of monoculture yields. Thus he proposed that a more appropriate index would be one that reflected the proportional impact of competition on plant performance, i.e.

relative competitive intensity (RCI) =

$$\frac{\text{performance in monoculture} - \text{performance in mixture}}{\text{performance in monoculture}}$$

De Wit and van den Bergh (1965) also pointed out that the intensity and course of competition between species in pasture could not be unambiguously quantified by simply comparing the performance of the species in the mixture. First, yields of individual species at particular times cannot be equated with others, i.e. 1 g of one is not necessarily the same as 1 g of another. They stressed the need for a dimensionless measure, such as the relative yield (yield of a species in mixture/yield of the species in monoculture). They also pointed out that reference to monoculture yields enables changes in growing conditions and varying lengths of growing period to be taken into account. If only differences between species in mixture are measured, it is difficult to determine whether these are due to differences in competitive ability or to differences in response to growing conditions. Including monoculture yields in the formula helps to account for the latter.

An important measure of competition, developed by de Wit (1960, 1961) is the relative crowding coefficient (k) (see also Sackville Hamilton, Chapter 2, this volume). This is a measure of 'competitive power', namely, the degree to which a stronger competitor crowds a weaker one. De Wit studied numerous field experiments in which barley and oats were grown both in monoculture and in mixtures, where various proportions of barley were replaced by the same proportions of oats (replacement design). The results showed that, in the mixtures, one species always crowded the other out of some of the space 'allotted' to it according to the composition of the sown mixture. Gains and losses were equivalent. Consequently, in terms of grain yield, the relative crowding coefficient of barley (k_b) with respect to oats was the reciprocal of the relative crowding coefficient of oats (k_o) with respect to barley, i.e. $k_b \times k_o = 1$. Furthermore, the relative yield total (RYT) = 1, where

$$\text{RYT} = \frac{\text{grain yield barley in mixture}}{\text{grain yield barley in monoculture}}$$
$$+ \frac{\text{grain yield oats in mixture}}{\text{grain yield oats in monoculture}}$$

In terms of competition theory, this means that the two species were crowding for the same 'space' or resources. In these circumstances, yields of mixtures cannot exceed the yield of the highest-yielding monoculture.

De Wit and van den Bergh (1965) and van den Bergh (1968) showed that the above concepts also apply to mixtures of pasture grasses. They found that yields of successive harvests provided an appropriate measure of plant performance for defining

the course of competition, in place of grain yields used for crops. The index to define the course of competition was called the relative replacement rate (ρ). Relative yields are used to define ρ of species a with respect to species b at the nth harvest with respect to the mth harvest by

$$^{nm}\rho_{ab} = \frac{^n r_a / {^m r_a}}{^n r_b / {^m r_b}}$$

If $\rho > 1$, species a is the strongest competitor. If $\rho < 1$, species b is the strongest. If ρ is plotted on a logarithmic scale against time, the angle of the line with the horizontal is a measure of the relative rate at which one species replaces another. The same course line may be obtained by plotting the ratio of relative yields at successive harvests (van den Bergh, 1968). This course line is very useful for judging the direction of competitive relationships over time but not for further quantitative analysis of the mutual interference. Van den Bergh conducted experiments to show the effects on the course of competition of various factors, e.g. plant density, plant nutrient treatments and pH levels. De Wit and van den Bergh also found that, almost invariably, grass species were mutually exclusive (RYT = 1), i.e. they were competing for the same resources and the relative replacement rate was independent of the relative frequency (sowing proportions) of the component species. This is an important ecological concept. While not yet explained, it helps illustrate and predict how stronger and weaker competitors interact when competing for the same set of resources.

In contrast to the situation with mixtures of grasses, de Wit et al. (1966) found that a grass and a legume were not mutually exclusive when the legume obtained N from symbiotic fixation. Their experiment was conducted with and without rhizobial inoculation of the legume. Without rhizobium and N fixation, the grass and legume were mutually exclusive (RYT = 1). With rhizobium, however, N fixation gave the legume a competitive advantage. RYT was greater than 1 and the species were not mutually exclusive because the legume had an additional source of N not available to the grass. When course lines of the ratio of relative yields were drawn, over seven harvests, the lines of mixtures of different sowing frequencies tended to converge and to approach equilibrium (no change, no one species winning competitively). These trends were attributed to a combination of N fixation (which favoured the legume competitively) and N transfer from legume to grass (which favoured the grass competitively). It is now widely assumed that mixtures of grasses and legumes, at least those based on white clover, have a capacity to regulate the N cycle in the pasture (Chapman *et al.,* 1996).

In later experiments in both field and glasshouse (Tow, 1993; Tow *et al.,* 1997), trends with time in the ratio of relative yields provided further evidence of the tendency for a dynamic equilibrium to occur, provided that: (i) one species did not remain dominant for too long; and (ii) growing conditions were generally favourable to the growth of the legume. As indicated above, such course lines show if and under what conditions grasses and legumes tend towards equilibrium, but do not provide a means of further analysing competitive interactions. A tendency for equilibrium should have a positive influence on stability of botanical composition and species persistence.

Where climatic conditions fluctuate over time, the course of competition may also fluctuate. This may result in breakdown in equilibrium. However, the work of Tow and his colleagues quoted above provides evidence that, as long as dominance is not too severe, there is a persistent tendency to equilibrium. Equilibrium between species, or at least coexistence, is often said to be due to the fact that they occupy different niches. In grass–legume mixtures, the legume occupies a different niche in the sense that it has an independent source of N.

The attainment of equilibrium or coexistence sometimes requires an input of management that assists towards reducing the dominance of strong competitors, e.g. grass over legume. Such management is of most benefit if it achieves competitive balance without loss of productivity and with benefit to the grazing animal (see Davies, Chapter 4; Kemp and King, Chapter 5; and Harris, Chapter 8, this volume).

Achieving a competitive balance is more complex than might be supposed. For instance, a general problem with white clover–grass pastures is the difficulty of maintaining the clover content of some 30% thought to be desirable (Martin, 1960). This might be simply a problem of reducing grass dominance by appropriate management. However, defining appropriate management of grass–legume competition and N relations has to take account of the spatial heterogeneity (patchiness) of clover content brought about by spatially random urine deposition. This keeps different areas in the field 'out of phase' with respect to surrounding grass or legume

dominance. Furthermore, white clover content in pastures is also subject to long-term fluctuations or cycles (Chapman *et al.*, 1996).

Renewed interest in white clover–grass pastures over the past 20 years (reflected in the increased number of relevant publications, for example, in *Grass and Forage Science*) is related to the belief that clover N, compared with fertilizer N will reduce costs, use of fossil energy and leaching of nitrate to groundwater. Further, Ennik (1981, 1982), examining experimental data in the literature, found that the DM yield of a mixed white clover–grass sward receiving N fertilizer at varying levels was always higher than that of a pure grass sward at the same rate of mineral N application. This was because, with increasing application of N, the gain in grass DM was higher than the loss of clover DM. He also estimated that the amount of fertilizer N needing to be applied to a pure grass sward to obtain an N yield equal to that of a mixed sward was about 80 kg N t^{-1} of clover in the mixture (after the first tonne). This linear relationship, accompanied by an inverse relationship between rate of fertilizer N input and clover content of the mixture led to the conclusion that most of the fertilizer N was taken up by the grass. Furthermore, he concluded that, while introduction of more competitive clover varieties into a mixed pasture may increase N yield of the mixture, it was unlikely to increase DM yield.

Improvement of clover yield, N_2-fixation and persistence have all been recent objectives of plant breeders, agronomists and modellers. (Caradus *et al.*, 1996; Chapman *et al.*, 1996; Evans *et al.,* 1996; Schwinning and Parsons, 1996). All agree that effective production and utilization of grass–clover pastures require understanding of the interactions of the two components. This becomes all the more important as attempts are made to achieve a combination of aims, such as: (i) increasing the yield of clover by breeding more competitive cultivars and cultivars with a higher capacity for N_2-fixation, while avoiding leakage of nitrate to groundwater; (ii) increasing total yield by the use of N fertilizer without losing clover content; (iii) managing grass–clover swards for optimal animal production; and (iv) assessing new cultivars of white clover under grazing conditions (see also Nurjaya and Tow, Chapter 3; and Davies, Chapter 4, this volume).

Experience with grass–clover mixtures provides a reminder that competition usually operates in conjunction with other factors that affect companion plants differentially. In such mixtures, the most important factors would probably be N$_2$-fixation, N transfer and selective grazing. Competition is sometimes distinguished from 'apparent competition', where reduced yield of one component of a mixture may be due to differential effects of another organism on that component, e.g. selective grazing of palatable species, leaving an unpalatable one in higher proportions; or the same effect by selective attack by an insect pest. Begon *et al.* (1996) quote an example discussed by Connell (1990) of an indirect effect of *Artemesia* bushes on the growth of associated herbs. The beneficial effect of removing the bushes on the growth of the herbs was initially attributed to reduced competition for water. It was then found that removal of *Artemesia* also discouraged deer, rodent and insect consumers of the herbs which used this plant as a source of both food and shelter. Figure 1.8 and accompanying commentary illustrate just how complex interspecies relationships can be.

In attempting to understand the mechanisms of competition and to predict the outcome, many researchers have identified morphological and physiological traits or characteristics of plants associated

Fig. 1.8. Plants of the annual legume rose clover (*Trifolium hirtum*) growing amongst a clump of a perennial, native speargrass, near Bukkulla, northern New South Wales, Australia. The relation between grass and legume may be quite complex. In summer, the grass clump may intercept seed pods of rose clover washed over the soil surface by heavy storm rains. In winter, the grass is almost dormant and poorly competitive, when the clover establishes from seed and makes much of its growth. In spring, as temperatures rise but soil moisture declines, plants of rose clover may survive only in the shelter of speargrass clumps, as in the photo. However, it must set seed quickly before the new growth of speargrass becomes too competitive.

with their competitive abilility. Such traits do not always define the mechanism involved, but they assist in explaining or predicting competitive outcomes (see Nurjaya and Tow, Chapter 3; Skarpe, Chapter 9; and Peltzer and Wilson, Chapter 10, this volume).

Succession

Succession, the change in botanical composition over time, is currently a subject of great importance in both natural and sown pastures (as illustrated by the contents of this volume). Such importance arises because of the many changes that have occurred over the past century, largely resulting from increasing intensification of pasture use. Succession has long been linked to competition. More than 70 years ago, Clements *et al.* (1929) concluded, from their North American research and experience, that competition 'is the controlling function in successional development, and it is secondary only to the control of climate in the case of climaxes'. They also concluded that the regular outcome of competition is dominance, the successful competitors coming to control the habitat more or less completely. Other components of the plant community face suppression or even extinction.

As a feature of cyclic changes, Clements and his colleagues envisage regular invasion of plant communities from species outside. Hence their assertion that:

> The [successful] invading community is in harmony with the changing climate, the one invaded is correspondingly handicapped by it, and is all the more readily replaced as a result of competition between them. The course of events in edaphic habitats where succession is occurring is much the same, but the advantage to the invaders arises from the changes brought about by the occupants, which serve as a progressive hindrance to possession.

They then see the climax as the mature stage 'in harmony with the climate' and yet exhibiting an 'annual departure in growth and numbers', due to climatic variation. In all these processes, Clements *et al.* regarded competition as having a leading role in determining the botanical structure of the vegetation. The competitive balance of various grass, herb and shrub types in grassland is disturbed by variations in rainfall and is also 'profoundly modified by grazing, burning or cutting'.

These and related conclusions were subsequently translated into a successional approach to rangeland management (the so-called range succession model) and a practical system of range classification (Westoby *et al.*, 1989; Laycock, 1991). As summarized by Westoby *et al.* (1989):

> the [range succession] model supposes that a given rangeland has a single, persistent state (the climax) in the absence of grazing. Succession towards this climax is a steady process. Grazing pressure produces changes which are also progressive and are in the opposite direction to the successional tendency. Therefore the grazing pressure can be made equal and opposite to the successional tendency, producing an equilibrium in the vegetation at a set stocking rate.

The main tool of range management for the range succession model is thus the level of stocking rate. However, 'vegetation changes in response to grazing have been found to be not continuous, not reversible or not consistent', particularly in arid and semi-arid areas. These observations have led to a general questioning of the range succession model.

In recent years, the need for an alternative model to describe and assess rangeland condition and dynamics has been discussed by many workers, e.g. Westoby *et al.* (1989), Friedel (1991), Laycock (1991) and Humphreys (1997). Particularly questioned has been the need to manage rangeland to achieve a single, climax state or at least some desirable, stable state in equilibrium with an economic stocking rate. A 'stable' system (in terms of botanical composition) returns to the original steady state after being disturbed or deflected. Some researchers and practitioners prefer a system to have 'resilience', namely, the capacity to adapt to change, without necessarily reverting to the original state. What is regarded as important would depend on both the economic and conservation goals of management and the opportunities and limitations set by the environment and available technology. Rangeland stability and resilience may each be important in particular situations and can be envisaged as dependent to some extent on interspecific competition.

The above authors thus favour a model of rangeland dynamics that caters for the occurrence of multiple states of vegetation structure, changing influences on these states and the need for flexibility of short-term aims and management. A model of this nature should also be appropriate for many other grasslands, where botanical structure has been or is being greatly modified by over-grazing, weed invasion, effects of climate change and an increasing range of technological inputs. The so-called state and transition model seems to satisfy these needs. It involves the concept of 'thresholds of environmental change', which cause 'transition' from one discrete or stable 'state' of the vegetation to another. Such transition requires the imposition of a threshold of stress or perturbation. The prediction or early detection of an impending threshold would allow management action to be taken to maintain or achieve desirable botanical structure and productivity levels.

Westoby *et al.* (1989) suggest that, for the effective use of the state and transition model, recorded information on particular areas of rangeland should include catalogues of possible alternative states, possible transition pathways, opportunities for positive management action and hazards which may produce an unfavourable transition. The experimental testing of hypotheses (e.g. opportunistically during the occurrence of isolated events or sequences of events) should be a regular feature of information gathering. This needs to be accompanied by the estimation of probabilities of occurrence of climatic circumstances relevant to particular transitions. Such information should also be of value for describing and managing other types of grassland, at least for long-term pastures. Similarly, the proposals of Friedel (1991) should be applicable to a wide range of grasslands and pastures. She argues for the need to monitor botanical composition and yield of arid and semi-arid rangelands in order to detect the approach of a 'threshold' of change from one state to another. She presents evidence that this is feasible from monitoring programmes and the use of multivariate analyses and ordination techniques. The research suggests that rangeland which is deteriorating may retain the capacity to recover up to a certain point, beyond which it cannot readily return to its former state. Some factor, such as drought, fire or flooding, usually coincides with excessive grazing to 'tip the balance'. Appropriate monitoring needs to be combined with an understanding of plant–environment relations to allow prediction of approaching thresholds, thereby enabling preventive action to be taken.

The role of competition in determining vegetation structure and succession has received little critical attention in the above debate. It may be that, in arid and semi-arid areas, the overriding influences on plant community structure and succession are management (e.g. stocking rate effects) and periodic climatic events. If so, competition may play a lesser role than that claimed by Clements *et al.* (1929). It may have an increasing effect on pasture plant community structure with increasing rainfall and the accompanying greater plant density and/or productivity. The role of competition may also vary with the level of soil nutrient availability. These suppositions have long occupied the attention of ecologists and continue to do so (see Peltzer and Wilson, Chapter 10, this volume). A clear understanding of such matters is needed for the effective use of a state and transition approach to pasture management, with its need for clear definition of thresholds of change and transitions between relatively stable states. Indeed, its effectiveness as a model will depend on an understanding of all relevant plant–environment, plant–plant and plant–animal relationships. These must be understood at the scale of both individual plants and the wider ecosystem.

As mentioned earlier in this chapter, the regulation of dominance and invasion by highly competitive weeds is important in managing succession or retaining a desirable stable or resilient botanical structure (see Kemp and King, Chapter 5; and Wolfe and Dear, Chapter 7, this volume). The management problems posed by the heterogeneity of vegetation and associated environments in rangelands and pastures, together with current concerns about loss of species diversity, are encouraging increasing investigation on these topics (see Clark, Chapter 6; and Chapman, Chapter 13, this volume).

In farming situations, a different form of complexity is provided by rotations of pastures with crops. For example, in the so-called South Australian ley farming system (Fig. 1.9), weeds of the annual legume-based pastures, if allowed to set seed, are likely to re-appear in and compete strongly with following crops (see Trenbath, Chapter 14, this volume). One principle that can be used in changing competitive relationships and botanical composition involves sowing mixtures of pasture species, which together, at appropriate densities, are more competitive against weeds than the legume alone (see Figs 1.2 and 1.3c; see also Harris, Chapter 8, this volume).

One attitude engendered by adoption of the state and transition model is that range management can be seen as a continuing 'game', the object of which is to seize opportunities and evade hazards as far as possible (Westoby *et al.*, 1989). This

Fig. 1.9. The annual legume *Medicago truncatula* establishing from self-sown seed in early winter, amongst the stubble of the previous season's cereal crop. This illustrates a phase in the traditional South Australian ley farming system.

encourages and frees the manager to use a wide range of information and management options to achieve goals of pasture production, botanical composition and long-term stability or resilience. Management based simply on opportunism is no basis for achieving long-term pasture outcomes. What is needed, for all types of pastures, is a system developed within a framework of clearly defined, long-term goals and an understanding of factors affecting competition and succession. Management decisions, incorporating various pathways and time periods, could then be taken to achieve the goals.

Conclusions

Ecologists and agronomists are currently making considerable efforts both to overcome and prevent degradation and to improve the long-term performance of natural and sown pastures. Achieving these objectives requires an understanding of the processes involved, e.g. processes leading to decline in productivity, loss of valuable species and ingress of weeds, as well as those associated with improved productivity and stability. Plant competition is an important factor controlling these processes. However, the nature of such competition is not yet fully understood. There are other factors – abiotic and biotic – which can have a major effect on the conditions under which competition occurs.

Parallel with research on the mechanisms of competition is work that has produced indices to measure outcomes of plant interactions. These indices enable comparisons to be made of the effects of genotype and environmental factors on such outcomes. Other investigations have resulted in the description and classification of the botanical structure of grasslands and any changes over time, thereby providing a basis for management decisions. Yet the precise role of competition in determining the botanical composition of pastures remains unclear. Even so, integrating the bank of information and improved understanding arising from the array of ecological and agronomic research should provide a real opportunity to develop management systems to achieve long-term use and stability of grasslands.

References

Begon, M., Harper, J.L. and Townsend, C.R. (1996) *Ecology: Individuals, Populations and Communities*, 3rd edn. Blackwell Science, Boston, USA.

Bolger, T.P. (1998) Aboveground and belowground competition among pasture species. In: *Proceedings of the 9th Australian Agronomy Conference, Wagga Wagga*. The Australian Society of Agronomy, Wagga Wagga, Australia, pp. 282–285.

Caradus, J.R., Woodford, D.R. and Stewart, A.V. (1996) Overview and vision for white clover. In: Woodford, D.R. (ed.) *White Clover: New Zealand's Competitive Edge*. Agronomy Society of New Zealand Special Publication No. 11/Grassland Research and Practice Series No. 6, Agronomy Society of New Zealand, Christchurch, New Zealand, and New Zealand Grassland Association, Palmerston North, New Zealand, pp. 1–6.

Chapman, D.F., Parsons, A.J. and Schwinning, S. (1996) Management of clover in grazed pastures: expectations, limitations and opportunities. In: Woodford, D.R. (ed.) *White Clover: New Zealand's Competitive Edge*. Agronomy Society of New Zealand Special Publication No. 11/Grassland Research and Practice Series No. 6, Agronomy Society of New Zealand, Christchurch, New Zealand, and New Zealand Grassland Association, Palmerston North, New Zealand, pp. 55–64.

Clements, F.E., Weaver, J.E. and Hanson, H.C. (1929) *Plant Competition: an Analysis of Community Functions*. Carnegie Institution, Washington DC, USA.

Connell, J.H. (1990) Apparent and 'real' competition in plants. In: Grace, J.B. and Tilman, D. (eds) *Perspectives on Plant Competition*. Academic Press, New York, USA, pp. 9–26.

de Wit, C.T. (1960) *On Competition*. Agricultural Research Reports 66(8), Centre for Agricultural Publishing and Documentation (PUDOC), Wageningen, The Netherlands.

de Wit, C.T. (1961) Space relationships within populations of one or more species. In: *Mechanisms in Biological Competition*. Symposia of the Society for Experimental Biology Number XV, Society for Experimental Biology, Cambridge, UK, pp. 314–329.

de Wit, C.T. and van den Bergh, J.P. (1965) Competition between herbage plants. *Netherlands Journal of Agricultural Science* 13, 212–221.

de Wit, C.T., Tow, P.G. and Ennik, G.C. (1966) *Competition between Legumes and Grasses*. Agricultural Research Reports 687, Centre for Agricultural Publishing and Documentation (PUDOC), Wageningen, The Netherlands.

Ennik, G.C. (1981) Grass–clover competition especially in relation to N fertilization. In: Wright, C.E. (ed.) *Plant Physiology and Herbage Production*. Proceedings of Occasional Symposium No. 13, British Grassland Society, Hurley, UK, pp. 169–172.

Ennik, G.C. (1982) De bijdrage van witte klaver aan de opbrengst van grasland. *Landbouwkundig Tijdschrift* 94(10), 363–369.

Evans, D.R., Williams, T.A. and Evans, S.A. (1996) Breeding and evaluation of new white clover varieties for persistency and higher yields under grazing. *Grass and Forage Science* 51, 403–411.

Friedel, M.H. (1991) Range condition assessment and the concept of thresholds: a viewpoint. *Journal of Range Management* 44(5), 422–426.

Goldberg, D.E. (1990) Components of resource competition in plant communities. In: Grace, J.B. and Tilman, D. (eds) *Perspectives on Plant Competition*. Academic Press, New York, USA, pp. 27–49.

Grace, J.B. (1995) On the measurement of plant competition intensity. *Ecology* 76, 305–308.

Grace, J.B. and Tilman, D. (eds) (1990) *Perspectives on Plant Competition*. Academic Press, New York, USA.

Grime, J.P. (1979) *Plant Strategies and Vegetation Processes*. John Wiley & Sons, Chichester, UK.

Harper, J.L. (1977) *Population Biology of Plants*. Academic Press, London, UK.

Humphreys, L.R. (1997) *The Evolving Science of Grassland Improvement*. Cambridge University Press, New York, USA.

Keddy, P. (1989) *Competition*. Chapman and Hall, London, UK.

Keddy, P. (1990) Competition hierarchies and centrifugal organisation in plant communities. In: Grace, J.B. and Tilman, D. (eds) *Perspectives on Plant Competition*. Academic Press, New York, USA, pp. 265–289.

Laycock, W.A. (1991) Stable states and thresholds of range condition on North American Rangelands: a viewpoint. *Journal of Range Management* 44(5), 427–433.

Martin, T.W. (1960) The role of white clover in grasslands. *Herbage Abstracts* 30, 159–164.

Milthorpe, F.L. (1961) The nature and analysis of competition between plants of different species. In: *Mechanisms in Biological Competition*. Symposia of the Society for Experimental Biology Number XV, Society for Experimental Biology, Cambridge, UK, pp. 330–355.

Radosevich, S., Holt, J. and Ghersa, C. (1997) *Weed Ecology: Implications for Management*, 2nd edn. John Wiley & Sons, New York, USA.

Rhodes, I. and Stern, W.R. (1978) Competition for light. In: Wilson, J.R. (ed.) *Plant Relations in Pastures*. CSIRO, Melbourne, Australia, pp. 175–189.

Schwinning, S. and Parsons, A.J. (1996) Analysis of the coexistence mechanisms for grasses and legumes in grazing systems. *Journal of Ecology* 84, 799–813.

Silvergrass Task Force (1998) *Report to the Woolmark Company on the Recommendation of the Silvergrass Task Force*. The Woolmark Company, Parkville, Victoria, Australia, 36 pp.

Society for Experimental Biology (1961) *Mechanisms in Biological Competition*. Symposia of the Society for Experimental Biology Number, XV, Cambridge University Press, Cambridge, UK.

Stern, W.R. and Donald, C.M. (1962) Light relationships in grass–clover swards. *Australian Journal of Agricultural Research* 13, 599–614.

Tilman, D. (1982) *Resource Competition and Community Structure*. Princeton University Press, New Jersey, USA.

Tilman, D. (1988) On the meaning of competition and the mechanisms of competitive superiority. *Functional Ecology* 1, 304–315.

Tilman, D. (1990) Mechanisms of plant competition for nutrients: the elements of a predictive theory of competition. In: Grace, J.B. and Tilman, D. (eds) *Perspectives on Plant Competition*. Academic Press, New York, USA, pp. 117–141.

Tilman, D. (1994) Competition and biodiversity in spatially structured habitats. *Ecology* 75(1), 2–16.

Tow, P.G. (1993) The attainment and disturbance of competitive equilibrium in tropical grass–legume mixtures. In: *Proceedings of the XVII International Grassland Congress, Palmerston North and Rockhampton*. New Zealand Grassland Association, Palmerston North, New Zealand and Rockhampton, Australia, pp. 1913–1914.

Tow, P.G., Lazenby, A. and Lovett, J.V. (1997) Relationships between a tropical grass and lucerne on a solodic soil in a sub-humid, summer–winter rainfall environment. *Australian Journal of Experimental Agriculture* 37, 335–342.

van den Bergh, J.P. (1968) *Analysis of Yields of Grasses in Mixed and Pure Stands*. Agricultural Research Reports 714, Centre for Agricultural Publishing and Documentation (PUDOC), Wageningen, The Netherlands.

Wedin, D. and Tilman, D. (1993) Competition among grasses along a nitrogen gradient: initial conditions and mechanisms of competition. *Ecological Monographs* 63, 199–229.

Westoby, M., Walker, B. and Noy-Meir, I. (1989) Opportunistic management for rangelands not at equilibrium. *Journal of Range Management* 42(4), 266–274.

Wilson, J.B. (1988) Shoot competition and root competition. *Journal of Applied Ecology* 25, 279–296.

Wilson, J.R. (ed.) (1978) *Plant Relations in Pastures*. CSIRO, Melbourne, Australia.

2 Measurement of Competition and Competition Effects in Pastures

N.R. Sackville Hamilton

Institute of Grassland and Environmental Research, Plas Gogerddan, Aberystwyth, Ceredigion, UK

Introduction

Competition is one of the most important concepts in ecology and the subject of a large literature and numerous reviews (e.g. Trenbath, 1978; Goldberg and Barton, 1992; Goldberg and Scheiner, 1993; Gibson *et al.*, 1999; Jolliffe, 2000). Yet it continues to cause confusion. Problems are manifold. Terms such as competitive ability, competition intensity and even competition are used in diverse, often conflicting or misleading senses (Harper, 1982; Sackville Hamilton, 1994). Many experiments have been poorly designed, analysed or interpreted (Cousens, 1991, 2000). The optimum design of experiments is debatable (Snaydon, 1991, 1994; Sackville Hamilton, 1994; Gibson *et al.*, 1999). It is suggested that the debate has become dogmatic and emotionally charged (Cousens, 1996; Jolliffe, 2000).

The debate still continues, but has generated two important generic conclusions. First, many different questions can be asked about competition and different experimental approaches are required to answer them (Sackville Hamilton, 1994; Cousens, 1996). Consequently, there is a need for rigour in defining the objectives of any experiment on competition and in matching experimental design and methodology to objectives. In this chapter we consider the kind of questions that can be asked and the approaches that can be used to answer them, based on a consideration of competition theory.

Secondly, it appears to be easy to misunderstand the concept of competition. Consequently it is common to fail to match experiment with objectives, to misuse competition experiments, to analyse data incorrectly and to interpret results incorrectly (Cousens, 2000). This chapter therefore also seeks to clarify the concept and highlight some of the major pitfalls in the study of competition.

Competition: the Concept

Competition is 'an interaction between individuals, brought about by a shared requirement for a resource in limited supply, and leading to a reduction in the survivorship, growth and/or reproduction of at least some of the competing individuals concerned' (Begon *et al.*, 1996). Plants can affect each other's survivorship, growth and reproduction in many ways, including ways that, ecological purists argue, do not constitute competition (Harper, 1982). Supposedly non-competitive interactions include allelopathy (Inderjit *et al.*, 1999), hemiparasitism (Matthies, 1996), the quasi-parasitic direct transfer of nutrients from one plant to another through mycorrhizal connections between root systems (Martins and Read, 1996; Simard *et al.*, 1997a, b) and the release of nutrients from living or decaying roots and shoots. In addition, the effect on survivorship, growth and/or reproduction depends not only on the competitive acquisition of limited resources, but on subsequent internal processes, such as developmental and allometric control of resource allocation, growth habit and patterns of dispersal. Moreover, the final outcome, in terms of changes in the composition of communities, depends not only on the effect of competition on reproduction and survival but also on other factors, such as the intrinsic maximum reproductive rate and the effects of pathogens and herbivores.

A variety of terms, such as 'crowding for biological space' (de Wit, 1960) and 'interference' (Harper, 1977), have been proposed in attempts to improve the conceptual rigour of studies on competition, although such terms have not gained wide acceptance. Indeed, 'interference', intended by plant ecologists as a more general term than competition, is used by animal ecologists as a narrower term, referring to one specific type of competition (Human and Gordon, 1996).

Most of the literature on competition among plants is concerned with quantifying the overall effect of one plant or population of plants on the survivorship, growth and/or reproduction of neighbouring plants and vice versa. Relatively little literature is concerned with: (i) identifying the resource in limited supply; (ii) demonstrating the role of a shared requirement for limited resources in the process of competition, as opposed to other mechanisms of interaction (such as allelopathy); or (iii) separating the process of interaction between individuals from its subsequent impacts. As such, most literature fails to identify the mechanism of competition or to distinguish competition *sensu stricto* from other types of interaction. Of necessity, therefore, most literature implicitly accepts the most general possible definition of competition between plants, as any effect of one plant on another. Even positive effects, such as the increase in grass growth following the transfer of fixed nitrogen (N) from clover, cannot be separated from negative ones if all we measure is the overall combination of all positive and negative effects. The implicit broad definition of competition is adopted in this book, although one of our objectives is to improve the ability to separate the different influences.

Competition: Population Theory

This section provides the minimum theoretical basis required for subsequent evaluation and understanding of the different approaches and indices used in the study of competition. Equations are presented only for the simplest mathematical forms of competitive interactions. It is often necessary to use more complicated equations, but the mathematical details of such complications do not affect the principles.

Population dynamics of pure stands

Before considering competition between species, we need first to consider competition between the individuals of one species growing in a monoculture. Classical competition theory is based on the logistic curve of population growth:

$$\frac{dN}{dt} = Nr(1 - qN) \qquad (2.1)$$

This equation describes how population density, N, changes with time; r is the maximum relative growth rate achieved in the absence of competition, and q is a competition coefficient, quantifying the effect of competition on population growth rate. The key concept underlying this equation is that increasing density increases the intensity of competition and so decreases relative growth rate. Growth rate is zero when $N = 1/q$ and negative (i.e. population density decreases) when $N > 1/q$. This means that $N = 1/q$ is the point of 'stable equilibrium': population density always changes towards this value, regardless of its initial density. Following any temporary impact that moves the population density away from this equilibrium value, the population will always tend to return to equilibrium.

Population dynamics of two-species[1] mixtures

Describing the dynamics of two-species mixtures appears more complicated because we need a separate equation for each species and some way of distinguishing the two species. We shall call the two species i and j and use subscripts i and j to distinguish corresponding terms in equations. For example, in Equation 2.1 for species i we write N_i, r_i and q_i instead of N, r and q.

However, the real increase in complexity is small, with just one new parameter p, similar to q but describing the effect of inter- rather than intraspecific competition. This gives the classical Lotka–Volterra equations (Volterra, 1926; Lotka, 1932):

$$\frac{dN_i}{dt} = N_i r_i(1 - q_i N_i - p_{ij} N_j)$$
$$\frac{dN_j}{dt} = N_j r_j(1 - q_j N_j - p_{ji} N_i) \qquad (2.2)$$

The equilibrium conditions are less simple than for the pure stand. We still have the pure stand situation, that growth rate of species i is zero when $N_i = 1/q_i$ and $N_j = 0$. However, from Equations 2.2 the condition for zero population growth rate of species i is no longer just $1 = q_i N_i$; it is now $1 = (q_i N_i + p_{ij} N_j)$. Thus, in terms of their effect on population growth of i, q_i plants of species j are equivalent to p_{ij} plants of species i, and we can also get zero growth rate for species i by replacing any number of plants of i with the equivalent number of plants of j. On a graph with the densities of species i and j on the axes (Fig. 2.1), this represents a straight line (a 'zero isocline') from $N_i = 1/q_i$ at $N_j = 0$ to $N_j = 1/p_{ij}$ at $N_i = 0$. The rate of change in pop-

ulation density of species i is zero at any point along this line. As in the case of pure stands, it is a line of stable equilibrium for species i, since the population density of species i always changes towards the line. If the initial joint density is below the line, $(q_i N_i + p_{ij} N_j) < 1$, population growth rate of i is positive and population density increases (solid arrows pointing to the right in Fig. 2.1). If the initial joint density is above the line, $(q_i N_i + p_{ij} N_j) > 1$, population growth rate of i is negative and population density decreases (solid arrows pointing to the left in Fig. 2.1).

However, it does not usually represent a line of stable equilibrium for the whole mixture, because the zero isoclines for species i and j will not usually

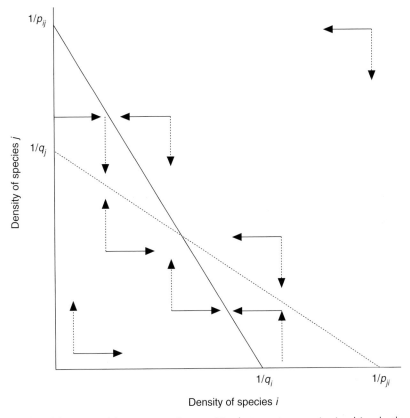

Fig. 2.1. Graphical depiction of the outcome of competition between two species i and j under the Lotka–Volterra equations. The solid and dashed lines are the 'zero isoclines' of species i and j respectively, i.e. the joint densities of i and j at which $dN_i/dt = 0$ and $dN_j/dt = 0$. The solid and dashed arrows show the direction of change in population densities of i and j, respectively; q_i, q_j, p_{ij} and p_{ji} are the coefficients for intra- and interspecific competition in Equations 2.2. The graph illustrates the situation for stable coexistence of the two species in mixture: regardless of the initial combination of densities, the joint densities of the two species tend towards the point at which the two isoclines cross over, which point thus represents the point of stable coexistence. Other cases are discussed in the text.

be coincident. That is, when the mixture is on the zero isocline for species i, the mixture will tend to move off that isocline because of changes in the density of j. The dashed arrows in Fig. 2.1 show corresponding changes in density of j.

The relative positions of the two zero isoclines determine the outcome of competition between the two species. If the two lines do not cross over, one species always outcompetes the other, leading ultimately to competitive exclusion and a monoculture of the superior competitor. Species i is the superior competitor if $(1/p_{ij}) > (1/q_j)$ and $(1/p_{ji}) < (1/q_i)$, i.e. if the effects of competition from i and from j are both less on i than on j. Conversely, species j is the superior competitor if $(1/p_{ij}) < (1/q_j)$ and $(1/p_{ji}) > (1/q_i)$.

If the two lines cross over with $(1/p_{ij}) > (1/q_j)$ and $(1/p_{ji}) > (1/q_i)$ – the case illustrated in Fig. 2.1 – any mixture will always tend to change towards the point of intersection between the two lines. This is easily envisaged by choosing any starting-point in Fig. 2.1 and following simultaneous changes in population densities of i, as indicated by the solid arrows, and of j, as indicated by the dashed arrows. Any starting-point above both lines leads to a reduction in density of both species. Any starting-point below both lines leads to an increase in density of both species. Any starting-point between the two lines leads to a reduction in density of one species and an increase in the density of the second, towards the point of intersection. Thus, regardless of the starting-point, repeated changes in population density lead gradually but invariably towards the point at which the two lines cross. The point of intersection thus represents a point of stable equilibrium, or stable coexistence of the two species in mixture. This represents the situation where, for both species, the competitive effect of each species on its own growth is greater than its competitive effect on growth of the other species, i.e. where intraspecific competition is more severe than interspecific competition.

This conclusion highlights a key feature of competition that may not be intuitively obvious. The outcome of competition between species depends not just on the competition between them but on their relative population responses to intra- and interspecific competition. Studies that fail to measure intraspecific competition, or at least the relationship between intra- and interspecific competition, provide no information on the outcome of interspecific competition.

Limitations of the Lotka–Volterra equations

The equations presented above suffer a number of limitations restricting their value for the study of competition. They do not always adequately describe the relationship between population growth rate and density. They provide population-level descriptions of the outcome of competition, and yet competition is a process involving individuals, which makes the equations unsuitable for elucidating the mechanisms of competition. They ignore the theory of competition for multiple resources (Tilman 1982, 1990). As population-level descriptions, they necessarily ignore stage-, age- and size-dependent effects.

They also ignore the effects of spatiotemporal heterogeneity of the environment. The immediate effect of an environmental change is to alter the values of parameters in the competition equations. Spatial heterogeneity therefore tends to produce a structured community, and temporal heterogeneity tends to change the equilibrium state so that a population may move away from, as well as towards, equilibrium. These effects, combined with spatially limited dispersal of individuals, the resulting spatial constraints on population dynamics and the time delay between imposition of a particular environment and the population response to that environment, can generate complex responses to environmental heterogeneity.

In addition, the conditions for coexistence outlined above, i.e. that intraspecific competition must be stronger than interspecific competition, have not been successful in explaining coexistence of plant species. Many reviews (e.g. Trenbath, 1974; Goldberg and Barton, 1992; Goldberg, 1996) conclude that, with the notable exception of legume/non-legume mixtures where the legume fixes N, competition within species is not usually stronger than competition between species of plant. This suggests that other issues are more important and that the limitations of the equations may be particularly severe for plants. Much of the difficulty with plants can be attributed to their relative immobility, which has numerous consequences, e.g.:

- Each plant interacts with few neighbours, so that the effective ecological population size (i.e. the number of individuals that interact with each other) is small.
- Larger units develop small-scale spatial patterns, comprising patches or subcommunities that do not compete with each other, and whose

dynamics are controlled by other processes, such as the chance dispersal of seed or vegetative propagules.

- The magnitude of interaction with each neighbour depends heavily on the extent of overlap of their zones of influence, which in turn depends mainly on the distance between and sizes of neighbouring plant pairs. Population-level summaries ignore this major factor.
- Plants respond to environmental heterogeneity mainly through their high phenotypic plasticity. It may therefore be expected that the effects of spatiotemporal heterogeneity of the environment on the dynamics of competition are particularly important for plants.
- Plant species differ in growth habit and mobility, ranging from plants with a highly compact tufted growth form, which makes them immobile, to plants with structures such as stolons and rhizomes, which make them relatively mobile. The dynamics of mixtures of plants depends heavily on the growth forms of the species present.

Thus, the Lotka–Volterra equations should not be viewed as providing a comprehensive theory of competition. Rather, they provide a baseline for understanding and developing the concept of competition.

Yield–density relationships

Yield is experimentally more tractable than population dynamics. Measuring the effect of density on yield has therefore been a more popular approach to analysis of competition in plants. This section presents the corresponding theory of yield–density relationships.

De Wit (1960) demonstrated that, if experimental populations of a species are grown in pure stands at a range of densities, Equation 2.1 generates an inverse-linear relationship between mean yield per plant w and density N:

$$w = \frac{1}{a + bN}$$
$$= \frac{1}{a(1 + \beta N)} \tag{2.3}$$

It is common to convert this to a linear relationship by using $1/w$, rather than w (Fig. 2.2a), or by using the inverse of total yield per plot, $Y = wN$ (Fig.

2.2c). In Equation 2.3, $1/a$ is the yield per individual extrapolated to zero density, i.e. without competition (Fig. 2.2b). To interpret the meaning of b, note that Equation 2.3 also describes a hyperbolic relationship between plot yield and density, $Y = N/(a + bN)$; $1/b$ is then the maximum yield per plot achieved at high density (Fig. 2.2d).

The decrease in mean plant size with increasing density (Fig. 2.2b) is attributed to the increase in intensity of competition associated with increasing density; $\beta = b/a$ is a key coefficient (the 'crowding coefficient' (de Wit, 1960)) that quantifies this effect of competition on plant size, in units of (area per individual), the inverse of density. Understanding the biological significance of β is fundamental to understanding many competition indices derived from yield–density relationships. It has numerous equivalent interpretations, all essentially describing the ability of plants to fill the available area, by increasing in size as density is decreased. It equals yield per isolated individual divided by asymptotic yield per unit area at high density, which at first sight may seem a rather strange ratio. An alternative way of thinking of it is as the size of plants grown without competition relative to their size at very high density, but standardized by dividing by that high density – a necessary standardization since plant size depends on the density used.

In graphical terms, β measures the convexity of the curves in Fig. 2.2(b and d), describing how rapidly yield per plant decreases and yield per plot increases as density increases. It is the spacing (1/density) at which plants are half the maximum size they attain without competition (Fig. 2.2b) and at which the yield per plot is half its maximum value (Fig. 2.2d). It is the slope at zero density of the curves in Fig. 2.2(b and d), standardized respectively for maximum plant yield and maximum plot yield. In other words, if in Fig. 2.2(b and d) yield per plant and yield per plot are expressed as proportions of their maximum values, such that $a = b = 1$, β is the slope at zero density.

Like the extension of Equation 2.1 to 2.2, Equation 2.3 may be extended for two-species mixtures simply by the inclusion of a coefficient, γ, to quantify interspecific competition in the same way as β quantifies intraspecific competition:

$$w_i = \frac{1}{a_i(1 + \beta_i N_i + \gamma_{ij} N_j)}$$
$$w_j = \frac{1}{a_j(1 + \beta_j N_j + \gamma_{ji} N_i)} \tag{2.4}$$

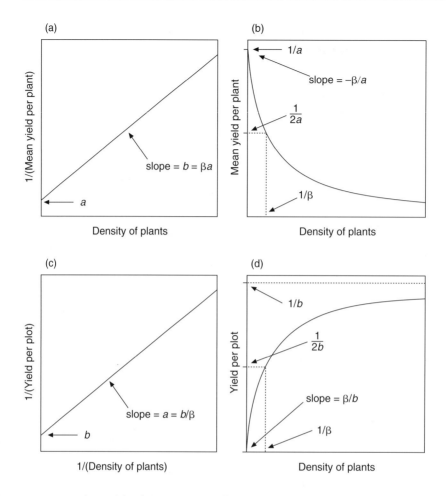

Fig. 2.2. Four equivalent graphical representations of yield–density relationships in pure stands described by Equation 2.3, highlighting the meanings of the parameters a, b and β. See text for further explanation.

In the same way as β describes the ability of plants to increase in size to fill the available space as their density is decreased, γ_{ij} describes the ability of plants of i to increase in size as the density of species j is reduced.

De Wit (1960) proved that Equations 2.4 are mathematically identical to the Lotka–Volterra Equations 2.2, at least when, in his terms, the species 'crowd for the same biological space'. Like the logistic and Lotka–Volterra equations, these equations do not always accurately describe yield–density relationships. Equation 2.3 often accurately describes total biomass of plants in pure stands, but not components of biomass, such as yield of seed or potato tubers (Willey and Heath,

1969; Harper, 1977). Generalized versions of the equations have been developed to handle more complex relationships: see, for example, Holliday (1960) and Jolliffe (1988) for a generalized version of Equation 2.3, and Firbank and Watkinson (1985), Law and Watkinson (1987), Menchaca and Connolly (1990) and Freckleton and Watkinson (2000) for generalized Equations 2.4.

Nevertheless, again like the Lotka–Volterra equations, Equations 2.4 provide an appropriate baseline for understanding and developing the concept of competition and for assessing the relative merits of the various approaches to the study of competition.

On the Choice and Interpretation of Competition Indices

The previous section presented equations describing the simplest form of competition. However, it is usually not sufficient, and indeed often not possible, to fit these equations to experimental data. It is necessary to devise competition indices that summarize competitive interactions in an appropriate way. Many such indices have been published (e.g. Trenbath, 1978; Mead and Riley, 1981; Connolly, 1986, 1987; Goldberg, 1996; Goldberg et al., 1999; Jolliffe, 2000). This section discusses the factors that affect the optimal choice of index.

Competition studies may address a range of different questions. For example, they may focus on the values of parameters in Equations 2.2 or 2.4, or on the importance of competition in the field. There are several potential pitfalls to be aware of when choosing competition indices. For example, indices may vary with density or frequency for several reasons, or may incorporate a size bias. Terms such as 'competitive ability' are used with many different meanings. Before embarking on any experiment on competition, it is essential to clarify objectives and to determine the relevance of these issues, and hence to determine their influence on design and analysis.

The biology and ecology of competition

Equations 2.2 and 2.4 provide information on the biology of competition in a given environment, in the sense that their parameters describe how the severity (Snaydon and Satorre, 1989) or intensity (Grace, 1995; Goldberg et al., 1999) of competition changes with density in that environment.

However, it is often desirable to determine the importance of interactions for ecosystem function by quantifying their strength (Laska and Wootton, 1998; Kokkoris et al., 1999). For example, how important is competition relative to herbivory in pastures (Reader, 1992; Rachich and Reader, 1999)? How does the importance of competition vary along productivity gradients (Goldberg et al., 1999)? In such cases, measurements of the intensity of competition per se provide the required information, with no underlying requirement to formalize how competition intensity varies with density.

The effect and outcome of competition

Gibson et al. (1999) emphasize the need to distinguish the effect of competition on the yield of plants (Equations 2.4) from the outcome of competition in terms of changes in the composition of mixtures (Equations 2.2). Both are valid topics for study. For example, in agricultural research, yield is often the primary target of study, and studies should focus on the effects of competition on yield. In studies on population ecology, the primary focus should be the outcome.

They are, however, distinct topics, requiring different units of measurement – yield per plant for the effect of competition, and density of plants for the outcome. The difference was explored in considerable mathematical and conceptual detail by de Wit (1960), who used the term 'crowding' to describe effects on yield per plant, and 'relative reproductive rate' for changes in mixture composition. He developed different graphical representations for the two phenomena – the replacement diagram for crowding effects, and the ratio diagram for changes in mixture composition – and presented equations formalizing the relationship between the two (e.g. de Wit, 1960, equations 3.7 and 9.4). He also demonstrated that, where two species are competitively neutral in terms of their effects on each other's yields (i.e. where $\beta_i = \beta_j = \gamma_{ij} = \gamma_{ji}$ in Equations 2.4), they may not be competitively neutral in terms of outcome. The 'winner' in such cases is the species with the higher reproductive rate (de Wit, 1960, p. 5) in the absence of competition.

Since they are distinct topics, studies on the effect of competition cannot always be used to deduce the outcome. Yet most studies focus on the effect, and few address the outcome. This criticism should be considered particularly important for the study of competition in permanent pastures. The long-term persistence and sustainability of the pasture depend on plant population dynamics. Improving our understanding of the outcome of competition is vital for understanding and improving pasture persistence and sustainability.

Size bias?

Indices describing effects on yield per plant can suggest an intrinsic competitive advantage of large plants that is not reflected in the outcome of competition between large and small plants (de Wit,

1960; Connolly, 1986; Gibson *et al.*, 1999). The nature of this phenomenon may be conceptualized using the 'thought experiment' of Connolly (1986). Imagine two species, L and S, identical except that we count two individuals of S as one of L, so that L 'plants' are twice the size of S plants. Clearly the composition of any mixture will not change with time: the ratio of numbers of L/S individuals produced will always equal their initial ratio. In this respect, L and S are competitively neutral.

However, by definition one plant of L has the same effect on its neighbours' yields as two plants of S. Consider also the meaning of the crowding coefficient β (Equation 2.3): it is the yield per isolated individual divided by the yield per plot at high density. Counting two individuals as one doubles the size of isolated individuals but, by definition, does not change asymptotic yield per plot at high density. Therefore, by definition, it also doubles the value of β, i.e. doubles the ability of plants to expand to fill available space at low density. It also accurately reflects the fact that the ratio of L/S biomass produced in a mixture will always be double their initial ratio of densities.

Thus, indices describing the effect of competition on yield correctly reflect the large competitive effect of large plants on the size of their neighbours. On the other hand, in relation to studies on the outcome of competition, the same indices incorporate a size bias that invalidates their use for such studies. This simply means that studies on the outcome of competition should use indices designed to address the outcome, while studies on the effect should use indices designed to address the effect.

Such a conclusion may seem almost trite. However, there is also a deeper, more philosophical issue here. Is it sensible to compare the effects of one plant of L with those of one plant of S if doing so gives answers that are misleading in terms of the outcome of competition? Answering this question is beyond the scope of this chapter.

Density and frequency dependence

Deviations from Equations 2.2 and 2.4 may be interpreted in terms of frequency- and density-dependent competition (Freckleton and Watkinson, 2000). Competition is frequency-dependent if the values of the parameters in these equations vary with the relative densities of the two species, and density-dependent if they vary with the combined density of both species regardless of their relative densities.

In situations where density is so low that there is no competition, yield per plant does not vary with density, i.e. in Equations 2.2 and 2.4, $p = q = \beta = \gamma = 0$. As density increases and competition starts to occur, the values of these parameters change. Experiments that include such low densities must be analysed using more complex equations. However, the additional coefficients required to describe this apparent density dependence are better interpreted as parameters describing the onset of competition as density increases, not as true density dependence.

Spurious density and frequency dependence

Most competition indices in common use are not based on the parameters in Equations 2.2 or 2.4. They therefore show spurious frequency and density dependence, i.e. their values depend on the densities used for the experiment even when there is no real frequency or density dependence.

Spurious frequency and density dependence may be demonstrated by expressing an index in terms of Equations 2.2 or 2.4 and noting whether the formula includes one or more density terms in addition to the parameter values (Connolly, 1986; Sackville Hamilton, 1994). The problem is most common with the simpler experimental designs that do not allow estimation of the parameter values or pure functions of them. It can also occur with the more comprehensive designs if the experimenter chooses inappropriate indices.

The usefulness of such indices is strictly limited to the densities used in the experiment. If possible they should be avoided in experiments where the choice of density is arbitrary. On the other hand, their use can be entirely acceptable if the densities have broader relevance – for example, in studies on natural communities or on crop mixtures where densities and management are defined by agronomic practice.

Outcome of competition

The direction of change in composition of a mixture may depend on the initial frequency or density of the species in mixture even when there is no frequency or density dependence in their underlying competitive interactions. As described above, Fig. 2.1 illustrates a form of such a frequency-dependent outcome, in which the species that increases in relative frequency is the one that was initially present at lower frequency. If the two lines in Fig. 2.1 were swapped (i.e. the line shown for species i was for species j and vice versa), then a different form of frequency-dependent outcome would occur, in which competition would always result in a pure stand of whichever species was initially most common. Such frequency-dependent outcomes of competition should not be confused with true density- or frequency-dependent competition.

Intensity of competition

Competition intensity varies with density. That is, it is intrinsically frequency- and density-dependent. It follows that measurements of competition intensity are useful *per se* only when the density used has broader relevance (as in field studies on existing communities) and is not just an arbitrarily chosen experimental state. Neither true nor spurious frequency and density dependence is an important issue for indices of competition intensity in such situations.

The meanings of competitive ability

Even under the simplest form of competition (Equations 2.2 and 2.4), six parameters are required to describe the competitive relationships between two species in a mixture, providing considerable scope for defining competitive ability in different ways. There is no single most acceptable definition.

Comparing the competitive abilities of two species implicitly requires definition of a reference point – ability to compete against what? Each other, or one or more other species? For example, the statement 'A is more competitive than B' may mean 'A performs relatively better than B in mixtures with C' or 'A performs relatively better than B in a mixture of A and B'. Within each of these meanings, the phrase 'performs relatively better' is also ambiguous. Performance can be defined in terms of the size of individuals (Equations 2.4) or of the dynamics of the population (Equations 2.2). As discussed above in the section on size bias, these two performance indicators have qualitatively different relationships with initial plant size.

In the first case, competitive ability may be defined in terms of response to competition (competitive response (Goldberg and Landa, 1991; Goldberg and Barton, 1992)) or in terms of effect on competitors (competitive effect), or a combination of both. For example, 'A is more competitive than B' may mean 'A has a greater effect than B on C' or 'A responds less than B to C'.

In the second case, several further options are possible. Competitive ability may be based on comparing the effect of and response to interspecific competition. 'A is more competitive than B' would then mean 'A has a greater effect on B than B has on A'.

Alternatively, it may be based on some comparison of the effects of inter- and intraspecific competition. 'A has a greater effect on B than on itself', 'B has a greater effect on itself than on A', 'A has a greater effect than B on B', or 'A has a greater effect than B on A' could all contribute to the statement 'A is more competitive than B'. However, it is unsatisfactory to use any one of these four senses on its own as a measure of competitive ability, since 'A is more competitive than B' would then not necessarily imply 'B is less competitive than A'. For example, A could have a greater effect on B than on itself, while at the same time B could have a greater effect on A than on itself. It is therefore necessary to combine the different senses in some way to form a satisfactory coefficient of competitive ability.

In experiments on three or more species, there is considerable choice over the reference species. Competitive abilities of each pair of species against each other may still be defined in terms of the relative effects of inter- and intraspecific competition within and between those two species. Alternatively, C may be a single reference species (which may be useful when comparing the abilities of several varieties of clover to compete with one variety of grass) or an average of several or all of the species under study. In the last case, competitive ability may be defined purely on the basis of the interspecific competition coefficients (p_{ij} or γ_{ij}), or it may also include the intraspecific competition coefficients.

As shown above, the outcome of competition depends on the relative effects of inter- and intraspecific competition. The main advantage of basing the measure of competitive ability on the relative effects of inter- and intraspecific competition is therefore the relevance of the measure to competition theory. A disadvantage is that, because the reference species is different for each pair of species, competitive hierarchies are not necessarily transitive; i.e. if A is more competitive than B and B more competitive than C, A is not necessarily more competitive than C (Keddy *et al.*, 1994). In contrast, the competitive hierarchy must be transitive if the same reference species (or set of reference species) is used for all species comparisons, although the rank order of species may change with different references and their rank orders for competitive response and competitive effect may differ.

When species compete for the same pool of resources and each species has the same effect on all competitors regardless of their identity (see the section below on resource complementarity), all these different meanings of competitive ability become equivalent. For example, competitive response is then the inverse of competitive effect. Conversely, the selected definition of competitive ability becomes important when these conditions are not met.

Summary

This section has discussed the importance of distinguishing between the intensity, effect and outcome of competition. It has also demonstrated the need to consider the relevance of frequency dependence, density dependence, size bias and the various possible meanings of competitive ability.

Once an experimenter has determined the objectives and relevant issues for an experiment, it is then possible to decide which indices of competition should be used: this is the subject of the next section.

Indices of Competition

This section presents some of the many indices that have been devised to summarize competitive interactions in terms of the intensity, effect and outcome of competition. The number of published indices is too great to review them all here. Rather, a few key indices have been chosen, which may be regarded as ideal for some purposes and which may be used to assess the merits of other indices.

Intensity of competition

The intensity of competition is the reduction in plant performance caused by competition. Several indices have been used, albeit with ambiguity over the qualifiers 'absolute' and 'relative'. Snaydon and Satorre (1989) and Snaydon (1991) use 'absolute

severity of competition' (they regard 'severity' as synonymous with, but preferable to, 'intensity') to describe the reduction in plant size relative to plants grown without neighbours:

$$I_0 = \log_{10}(w_0/w_n) \qquad (2.5)$$

where w_0 = size of plants grown without neighbours and w_n = size of plants grown with neighbours. In practice, as Snaydon (1991) observed, few studies on competition include treatments where plants are grown without neighbours and so it is rarely possible to measure this index. Most studies have some competition occurring in all treatments, and so at best can only measure the additional intensity of competition occurring in treatments with additional neighbours:

$$I_{ij} = \log_{10}(w_{ii}/w_{ij}) \qquad (2.6)$$

where w_{ii} = size of plants of species i grown without species j and w_{ij} = size of plants of i grown with j. The term w_{ii} may refer to the size of i in a monoculture of i or to the size of i in a multispecies mixture containing all the species of a community except j: all that matters is that only the density of j differs between w_{ii} and w_{ij}.

Snaydon and Satorre (1989) and Snaydon (1991) describe this index as the 'relative severity of competition'. However, to avoid confusion with other indices, especially those describing the relative severity of inter- and intraspecific competition, it is preferable to describe the index as the 'additional intensity of competition'. Besides avoiding the ambiguity of 'relative', this correctly describes what the index measures – the additional intensity of competition caused by the presence of species j at a particular density, on top of the lower intensity of competition experienced in its absence.

The index is identical to the log response ratio used in many field studies (e.g. Goldberg *et al.*, 1999), except that the latter uses natural logarithms. Goldberg *et al.* (1999) found that it was more effective in detecting significant differences than the conceptually similar 'relative competition intensity' recommended by Grace (1995).

Effect of competition

The greatest number of competition indices is found in studies on the effect of competition on the growth, reproduction or survival of plants. This section presents some indices that are 'ideal' for measuring the effect of competition, in the sense that they are based on the parameters of Equations 2.4. As such, they can be readily interpreted in terms of competition theory, and they provide required summaries about competitive relationships with no spurious frequency or density dependence.

Types of index

There are three general approaches to deriving competition indices. The first two involve estimating coefficients independently for each species – at least independently in a mathematical sense, although the estimates are not usually statistically or biologically independent. The first focuses on interspecific competition: for each species, estimates are obtained for its interspecific competitive response to other species and for its interspecific competitive effect on other species. The second approach focuses on the relationship between inter- and intraspecific interactions, as the relative response of each species to intraspecific and interspecific competition (relative competitive response) or the competitive effect of each species on other species relative to their intraspecific effect (relative competitive effect).

The third approach summarizes the two-way interaction between pairs of species rather than the contribution of each species to the interaction, using two indices. Resource complementarity (a term introduced by Snaydon and Satorre (1989), but with two distinct meanings, as discussed later) measures the extent to which competing species differ in their resource requirements. Competitive ability measures the ability of one species to compete with another in a two-species mixture.

Interspecific competitive response and effect

The γ_{ij} of Equations 2.4 is the interspecific competitive response of species i to species j. It is also the interspecific competitive effect of j on i. In practice, it is more common to measure interspecific competitive response and effect using density- and frequency-dependent functions of γ. Since the same coefficient may be regarded as an effect or as a response, it is meaningful to distinguish response and effect only in experiments involving many species, and an average response or effect of each species is estimated. Such experiments may involve multispecies mixtures or as a series of two-species

mixtures. In either case, Equations 2.4 must be extended. In the case of a series of two-species mixtures, there must be one equivalent pair of equations for each pair of species grown in a mixture. In the case of a multispecies mixture, there must be one equation for the yield of each species, but now each equation includes a separate γ for each other species in the mixture. In both cases, if we have n species in total, we have $(n^2 - n)$ different γ_{ij}, one for every combination of i and j where $i \neq j$.

The average interspecific competitive response of each species i (i = 1 to n) to all other species j (j = 1 to n, $j \neq i$) is then the average of γ_{ij} over all values of j. Similarly, the average interspecific competitive effect of each species i on all other species j is the average of γ_{ji} over all values of j.

The advantage of this approach is that it enables comparison of interspecific competitive interactions directly, without the additional complexities introduced by comparing intra- and interspecific effects of species pairs. In particular, it simplifies the identification and interpretation of competitive hierarchies in natural ecosystems. As such, it is particularly common in field experiments (e.g. Goldberg and Barton, 1992; Goldberg, 1996).

Relative competitive response and effect

The substitution rates (Connolly, 1987), S_{ij} and S_{ji}, of species j for species i and of species i for species j are:

$$S_{ij} = \gamma_{ij}/\beta_i$$
$$S_{ji} = \gamma_{ji}/\beta_j$$

(2.7)

where β and γ are coefficients from Equations 2.4, respectively quantifying the effects of intra- and interspecific competition; S_{ij} is thus the number of plants of species i that have the same effect as one plant of species j on the mean yield per plant of species i. For example, when S_{ij} = 1, one plant of j has the same effect as one plant of i on the yield per plant of i; and, when S_{ij} = 2, one plant of j has the same effect as two plants of i on the yield per plant of i. As such it is also a measure of relative competitive response, i.e. the response of species i to competition from species j relative to that from itself. The null hypothesis addressed by the substitution rate is that each species responds to interspecific competition in the same way as it responds to intraspecific competition, i.e. that $S_{ij} = S_{ji} = 1$.

Although it is rarely done, it is also possible to derive a corresponding pair of coefficients for relative competitive effect:

$$E_{ij} = \gamma_{ij}/\beta_j$$
$$E_{ji} = \gamma_{ji}/\beta_i \qquad (2.8)$$

Here, E_{ij} is the effect of species j on i relative to the effect of species j on itself. The null hypothesis addressed by these coefficients, $E_{ij} = E_{ji} = 1$, is that each species has the same competitive effect on its own yield as it has on the yield of other competing species.

De Wit (1960) argued that the tendency of a plant to acquire resources is a characteristic of the plant. As such, it might be expected that each species would have the same competitive effect on itself as on other species. For example, if the intraspecific effect β_i of species i is greater than the intraspecific effect β_j of species j, then the interspecific effect γ_{ji} of species i should be correspondingly greater than the interspecific effect γ_{ij} of species j. The null hypothesis $E_{ij} = E_{ji} = 1$ may therefore be biologically more appropriate than the null hypothesis $S_{ij} = S_{ji} = 1$. De Wit (1960) demonstrated that this was true in the case of an experiment on competition between peas and oats, but few other studies have examined the hypothesis.

The possibility that $E_{ij} = E_{ji} = 1$ is of considerable interest, since in this case interspecific interactions can be deduced purely from a knowledge of intraspecific competition (replacing γ_{ji} with β_i and γ_{ij} with β_j). Moreover, deviations from $E_{ij} = E_{ji} = 1$ would require interpretation purely in terms of interspecific competition, as opposed to a general effect on neighbours. It is therefore unfortunate that so little literature measures E.

Resource complementarity

Of particular interest, both ecologically and agriculturally, is the overall relationship between intraspecific and interspecific competition. As shown above from the Lotka–Volterra equations, stable coexistence of two species in an intimate mixture in a uniform environment requires that on average intraspecific competition must be stronger than interspecific competition. This situation implies a degree of niche separation between the species, such that individuals from different species are not competing for exactly the same set of resources (hence the term 'resource complementarity').

Conversely, mutual exclusion occurs if on average intra- and interspecific competition have the same effects (no resource complementarity). An alternative way to express the same condition is that the relative effects of i and j on i should be the same as their relative effects on j, i.e. that the ratio β_j/γ_{ij} is the same as the ratio γ_{ji}/β_j. This condition of no resource complementarity is represented by the situation $R_{ij} = 1$ in the following equation, which defines the recommended coefficient of resource complementarity:

$$R_{ij} = \frac{\beta_i\beta_j}{\gamma_{ij}\gamma_{ji}} = \frac{1}{S_{ij}S_{ji}} = \frac{1}{E_{ij}E_{ji}} \qquad (2.9)$$

The situation $R_{ij} = 1$ was described by de Wit (1960) as 'crowding for the same biological space', where the species are competing for the same pool of resources. In this situation, the effects of competition are particularly simple, with $S_{ij} = 1/S_{ji}$ and $E_{ij} = 1/E_{ji}$. That is, the relative competitive effect and relative competitive response of j are the inverse of those of i.

The possibility of $R_{ij} > 1$ introduces a consequence of additional agricultural interest, that a mixture of the two species will obtain more resources, and so give a higher relative yield, than the same plants grown over the same area but divided into two pure stands. De Wit and van den Bergh (1965) introduced the concept of relative yield total (RYT) to quantify this effect, and described the conditions under which the absolute yield of a mixture would exceed the pure stands.

It has become popular to use RYT as a measure of resource complementarity, although for this purpose it has undesirable properties (Snaydon, 1991; Sackville Hamilton, 1994). De Wit (1960, pp. 74–80) had earlier introduced a more appropriate measure that does not suffer the drawbacks of RYT, namely the product of the two relative crowding coefficients. At high density this product equals R_{ij} (Sackville Hamilton, 1994). In choosing between RYT and R_{ij}, it is necessary to determine whether the hypotheses being addressed relate to yield advantages of mixtures (if so, use RYT) or to biological resource complementarity of the species (if so, use R_{ij}).

Snaydon and Satorre (1989) introduced a second distinct concept of resource complementarity based on additive mixtures, in which each species is sown in pure stand at the same density as in the mixture. This coefficient is affected by the physical proximity of plants: the coefficient indicates complete 'complementarity' when plants are so far apart that they do not interact. That is, it measures the tendency of each plant to acquire molecules or photons of resource that are physically beyond the reach of any other plant. In contrast, R_{ij} measures the tendency of species to compete for the same type of resource, such that, regardless of the physical proximity of their plants, species with identical resource requirements will always have $R_{ij} = 1$, and $R_{ij} \neq 1$ only if the species are biologically different. To distinguish between these two different concepts, Sackville Hamilton (1994) suggested the terms 'physical resource complementarity' for the Snaydon and Satorre (1989) coefficient, and 'biological resource complementarity' for R_{ij}.

The coefficient of biological resource complementarity, R_{ij}, is essential for questions relating to the biological similarity of species, such as niche overlap, coexistence and yield benefits of mixtures. The coefficient of physical resource complementarity is more appropriate where it is desirable to combine the effects of biological similarity and physical proximity in a single coefficient.

Competitive ability

The following definition of competitive ability is orthogonal on a multiplicative (logarithmic) scale to resource complementarity as defined in Equation 2.9:

$$C_{ij} = \frac{\beta_i \gamma_{ji}}{\beta_j \gamma_{ij}} = \frac{S_{ji}}{S_{ij}} \qquad (2.10)$$

which is the ratio of the two substitution rates, which is also the ratio of the intraspecific effects of the two species multiplied by the ratio of their interspecific effects. Thus C_{ij} measures the competitive ability of species i against species j, such that, if $C_{ij} > 1$, species i is more competitive than species j.

The null hypothesis usually tested is that $C_{ij} = C_{ji} = 1$, i.e. that the two species are equally competitive. However, it was suggested above that $E_{ij} = E_{ji} = 1$ may be a biologically more realistic null hypothesis than $S_{ij} = S_{ji} = 1$. Equivalently, $C_{ij} = \beta_i/\beta_j$ may be a biologically more realistic null hypothesis than $C_{ij} = C_{ji} = 1$, i.e. that the relative abilities of two species to compete against each other equal their relative responses to intraspecific competition.

Other indices

The competition indices presented above are only a few of many published indices. Additional indices are needed where Equations 2.4 do not accurately describe competitive relationships. In such cases, competitive interactions depend on density or frequency of the species in the mixture, and the additional indices are needed to describe these effects.

Some competition indices are simple transformations of those presented above. For example, Snaydon and Satorre (1989) note that transformation of the indices to logarithms would facilitate interpretation by making values additive and symmetrical about zero, rather than ratios, which are multiplicative and asymmetrically distributed about one.

Most studies on competition use simplified designs that do not permit estimation of parameter values for Equations 2.4, and therefore usually do not allow estimation of the indices in Equations 2.7 to 2.10. Many alternatives have been developed for these designs, but all share the problem of spurious frequency and density dependence. This problem applies to most of the most popular indices, such as relative crowding coefficient, relative competition intensity, relative yield, aggressivity, competitive ratio and so on (for a summary of additional coefficients, see Jolliffe, 2000). Detailed assessment of these and other indices in common use is beyond the scope of this chapter. We merely caution the reader to be aware of their limitations, and discourage the uncritical use of any index without first considering its properties in the context of the equations presented here.

Outcome of competition

The above indices, which quantify the effect of competition on growth, reproduction or survival, are not generally suitable for studies on the outcome of competition, i.e. on long-term changes in composition of a mixture. As discussed above in the section on size bias, they can suggest an intrinsic competitive advantage of large size that is not reflected in the outcome of competition. By applying the 'thought experiment' of Connolly (1986), it can be shown that the size bias applies to β, γ, I, S and C (Equations 2.4, 2.6, 2.7 and 2.10) but not to E or R (Equations 2.8 and 2.9).

Despite the absence of a size bias, R has additional problems limiting its value for deducing coexistence. If $R > 1$, the two zero isoclines in

Fig. 2.1 have different slopes. This is necessary but not sufficient for the crossover of zero isoclines. Thus, since stable coexistence in a uniform environment requires crossover of the zero isoclines, $R > 1$ is necessary but not sufficient for coexistence.

The size bias in β, γ, I, S and C can sometimes be eliminated by measuring initial biomass, regarding initial biomass as a measure of functional density and replacing density with initial biomass in Equations 2.4. However, initial biomass is difficult to control directly. With experimental designs where data can be analysed by regression of yield on density, it is sufficient to control initial density and measure initial biomass (J. Connolly, Dublin, 1999, personal communication). This approach fails when using the simpler experimental designs that require initial density to be fixed at a predefined value. For these designs the outcome of competition can be assessed only by using indices devised specifically for the purpose.

Relative reproductive rate

The relative reproductive rate α of de Wit (1960) is a key index for describing changes in mixture composition:

$$\alpha_{ij} = \frac{\left(\dfrac{O_i}{O_j}\right)}{\left(\dfrac{N_i}{N_j}\right)} \qquad (2.11)$$

$$= \frac{o_i}{o_j}$$

In this equation, O_i and O_j are the number of seed produced per unit area produced by species i and j, in the same units as their initial densities N_i and N_j. That is, the relative reproductive rate is the ratio of the seed yields per unit area, relative to the ratio of initial number of seed per unit area. If these two ratios are the same, then the relative abundance of the two species in the mixture does not change and the relative reproductive rate is 1. Since $O_i/N_i = o_i$ is the mean number of seed produced per plant, the relative reproductive rate is more simply expressed as $\alpha_{ij} = o_i/o_j$, the ratio of the number of seed produced per plant of the two species.

The use of relative reproductive rate based on seed counts can be problematic for pastures, especially in pastures of perennial species where clonal dynamics (e.g. production and loss of tillers) are often a major component of the dynamics and persistence of plants. The number of seed produced is then often not an appropriate unit for assessing mixture composition. The use of number of vegetative units such as tillers, stolons or rooted units can be an appropriate alternative. To obtain a valid estimate of α, it is essential to use the same units for both initial (N) and final (O) densities: if O is measured as density of tillers, then so must N.

Counting vegetative units can also be questionable because of the variation in size of tillers produced. Moreover, it is often extremely laborious. Biomass can be a more appropriate and more convenient measure. Again, the caveat applies that the same units must be used for both O and N. The resource efficiency index of Connolly (1987) is the appropriate index, effectively identical to relative reproductive rate, except that it measures the relative biomass production per unit initial biomass. Since the two coefficients are identical except in the units of measurement, they will hereafter be discussed jointly and referred to, using de Wit's term, as α.

The value of α varies with density and frequency of the species in the mixture. The pattern of frequency and density dependence is crucial to the final outcome. It is therefore necessary to explore these patterns. Figure 2.1 shows a simple example of frequency dependence in α even without density- or frequency-dependent competition. The pattern can often be more complex, because Equations 2.2 and 2.4 often do not accurately describe variation in seed production. However, for simplicity we shall explore only the simple case, where Equations 2.4 are accurate.

To ensure that we use the same units (we shall use numbers of seed, but could equally use biomass) on both sides of the equation, we put o instead of w on the left-hand side of Equations 2.4. We then substitute o_i and o_j in Equations 2.11 with the right-hand side of Equations 2.4. To enable a distinction to be made between frequency and density dependence, we define D as the combined effective density of i and j, $D = \beta_i N_i + \gamma_{ij} N_j$. This is the effective density for i, since, at a given value of D, the mean plant size of i is constant regardless of the relative densities of i and j. It is also the effective density for j and for the whole mixture if $R_{ij} = 1$ (i.e. if the average severity of intraspecific competition equals the average severity of interspecific competition). Then, by expressing α in terms of N_i and D instead of N_i and N_j, density dependence appears as variation with D and frequency dependence as variation with N_i (since, at a given value

D, any increase in N_i must be accompanied by a corresponding decrease in N_j). This gives:

$$a_{ij} = \frac{o_i}{o_j}$$

$$= \frac{a_j}{a_i} \left[\frac{1 + E_{ji}\left(1 - R_{ij}\right)\beta_i N_i + E_{ji} R_{ij} D}{1 + D} \right] \quad (2.12)$$

This analysis reveals several important characteristics of α. First, in the absence of frequency- and density-dependent competition, α is frequency-dependent only if $R_{ij} \neq 1$ (if $R_{ij} = 1$, $1 - R_{ij} = 0$ so there is no variation with N_i). If $R_{ij} > 1$, any gain to i decreases as the relative frequency of i increases, i.e. the rarer species tends to be favoured. This is the result discussed earlier – that stable coexistence in a uniform environment requires $R_{ij} > 1$.

Secondly, in the absence of frequency- and density-dependent competition, α varies with the effective combined density D even if $R_{ij} = 1$, unless $E_{ij} = (1/E_{ji}) = 1$. That is, α is density-dependent if each species has interspecific competitive effects that do not simply reflect how they crowd for space in monoculture.

Thirdly, at all densities if $E_{ij} = E_{ji} = 1$, or just at low densities if $E_{ij} = (1/E_{ji}) \neq 1$, we have quite simply $\alpha_{ij} = a_j/a_i$. Remembering that $1/a_i$ is the yield (in this case measured as number of seed) produced by a single isolated plant of i growing without competition, α is then simply the relative reproductive rate of spaced plants of i and j. This is a key point, emphasizing again the difference between the effect and outcome of competition. The 'outcome of competition' is basically the relative reproductive rate of plants grown without competition. In one sense, this clearly has nothing to do with competition, but, in another sense, it correctly reflects the fact that the winner of any competition is simply the fastest, regardless of whether competition is actually occurring. The 'effect of competition' is then to modify that outcome to generate more complex solutions.

Fourthly, at high density with $E_{ij} = (1/E_{ji}) \neq 1$, the equation simplifies to $\alpha_{ij} = E_{ji} a_j/a_i$, that is, α_{ij} is E_{ji} times its basic value of a_j/a_i. This again is a key feature. It shows that the effect of interspecific competition that fundamentally determines the output of competition is not the overall effect of interspecific competition or any of the popular measures of competitive ability. Rather, it is the special additional effects of interspecific compe-

tition on top of the basic effects of crowding for space in monoculture. That is, i has a competitive advantage over j if its effect, γ_{ji}, on the yield of j is greater than its effect, β_i, on its own yield.

Of the three key components of α (resource complementarity, R, relative competitive effect, E, and spaced plant reproductive rate, $1/a$), only one (R) is commonly measured in competition studies. There is a good body of information on interspecific variation in $1/a$, at least in terms of maximum relative growth rate of seedlings (e.g. Cornelissen et al., 1996), although it has not been obtained as part of competition experiments. It is suggested that in future higher priority be attached to enhancing competition studies by measuring relative competitive effect and the reproductive rates of plants growing without competition.

Summary

In this section we have discussed a range of key indices used to describe the intensity, effect and outcome of competition. No attempt has been made to discuss all available indices, regardless of their current popularity. The indices discussed were chosen because they are considered ideal for the majority of purposes, including the purpose of evaluating the merits and demerits of other competition indices.

It is argued that most studies have failed to measure certain key indices, despite their central importance for understanding competition. In particular, the relative competitive effect, E (Equation 2.8) is a key index for both the effect and the outcome of competition. The reproductive rate or relative growth rate, $1/a$ (Equations 2.3 and 2.4), of plants grown without competition is a key index for the outcome of competition. It defines the baseline expectation for the outcome of competition between species with neutral competitive effects.

Methodologies

The previous section presented a range of indices for different purposes. Having chosen the most appropriate indices for a particular study, an experimenter can then proceed to determine the most appropriate methodologies for estimating values of the chosen indices. A description of possible methodologies is the subject of this section.

Experimental approaches

Numerous designs have been developed to study different aspects of competition (e.g. Harper, 1961; Begon *et al.*, 1996). No single approach is suitable for all purposes. The experimental study of competition requires experimental manipulation of the severity of competition. At its simplest, this may just involve comparing a 'with competition' treatment and a 'without competition' control. Alternatively, it may involve establishing a series of treatments varying quantitatively in the severity of competition.

The severity of competition is usually manipulated indirectly, by controlling the initial density or size of individuals in a plot. Average density or size may be controlled over whole plots, or systematic designs may be used that specify the size and position of individual plants. The assumption is that competition is more severe with more closely spaced and/or larger plants, as described at its most simple by Equations 2.2 and 2.4. Measured effects of initial density or size on growth, reproduction or survival can then be interpreted in terms of competition.

The severity of competition may also be manipulated directly, by controlling the supply of resources. The advantage of this approach is that it enables competition for each resource to be studied independently. For example, competition for P may be studied by restricting the P supply while supplying abundant light, water and other minerals. However, this approach is not sufficient on its own, because it does not enable separation of the effects of competition from the non-competitive effects of variation in nutrient supply. The approach is useful for studies on competition when combined with the experimental manipulation of density or size.

Studies on the outcome of competition require comparison of mixture composition before and after a period of competition. At its simplest, this can be done with only one initial mixture composition. However, experimental manipulation of the severity of competition is central to assessment of whether the outcome of competition depends on its initial composition.

There is a fundamental, intrinsic problem with the experimental manipulation of competition. The outcome of competition is a change in the very attributes (density, size and resource supply) that need to be controlled in order to manipulate the severity of competition, leading to uncontrolled,

self-driven changes in severity with time. The consequence of this is that treatment effects can be directly associated with competition effects only over a single period starting at the moment of manipulation. Assessment of changes in mixture composition over several periods of time is an essential part of the suite of tools for analysing the outcome of competition (e.g. de Wit *et al.*, 1966; Tow *et al.*, 1997), but attributing changes in mixture composition to competition effects can be problematic.

Artificial plots

The use of artificial plots, in glasshouse or field, has been one of the most popular experimental approaches. The main advantage of the approach is its ease. Plots can be set up with full control over initial densities in mixture and in pure stand, and with good control over environmental heterogeneity.

One disadvantage of the approach, as with all artificial plots, is the questionable relevance to the field. Indeed, the existence and importance of competition in the field have long been debated in the ecological literature, despite a very large literature demonstrating competition in artificial plots (see, for example, Connell, 1983; Schoener, 1983).

In addition, the above-mentioned problem – that treatment effects can be directly associated with competition only over a single period starting when the plots are established – causes a particular difficulty for studying competition in permanent pastures based on perennial forages. It is impossible to exclude the effects of competition during the period of establishment of the plots. Since most studies involve sowing seed, they inevitably address seedling competition. Most studies are of short duration, and so competition effects are dominated by seedling competition. These may be of little relevance to competition among mature plants and therefore to the long-term persistence and stability of mixtures in established pastures. Even in studies that are continued for several years, results still depend partly on competition during the initial establishment phase and can be difficult to interpret.

A further problem is a logistical one – that experiments on competition in artificial plots can be so large that it is often impractical to study multispecies mixtures. The majority of published experiments deal only with two-species mixtures (Gibson *et al.*, 1999).

Field studies

Density in pre-existing communities in the field may be manipulated by adding new plants or by removing existing ones (e.g. Goldberg, 1996; Reader *et al.*, 1994). The approach has not been as popular as experiments with artificial plots because of the greater difficulty of manipulating the severity of competition. However, it has a number of major advantages. Obviously it generates results that are directly relevant to competition in the field. Unlike glasshouse experiments, measures of the intensity of competition are of direct interest *per se*. In addition, most experiments deal with multispecies mixtures. Finally, of particular importance for pastures of perennial forages, the approach does not depend on establishing new plots and therefore is readily applicable to the study of competition in mature communities, ignoring seedling competition. Because of these advantages, further use of this approach should be encouraged.

Difficulties associated with the approach include the high level of environmental heterogeneity in most field communities, necessitating high replication, simple designs and special consideration of spatial heterogeneity (Goldberg and Scheiner, 1993; Ives, 1995).

In addition, manipulating density by adding or removing plants presents the major practical difficulty of how to avoid causing other environmental changes, so that the effects of manipulation can be attributed to competition rather than to the environmental side-effects. For example, removal of plants by systemic herbicides increases the organic matter content of the soil. Mechanical removal of above-ground parts may leave plants able to re-establish from below-ground parts. Mechanical removal by cultivation to kill below-ground parts causes major changes in soil structure, nutrient cycling and the soil microbial community. All of these changes will affect nutrient uptake by the plants and so may affect competition.

Likewise, addition of plants changes the soil during the process of planting, and it may be very difficult to add plants that are equivalent to those already present. These difficulties can be overcome by adding seed to the seed rain instead of adding adult plants, but this gives relatively poor control over the target density. It may work relatively well in self-seeding pastures of annual forages, but is likely to be relatively ineffective in temperate permanent pastures of perennial forages, where seedling establishment is low in mature pastures.

Measurement

Level of measurement

Measurements are often made on whole plots, giving data that can be used to fit whole-population equations, such as Equations 2.2 or 2.4, or to estimate other forms of whole-population competition indices.

An alternative to measuring whole-plot variables is the 'neighbourhood' approach. In this approach, measurements are made on the growth, reproduction or survival of individual plants, and the measurements are analysed in relation to the size of the plant and the size and proximity of their neighbours (Ives, 1995; Bi and Turvey, 1996; Oosthuizen *et al.*, 1996; Wagner and Radosevich, 1998; Grist, 1999). This is a more recent approach than whole-population studies, and a range of different regression algorithms have been tried in different situations. It is premature to propose ideal designs and analyses for such experiments. However, it should be emphasized that, being based on the individual plant, they provide a more promising way forward in terms of understanding the process and mechanisms of competition.

Choice of experimental units for measurement

In many published studies, all experimental units (i.e. all the plots or plants grown in the experiment) are measured. The same plants are used both to cause a competitive effect and to measure a competitive response. In two situations the experimenter may choose to separate plants into two groups, using the response of one group of plants to measure the competitive effect of another group.

Firstly, all experiments on mixtures raise a statistical difficulty, namely, that the yields of all species within one mixture will show correlated errors, whereas conventional statistical analyses assume independent measurements of yield. For some designs, statistical solutions to this difficulty have been published in which the correlation is estimated and allowed for (e.g. Machin and Sanderson, 1977). Alternatively, the problem can be avoided by using more plots (doubling the number of mixed plots in the case of competition between two species) and measuring only one of the species in each plot, which generates truly independent sets of measurements for each species (e.g. Law and Watkinson, 1987).

Secondly, in field experiments plants may be sown as phytomers, i.e. used solely to measure the competitive effect of neighbouring plants. Sown at low density, they will have minimal effect on the severity of competition experienced by neighbouring plants. The advantage of using phytomers rather than existing plants is the opportunity it affords to standardize the initial size of the phytomers, thus reducing the high error terms associated with field experiments.

Experimental Design and Analysis

The previous section discussed generic approaches in terms of working on artificial plots or field communities and of what to measure. Having chosen the most appropriate approach, the experimenter can then proceed to the subject of this section, which is to determine the most appropriate experimental design.

This section presents a range of different designs and considers the advantages and disadvantages of each. For simplicity, specific designs will be discussed only for competition between two species, although the principles generalize readily to more species. Similarly, and again without loss of generality, discussion will be restricted to random mixtures, in which plants of the two species are mixed and distributed at random. The same principles apply to non-random mixtures, in which experimental design defines the precise position of each plant in the mixture. Since the key issue is to match experimental design to objectives, designs will be discussed primarily in terms of what can and cannot be achieved with them.

Implicit in Equations 2.4 is that an experiment on competition between two species ideally requires both species to be present initially at a wide range of densities, in mixture and in pure stand. Plotting the initial densities of the two species on orthogonal axes, the 'treatments' should form a complete two-dimensional plane of densities. Growth, reproduction and/or survival of both species should be measured, giving two measured response surfaces to the two-dimensional density plane.

Quantifying this double response surface is 'ideal' in the sense that it provides the most complete description of the effects of competition in one set of conditions. However, it requires exceptionally large experiments. Quantifying the full double response surface may be very far from ideal

if the resulting size of the experiment prevents other important factors from being addressed. The design of competition experiments will always involve a compromise, in which some factors are ignored, despite their importance for competition, in order to make it possible to study other factors. It is up to each experimenter to ensure that the factors ignored and the factors studied are appropriate to the hypotheses at hand.

Figure 2.3 illustrates four experimental designs that allow estimation of response surfaces. Figure 2.4 illustrates four simplified designs that provide less complete information on competition. Table 2.1 summarizes the advantages and disadvantages of each of these designs, as discussed below.

Response-surface designs

All designs in Fig. 2.3 enable estimation of all parameters in Equations 2.4, and therefore enable estimation of the ideal competition indices presented in Equations 2.7 to 2.10. In all cases, analysis should involve regression of the final state of each species on the initial state of the composition, with due statistical care for issues such as normality, homogeneity, correlated errors, tests for goodness of fit to the equations and tests of null hypotheses. In all cases, size bias is readily allowed for by replacing initial density in Equations 2.4 with initial biomass.

Snaydon and Satorre (1989) recommend a 'bivariate factorial design' (Fig. 2.3a), although the design is not factorial (there can be no data from a 'plot' with zero density of both species) and should not be analysed as a factorial design. The design is sometimes considered the most comprehensive of all. However, since it includes mixtures at twice the density of the highest pure-stand density, it involves using unrealistically high mixture densities and/or failing to include realistically high pure-stand densities (J. Connolly, Dublin, 1999, personal communication).

Figure 2.3b shows the most complete recommendable design that does not suffer the conceptual difficulty of the design in Fig. 2.3a. The design enables complete coverage of the entire biologically realistic part of the density plane. It is a large design, justifiable only when the hypotheses under consideration require quantification of the nature of frequency- and density-dependent variation. The design is particularly important for stud-

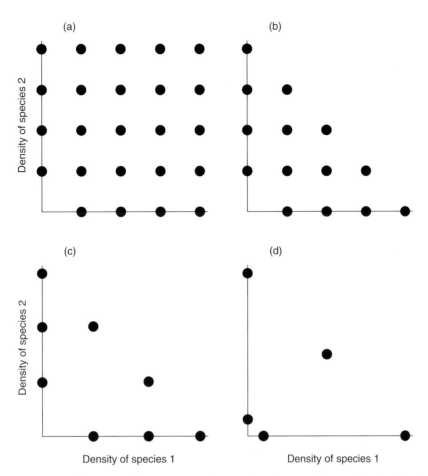

Fig. 2.3. Experimental designs enabling response-surface analysis of competition: (a) bivariate factorial (Snaydon and Satorre, 1989); (b) generalized response surface; (c) minimal response surface for detecting frequency and density dependence; (d) minimal response surface ignoring frequency and density dependence.

ies on the outcome of competition, since they require measurement of numbers of offspring produced, and, as noted above, numbers of offspring are not usually described adequately by Equations 2.4. Addressing other important factors, such as variation in soil, management, etc., at the same time as quantifying frequency- and density-dependent variation may make an experiment impossibly large.

Figure 2.3c is the simplest possible design enabling detection of both true density-dependent and true frequency-dependent variation. Because of its smaller size, it should be used in preference to the design in Fig. 2.3b if an objective is to be able to detect true density and frequency dependence but not its precise algebraic form.

Figure 2.3d is the simplest possible design that enables estimation of all the competition indices in Equations 2.7 to 2.10, with no spurious frequency or density dependence and with easy correction for size bias. With this design it is not possible to test the goodness of fit to Equations 2.4. As such, it assumes that there is no true density- or frequency-dependent variation. This may be considered a disadvantage: if so, the more complex designs must be used. On the other hand, it has major advantages over the simpler designs adopted and accepted by many authors. As a compromise that achieves relative simplicity while still permitting complete unbiased estimation of all relevant competition coefficients, the design thus has much to recommend it.

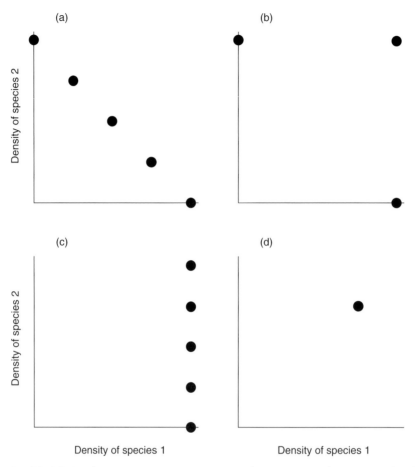

Fig. 2.4. Simplified designs for competition experiments: (a) replacement series (deWit, 1960); (b) additive mixture (Snaydon, 1991); (c) target-neighbour (Gibson *et al.*, 1999); (d) simple pairwise mixture (Gibson *et al.*, 1999).

Cross-sectional designs

The designs in Fig. 2.4 represent cross-sections through the two-dimensional density plane. As such, they all provide only limited information on competition and they all share several problems. First, the designs are of relatively little value for studies on the outcome of competition. Secondly, they all confound density and frequency in some way. Thirdly, none of them enables estimation of all the parameters in Equations 2.4, and so almost all derived competition indices show spurious density and frequency dependence. Fourthly, as noted above, this makes it difficult to allow for size bias when biomass is measured. Fifthly, none of the cross-sectional designs can be used to estimate the relative competitive effect, E (Equation 2.8).

These disadvantages are counterbalanced by one important advantage: their small size allows other factors to be addressed. It is crucial to ensure that the information obtained is relevant, and that the information disregarded is not relevant, to the hypotheses being addressed. If the experimenter fails to appreciate the biological significance of the aspects of competition ignored by each design, they are all easy to misinterpret.

Much of the debate over optimal designs for competition experiments centres on comparing 'replacement' or 'substitutive' designs with 'additive' designs.[2] This aspect of the debate is only relevant for cross-sectional designs that do not allow estimation of the complete response surfaces. However, the term 'additive design' has been used in many different ways, including the designs shown in

Table 2.1. Summary of the advantages and disadvantages of a range of experimental designs for competition studies.

Feature of design	Experimental design (illustrated in Figs 2.3 and 2.4 as indicated in parentheses)						
	Bivariate factorial (Fig. 2.3a)	General response surface (Fig. 2.3b,c)	Minimal response surface (Fig. 2.3d)	Replacement series (Fig. 2.4a)	Additive mixture (Fig. 2.4b)	Target-neighbour (Fig. 2.4c)	Simple pairwise mixture (Fig. 2.4d)
Estimate β? (Equation 2.4)	Yes	Yes	Yes	No	No	No	No
Estimate γ? (Equation 2.4)	Yes	Yes	Yes	No	No (but can estimate $a\gamma$)	No (but $a\gamma$ for one species)	No
Estimate I? (Equation 2.6)	Yes	Yes	Yes	No	Yes	Yes	No
Estimate S? (Equation 2.7)	Yes	Yes	Yes	Yes at high equal density	No	No	No
Estimate E? (Equation 2.8)	Yes	Yes	Yes	No	No	No	No
Estimate R? (Equation 2.9)	Yes	Yes	Yes	Yes at high density	No	No	No
Estimate C? (Equation 2.10)	Yes	Yes	Yes	Yes at high equal density	No	No	No
Estimate α? (Equation 2.11)	Yes	Yes	Yes	Yes	Yes	Yes	Yes
Easy to correct for size bias of β, γ, I, S, C?	Yes	Yes	Yes	No	No (yes for $a\gamma$)	No (yes for $a\gamma$)	No
Detect true density and frequency dependence?	Yes	Yes	No	Yes, but density and frequency confounded	No	Yes for γ of one species, but confounded	No
Spurious density dependence?	No	No	No	At low density	At low density (No for $a\gamma$)	At low density Yes (No for $a\gamma$)	Yes
Spurious frequency dependence?	No	No	No	Yes	Yes	Yes	Yes
Size of experiments	Large	Large	Medium	Small	Small	Small	Very small
Preferred usage	Studies on outcome of competition	Studies on outcome of competition	Crop mixtures with E and no size bias	Crop mixtures; smallest design for S, R, C. Mechanical diallels	Field studies Mechanical diallels	Weed–crop studies	Mechanical diallels

Fig. 2.3(a) (Snaydon and Satorre, 1989), 2.3(b–d) (Gibson et al., 1999), 2.4(b) (Goldberg and Scheiner, 1993), 2.4(c) (Snaydon, 1991) and 2.4(d) (Austin et al., 1988). Figures 2.3(a–d) also incorporate replacement designs, so the potential for confusion is clear and the term 'additive' should be used with caution.

Replacement designs

Since 1960, replacement designs (Fig. 2.4a) have been, and remain, the most popular design for competition experiments, yet they have been heavily criticized for many reasons by many authors throughout the 1980s and 1990s (Gibson et al., 1999). The criticisms cover not only the design but also the large number of publications that incorrectly use, analyse or interpret results. Gibson et al. (1999) conclude: 'The tendency to misuse the method is so pervasive that its continued use should be discouraged.' Like all cross-sectional designs, the replacement design has value for certain purposes, but the value is limited and the flaws necessitate great caution in its application (Jolliffe, 2000).

The value of de Wit's (1960) 'relative crowding coefficient' (which he denoted k) depends on both the absolute and relative densities used for the two pure stands. It is density-dependent only at low densities; as density is increased, it tends towards a density-independent value. In a replacement design with high, equal, pure-stand densities, k_{ij} equals the inverse of the substitution rate S_{ij} (Equation 2.7) of Connolly (1987) (Sackville Hamilton, 1994). Therefore the design can also be used to estimate resource complementarity, R_{ij} (Equation 2.9), and competitive ability, C_{ij} (Equation 2.10). It is the most size-efficient design available for estimating these coefficients, which is its major advantage over other designs. Conversely, it cannot be used to estimate these coefficients if the densities of the two pure stands are unequal or not high enough.

Like S_{ij}, de Wit's (1960) relative crowding coefficient measures relative responses to intra- and interspecific competition. Other cross-sectional designs fail to address this aspect of competition. Conversely, the replacement design has one major conceptual limitation that is overcome with other cross-sectional designs: it fails to estimate the intensity of competition, either overall or of intra- or interspecific competition.

The problems of size bias, spurious frequency and density dependence, difficulty of analysing and interpreting, failure to estimate relative competitive effect, E, the limited value of estimates of relative replacement rate, α, and the lack of relevance of results to the outcome of competition are the same as for other cross-sectional designs. Density and frequency are confounded, but in a qualitatively different way from other cross-sectional designs: density- and frequency-dependent variation both appear as frequency-dependent variation in S_{ij}.

Replacement designs can also be constructed with multispecies mixtures, although this is rare in the literature. It is more often used as the basis for comparing many species in all possible paired combinations (a physical or mechanical diallel (Harper, 1977)). Usually this is done only with a single 1 : 1 mixture of each pair of species. The value of replacement mechanical diallels is best conceptualized by thinking of one of the species in a mixture as the experimental unit and the second species as the treatment. The pure stands are then simply another treatment, at a density such that conceptually half the plants in the pure stand correspond to the measured experimental unit and half to the treatment. That is, 'competing species' is the treatment design, with one level of the factor for each species in the experiment, and the design simply compares the effects of all species on the yield of their neighbours.

Additive mixtures

Additive mixtures (Fig. 2.4b) ignore intraspecific competition. For hypotheses not concerned with intraspecific competition, additive mixtures are preferable to replacement mixtures. Adding or removing plants enables estimation of the intensity of competition caused by those plants, which overcomes the main conceptual limitation of replacement designs.

Like other simplified designs, the competition indices in use for additive designs show spurious density and frequency dependence. One possible index shows no spurious density and frequency dependence, but it appears not to have been used. Specifically, linear regression of $1/w$ of one species on the density of the other species estimates the products $a_i\gamma_{ij}$ and $a_j\gamma_{ji}$ of Equations 2.4, i.e. the effects of interspecific competition not standardized for the size of plants grown without competition. Because it is not standardized for size, the coefficient has little value except in ranking several species j in terms of their competitive effect on one species i, for which purpose all the $a_i \gamma_{ij}$ share the same value, a_j, so that the species are ranked by γ_{ij}.

Since the design has only one mixture, neither true nor spurious density and frequency dependence can be detected or estimated, so confounding of density and frequency is complete.

Like the replacement design, additive mixtures can also be used to study many species with a mechanical diallel of all possible pairs of 1 : 1 mixtures. The only design difference between additive and replacement mechanical diallels is that the additive pure stands are at half the density, but this small design difference has large implications for analysis and interpretation. Pure stands of additive mechanical diallels represent 'low competition controls', used to assess the increase in severity of competition caused by adding interspecific competition. Thus only the interspecific competitive effects and responses of each species are measured.

Essentially the same design is also commonly used as the basis for field experiments where density is manipulated by removing plants (e.g. Reader *et al.*, 1994). In removal experiments in the field, the basic design is the full multispecies community, and each 'pure stand' depicted in Fig. 2.4b is actually a mixture of all the species except the one species removed. Then the graph is conceptually a hypervolume, with one axis for each species in the community. The analysis of data can be essentially identical. The popular use of this design for field experiments adds to the value of using the approach in the glasshouse also, since results from glasshouse and field experiments are then directly comparable.

Target-neighbour designs

Target-neighbour designs (Fig. 2.4c) are appropriate when the primary interest is in the one-directional competitive effect of one species (typically a weed) on a second (typically a crop). In these cases, the performance of the weed in pure stand may be of no interest and the density of the crop may be fixed by agronomic practice. All relevant information can then be obtained by varying the density of the weed at a single crop density, including estimation of $a_c \gamma_{cw}$ (using subscripts c and w for crop and weed). Density and frequency are again confounded, but in this case density and frequency dependence both appear as density-dependent variation in γ_{cw}.

Simple pairwise mixtures

A single pairwise mixture (Fig. 2.4d) on its own provides no useful information on the effects of competition. A series of repeated measurements on the composition of one such mixture is sufficient for a descriptive study of the outcome of competition from a single starting-point, but cannot be interpreted in terms of competitive effects or other factors affecting the outcome.

The main use of these mixtures for assessing the effects of competition is in comparing many species in all possible combinations in the smallest possible design. This mechanical diallel without pure stands is analogous to a genetic diallel cross without parents. Its main benefit is its minimal size. Like the replacement and additive diallels, it enables estimation of interspecific competitive response, competitive effect and physical combining abilities, using indices with all the disadvantages of the other cross-sectional designs. Additional disadvantages are that it fails to estimate either the severity of interspecific competition or the relative severity of intra- and interspecific competition.

Summary

This section considers the advantages and disadvantages of a range of experimental designs for the study of competition. Optimal design depends on the choice of experimental approach and competition index, both of which depend on objectives. Selection of the optimal design for a given objective therefore logically represents the last stage in planning a competition study.

However, before concluding the chapter, we shall briefly consider the implications of some of the special features of grass–legume dynamics for studying competition in pastures.

Dynamics of Grass–Legume Mixtures

The purpose of this section is to indicate, without duplicating the detail in the rest of this book, how the special features of grass–legume dynamics may influence results. It seeks thereby to clarify interpretation of some competition indices, especially with regard to how they change with time in experiments with repeated harvests. In addition, it highlights some important questions specific to grass–legume mixtures, and so aims to guide future experimental objectives and help improve future experimental designs for competition studies.

Mixtures of N_2-fixing legumes with non-legumes are among the very few cases where competition experiments consistently show yield advantages of mixtures (de Wit *et al.*, 1966; Trenbath, 1974) and, by implication, niche differentiation, and a possible equilibrium frequency and possible coexistence through the Lotka–Volterra mechanism illustrated in Fig. 2.1. This is attributed to N_2-fixation by the legume, which gives it access to a N source not accessible to the grass – which, after all, is one of the most important reasons for using legumes in pastures and other cropping systems.

Determining the existence and characteristics of an equilibrium state are important issues, especially in the context of sustainable pastures. If there is an equilibrium state, a mixture will change in frequency and density towards the equilibrium regardless of the starting-point. At its simplest, this may be demonstrated in a single-harvest experiment at a wide range of frequencies and densities.

However, there is also a need for confirmation in the field that initial frequency- and density-dependent relative replacement rates do indeed result in convergence on a stable end-point. This requires repeated harvests and a demonstration that widely different initial densities and frequencies finally converge on the same value. If there is a stable equilibrium, initial plant density or frequency will not affect the final state, although it will change the speed of approach and the path towards that final state. Similarly, the use of additive or replacement mixtures will not change the final state, although it will change the speed of approach and the path towards that final state.

A consequence of the approach to a single final equilibrium point is a progressive reduction in frequency- and density-dependent variation in indices such as relative replacement rate, relative yield and relative intensity of competition. This necessarily occurs because of the progressive reduction in the range of frequencies and densities in different treatments. However, it does not necessarily imply a reduction in the underlying patterns of frequency and density dependence of these indices. Examination of any possible temporal changes in these underlying patterns would require repeated experimental manipulation of density to re-establish differences in density and frequency.

A further consequence is a reduction in difference between additive and replacement mixtures. For example, the difference between replacement relative yield and additive relative yield, measured for one harvest interval, will decrease with time. In a sense, this is an artefact, because the relevant densities of the mixture and monocultures are their densities at the beginning of the harvest interval in question; and, as already indicated, they become progressively more similar, regardless of whether they started as additive or replacement mixtures.

It should be noted that convergence is towards the 'replacement' version of competition indices. As time passes, mixtures and monocultures both tend towards their equilibrium (maximum) yields and densities per plot. This means that the additive mixture is no longer increasing the intensity of competition over that experienced in the monoculture; instead, it tends towards the replacement mixture, where the total intensity of competition in monoculture and mixture is the same. Thus, for repeat harvests, additive mixtures cannot be used for their primary purpose, which is to measure the additional intensity of interspecific competition.

The above discussion applies primarily to grass–legume dynamics in the simplest case, with N_2-fixation by legumes and N-limited mixture yield. In this case, Equations 2.2 and 2.4 may be accurate and Fig. 2.1 may give an accurate summary of dynamics and stable coexistence. Three additional complications may arise.

First, mixture growth rate may not be N-limited. If the soil is fertile with abundant N, growth rate may be limited by pH, water-supply or other nutrients or, if cutting and grazing are sufficiently infrequent, by light. In this case, N_2-fixation by the legume cannot increase mixture yield. However, provided grass is a superior competitor for soil mineral N, in theory this would result in frequency-dependent competition, with deviations from Equations 2.2 and 2.4 favouring the rarer component. This would again result in the existence of a stable equilibrium and convergence of all mixtures on that final equilibrium state, regardless of their initial composition, although the mathematical description of convergence would be more complicated and the precise speed and paths to convergence would differ. There is little literature empirically testing this theoretical effect of soil N status on the adequacy of Equations 2.2 and 2.4 to describe grass–legume competition.

A second complication arises from the transfer of fixed N from legume to grass. This would also result in frequency-dependent competition, with deviations from Equations 2.2 and 2.4 favouring the rarer component – again provided that the grass is a superior competitor for soil mineral N. However, there is one major difference from the previous case. Frequency dependence resulting from mixture yield not being N-limited would occur immediately. In contrast, frequency dependence resulting from transfer of fixed N from legume to grass would not occur immediately because of the time delay involved in N transfer. So, in this case, Equations 2.2 and 2.4 would initially suffice and competition would change to being frequency-dependent only after sufficient time has elapsed for N transfer to occur. Harris and Thomas (1973) and Harris (1974) demonstrated an increase with time in the frequency dependence of competition between white clover and grass. It is supposed that N transfer is the cause of their results.

The third and biggest complication is a direct consequence of the time delay involved in N transfer from legume to grass. The equilibrium state may then not be a single mixture composition but rather an equilibrium cycle, involving indefinitely repeated fluctuations in mixture composition. The reason, proposed by Turkington and Harper (1979) as a possible explanation of the observed dynamics of grasses and white clover in permanent temperate pastures, is as follows. Ryegrass is a superior competitor at high soil mineral N. White clover is the superior competitor at low soil mineral N because of its ability to use atmospheric N_2 as its source of N. A ryegrass-dominant patch will reduce soil mineral N level because of uptake by the grass, and so will alter the competitive balance in favour of white clover. A clover-dominant patch will increase soil mineral N because of N_2-fixation and subsequent release of fixed N by leakage from living tissue or by decay of dead tissue, and so will alter the competitive balance in favour of ryegrass. Because of the time delay involved in these processes, the mixture will never converge on a single equilibrium state. Instead, it will converge on an equilibrium cycle, in which patches alternate between a grass-dominant and clover-dominant state. This mechanism was proposed as an explanation of the patchy distribution (in space and time) of legumes in sustainable pastures, and has been mathematically formalized by Schwinning and Parsons (1996a, b).

Experimental study of such equilibrium cycles clearly requires still more sophisticated methodologies. They must include studies of the effect of each species on its immediate environment, the feedback effect of that environmental change on the competitive balance between grass and legume and the time course of the effects of plant on environment and of environment on plant. In addition, consequences for spatial patchiness will depend on the growth habit of the species involved, in particular on patterns of clonal or sexual dispersal through the field.

Thus grass–legume dynamics can involve some of the most complicated forms of competitive interaction known between plants. Combined with their central importance for pasture sustainability, this further emphasizes the need for high-quality studies on competition where experimental methods are carefully matched to objectives.

Conclusion

Probably the greatest difficulty with studies on competition stems from the wide range of questions that can be asked. Different experimental approaches are required to answer different questions. This chapter provides a guide to the diversity of questions that can be asked, to the pitfalls to be avoided and to the selection of the most appropriate experimental approach for each purpose. It makes no attempt to review all the experimental designs or competition indices that have been used. Instead it provides a theoretical framework that can be used to assess the merits of other designs and indices, and it highlights aspects of competition that have been inadequately studied to date, particularly in the context of the special features of grass–legume dynamics.

The chapter identifies four main categories of competition study requiring different methodologies. One is concerned with measuring the intensity of competition in the field. The value of such studies lies in quantifying the importance of competition relative to other ecological factors as forces driving community structure. The second is concerned with measuring the effects of competition on the growth, reproduction and survival of plants, and is the subject of the vast majority of competition studies. The third is concerned with measuring the outcome of competition in terms of changes in community structure. Such studies are particularly important for understanding the long-term persistence and sustainability of pastures. The fourth is concerned with identifying the mechanisms of competition. This category has seen relatively little progress to date, and has involved a diversity of methodologies too great to be reviewed here. Identifying mechanisms of competition, especially at the molecular level, is left as the biggest challenge for the future.

Acknowledgements

I am grateful to John Connolly for valuable discussions while preparing this chapter, and to Alison Davies for comments on the draft.

Notes

1. The term 'species' is used in this chapter for simplicity. All the theory and methodology presented can be applied not only to competition between species but also to competition between varieties, populations, ecotypes, genotypes, or any other definable genetic or taxonomic entity, or between different developmental states of the same genetic or taxonomic entity.

2. A replacement or substitutive mixture between two species is formed by substituting a proportion of plants from the pure stand of one of the species with the same proportion of plants from a pure stand of the second species (Fig. 2.4a). An additive mixture is formed by adding plants without removing others; in some usages (e.g. Snaydon, 1991) both species in the mixture should be present in pure stand at the same density as in the mixture (Fig. 2.4b).

References

Austin, M.P., Fresco, L.F.M., Nicholls, A.O., Groves, R.H. and Kaye, P.E. (1988) Competition and relative yield: estimation and interpretation at different densities and under various nutrient concentrations using *Silybum marianum* and *Cirsium vulgare*. *Journal of Ecology* 70, 559–570.

Begon, M., Harper, J.L. and Townsend, C.R. (1996) *Ecology: Individuals, Populations and Communities*, 3rd edn. Blackwell Science, Oxford, UK.

Bi, H.Q. and Turvey, N.D. (1996) Competition in mixed stands of *Pinus radiata* and *Eucalyptus obliqua*. *Journal of Applied Ecology* 33, 87–99.

Connell, J.H. (1983) On the prevalence and relative importance of interspecific competition: evidence from field experiments. *American Naturalist* 122, 661–696.

Connolly, J. (1986) On difficulties with replacement-series methodology in mixture experiments. *Journal of Applied Ecology* 23, 125–137.

Connolly, J. (1987) On the use of response models in mixture experiments. *Oecologia* 72, 95–103.

Cornelissen, J.H.C., Diez, P.C. and Hunt, R. (1996) Seedling growth, allocation and leaf attributes in a wide range of woody plant species and types. *Journal of Ecology* 84, 755–765.

Cousens, R. (1991) Aspects of the design and interpretation of competition (interference) experiments. *Weed Technology* 5, 664–673.

Cousens, R. (1996) Design and interpretation of interference studies: are some methods totally unacceptable? *New Zealand Journal of Forestry Science* 26, 5–18.

Cousens, R. (2000) Greenhouse studies of interactions between plants: the flaws are in interpretation rather than design. *Journal of Ecology* 88(2), 352–353.

de Wit, C.T. (1960) On competition. *Verslagen van Landbouwkundige Onderzoekingen* 66, 1–82.

de Wit, C.T. and van den Bergh, J.P. (1965) Competition between herbage plants. *Netherlands Journal of Agricultural Science* 13, 212–221.

de Wit, C.T., Tow, P.G. and Ennik, G.C. (1966) Competition between legumes and grasses. *Verslagen van Landbouwkundige Onderzoekingen* 687, 3–30.

Firbank, L.G. and Watkinson, A.R. (1985) On the analysis of competition within two-species mixtures of plants. *Journal of Applied Ecology* 22, 503–517.

Freckleton, R.P. and Watkinson, A.R. (2000) Designs for greenhouse studies of interactions between plants: an analytical perspective. *Journal of Ecology* 88(3), 386–391.

Gibson, D.J., Connolly, J., Hartnett, C.D. and Weidenhamer, J.D. (1999) Designs for greenhouse studies of interactions between plants. *Journal of Ecology* 87, 1–16.

Goldberg, D.E. (1996) Competitive ability: definitions, contingency and correlated traits. *Philosophical Transactions of the Royal Society of London Series B (Biological Sciences)* 351, 1377–1385.

Goldberg, D.E. and Barton, A.M. (1992) Patterns and consequences of interspecific competition in natural communities: a review of field experiments with plants. *American Naturalist* 139, 771–801.

Goldberg, D.E. and Landa, K. (1991) Competitive effect and response – hierarchies and correlated traits in the early stages of competition. *Journal of Ecology* 79, 1013–1030.

Goldberg, D.E. and Scheiner, S.M. (1993) ANOVA and ANOCOVA: field competition experiments. In: Scheiner, S.M. and Gurevitch, J. (eds) *Design and Analysis of Ecological Experiments*. Chapman and Hall, New York, USA, pp. 69–93.

Goldberg, D.E., Rajaniemi, T., Gurevitch, J. and Stewart-Oaten, A. (1999) Empirical approaches to quantifying interaction intensity: competition and facilitation along productivity gradients. *Ecology* 80, 1118–1131.

Grace, J.B. (1995) On the measurement of plant competition intensity. *Ecology* 76, 305–308.

Grist, E.P.M. (1999) The significance of spatio-temporal neighbourhood on plant competition for light and space. *Ecological Modelling* 121, 63–78.

Harper, J.L. (1961) Approaches to the study of plant competition. In: Milthorpe, F.L. (ed.) *Mechanisms in Biological Competition*. Symposia of the Society for Experimental Biology, 15, Cambridge University Press, Cambridge, UK, pp. 1–39.

Harper, J.L. (1977) *Population Biology of Plants*. Academic Press, London, UK.

Harper, J.L. (1982) After description. In: Newman, E.I. (ed.) *The Plant Community as a Working Mechanism*. Special Publication of the British Ecological Society, 1, Blackwell Scientific Publications, Oxford, UK, pp. 11–25.

Harris, W. (1974) Competition among pasture plants: V. Effects of frequency and height of cutting on competition between *Agrostis tenuis* and *Trifolium repens*. *New Zealand Journal of Agricultural Research* 17, 251–256.

Harris, W. and Thomas, V.J. (1973) Competition among pasture plants: III. Effects of frequency and height of cutting on competition between white clover and two ryegrass cultivars. *New Zealand Journal of Agricultural Research* 16, 49–58.

Holliday, R. (1960) Plant population and crop yield. *Nature (London)* 186, 22–24.

Human, K.G. and Gordon, D.M. (1996) Exploitation and interference competition between the invasive Argentine ant, *Linepithema humile*, and native ant species. *Oecologia* 105, 405–412.

Inderjit, Asakawa, C. and Dakshini, K.M.M. (1999) Allelopathic potential of *Verbesina encelioides* root leachate in soil. *Canadian Journal of Botany* 77, 1419–1424.

Ives, A.R. (1995) Measuring competition in a spatially heterogeneous environment. *American Naturalist* 146, 911–936.

Jolliffe, P.A. (1988) Evaluating the effects of competitive interference on plant performance. *Journal of Theoretical Biology* 130, 447–459.

Jolliffe, P.A. (2000) The replacement series. *Journal of Ecology* 88(3), 371–385.

Keddy, P.A., Twolanstrutt, L. and Wisheu, I.C. (1994) Competitive effect and response rankings in 20 wetland plants – are they consistent across 3 environments? *Journal of Ecology* 82, 635–643.

Kokkoris, G.D., Troumbis, A.Y. and Lawton, J.H. (1999) Patterns of species interaction strength in assembled theoretical competition communities. *Ecology Letters* 2, 70–74.

Laska, M.S. and Wootton, J.T. (1998) Theoretical concepts and empirical approaches to measuring interaction strength. *Ecology* 79, 461–476.

Law, R. and Watkinson, A.R. (1987) Response-surface analysis of two-species competition: an experiment on *Phleum arenarium* and *Vulpia fasciculata*. *Journal of Ecology* 75, 871–886.

Lotka, A.J. (1932) The growth of mixed populations: two species competing for a common food supply. *Journal of the Washington Academy of Sciences* 22, 461–469.

Machin, D. and Sanderson, B. (1977) Computing maximum-likelihood estimates for the parameters of the de Wit competition model. *Applied Statistics* 26, 1–8.

Martins, M.A. and Read, D.J. (1996) The role of the external mycelial network of arbuscular mycorrhizal (AM) fungi. 2. A study of phosphorus transfer between plants interconnected by a common mycelium. *Revista de Microbiologia* 27, 100–105.

Matthies, D. (1996) Interactions between the root hemiparasite *Melampyrum arvense* and mixtures of host plants: heterotrophic benefit and parasite-mediated competition. *Oikos* 75, 118–124.

Mead, R. and Riley, J. (1981) A review of statistical ideas relevant to intercropping research. *Journal of the Royal Statistical Society (A)* 144, 462–487.

Menchaca, L. and Connolly, J. (1990) Species interference in white clover–ryegrass mixtures. *Journal of Ecology* 78, 223–232.

Oosthuizen, M.A., van Rooyen, M.W. and Theron, G.K. (1996) Neighbourhood analysis of competition between two Namaqualand ephemeral plant species. *South African Journal of Botany* 62, 231–235.

Rachich, J. and Reader, R. (1999) Interactive effects of herbivory and competition on blue vervain (*Verbena hastata* L.: Verbenaceae). *Wetlands* 19, 156–161.

Reader, R.J. (1992) Herbivory, competition, plant mortality and reproduction on a topographic gradient in an abandoned pasture. *Oikos* 65, 414–418.

Reader, R.J., Wilson, S.D., Belcher, J.W., Wisheu, I., Keddy, P.A., Tilman, D., Morris, E.C., Grace, J.B., Mcgraw, J.B., Olff, H., Turkington, R., Klein, E., Leung, Y., Shipley, B., Vanhulst, R., Johansson, M.E., Nilsson, C., Gurevitch, J., Grigulis, K. and Beisner, B.E. (1994) Plant competition in relation to neighbor biomass – an intercontinental study with *Poa pratensis*. *Ecology* 75, 1753–1760.

Sackville Hamilton, N.R. (1994) Replacement and additive designs for plant competition studies. *Journal of Applied Ecology* 31, 599–603.

Schoener, T.W. (1983) Field experiments on interspecific competition. *American Naturalist* 122, 240–285.

Schwinning, S. and Parsons, A.J. (1996a) Analysis of the coexistence mechanisms for grasses and legumes in grazing systems. *Journal of Ecology* 84, 799–813.

Schwinning, S. and Parsons, A.J. (1996b) A spatially explicit population model of stoloniferous N-fixing legumes in mixed pasture with grass. *Journal of Ecology* 84, 815–826.

Simard, S.W., Jones, M.D., Durall, D.M., Perry, D.A., Myrold, D.D. and Molina, R. (1997a) Reciprocal transfer of carbon isotopes between ectomycorrhizal *Betula papyrifera* and *Pseudotsuga menziesii*. *New Phytologist* 137, 529–542.

Simard, S.W., Perry, D.A., Jones, M.D., Myrold, D.D., Durall, D.M. and Molina, R. (1997b) Net transfer of carbon between ectomycorrhizal tree species in the field. *Nature* 388, 579–582.

Snaydon, R.W. (1991) Replacement or additive designs for competition studies? *Journal of Applied Ecology* 28, 930–946.

Snaydon, R.W. (1994) Replacement and additive designs revisited: comments on the review paper by N.R. Sackville Hamilton. *Journal of Applied Ecology* 31, 784–786.

Snaydon, R.W. and Satorre, E.H. (1989) Bivariate diagrams for plant competition data – modifications and interpretation. *Journal of Applied Ecology* 26, 1043–1057.

Tilman, D. (1982) *Resource Competition and Community Structure*. Princeton University Press, Princeton, New Jersey, USA.

Tilman, D. (1990) Mechanisms of plant competition for nutrients: the elements of a predictive theory of competition. In: Grace, J.B. and Tilman, D. (eds) *Perspectives on Plant Competition*. Academic Press, New York, USA, pp. 117–141.

Tow, P.G., Lazenby, A. and Lovett, J.V. (1997) Relationships between a tropical grass and lucerne on a solodic soil in a subhumid, summer–winter rainfall environment. *Australian Journal of Experimental Agriculture* 37, 335–342.

Trenbath, B.R. (1974) Biomass productivity of mixtures. *Advances in Agronomy* 26, 177–210.

Trenbath, B.R. (1978) Models and the interpretation of mixture experiments. In: Wilson, J.R. (ed.) *Plant Relations in Pastures*. Commonwealth Scientific and Industrial Research Organization, pp. 145–162.

Turkington, R. and Harper, J.L. (1979) The growth distribution and neighbour relationships of *Trifolium repens* in a permanent pasture. 1. Ordination, pattern and contact. *Journal of Ecology* 67, 201–281.

Volterra, V. (1926) Variations and fluctuations of the numbers of individuals in animal species living together. In: Chapman, R.N. (ed.) *Animal Ecology* (1931). McGraw Hill, New York, USA, pp. 409–448.

Wagner, R.G. and Radosevich, S.R. (1998) Neighborhood approach for quantifying interspecific competition in coastal Oregon forests. *Ecological Applications* 8, 779–794.

Willey, R.W. and Heath, S.B. (1969) The quantitative relationships between plant population and crop yield. *Advances in Agronomy* 21, 281–348.

3 Genotype and Environmental Adaptation as Regulators of Competitiveness

IG.M.O. Nurjaya[1] and P.G. Tow[2]

[1]*Department of Biology, Faculty of Mathematical and Natural Sciences, Udayana University, Denpasar, Bali, Indonesia;* [2]*Department of Agronomy and Farming Systems, University of Adelaide, Roseworthy Campus, Roseworthy, Australia*

Introduction

Plants growing together in pasture swards must be able to adapt to changes in their environment and to capture growth resources in the presence of competitors, in order to contribute usefully to pasture botanical composition and yield. This depends not only on their morphological and physiological characteristics but also on their ability to make morphological and physiological adjustments, i.e. on their phenotypic plasticity (Schlichting, 1986).

Competitiveness, at least for agronomic purposes, is often defined by relative yield, i.e. the ratio of yield in mixture to that in monoculture. In mixed swards, the genotype with the higher relative yield is regarded as the more competitive. By this definition, competitiveness is the ability of plants of one genotype to capture growth resources in the presence of another genotype, relative to their ability to capture such resources in pure stand. Preempting of growth resources by one plant to the disadvantage of another may result in dominance of the stronger competitor. In pastures comprising mixtures of genotypes, management is often directed to preventing or correcting such dominance, which may occur at particular times of the year or in particular circumstances of production.

Plant competitiveness in swards is thus likely to be affected by genotype and by environmental and management factors (van den Berg, 1968; Grime, 1979; Berendse and Elberse, 1990). The response to these factors is mediated through physiological and morphological characteristics or traits and their adjustment. First, in this chapter, those traits that characterize individual pasture genotypes and affect competitiveness through the capture and utilization of light, water and nutrients will be reviewed. This will be followed by the broader question: 'How far does genotype regulate or influence competitiveness and competitive interrelationships in pasture mixtures?'

Environmental factors, such as temperature, moisture and day length, affect growth of different genotypes in different ways and thus may affect their relative competitiveness. Similarly, soil physical properties (porosity, texture, moisture-holding capacity) and chemical properties (pH, nutrient status and ion-exchange capacity) may affect the relative competitiveness of species in mixture (Harper, 1977). Understanding genotypic adaptation to a set of environmental conditions can assist pasture management decisions (the choice of cultivars, fertilization, irrigation, defoliation regimes, grazing, utilization and pest control). Therefore it will also be important to know if the relative competitiveness of associated genotypes depends on how well they are adapted to the environment. This question will be considered finally in this chapter.

© CAB *International* 2001. *Competition and Succession in Pastures*
(eds P.G. Tow and A. Lazenby)

Morphological and Physiological Traits Affecting Competitiveness

Morphological and physiological characteristics have been suggested as affecting competitiveness where there are significant correlations between these factors and measurable success in mixture.

Shoot characteristics

Plant size has been shown to be highly correlated with competitive ability. Thus, Gaudet and Keddy (1988) in a comparison of 44 wetland test species for their short-term effect on the biomass of a phytometer (plant on which the competitive effect of a test plant is measured), showed that competitive ability of the test plants relative to the phytometer was highly correlated with their above-ground biomass and associated traits, such as plant height and canopy area. This correlation applied to test plants grown both alone and in mixture.

Rösch et al. (1997) also obtained high correlations between the competitive effect of some pioneer desert test plants on phytometer biomass and traits measured on these test plants grown singly. Again, plant traits of significance were those associated with plant size and leaf area, such as leaf area ratio (LAR) (leaf area per unit whole plant or shoot mass) and specific leaf area (SLA) (leaf area per unit leaf mass), rather than root production and plant reproduction. An equation comprising only maximum shoot mass and maximum SLA accounted for about 83% of the competitive effect.

In pastures, it is common to find that grasses supplied with optimum N are more highly productive than pasture legumes. For example, Davidson and Robson (1986) quote maximum annual values of 12 t ha^{-1} and 8 t ha^{-1} for grass and clover, respectively, in the UK. This difference in potential is probably at least partly responsible for the commonly experienced suppression of legumes by productive grasses supplied with high levels of N. Mahmoud and Grime (1976) found the same effect with mixtures of taller- and shorter-growing grasses, the taller dominating the shorter through shading.

It would seem likely that a high relative growth rate (RGR) (the increase in plant mass per unit of time per unit of plant mass) at an early stage of growth and competitive interference would confer a competitive advantage on a plant (Grime, 1979). However, this was not found by Rösch et al. (1997). RGR usually declines with time, but the decline may

be less in a plant that maintains a high LAR and SLA. Furthermore, factors other than RGR may operate to determine competitive success. Thus Cocks (1974) found that, in mixtures of the annual grasses Hordeum leporium (barley grass) and Lolium rigidum (annual ryegrass), while ryegrass had the higher RGR and was usually the stronger competitor, barley grass competed successfully when it had a relative high density. This was because it germinated more quickly, had larger seedlings and was a superior competitor for nitrogen (N) in the early stages of growth. The last feature suggests the importance of rapid early root growth in conferring a competitive advantage.

Root characteristics

Several authors have related differences between grass species in competitive ability to traits of their root systems. Many of these studies have been done in pots and so are more relevant to the establishment phase of a pasture. Rhodes (1968a) grew Festuca arundinacea, Dactylis glomerata, L. rigidum and Phalaris arundinacea in monoculture and in substitutive mixtures with H1 ryegrass and with Phalaris coerulescens at two total densities. After 9 weeks, yield per plant was higher for plants of all species in association with P. coerulescens than in association with the ryegrass. That is, P. coerulescens was a weaker competitor than H1 ryegrass. Subsequent detailed studies of monocultures (Rhodes, 1968b) showed that, while the initial development of seminal and nodal roots was faster in P. coerulescens than in H1 ryegrass, after 2–3 weeks the ryegrass became superior in number and length of both seminal and nodal roots and in dry matter (DM) yield of roots and shoots. The relatively low seedling competitive ability of D. glomerata and F. arundinacea was also attributed to relatively slow initial root growth. These results were in keeping with the outcome of another study, in which F. arundinacea and P. coerulescens were grown with H1 ryegrass, at two densities and with separation of root and shoot competition. This study confirmed the importance of root competition in determining the superior competitive ability of H1 ryegrass, especially at the higher density. However, at the lower density, P. coerulescens was not inferior to H1 ryegrass, a result attributed to the good start to phalaris given by its earlier root development, before contact was made with roots of ryegrass. This example illustrates the transitory effect some plant traits may have on competitive ability and the importance of interaction with other factors, such as density, at least in the establishment phase of a pasture.

Nevertheless, species differences in competitive ability are often more persistent. For example, when seed of *P. coerulescens*, *L. rigidum* and H1 ryegrass were sown in association with established plants of each of the same species, both defoliated and undefoliated, all seedings grew better in association with *P. coerulescens* (Rhodes, 1968a). Thus, as before, *P. coerulescens* was the least competitive and regarded as the species most open to invasion. An unidentified trait of the root system was again indicated as contributing to the species differences, but this time it was the established root system (10 weeks of age onwards) that was implicated.

Root production is not often measured in competition studies, but available evidence indicates that it is positively related to competitiveness. Baan Hofman and Ennik (1982) found that size of root system was consistently and positively correlated with differences in competitive ability (as measured by relative reproductive rate and relative replacement rate) between six perennial ryegrass clones, even though shoot production was similar in all clones. Whilst it could not be concluded that the relationship was causal, it is pertinent that the root system of the stronger competitor contributed substantial net root production (measured in monoculture) over the four harvest months of the experiment, while root yield of the weakest competitor was at a standstill. Other clones were intermediate in competitiveness and root yield. These differences were reflected in differing root/shoot ratios, which were considered to be genetically based. The results suggest that differences in competitive ability were related to root activity as measured by net root growth. Using additional data, Ennik and Baan Hofman (1983) again concluded that competitive ability of their ryegrass clones was positively related to root mass, as measured in monocultures. They also used the technique of root partioning in mixtures of clones to further investigate root competition. In the later cuts of the experiment, the yield of the more competitive clone was larger where its roots were allowed to grow among those of the weaker competitor than where the root systems were kept separate; in contrast, the roots of the weaker competitor were further restricted and its yield reduced by competition from the stronger competitor. The differences in root growth were concluded to be genetic and occurred over a wide range of levels of applied N. However, the differences were greatest at intermediate N levels. At the highest N levels, root growth of both clones was lower and it seems likely that, in such a situa-tion, competition becomes more closely related to shading by dense shoot growth than to root interference. This illustrates the well-known shading effect of applying high rates of N to pastures. The ryegrass clones were also grown with another species (*Elytrigia repens*, couch grass). In agreement with the previous results, the higher the root mass of ryegrass (measured in monoculture), the lower the yield of couch grass.

Root mass was also correlated with competitive ability in an experiment to test Tilman's hypothesis that the strongest competitor is the one that can reduce resources to the lowest level. Tilman and Wedin (1991) found that root mass accounted for 73% of the observed variance in soil nitrate levels, in studies to compare the ability of five grasses to reduce soil nutrient levels in a N gradient experiment. The late successional species (more successful competitors) reduced soil-solution N to the lowest level.

Cahill and Casper (2000) designed an experiment specifically to measure the influence of neighbour root biomass, in a field dominated by grasses and thistles, on the strength of root competition experienced by an individual target plant of *Amaranthus retroflexus*. Target plants were transplanted into neighbour vegetation, either with full access from neighbour roots or with neighbour root access to the target plant impeded to varying degrees by root exclusion tubes having varying numbers of wall perforations to allow root entry. Shading of target plants by neighbours was prevented by tying back the shoots of the neighbour plants. Soil volumes allocated to each target plant were excavated and refilled with the same soil prior to transplanting, to ensure that no living neighbour roots were present at the start. Over 10 weeks up to harvest, target plant growth was closely related to the degree of root exclusion from these volumes. Below-ground competition intensity, defined as final target plant biomass reduction due to root interactions, relative to growth in the absence of neighbour roots, was highly correlated with neighbour root abundance in a quadratic relationship. These results, under natural field conditions, agree with those from less natural situations, mentioned above, examining the role of root growth and biomass in competition.

Combination of root and shoot characteristics

Considerable light has been shed on the relative importance of various plant traits for competitive ability by the experiment of Aerts *et al.* (1991) with the evergreen dwarf shrubs *Erica tetralix* and *Calluna vulgaris* (dominant on nutrient-poor heathland soils) and the perennial, deciduous grass *Molinia caerulea* (dominant on nutrient-rich heathland). They grew these species in field plots in monoculture and pairwise mixtures at low and high nutrient levels and with and without separation of shoot and root competition.

Molinia was the only species whose root system penetrated the soil volume of the other species and it had a much higher allocation of its biomass to the root system than did the shrubs. Yet, in the low-nutrient treatments, *Molinia* had no significant effect on the total biomass per plant of either *Erica* or *Calluna*. It was concluded that, at low nutrient levels, the more aggressive root growth (and presumably nutrient uptake) of *Molinia* was balanced by the higher nutrient retention capacity of the shrubs (in contrast to the deciduous nature of *Molinia*). In another experiment quoted by Aerts and his colleagues, it had been found that the shrubs could outcompete *Molinia* if they were at a high enough density to reduce light interception by the grass at an early stage of its growth.

At higher nutrient level, the total biomass per plant of each shrub was significantly reduced by root and/or total plant competition from *Molinia*. Concurrently, the biomass per plant of *Molinia* was significantly increased. This effect was attributed both to the aggressive root growth of the grass and to its morphological plasticity, which, in association with the shrubs, enabled it to position its leaves higher than in the monoculture. No such plasticity occurred in the shrubs. Thus, in spite of a relatively low shoot : root ratio at the high nutrient level, *Molinia* was more competitive for light.

This experiment illustrates how various plant traits, probably having a genetic basis, including shoot and root morphology, acted differentially to determine the outcome of interspecies competition, depending on other factors, such as plant density and available nutrient levels. It was also pointed out (with respect to other work quoted by Aerts *et al.*) that the lower allocation of biomass to the leaves in *Molinia* as compared with *Erica* and *Calluna* is compensated for by its higher SLA. On the other hand, the lower biomass allocation to the roots of the shrubs compared with the grass is compensated for by their higher specific root length (SRL) (length per unit of root mass).

Other work that points to the importance of such shoot and root traits includes that of Svejcar (1990), who proposed that *Bromus tectorum* was more competitive than *Agropyron desertorum* because it was more efficient (per unit of biomass) in producing leaf area and root length. In the period 40–60 days from sowing, *B. tectorum* had 12% more root DM and 56% more shoot DM than *A. desertorum*, while having more than twice the root length and leaf area. Genetic variation for root growth has also been observed in populations of white clover (Ennos, 1985). In mixtures during drought, white clover with short roots that remain in the surface soil layer are more affected than types with longer roots that explore deeper soil layers.

Laurenroth and Aguilera (1998), in a review of plant–plant interactions in grasslands quoted cases where differential competitive ability of grasses was associated with ability to extend roots rapidly and deplete soil water and nutrients quickly or from deeper soil layers, or to invade root-free gaps and nutrient-rich patches, i.e. to pre-empt acquisition of soil resources. However, they found that such traits were not always present to explain differences in competitive ability.

Analysis of competition between species

The concepts discussed above have been used by Berendse (1994) to develop an analytical model for competition between perennial species at low and high nutrient levels. The model is designed in such a way that it integrates the forms of the de Wit, Lotka–Volterra and Tilman models (see Sackville Hamilton, Chapter 2; and Peltzer and Wilson, Chapter 10, this volume). The aim of the author was to 'introduce a theory that is sufficiently simple to allow analytical solutions, but nevertheless produces qualitative predictions about the effect of changes in nutrient supply that agree with experimental results'.

Given a plant density adequate for complete uptake of a limiting resource, the uptake of this resource by each competing plant population is taken as proportional to the fractional biomass contribution of each species to total biomass. For low-nutrient situations, a function is introduced to convert the biomass of each species into its root length, its root surface or any other variable that determines the fraction of available nutrients that it can absorb. Another term, relative nutrient loss rate, is also introduced into the formula to account for losses of nutrients by abscission or death of plant parts and removal by grazers and parasites. Furthermore, the amount of nutrients in the plant is derived from nutrient concentration and biomass. From the equation, zero growth lines can be drawn (see Sackville-Hamilton, Chapter 2, this volume) to predict the circumstances when one species replaces another. For example, the equation predicts that the species with the lower relative nutrient requirement will become dominant if the larger nutrient losses from an associated species are not sufficiently compensated for by greater competitive ability for absorption of nutrients.

In the contrasting situation where nutrient availability is high and growth is regulated by the amount of irradiation captured by each competitor, the starting-point of the formula is an expression of the potential growth rate of the monoculture. Since the potential growth rate of each species in a mixture depends on the fraction of the total radiation it can intercept, the next most simple step for an analytical solution of the equation is the inclusion of terms to convert biomass into leaf area (via the LAR) and for losses of biomass (via relative loss rate). The conditions for a change in species dominance can be worked out. For example, in environments where the nutrient supply is sufficiently large, the species that was the weaker competitor at a low nutrient supply would become dominant despite greater losses, if these losses are compensated for by a higher potential growth rate. In other words, the difference between the potential growth rates overrides the difference between the relative loss rates of the two species. The model might be extended to determine the intermediate nutrient supply at which the competitive balance between the two species is reversed and at which coexistence is possible. Such determinations would have considerable practical value in pastures where the balance between species is commonly affected by regular inputs of nutrients (either natural or applied) and by other factors, such as temperature, rainfall and grazing, which may affect which species is dominant (see later).

Berendse (1994) relates his model to the example discussed earlier in this chapter (Aerts *et al.*, 1991), where situations of *Erica* and *Molinia* dominance in heathland are analysed. Given the differences between the species in potential growth rate and relative nutrient losses, Berendse concludes that competition between these species appears to be a characteristic case, where the slow-growing species

with the lowest loss rate (*Erica*) is superior under nutrient-poor conditions, whereas the species with the faster potential growth rate (*Molinia*) is superior in relatively nutrient-rich environments, in spite of its higher loss rate. Such a model, while very simple, provides a framework for analysing competitive situations. The plant traits chosen for use in the model, such as root length per unit biomass, root diameter, LAR, leaf thickness, relative nutrient loss rates and maximum growth rate, are measurable and appear appropriate for explaining the change in species dominance in heathlands. Such models can be used: (i) to provide qualitative predictions about which plant features lead to dominance of plant populations in environments with different nutrient supplies; and (ii) more specifically, to estimate the contribution of various plant traits to competitiveness and to define the conditions under which plants tend to dominance in mixtures with other plants. Thus, such models go hand in hand with and guide research on competitive relationships.

Response to grazing and cutting

In grazed pastures, grazing, both non-selective and selective, may change a trend to dominance by one component. Residual leaf area after grazing becomes important in determining which species in a mixture initially intercepts more light for regrowth. However, Nassiri and Elgersma (1998) concluded that residual leaf area was less important than both the rate of increase of leaf area index (LAI) (growth rate) and SLA in determining the composition and growth of perennial ryegrass–white clover mixtures. They compared binary mixtures of three clover cultivars and two ryegrass cultivars, cutting to 5 cm above ground level. 'Aggressivity' was expressed by the slope of the linear regression between weekly estimates of ryegrass growth and clover growth. Two of the three clover cultivars (Alice and Gwenda) were more aggressive than ryegrass, even though ryegrass had the greater residual (stubble) leaf area after cutting. Clover increased its content of the mixture by means of a higher relative rate of increase of LAI and probably also by means of higher SLA than the grass. A third clover cultivar (Retor) had a higher residual leaf area than the other two but a lower SLA and higher pest damage. Hence, its light interception and aggressivity were lower.

Because of: (i) the lack of monoculture measurements in Nassiri and Elgersma's experiment to compare with the mixtures; and (ii) the influence of both soil N and N_2-fixation on growth and competitive ability, it is uncertain how far the plant traits mentioned are directly responsible for differences in aggressivity. However, the results are consistent with other data discussed earlier in this chapter, which show that: (i) traits such as rate of increase in leaf area and SLA values are closely related to competitive ability; and (ii) there are genetic differences both between and within species. The results also showed seasonal effects in DM and leaf area composition which varied between and within species. These may have been due to different seasonal growth cycles and temperature responses.

Variation in residual leaf area was also suggested as a reason for differences in competitive ability among three types of the subtropical grass *Eragrostis curvula*, which invades temperate pasture grasses on the northern tablelands of New South Wales, Australia. However, in a field experiment (Robinson and Whalley, 1991), these differences developed over time, while variations in height and yield potential also seemed related to competitive ability. Furthermore, competitive relations between *E. curvula* and the temperate grasses varied throughout the year, because of temperature restrictions on the subtropical species in the cooler months and on the temperate species in summer. Preferential grazing of temperate grasses was also thought to reduce their ability to compete with *E. curvula*; on one occasion, severe defoliation adversely affected *D. glomerata* more than *F. arundinacea*. Thus a range of plant characteristics can affect or modify competitive relationships in field situations.

Several examples given in this chapter show how, in short-term, controlled experiments, a particular plant trait stands out in importance from others in determining competitive ability. This is especially obvious in the experiments of Black (1960, 1961) comparing two cultivars of *Trifolium subterraneum* (subterranean clover) – Tallarook, with short leaf petioles, and Yarloop, with much longer petioles. The cultivars, both with prostrate stems, were grown in monocultures and replacement series mixtures at a high density, ensuring maximum sward growth rate and competitive interaction. Four serial harvests at 10-day intervals in the second month after sowing provided data on DM production and detailed sward profile information on leaf area of each cultivar and its light interception. The two cultivars held their leaf canopies in distinct bands at different heights. At all relative densities from harvest 2 onwards, Yarloop shaded and progressively suppressed Tallarook by means of its longer petioles. The competitive advantage of long petioles in providing a superior leaf profile was clearly shown. Plotting the DM yields of each species against their plant frequencies gave the 'mirror image' hyperbolic curves typical of species which are mutually exclusive (*sensu* de Wit). That is, they were competing for the same set of resources, as would be expected of plants so closely related. Because the experiment was very short-term and conducted under highly controlled conditions, it may be expected that the effects of other important traits may not have had the opportunity to appear.

On the basis that defoliation may reduce the shading by Yarloop and allow Tallarook to survive, Black (1963) conducted defoliation experiments with these (and other) cultivars. Recovery from complete defoliation to the leaf base was more rapid in Tallarook monoculture than in Yarloop, because of a much higher rate of appearance of new (though smaller) leaves. Furthermore, 26% of Yarloop plants died compared with 5% of Tallarook. A mixture of the two cultivars was subjected to three defoliation treatments. With no defoliation, Tallarook was at an extreme disadvantage, as before (Black, 1960), and failed to survive. With the removal of the Yarloop leaf canopy over Tallarook (twice) but not of young leaves, an improved light regime and recovery of Tallarook was short-lived. Yarloop again suppressed it because of rapid elongation of petioles left below the cutting height. A similar result occurred when canopies of both cultivars were removed.

Thus the relation between the two cultivars in the mixture depended on the light energy available to each, as determined by relative heights of leaf canopies. Whilst complete defoliation may have allowed the shorter-leaved Tallarook to become dominant because of its faster rate of leaf production, less severe defoliation allowed Yarloop to dominate because of its capacity for rapid elongation of petioles of developing leaves. This provides a good example of a plant trait that may have an overriding effect on competitive relationships, unless measures are taken to reduce its effects, bringing another trait to the fore or perhaps avoiding the interplant reaction altogether. Thus, when Yarloop was found to have a high content of oestrogen, which reduced the fertility of ewes grazing the cultivar, the strategy recommended to farmers to replace it with a non-oestrogenic cultivar was to prevent seed set for a couple of years to reduce soil seed reserves and regeneration capacity, before sowing the new cultivar.

Petiole length is important in subterranean clover because of the plant's prostrate growth habit and planophile (horizontal) leaf blade positioning. In mixtures of clovers with grasses, shorter petiole clovers may still be able to position leaves to receive light that passes between grass leaves, as found by Woledge (1988) for white clover in tall-growing grass. Legumes with an upright habit, such as lucerne and red clover, have a greater capacity to compete for light than those with a prostrate habit (Rhodes and Ngah, 1983).

Lateral spread by stolons and rhizomes enables plants to explore unoccupied space ahead of a competitor. This 'foraging' capacity and any competitive advantage will depend on the length and density of these organs. Cultivars of white clover have been shown to differ in these respects. Rhodes and Evans (1993) found that stolon density in spring varied from about 20 to 120 m m^{-2}. Over this range, the greater the length of stolon, the greater the annual yield of white clover. Stolons exposed to light are capable of photosynthetic activity of 12–22% (on a unit area basis) of that in the leaves (Chapman and Robson, 1992).

Plants with larger stolon or rhizome biomass tend to be more persistent and more competitive than plants with smaller organs (Thom *et al.*, 1989). Thomas (1984) compared the competitive ability of the white clover cultivars Olwen (large leaf) and S 184 (small leaf) growing with perennial ryegrass. S 184 grew below the cutting height and this resulted in almost double the stolon weight compared with the uncut control. This larger stolon weight enhanced the rate of growth of S 184 during recovery from drought.

ences in N-fixing capacity in legumes can be associated directly with differences in competitive ability with respect to grasses. Such differences in N$_2$-fixation exist. For instance, Goodman and Collison (1986) found that Olwen white clover produced more DM and fixed more N than S 184 white clover and that the associated grass assimilated more N when grown with S 184 than with Olwen. One objective of plant breeding is to increase the competitiveness of pasture legumes (see Tow and Lazenby, Chapter 1, this volume). The question arises as to whether this can be done simply by selecting for higher legume productivity or whether efficiency of N$_2$-fixation must also be considered. Complexity is increased by the exchange of N between legume and grass (see section on Genotypes as Regulators of Competitiveness in Pasture Plants).

Another source of complexity in the determination of traits related to competitiveness is the occurrence of vesicular arbuscular mycorrhizas (VAM) in plant roots. Mycorrhizae occur widely in the field but infection with these fungi varies among species. For example, Goodman and Collison (1981) found that *Lolium perenne* cv. S23 had more mycorrhizae than *Trifolium repens* cv. Olwen roots, resulting in ryegrass absorbing more ^{32}P than clover, especially from deeper soil layers. Cultivar differences in ability to become infected by VAM are also known. For example, Hall *et al.* (1977) found that, when they inoculated white clover cultivars with VAM, cv. Tamar had more of its roots converted to mycorrhizae and was more mycotropic than cv. Huia. Inoculation with VAM stimulated the uptake of soil phosphorus (P) and enhanced the growth of the clovers at low levels of P, an effect which might be expected to improve the competitiveness of the clovers.

Effects of symbiotic N$_2$-fixation and mycorrhizal infection

Symbiotic N$_2$-fixation confers greater competitiveness on legumes in respect of associated grasses. This was clearly illustrated by de Wit *et al.* (1966) in an experiment where the legume was sown with and without rhizobial inoculation and nodulation. The results also suggested that, over a period of seven harvests, there was a transfer of N from legume to grass, which helped maintain the competitiveness of the grass. Results from this and other experiments raise the question of whether differ-

Carbon dioxide fixation pathways

Plants possessing the C$_4$ carbon fixation pathway have the potential to be more competitive than those with a C$_3$ pathway, because of higher rates of photosynthesis and higher water use efficiency (Hatch and Slack, 1970; Ludlow and Wilson, 1972). However, Pearcy *et al.* (1981) found that a C$_4$ agricultural weed, *Amaranthus retroflexus*, and a C$_3$ weed, *Chenopodium album*, had different photosynthetic temperature response curves, so that, while *Amaranthus* had higher rates of photosynthesis than

Chenopodium at day/night growth temperatures of 34/28°C, their rates were similar at 25/18°C, and *Chenopodium* had higher rates than *Amaranthus* at 17/14°C. The ability of *Amaranthus* to maintain high mesophyll conductances due to the presence of the C_4 pathway accounted for its photosynthetic advantage over *Chenopodium* at higher temperatures. Competitive abilities of the two plants in mixtures were measured using de Wit (1960) replacement series and diagrams of the resulting DM yield at 52–60 days. The shift in relative competitive abilities with growth temperature showed a very close parallel to the photosynthetic responses: competitive abilities were about the same at the growing temperatures of 25/18°C. *Chenopodium* was by far the stronger competitor at 17/14°C and *Amaranthus* at 34/28°C. Those competitive outcomes were determined primarily by differences in relative growth rates, which were visibly discernible prior to canopy closure at 2–3 weeks of age. Seedlings of the two species were the same age and size at the beginning of the experiment.

In contrast to the temperature and photosynthetic effects, the imposition of water stress in another experiment did not favour the C_4 species competitively. Other investigators have shown the dominating effect of temperature on the competitive relationships between C_3 and C_4 grasses. For example, Christie and Detling (1982) showed that soil N supply had little effect on relative crowding coefficients compared with the overriding effect of temperature. Other examples of temperature effects on competition between C_3 and C_4 grasses are discussed in relation to temperature adaptation (see section on Environmental Adaptation as a Regulator of Competitiveness – Temperature).

Carbohydrate reserves

The amount of carbohydrate reserves in storage organs has been related to persistence of plants in response to environmental and competitive stress. Persistence may vary among genotypes subjected to adverse environmental and management conditions. It is important to know how persistence is affected by the additional stress of interspecific competition. Smith *et al.* (1992) compared the persistence and productivity of four cultivars of *Medicago sativa* (lucerne/alfalfa) under continuous grazing, a management that is generally detrimental to lucerne persistence because of depletion of car-

bohydrate reserves in the root and crown. The cultivars selected for testing were expected to vary in their tolerance of continuous grazing and were grown in monoculture and mixture with tall fescue. After 3 years of continuous grazing, the mean density of the cultivars was 57, 41, 5 and 4 plants m^{-2} in pure stand and 17, 9, 0 and 1, respectively, in mixture. Superior persistence was associated with the maintenance of higher levels of total non-structural carbohydrates (TNC), as well as some other factors, such as disease tolerance. Overall, it was found that the traits that assisted two of the cultivars to persist better under the stress of continuous grazing also conferred superior persistence under the additional stress of competition from tall fescue.

Accumulation of carbohydrate reserves and their mobilization after removal of stress may also confer a competitive advantage over a plant with fewer reserves. This appears to have been the case in mixed pasture of the subtropical grass *Paspalum dilatatum* and the temperate grass *L. perenne* in the temperate environment of the North Island of New Zealand (Thom *et al.*, 1989). The survival of paspalum through the cold winters was aided by underground rhizomes, which increased in biomass from 1.6 t ha^{-1} DM in December to 4.4 t ha^{-1} DM in May, while its non-structural carbohydrate (NSC) content increased from 6% to 19% (much higher than that of ryegrass). N accumulation in rhizomes was also high. Levels of biomass, TNC and N in rhizomes were high enough to ensure survival of paspalum in winter and give it a good start in spring, and it tolerated summer temperatures and moisture stress better than ryegrass.

The accumulation and use of NSC may vary between cultivars of the same species and this may affect competitive ability. Nurjaya (1996) found that the white clover cultivar Huia accumulated NSC particularly in stolons, while cv. Olwen accumulated more in roots. The higher competitiveness of Olwen with perennial ryegrass seemed to be related to its mobilization of greater amounts of NSC from its roots.

Genotypes as Regulators of Competitiveness in Pasture Plants

The discussion in previous sections of this chapter indicates that many heritable traits may be involved in determining the competitiveness of one genotype with respect to another. Some of such traits vary within as well as between species. For instance, in perennial ryegrass, one of the most intensively studied pasture species, clones within varieties have been shown to vary markedly in response to N (Lazenby and Rogers, 1965) and in root growth (Ennik and Baan Hofman, 1983), features likely to be closely related to competitive ability. As discussed earlier in this chapter, more than one trait may affect competitive ability, especially over the development of a plant from seedling stage to maturity. Thus, while certain morphological or physiological features may be characteristic of particular genotypes, their influence on competitiveness may not be constant. The question arises as to how far competitive relations in pastures are regulated by the particular mix of component genotypes, either naturally in 'natural' pastures or by choice of cultivars in sown pastures.

Aarssen (1983) has argued that genetic variation will occur in populations, for relative competitive ability, in terms of either 'ability of an individual to reduce the availability of contested resources to another' or 'the ability to tolerate reduction in contested resource availability by another'. She also argued for genetic variation in niche requirements. This would allow partial avoidance of competition between individuals, which is generally regarded as a requirement for coexistence in different genotypes. Aarssen regards the number of genes controlling the above characteristics as very large and envisages a continuum of selection processes, recombinations, mutations and gene flows in populations, which will regulate competitive measures among constituents.

This process will produce cases of both coexistence and competitive exclusion. Given that one species in a population starts as a stronger competitor than another, the selection pressure of competition over time may result in avoidance of competitive interaction through niche differentiation (termed selection for 'ecological combining ability'). Here, the two species are not competing for the same 'space', so have the opportunity to coexist and achieve a relative yield total (RYT) greater than 1 (sensu de Wit). Alternatively, selection may result in increased competitive ability in surviving plants of the initially weaker competitor. This could be classed as a case of genotypes regulating competitiveness. Such regulation requires in a weaker competitor the potential to generate and propagate new genetic variants with increased com-

petitive pressures. The trend towards less difference in competitiveness between species is seen as giving more chance of coexistence ('competitive combining ability'). Reciprocal selection may also be involved in the maintenance of near-equal competitive pressures between genotypes of species in a population.

This theory has been supported by further research of Aarssen (1989) and Aarssen and Turkington (1985). Research undertaken with mixtures of naturally occurring genotypes of *T. repens* and *L. perenne* by a number of people has indicated that natural selection leads to the adaptation of *T. repens* to site effects and to interspecific competitors (Lüscher *et al.*, 1992; see also Peltzer and Wilson, Chapter 10, this volume). Lüscher *et al.* (1992), in a competition experiment designed to eliminate weaknesses of former experiments, found that the *Lolium* genotype × *Trifolium* genotype combinations that were sampled as immediate neighbours in the field had a higher percentage of clover than those combinations in which the two components came from different neighbourhoods. A higher percentage of clover indicated an increasing competitive ability of *Trifolium* relative to *Lolium* and appeared to be in agreement with Aarssen's theory (see above) that natural selection may result in an increase in the competitive ability of the inferior competitor, since the superior one acts as a selective agent. In this natural pasture, *Trifolium* was in the minority quantitatively and thus regarded as the inferior competitor.

Understanding those characteristics which confer greater competitiveness on legumes would be useful for breeding and selection of genotypes that are higher-yielding and more persistent in association with grass. However, this is a complex matter and it is necessary to appreciate the implications of having more competitive legumes. For instance, Goodman and Collison (1986) compared the clover varieties Olwen and S184, using ammonium and nitrate sources of ^{15}N to measure N recovery from fertilizer and soil, N_2-fixation and N transfer from clover to grass. At a lowland site, Olwen produced more DM, took up more ^{15}N and fixed more N than did S184. In contrast, soil N uptake by the companion grass with Olwen was less than in the grass growing with S184. This difference was attributed to a greater competitive ability of Olwen for mineral nutrients. Evans *et al.* (1990) also found in a rotationally grazed experiment that, while the yields of cultivars Olwen and Nesta were about the same in each of 3 years, the yield of the associated grass was consistently higher in Nesta, leading to higher total mixture yields. They stressed the importance, for low-input systems, of a choice of clover variety that can increase grass production and improve spring growth of swards without input of N fertilizer. This suggests that highly competitive clover cultivars may not be appropriate for such a system, i.e. where soil N levels are low.

Collins and Rhodes (1989) also found large interactions between clover and grass genotypes for clover yield in the mixture. Similar results were also reported by Widdup and Turner (1983) for comparisons of four white clover cultivars grown with perennial ryegrass under grazing. The lowest-producing clover cultivar in both monoculture and mixture (Kent) was associated with the highest yield of associated grass; the opposite occurred with the highest-yielding clover (Pitau). In this experiment, there was little, if any, improvement in total herbage yield in the most grassy mixture, although a high proportion of grass could be regarded as beneficial in ensuring a reliable distribution of herbage through the year in low-fertility environments. Pitau white clover, in contrast, could be more valuable in high-fertility areas (being also adapted to high grazing frequency).

Another example of the effect of genotype on the proportions of grass and legume was provided by McCloud and Mott (1953). They grew a single cultivar of lucerne with a number of temperate grasses separately. Yields of the mixtures were similar, but grass : legume ratios varied widely. The grass : legume ratio may vary not only according to relative competitive abilities and net transfer of N from legume to grass, but also to relative seasonal growth potentials (presumably determined by temperature responses), as found by Ledgard et al. (1990) for New Zealand white clover cultivars.

The relation between net N transfer and relative competitive abilities of legumes and grass may also be modified by environment. Tow et al. (1997a, b) found that a net transfer of N from lucerne to the tropical grass Digitaria eriantha and/or a higher DM yield from the mixture than from either monoculture occurred at summer temperatures, where lucerne was at a relative disadvantage competitively. A review of literature showed that a net transfer of N from lucerne to a tropical grass had occurred in some experiments but not others. The various results suggested that the grass benefits from a transfer of N when lucerne aggres-

siveness is somewhat reduced, but that soil N mineralization and grass growth are increased by high temperature and adequate moisture. Under conditions of low soil N, the mixture then outyields both monocultures if the improvement in grass from increased N outweighs any restriction in lucerne growth.

As foreshadowed by discussion earlier in this section, grass genotypes within species also vary in competitive ability. Thus, when Evans et al. (1985) found that their white clover genotypes, selected from a range of European natural pastures, were more productive growing with particular ryegrass genotypes associated with them in those pastures than when grown with two local Welsh ryegrass genotypes, they also found that clover yields differed markedly according to which Welsh genotype was associated. As emphasized by Aarssen (1983), in agreement with the terminology of de Wit, competitive ability has meaning only in a context relative to other competitors. This does not preclude the identification of legume genotypes with generally superior competitive ability. However, when defining the role of genotype in determining or regulating competitiveness and competitive relations, it should be more meaningful to deal with combinations of genotypes rather than individuals. In the practical sense, this accords with the conclusion of Collins and Rhodes (1989) that clover genotypes should be tested in mixture with a range of companion grasses and that particular grass : legume mixtures should be tested against a range of standard mixtures.

Environmental Adaptation as a Regulator of Competitiveness

Environmental factors (climatic and edaphic) can be expected to modify the expression of morphological and physiological plant traits that determine competitive relations between genotypes. Experiments that relate plant response to environmental factors to competitive effects are relatively few. Where the effect of environmental factors on competition between two species is being assessed, it is at least necessary to grow the plants in monocultures, as well as in mixture. Only then can both a measure of the response of each species to environmental change and the indices of competition (e.g. relative crowding coefficient, aggressivity or relative replacement rate (RRR)) be determined.

Temperature

As indicated in the section on Morphological and Physiological Traits Affecting Competitiveness, temperature can have a marked effect on the competitive relations between genotypes. The temperature effects are of particular relevance where the occurrence of rainfall in both the cooler and warmer times of the year and the absence of killing extremes of temperature allow both cool- and warm-temperature plants to grow, albeit at different times of the year. Such conditions occur, for example, in parts of New South Wales (Australia), the North Island of New Zealand and the southern USA. If the two types of plant can exist in the same mixed pasture, their complementary temperature responses can facilitate the provision of year-round feed and more stable pasture yields (Harris *et al.*, 1981).

Harris *et al.* (1981) studied competition between *P. dilatatum* (paspalum) and *L. perenne* (perennial ryegrass) over 12 combinations of temperature and cutting treatments, using monocultures and mixtures and 2- and 4-weekly harvests. The different temperature responses of the two species were reflected in their relative competitive abilities. Overall, paspalum was the stronger competitor at day/night temperatures of 24/18°C and ryegrass at 14/8°C. However, infrequent cutting allowed ryegrass to become more competitive at the higher temperatures. Transfer from high to low temperatures strengthened ryegrass dominance, especially under frequent cutting. With transfer from low to high temperature, paspalum recovered rapidly from earlier suppression only when cutting was frequent.

The study shows that, although temperature adaptation has a dominant role in regulating competition between these species with different temperature response curves, these patterns of competitive relationships may be modified by other factors – here defoliation frequency. This is also illustrated by the multifactorial experiments of Cook *et al.* (1976), comparing the responses of *L. perenne* and *Bothriochloa macra* (a summer-growing, Australian native grass) to temperature, moisture, nutrient supply and defoliation. Day/night temperature regimes were 16/10°C, 23/17°C and 31/25°C. Competition was calculated as the RRR of *Lolium* with respect to *Bothriochloa* between two harvests. Temperature responses of monoculture yields and RRR of *Lolium* with respect to *Bothriochloa* followed the same pattern. *Lolium* was superior to *Bothriochloa* at 16/10°C, *Bothriochloa* was far superior at 31/25°C

and the two were similar at 23/17°C. However, moisture and soil fertility interacted strongly, in terms of RRR, *Lolium* gaining relative to *Bothriochloa* under high fertility and in the absence of moisture stress. *Bothriochloa* was at a competitive advantage under the combined conditions of moisture stress and low fertility. These moisture–fertility effects were similar in each temperature regime. Thus moisture stress modified the effects of temperature on competition between the two species.

Similar findings were reported by Tow *et al.* (1997c) for relationships between *Digitaria eriantha* (digitaria, a tropical C_4 grass) and *M. sativa* (lucerne, a C_3 legume). The species were sown in monoculture and a 50 : 50 mixture in all combinations of two day/night temperature regimes (spring, 24/12°C, and summer, 33/19°C), three moisture regimes (periodically droughted, intermediate/nonlimiting and periodically flooded) and two levels of N (with and without N application). Digitaria DM production was much higher at summer temperatures and lucerne at spring temperatures. However, there was a significant species × temperature × moisture × N interaction. Of particular relevance here is the fact that lucerne was more adversely affected by droughting and flooding than digitaria, especially at summer temperatures. These differences in environmental adaptation were translated into corresponding differences in relative competitiveness, as determined by trends in relative yield ratios over six harvests (Fig. 3.1). The outcome was that, overall, digitaria became increasingly competitive over time at summer temperatures but the two species tended towards equilibrium at spring temperatures, especially in the intermediate moisture regime. Thus, conditions that favoured lucerne competitively and the development of equilibrium were those to which lucerne was best adapted.

Soil chemical and physical conditions

It seems reasonable to expect that plants growing in edaphic habitats to which they are adapted may be more competitive than associated plants that are less well adapted. Botanists have long quoted examples of such principles. For instance, Clements *et al.* (1929) give examples of Nageli's 1865 and 1874 writings: 'competition favours those adapted to a particular habitat'.

It is possible for a species to dominate in a soil of a particular holard, but in another to be suppressed or

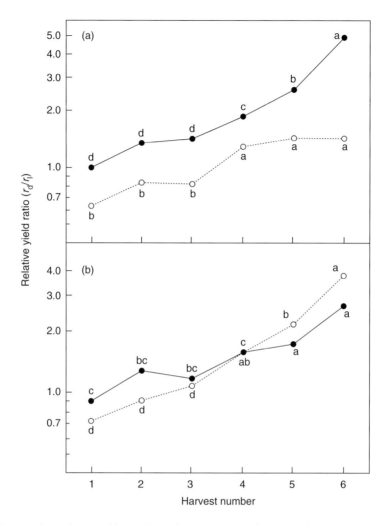

Fig. 3.1. Time trends in relative yield ratios for replacement series relationships between digitaria (d) and lucerne (l), over six harvests, for (a) the temperature × harvest interaction (○, spring temperatures; ●, summer temperatures) and (b) moisture × harvest interaction (○, droughted regime; ●, intermediate moisture regime). Values in the graphs associated with the same lower-case letters within temperature or moisture treatments are not significantly different at $P = 0.05$. (Permission from the *Australian Journal of Experimental Agriculture*.)

excluded. This is the situation with *Primula officinalis* and *Primula elatior*. When both occur together, they sometimes exclude each other almost completely, the one preferring the drier and the other the moist areas. Each is the more dominant in its own habitat, where it is able to suppress its competitor, but when one alone is present, it is much less exacting.

Clements *et al.* (1929) and also Begon *et al.* (1996) quote Tansley's (1917) experiments with *Galium silvestre* (now *Galium pumilum*) and *Galium saxatile* (now *Galium hercynicum*) on cal-

careous and acidic soil types. When grown separately, both species grew well on both types of soil. However, when grown together, competition between the two resulted in *G. pumilum* being confined to the calcareous soil and *G. hercynicum* being confined to the acidic soil. In this example, differences in the distribution of the two species were attributed to differences in adaptation to soil conditions, which only became apparent when the two species were grown together. Because monocultures of both species appeared to grow equally well on each soil type, the results do not necessarily lead to the conclusion that the stronger competitor in each soil environment was the better adapted to that environment. However, from a practical viewpoint, they signify the value of evaluating pasture cultivars for particular soil conditions in association with the plants which will be grown with them. This has proved particularly important in testing new legumes to grow with grasses.

Snaydon (1971) also found that differences in competitive ability between white clover strains growing on calcareous and acidic soils varied with the origin of the strains. He grew monocultures and mixtures of white clover strains originating from the calcareous and acidic soils and measured growth to a single harvest. The effect of root competition upon the plant yield of the populations indicated that the strain from acid soil had greater root competitive abilities on acid soils, while the strain from calcareous soil had greater root competitive abilities on calcareous soil. This is in keeping with the premise that competition favours plants best adapted to the environment.

The question of adaptation to acidity or alkalinity (soil pH) was examined by van den Berg (1968) for a comparison of common bent (*Agrostis tenuis*), an indicator of poor, acid soils, and cocksfoot (*D. glomerata*), an indicator of fertile, slightly alkaline soils. The pH of the poor, acid, sandy soil was adjusted with $CaCO_3$ to give pH values of 4.2, 6.2 and 6.7. These were combined with two NPK levels. Competition was studied by measuring yields of monocultures and mixtures in replacement series, over 16 harvests. The effect of the treatments over this period of time was more complex than expected. The considerable response of monocultures of both species to $CaCO_3$ addition at the low NPK level disappeared at high NPK, indicating that pH within this range affects growth only under low-fertility conditions. Cocksfoot was consistently the stronger competitor at high fertility, for all pH levels, and it suppressed the bent completely after harvest 5. The rate of suppression was increased by a fungal infection in the bent, not apparent at low fertility. At low fertility, for a few harvests, cocksfoot tended to lose competitively to bent, especially at pH 4.2, but it gained competitively over the remaining harvests. Thus, the adverse effect of low pH on the competitiveness of cocksfoot was confined to the establishment phase.

Van den Berg (1968) also showed the importance of pH in the establishment phase in an experiment to study competition between perennial ryegrass and common bent: ryegrass was initially more competitive at higher than at lower pH values, but, during the first year, common bent gained competitively on ryegrass, regardless of pH, and this trend continued in the second year. Competition between ryegrass and four other grasses was also found to be influenced by pH only in the early, establishment phase of the pasture.

Although the influence of pH on competitiveness may operate only in the early stages of a pasture, its effect on relative population densities may persist for longer. Van den Berg (1968) found that the results of the above experiments were consistent with the proportions of various grasses in the field at high and low pH. The application of lime with seed at sowing may preclude the short-term adverse effects of low pH on competition between pasture species. Favouring desirable species that are more competitive at high levels of soil fertility (e.g. cocksfoot) should be achieved by the regular application of nutrients.

Several other examples exist of correlations between soil pH and the composition of pastures, e.g. Robinson *et al.* (1993) for a mixture of exotic and native grasses and Simpson *et al.* (1987) for mixtures of *T. repens*, *L. perenne* and *A. tenuis*. Some effects of pH have been traced to toxic levels of available manganese and aluminium at low pH levels. Thus Scott and Lowther (1980) found that the superior competitiveness of *Lotus pedunculatus* over Huia white clover in acid, infertile soil was due to the intolerance of Huia to high levels of aluminium in the soil solution and its lower ability than the lotus to absorb phosphorus. The competitive ability of white clover relative to lotus was improved by additional application of phosphorus fertilizer.

The effect of level of adaptation to low pH and accompanying high aluminium levels in strains of *Phalaris aquatica* and *D. glomerata* on competitive interaction on different soil types is discussed by Wolfe and Dear (see Chapter 7, this volume).

Many other experiments have suggested that competitiveness of one pasture species in respect of another varies with level of soil nutrients, although relatively few experiments are designed to measure an index of competition.

Van den Berg (1968) compared the competitive relations of perennial ryegrass and couch grass (*E. repens*) at four levels of NPK fertilizer over four harvests, using a replacement series. At the lowest level, ryegrass was clearly the stronger competitor, but couch grass improved its competitiveness with increasing fertilizer, until, at the highest rate, the two species were equally matched. The increasing competitiveness of couch grass with rising fertility was associated with a greater DM increase (in both absolute and percentage terms) in couch grass monoculture than in ryegrass monoculture. Yet the level of DM yield was consistently higher in ryegrass than in couch grass at all levels of fertility. Thus, to define adaptation to soil nutrient conditions for comparison with competitiveness, DM responses over a range of nutrient levels should be more useful than absolute levels of yield of each species. In practical terms, couch and ryegrass should have the best chance of coexisting where soil fertility levels are high. Allowing nutrient levels to decline in a pasture is very likely to change the relative competitiveness of the components.

The level of a single nutrient may influence the relative competitiveness of pasture species. This is illustrated by the pot experiment of Hall (1974, 1978), in which he studied the influence of potassium (K) level on competition between the tropical grass *Setaria anceps* cv. Nandi and the tropical legume *Desmodium intortum* cv. Greenleaf. Monocultures and mixtures in a replacement series were grown at two levels of K. The DM and nutrient yields were presented as the totals for two harvests. The addition of K produced only a small response in DM yield in the monocultures of both species, but had a marked effect on their relative competitiveness in mixture

(Fig. 3.2). At low K, desmodium was a relatively poor competitor for the uptake of N, P and K, especially K, and thus also in terms of DM. The product of the relative crowding coefficients and the RYT for K was about 1, signifying that the two species were mutually exclusive (*sensu* de Wit) in respect of competition for K. However, values for P and DM were a little higher than 1 and for N even higher. This indicated that symbiotic N_2-fixation supplied at least some of this element. In contrast, at high K, desmodium was much more competitive in terms of N, P, K and DM. The relative crowding coefficient of setaria also remained quite high at low N. The outcome, therefore, was that RYTs were much higher than 1, which was probably due to a higher level of symbiotic N fixation in the legume at the higher level of K.

As far as this short-term pot experiment can show, monoculture DM yields did not indicate differences in adaptation of the two species to low or high levels of K. Differences were only revealed by comparison of relative crowding coefficients. Differences in relative yield totals also showed that N fixation was improved, under competitive stress, by the application of additional K. These results do not show whether the benefit of K was to the plant or to the plant–rhizobium relationship. The uptake of soil mineral N by the legume and the transfer of N from legume to grass may also have influenced competitive relationships. Thus, while it can be said that setaria was strongly competitive at both low and high levels of K and desmodium was much

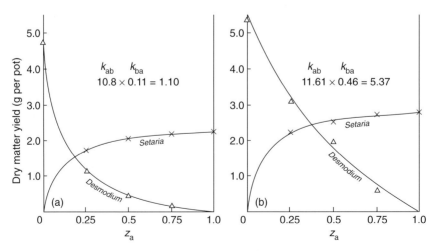

Fig. 3.2. Replacement series diagrams showing the influence of potassium addition on the growth of Nandi *Setaria* (species a) and Greenleaf *Desmodium* (species b) at various relative plant frequencies (*z*). Data for harvests 1 and 2 combined. (a) No potassium; (b) plus potassium. Relative crowding coefficients (*k*) for each species with respect to the other are also shown. (Permission from the *Australian Journal of Agricultural Research*.)

more competitive at high than at low K, it is not clear how this can be related to the concept of environmental adaptation. From a practical point of view, however, stable grass–legume relationships will only be achieved where the competitive outcome of varying nutrient levels is known and acted on.

A similar type of study was conducted by Gillard and Elberse (1982). They compared the effects of low and high levels of P and N on competition between the tropical grass *Cenchrus biflorus* and the tropical legume *Alysicarpus ovalifolius*. The species were grown in monocultures and mixtures in a replacement series and were harvested once only (Fig. 3.3). At low N, addition of P improved the monoculture productivity of *Alysicarpus*, but the improvement in its competitive ability relative to *Cenchrus* was quite small. The legume maintained a good competitive status (relative crowding coefficient 0.8) at low P by being able to function with low levels of P in its tissues (0.12%), similar to those of the grass. Even at low P, RYTs were greater than 1, indicating good N fixation in competition with *Cenchrus*. In the high-N treatments, *Alysicarpus* was less competitive than at low N, because of shading by the tall growth of *Cenchrus*. The grass had its highest values for relative crowding coefficient (16.5) at high N/high P. Yet, as occurred at low N, the relative crowding coefficient of the legume at high N was almost the same at low and high P.

Thus, in terms of competitive ability with respect to *Cenchrus*, *Alysicarpus* was adapted to be almost equally competitive at high and low P, at both high and low N levels. It was less competitive overall at the high N level, where *Cenchrus* was

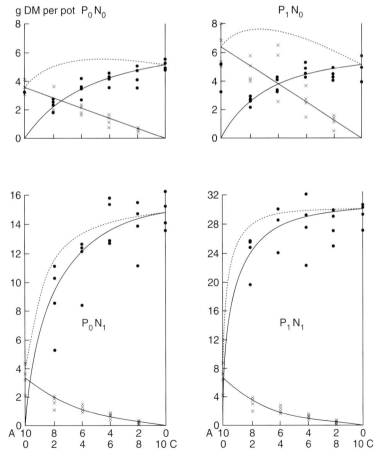

Fig. 3.3. Replacement diagrams presenting dry matter of *Alysicarpus* (A: ×) and *Cenchrus* (C: ●) and total yield (---), for four treatments combining low P (P_0), high P (P_1), low N (N_0) and high N (N_1). Notice the different scale for the yield of P_1N_1. (Permission from the Royal Society of Agricultural Sciences, The Netherlands (KLV).)

more productive and competitive. However, even in this situation favouring the grass, reduction of shading by defoliation would probably have assisted the legume competitively. A legume able to compete satisfactorily with a grass over a range of available N and P levels would be a useful component of a pasture subject to variable inputs of plant nutrients.

The above experiments show how variations in plant nutrient status affect the competitive ability of one species in respect of another. From such experiments, however, it is difficult to identify an adaptive feature or mechanism in the plants, related to their nutrient status or their response to nutrients, which may be regarded as regulating competitiveness.

With grass–legume relations, it is also difficult to separate the effects on competition of the response of the plant to added nutrients and the response of the symbiotic system. Progress can be made if the legume is grown with and without rhizobial inoculation and N_2-fixation, as shown by de Wit *et al.* (1966). They grew a tropical grass, *Panicum maximum* var. *trichoglume* cv. Green panic, and a legume, *Glycine javanica* (now *Neonotonia wightii*) cv. Tinaroo, in monocultures and mixtures in a replacement series over seven harvests. Figure 3.4 summarizes the treatment differences that developed consistently over the experiment The with- and without-rhizobium treatments were applied at low and high levels of N. Without rhizobium and at low N, *Neonotonia* made very little growth and competed very poorly. With rhizobium at low N, the legume grew well and was strongly

competitive. At high N, even without rhizobium, the legume grew well in both mixture and monoculture. Although the weaker competitor, it was not severely suppressed and it gained competitively with time. With both rhizobium and high N, the legume was strongly competitive. This experiment separated the effects of rhizobial N and mineral N on the competitiveness of the legume in relation to the grass. It shows that *Neonotonia* is more competitive at high levels of available mineral N, even in the absence of rhizobium, although the reason for this remains to be determined. The monoculture yields of DM and of N (apart from the treatment without rhizobium at low N) were not very different from each other, and so could not be used to predict the outcomes of competition.

The studies discussed in this section assist in defining how competition is influenced by changes in soil conditions and thus in making management decisions, but specifically designed experiments are needed to explain competition effects in terms of plant–soil relationships.

Conclusions

Many plant morphological and physiological traits have been shown, experimentally, to be associated with the superior competitiveness of particular genotypes relative to others. These characteristics, which include potential growth rate, LAR, SLA, root mass, rapid or prolonged root growth, SRL and recovery from defoliation, all point to the linking of superior competitive ability with ability to

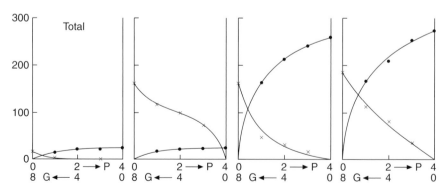

Fig. 3.4. Total DM yields (g, means of four replicates) over seven harvests of green panic (P) and glycine (G) in replacement series in four combinations of rhizobium (R) and nitrogen (N) treatments. Numbers on the *x* axis refer to the number of plants per pot of each species. From left to right: without R, without N; with R, without N; without R, with N; with R, with N. (Permission from PUDOC, Wageningen, The Netherlands.)

pre-empt resources of light, nutrients and water. However, the absolute or relative importance of various traits and even the outcome of competition may change with: (i) stage of plant development; (ii) plant population density; (iii) the degree to which a genotype retains mineral nutrients or carbohydrate reserves in its tissues for use in times of environmental stress; and (iv) morphological and physiological response to environmental change (especially temperature), grazing (defoliation) and soil nutrient status (e.g. N).

Genetic variation has been found to occur both between and within species for relative competitiveness. In natural or long-established pastures, such competitiveness may be modified over time by gene flows and natural selection, which may facilitate competitive exclusion in some cases and coexistence in others. Experiments with a range of grass and clover genotypes (especially ryegrass and white clover) suggest that they can be selected for coexistence, with higher or lower proportions of either component as required. The selection of 'balanced' or 'compatible' grass–legume associations depends on relative competitiveness, along with the capacity for N_2-fixation in the legume and the 'transfer' of N for use by the grass.

Differences between pasture species in their DM response to temperature (e.g. tropical vs. temperate) have been translated to parallel differences in relative competitiveness in mixtures of these species over a range of temperatures. There is also evidence for similar parallelism between competitiveness and adaptation to high and low extremes of moisture and temperature × moisture interactions.

However, few experiments show parallelism between relative competitiveness of genotypes and adaptation to soil conditions, in spite of striking differences in relative competitive ability under different soil conditions.

The lack of parallel adaptation to pH may be because the effect of pH is restricted to the establishment phase of some species or is confounded with soil nutrient status or levels of potentially toxic elements, such as aluminium and manganese. Relative competitivenes has also been shown to vary with level of 'soil fertility', i.e. with general level of all nutrients and with level of availability of particular nutrients (e.g. potassium). The experiments reviewed do not show a consistent parallelism of relative DM responses to nutrient supply and relative competitiveness. Some increased understanding is gained by measuring nutrient content and concentration in both monocultures and mixtures and by conducting the investigation over several harvests. In the case of grass–legume competition, growing legumes with and without rhizobial inoculation allows a separation of effects on competition of the response of the plant to added nutrients and the response of the symbiotic system.

Research reviewed in this chapter provides many examples where competitiveness is closely linked to genotype (with associated traits) and environmental characteristics. This information assists in regulating competition in pastures. However, more purposefully designed experiments are required to enable the roles of genotype and environmental adaptation in regulating competitiveness to be more clearly defined.

References

Aarssen, L.W. (1983) Ecological combining ability and competitive combining ability in plants: toward a general evolutionary theory of coexistence in systems of competition. *The American Naturalist* 122, 707–731.

Aarssen, L.W. (1989) Competitive ability and species co-existence: a 'plant's eye' view. *Oikos* 56, 386–401.

Aarssen, L.W. and Turkington, R. (1985) Biotic specialization between neighbouring genotypes in *Lolium perenne*, and *Trifolium repens* from a permanent pasture. *Journal of Ecology* 73, 605–614.

Aerts, R., Boot, R.G.A. and van der Aart, P.J.M. (1991) The relation between above- and below-ground biomass allocation patterns and competitive ability. *Oecologia* 87, 551–559.

Baan Hofman, T. and Ennik, G.C. (1982) The effect of root mass of perennial ryegrass (*Lolium perenne* L.) on the competitive ability with respect to couchgrass (*Elytrigia repens* (L.) Desv.). *Netherlands Journal of Agricultural Science* 30, 275–283.

Begon, M., Harper, J.L. and Townsend, C.R. (1996) *Ecology: Individuals, Populations, Communities*, 3rd edn. Blackwell Science, Boston, USA.

Berendse, F. (1994) Competition between plant populations at low and high nutrient supplies. *Oikos* 71, 253–260.

Berendse, F. and Elberse, W.Th. (1990) Competition and nutrient availability in heathland and grassland ecosystem. In: Grace, J.B. and Tilman, D. (eds) *Perspectives on Plant Competition*. Academic Press, San Diego, USA, pp. 93–116.

Black, J.N. (1960) The significance of petiole length, leaf area, and light interception in competition between strains of subterranean clover (*Trifolium subterraneum* L.) grown in swards. *Australian Journal of Agricultural Research* 11, 177–191.

Black, J.N. (1961) Competition between two varieties of subterranean clover (*Trifolium subterraneum* L.) as related to the proportions of seed sown. *Australian Journal of Agricultural Research* 12, 810–820.

Black, J.N. (1963) Defoliation as a factor in the growth of varieties of subterranean clover (*Trifolium subterraneum* L.) when grown in pure and mixed swards. *Australian Journal of Agricultural Research* 14, 206–225.

Cahill, J.F. and Casper, B.B. (2000) Investigating the relationship between neighbor root biomass and below ground competition: field evidence for symmetric competition below ground. *Oikos* 90, 311–320.

Chapman, D.F. and Robson, M.J. (1992) The physiological role of old stolon material in white clover (*Trifolium repens* L.). *New Phytologist* 122, 53–62.

Christie, E.K. and Detling, J.K. (1982) Analysis of interference between C_3 and C_4 grasses in relation to temperature and soil nitrogen supply. *Ecology* 63, 1277–1284.

Clements, F.E., Weaver, J.E. and Hanson, H.C. (1929) *Plant Competition: an Analysis of Community Functions*. Carnegie Institution of Washington, Washington, USA.

Cocks, P.S. (1974) The influence of density and nitrogen on the outcome of competition between two annual pasture grasses (*Hordeum leporinum* Link and *Lolium rigidum* Gaud.). *Australian Journal of Agricultural Research* 25, 247–248.

Collins, R.P. and Rhodes, I. (1989) Yield of white clover populations in mixture with contrasting perennial ryegrass. *Grass and Forage Science* 44, 111–115.

Cook, S.J., Lazenby, A. and Blair, G.J. (1976) Comparative responses of *Lolium perenne* and *Bothriochloa macra* to temperature, moisture, fertility and defoliation. *Australian Journal of Agricultural Research* 27, 769–778.

Davidson, I.A. and Robson, M.J. (1986) Effect of temperature and nitrogen supply on the growth of perennial ryegrass and white clover. 2. A comparison of monocultures and mixed swards. *Annals of Botany* 57, 709–719.

de Wit, C.T. (1960) *On Competition*. Agricultural Research Reports 668, PUDOC, Wageningen, The Netherlands.

de Wit, C.T., Tow, P.G. and Ennik, G.C. (1966) *Competition Between Legumes and Grasses*. Agricultural Research Reports 687, PUDOC, Wageningen, The Netherlands.

Ennik, G.C. and Baan Hofman, T. (1983) Variation in the root mass of ryegrass types and its ecological consequences. *Netherlands Journal of Agricultural Science* 31, 325–334.

Ennos, R.A. (1985) The significance of genetic variation for root growth within natural populations of white clover (*Trifolium repens*). *Journal of Ecology* 73, 615–624.

Evans, D.R., Hill, J., William, T.A. and Rhodes, I. (1985) Effects of coexistence on the performance of white clover–perennial ryegrass mixtures. *Oecologia* 66, 536–539.

Evans, D.R., William, T.A. and Mason, S.A. (1990) Contribution of white clover varieties to total swards production under typical farm management. *Grass and Forage Science* 45, 129–134.

Gaudet, C.L. and Keddy, P.A. (1988) A comparative approach to predicting competitive ability from plant traits. *Nature* 334, 242–243.

Gillard, P. and Elberse, W.Th. (1982) The effect of nitrogen and phosphorus supply on the competition between *Cenchrus biflorus* and *Alysicarpus ovalifolius*. *Netherlands Journal of Agricultural Science* 30, 161–171.

Goodman, P.J. and Collison, M. (1981) Uptake of ^{32}P labeled phosphate by clover and ryegrass growing in mixed swards with different nitrogen treatments. *Annals of Applied Biology* 98, 499–506.

Goodman, P.J. and Collison, M. (1986) Effect of three clover varieties on growth, ^{15}N uptake and fixation by ryegrass/white clover mixtures at three sites in Wales. *Grass and Forage Science* 41, 191–198.

Grime, J.P. (1979) *Plant Strategy and Vegetation Processes*. Wiley, Chichester, UK.

Hall, I.R., Scott, R.S. and Johnstone, P.D. (1977) Effect of vesicular-arbuscular mycorrhizas on response of 'Grasslands Huia' and 'Tamar' white clovers to phosphorus. *New Zealand Journal of Agricultural Research* 20, 349–355.

Hall, R.L. (1974) Analysis of the nature of interference between plants of different species. II Nutrient relations into a Nandi *Setaria* and Greenleaf *Desmodium* association with particular reference to potassium. *Australian Journal of Agricultural Research* 25, 749–756.

Hall, R.L. (1978) The analysis and significance of competitive and non-competitive interference between species. In: Wilson, J.R. (ed.) *Plant Relations in Pasture*. CSIRO, Melbourne, Australia, pp. 163–174.

Harper, J.L. (1977) *Population Biology of Plants*. Academic Press, London, UK.

Harris, W., Forde, B.J. and Hardacre, A.K. (1981) Temperature and cutting effects on the growth and competitive interaction of ryegrass and paspalum. I. Dry matter production, tiller numbers and light interception. *New Zealand Journal of Agricultural Research* 24, 299–307.

Hatch, M.D. and Slack, C.R. (1970) Photosynthetic CO_2 fixation pathways. *Annual Review of Plant Physiology* 21, 115–141.

Laurenroth, W.K. and Aguilera, M.O. (1998) Plant–plant interrelations in grasses and grasslands. In: Cheplick, G.P. (ed.) *Population Biology of Grasses*. Cambridge University Press, New York, USA, pp. 209–230.

Lazenby, A. and Rogers, H.H. (1965) Selection criteria in grass breeding. V. Performance of *Lolium perenne* genotypes grown at different nitrogen levels and spacings. *Journal of Agricultural Science* 65, 79–89.

Ledgard, S.F., Brier, G.J. and Upsdell, M.P. (1990) Effect of clover cultivar on production and nitrogen fixation in clover–ryegrass swards under dairy cow grazing. *New Zealand Journal of Agricultural Research* 33, 243–249.

Ludlow, M.M. and Wilson, G.L. (1972) Photosynthesis of tropical pasture plants. IV Basis and consequences of differences between grasses and legumes. *Australian Journal of Biological Sciences* 25, 1133–1145.

Lüscher, A., Connolly, J. and Jacquard, P. (1992) Neighbour specificity between *Lolium perenne* and *Trifolium repens* from a natural pasture. *Oecologia* 91, 404–409.

McCloud, P.E. and Mott, G.O. (1953) Influence of association upon the forage yield of legume–grass mixtures. *Agronomy Journal* 45, 61–65.

Mahmoud, A. and Grime, J.P. (1976) An analysis of competitive ability in three perennial grasses. *New Phytology* 77, 431–435.

Nassiri, M. and Elgersma, A. (1998) Competition in perennial ryegrass–white clover mixtures under cutting. 2. Leaf characteristics, light interception and dry-matter production during regrowth. *Forage Science* 53, 367–379.

Nurjaya, IG.M.O. (1996) Studies on the competitive ability of white clover (*Trifolium repens* L.) in mixtures with perennial ryegrass (*Lolium perenne* L.): the importance of non-structural carbohydrates and plant traits. PhD thesis, University of Adelaide, South Australia, Australia.

Pearcy, R.W., Tumosa, N. and Williams, K. (1981) Relations between growth, photosynthesis and competitive interactions for a C_3 and a C_4 plant. *Oecologia* 48, 371–376.

Rhodes, I. (1968a) The growth and development of some grass species under competitive stress. 1. Competition between seedlings, and between seedlings and established plants. *Journal of the British Grassland Society* 23, 129–136.

Rhodes, I. (1968b) The growth and development of some grass species under competitive stress. 3. The nature of competitive stress, and characteristics associated with competitive ability during seedling growth. *Journal of the British Grassland Society* 23, 330–335.

Rhodes, I. and Evans, D. (1993) White clover: breeders models. In: *Science for Agriculture and Environment 1993 Annual Report*. Institute of Grassland and Environmental Research, Hurley, UK, pp. 26–27.

Rhodes, I. and Ngah, A.W. (1983) Yielding ability and competitive ability of forage legumes under contrasting defoliation regimes. In: Jones, D.G. and Davies, D.R. (eds) *Temperate Legumes. Physiology, Genetics and Nodulation*. Pitman Advanced Publishing Program, Boston, USA, pp. 77–88.

Robinson, G.G. and Whalley, R.D.B. (1991) Competition among three agronomic types of the *Eragrostis curvula* (Schrad.) Nees complex and three temperate pasture grasses on the northern tablelands of New South Wales. *Australian Journal of Agricultural Research* 42, 309–316.

Robinson, J.B., Munnich, D.J. Simpson, P.C. and Orchard, P.W. (1993) Pasture associations and their relation to environment and agronomy in the Goulburn district. *Australian Journal of Botany* 35, 283–300.

Rösch, H., Van Rooyen, M.W. and Theron, G.K. (1997) Predicting competitive interactions between pioneer plant species by using plant traits. *Journal of Vegetation Science* 8, 489–494.

Schlichting, C.D. (1986) The evolution of phenotypic plasticity in plants. *Annual Review of Ecology and Systematics* 17, 667–693.

Scott, R.S. and Lowther, W.L. (1980) Competition between white clover 'Grasslands Huia' and *Lotus pedunculatus* 'Grasslands Maku'. *New Zealand Journal of Agricultural Research* 23, 501–507.

Simpson, D., Wilman, D. and Adams, W.A. (1987) The distribution of white clover (*Trifolium repens* L.) and grasses within six sown hill swards. *Journal of Applied Ecology* 24, 201–206.

Smith, S.R., Bouton, J.H. and Hoveland, C.S. (1992) Persistence of alfalfa under continuous grazing in pure stand and in mixtures with tall fescue. *Crop Science* 32, 1259–1264.

Snaydon, R.W. (1971) An analysis of competition between plants of *Trifolium repens* L. populations collected from contrasting soils. *Journal of Applied Ecology* 8, 687–697.

Svejcar, T. (1990) Root length, leaf area and biomass of crested wheat grass and cheatgrass seedlings. *Journal of Range Management* 43, 446–448.

Thom, E.R., Sheath, G.W. and Bryant, A.M. (1989) Seasonal variations in total nonstructural carbohydrate and major element levels in perennial ryegrass and paspalum in a mixed pasture. *New Zealand Journal of Agricultural Research* 32, 157–165.

Thomas, H. (1984) Effects of drought on growth and competitive ability of perennial ryegrass and white clover. *Journal of Applied Ecology* 21, 591–602.

Tilman, D. and Wedin, D. (1991) Plant traits and resource reduction for five grasses growing on a nitrogen gradient. *Ecology* 72, 685–700.

Tow, P.G., Lovett, J.V. and Lazenby, A. (1997a) Adaptation and complemetarity of *Digitaria eriantha* and *Medicago sativa* on a solodic soil in a subhumid, summer–winter rainfall region. *Australian Journal of Experimental Agriculture* 37, 311–322.

Tow, P.G., Lazenby, A. and Lovett, J.V. (1997b) Effect of environmental factors on the performance of *Digitaria eriantha* and *Medicago sativa* in monoculture and mixture. *Australian Journal of Experimental Agriculture* 37, 323–333.

Tow, P.G., Lazenby, A. and Lovett, J.V. (1997c) Relationships between a tropical grass and lucerne on a solodic soil in a subhumid, summer–winter rainfall environment. *Australian Journal of Experimental Agriculture* 37, 335–342.

van den Bergh, J.V. (1968) *An Analysis of Yields of Grasses in Mixed and Pure Stands*. Agricultural Research Reports 714, Institute for Biological and Chemical Research on Field Crops and Herbage, PUDOC, Wageningen, The Netherlands.

Widdup, K.H. and Turner, J.D. (1983) Performance of 4 white clover populations in monoculture and with ryegrass under grazing. *New Zealand Journal of Experimental Agriculture* 11, 27–31.

Woledge, J. (1988) Competition between grass and clover in spring as affected by nitrogen fertilizer. *Annals of Applied Biology* 112, 175–186.

4   Competition between Grasses and Legumes in Established Pastures

Alison Davies

Institute of Grassland and Environmental Research, Plas Gogerddan, Aberystwyth, Ceredigion, UK

Introduction

The success of grasses and clovers as pasture species is largely attributable to their morphology, which ensures that a substantial proportion of growing points remain below the level at which animals normally graze. Clover and other legumes also increase fertility by fixing nitrogen (N), which is transferred by the grazing animal (in the form of dung and urine) to the soil and made available to companion species, such as high-N-demanding forage ryegrasses.

The obvious advantages of a grass–legume relationship are not, however, easily realized to the full in agricultural practice because it can be difficult to maintain a favourable balance between the two species. The selective effects of differential defoliation, fertilizer applications (particularly N), animal excreta, environmental stresses (such as low temperature and drought) and pests and diseases may all have an impact on the relationship.

The purpose of this chapter is to examine the factors most likely to influence the dynamic grass–legume relationships in established pastures. Attention has necessarily been focused on the economically important ryegrass–white clover mixtures and on work conducted in the British Isles and New Zealand (with their different seasonal growth patterns and management practices). Reference has also been made to supporting Australian observations on subterranean clover (a winter-growing annual) where this is appropriate. The intention is to use these studies to show how the results

obtained may provide answers not only to the essentially agronomic questions originally raised but also to more fundamental questions about the factors that control grass–legume relationships and, in particular, about the role of N in this relationship.

White clover (*Trifolium repens* L.) probably originates from the eastern Mediterranean (Duke, 1981) and seems to have been commonest in fairly open habitats, while perennial ryegrass (*Lolium perenne* L.), its commonest companion grass in agricultural practice, probably originated in southern Europe (Terrell, 1968). The earliest written record of its use in UK pastures comes from the latter half of the 17th century (Beddows, 1967). White clover produces prostrate branches (stolons), which, in suitably moist conditions, root at the nodes. Seed of Kentish white clover was being produced for use in UK pastures, again by the latter half of the 17th century (Whitlock, 1983), while the presence of introduced white clover in Australian pastures was recorded in Sydney in 1857 (Davidson and Davidson, 1993). Subterranean clover (*Trifolium subterraneum* L.) is a winter-growing annual that also produces prostrate branches, but these do not root at the nodes. It has a well-documented history, having first been observed growing in the Mount Barker region of South Australia in 1887 by Amos Howard, who, noting that it was well-nodulated and that stock were prepared to eat it dry or green, proposed its introduction into the drier areas of the continent (Davidson and Davidson, 1993). It is most commonly grown with annual grasses, such as *Lolium rigidum* (Wimmera ryegrass)

© CAB *International* 2001. *Competition and Succession in Pastures*
(eds P.G. Tow and A. Lazenby)

and *Bromus mollis*, or with volunteer species, such as *Bromus rigidus*, *Hordeum leporinum*, *Vulpia myuros*, *Erodium botrys* and capeweed (*Arctotheca calendula*) (Smith *et al.*, 1972). The perennial grass *Phalaris aquatica* also shows promise as a companion grass, but needs to be heavily grazed to reduce its soil moisture demand when subterranean clover is establishing (Dear *et al.*, 1998).

Ryegrass and white clover are both potentially fast-growing species characteristic of fertile situations. Tissue turnover techniques (Davies, 1993) indicate that the comparatively low percentage of white clover present in swards in spring (Evans and Williams, 1987) is not necessarily the result of a growth rate slower than that of ryegrass. Gross crop growth rates in pure white clover swards supplied with an élite mixture of *Rhizobium* strains reached just less than 200 kg ha^{-1} day^{-1} in the May–June period in the UK (Davies and Evans, 1982). Maximum growth rates in well-fertilized perennial ryegrass swards in August–September were 190 kg ha^{-1} day^{-1} (Davies, 1971), the difference between the two rates being very much in accordance with the expected differences in the daily radiation receipts. This serves to illustrate the very high growth potential of white clover, which in Western European practice is often seen (somewhat mistakenly) as the poor relation. Later sections will show that management, the location of the grass and clover leaves in the canopy and the overwintering capacities of grass and clover have more to do with poor clover performance in spring than their respective growth rates. Increased reliance on white clover as a N source in the temperate northern hemisphere has been greatly retarded by the lack of a sufficient understanding of the factors that control the grass–clover relationship and confused by the success of apparently similar management systems in the southern hemisphere.

Effects of Temperature and Moisture on the Growth of Grass and Clover

It is widely accepted that white clover grows less well than ryegrass at low temperatures, a conclusion based partly, at least, on the pioneering controlled growth room work of Mitchell and Lucanus (1962). They found that white clover grew less well than ryegrass below 20°C and better at higher temperatures. Work of Woledge and Dennis (1982) has since shown that

the photosynthetic responses of leaves of ryegrass and white clover to temperature are very similar, and that photosynthesis in both species was twice as high at 15°C as at 5°C. Leaf area expansion has, however, been found to be higher in ryegrass than in white clover when growing in simulated swards in a 10°C day/8°C night regime, rather than in a 20°C/15°C regime (Davidson and Robson, 1986b). Clover seedlings both photosynthesize and grow very poorly in an 8°C/4°C day/night regime (Woledge and Calleja, 1983). Such findings differ from the observations of Fukai and Silsbury (1977), who, working with subterranean clover, showed that the photosynthesis of box swards was relatively temperature-insensitive; but concur with the observations of Woledge and Parsons (1986), which showed that photosynthesis increased with temperature in ryegrass swards. The latter authors, however, also demonstrated that photosynthesis was only sensitive to temperature when the saturation deficit of water vapour was minimized, and concluded that photosynthesis was affected by relatively small saturation deficits. The temperature reactions of grass and clover are thus intimately related to their respective states of hydration, and further studies will need to focus on this aspect of grass and legume physiology.

Clover is at greater risk of desiccation than its usual companion grasses in dryland conditions, in which its herbage production was found to be inferior to that of tall fescue and phalaris (Johns and Lazenby, 1973). Its ability to control leaf hydration is also poorer than that of ryegrass and tall fescue, and a tendency towards incomplete stomatal closure has been noted in clover by Hart (1987). Clark *et al.* (1999) showed that withholding water reduced the photosynthesis and growth in white clover to a greater extent than in ryegrass, and that ryegrass also showed a more positive response to rewatering. Clover can also exhibit morphological changes in relation to water stress: Thomas (1984) observed that, although drought reduced the amount of clover herbage harvested, it increased growth below cutting level, particularly in S184, a variety with small leaves and short petioles, in which regrowth after drought was greatly enhanced by this reaction. A similar reaction may account for the increase in clover content observed in untreated lawns in a dry summer. Thomas (1984) also noted that white clover root systems were more affected by drought when growing in mixture with ryegrass than they were in monoculture.

Further studies will need to take careful account of interactions between temperature, relative humidity and water-supply. There is also, in general, a pressing need for further side-by-side growth room comparisons both of grass and clover plants and of well-established grass and clover communities at differing levels of humidity, temperature and light (including intensities higher than those which were possible in the earlier studies). Naturally occurring ecotypes of similar provenance are likely to be of particular interest in such comparisons.

Drought can affect N_2-fixation as well as root growth, and experiments on soybean (Djekoun and Planchon, 1991) showed that water reduction can bring about a marked limitation in the yield of soybean by impairing N_2-fixation as well as photosynthesis, but recovery was slower in the former and seemed to be related to nodule mass. Later work, also in soybean, has indicated a significant correlation between sucrose synthase activity and apparent nitrogenase activity, which suggests that a stress-related decline in N_2-fixation could well be caused by a reduction in sucrose flow to the nodule through the phloem (Gordon et al., 1997). A similar link between a reduced water-supply and reduced N_2-fixation may well be found in clover.

Desiccation (or adaptation to an increased risk of desiccation) may also be important in relation to winter performance of clover relative to that of grass. This is evident in the report of Harris et al. (1983) of an instance in which cold winds, rather than low temperature, caused damage to white clover, but not to the associated ryegrass. Avoidance of desiccation may also, at least partly, account for the low and relatively protected position which clover comes to occupy in a mixed sward in winter (see section on seasonal growth). A full account of drought resistance and drought avoidance in white clover varieties of different origin can be found in Collins (in press).

Cold-induced injury in most plants is the consequence of severe cellular dehydration. Freezing tolerance, which involves membrane stabilization, is induced in response to non-freezing temperatures below about 10°C (Thomashow, 1998), a process known as hardening. It is, however, only one aspect of the differences in winter survival that can be observed between hardy and susceptible varieties: others include resistance to wind and to snow cover and susceptibility to attacks from low-temperature pathogens (Eagles et al., 1997).

Susceptible cultivars of both ryegrass and clover are often characterized by a loss of hardening during warmer periods, which may occur relatively quickly in comparison with the hardening process. Hardier varieties of grass and clover often have additional photoperiodic requirements that enhance hardening. For example, Juntilla et al. (1990) noted that hardening was enhanced by short photoperiods, while Eagles (1994) noted that the increased photoperiodic requirement of hardy varieties of clover prevented premature dehardening in response to periods of raised winter temperatures. Such mechanisms underlie the differences in stolon survival and in the capacity to expand leaves during milder periods in winter that were observed by Rhodes et al. (1994) in clover varieties of different provenances.

Canopy Development

The extent to which the grass–clover relationship is influenced by temperature and N is strongly dependent on the stage of development of the canopy. This is best considered as comprising three stages of variable duration:

1. Active increase in light capture.
2. Light capture and growth rate reach a maximum.
3. Maturation, during which processes such as self-thinning (Kays and Harper, 1974) and/or programmed leaf senescence (Davies, 1971) are initiated and net dry matter increase terminates. Annual plants (such as subterranean clover) set seed and die.

The effects of temperature and N (both of which affect leaf area expansion) depend on the stage of canopy development. Before the canopy closes, the species with the highest rate of leaf area expansion will increase its share of the light intercepted at the expense of its competitor(s), but, once a closed canopy is present, temperature and N may cease to affect the share of radiation received or the crop growth rate. This principle is well illustrated in studies of subterranean clover communities conducted by Cocks (1973), which show that crop growth rates increased with temperature when LAI (leaf area index) was less than 3 cm^2 cm^{-2} of ground area, but not thereafter (the response to raised temperature actually became negative at LAI > 5.5, since the plants matured and set seed faster at the higher temperatures). Fukai and Silsbury (1976), also working on subterranean clover, found that temperature ceased to affect the rate of dry matter production when the canopy closed (before canopy closure, high temperatures increased leaf expansion and photosynthesis, so that maximum crop growth rate was attained sooner). Net carbon dioxide exchange at 250 W m^{-2} light intensity increased only up to LAI = 3 (Fukai and Silsbury, 1977) and daily net production (at low light levels characteristic of midwinter) was better at low temperatures. Relative humidity effects cannot, however, be excluded.

The same temperature insensitivity was observed in closed canopies of ryegrass and white clover in controlled conditions. Measurements of photosynthesis made by Davidson and Robson (1984) confirmed that clover fixed at least as much carbon per unit leaf area as ryegrass and that it was able to maintain its favourable position in the canopy in both warm (20/15°C) and cool (10/8°C) conditions. Point quadrat measurements demonstrated that clover leaves were concentrated in the upper sward layers and radiocarbon studies (Dennis and Woledge, 1985) showed that clover photosynthesized at least as well as ryegrass at equivalent height.

Investigations on the effects of N on the grass–clover balance in the established canopy have similarly revealed that photosynthesis per unit leaf area in white clover was at least as high as in ryegrass, even at higher N levels (Davidson et al., 1986). Clover dry weight was the same in low- and high-N situations, and the clover percentage in the low-N treatment was actually increasing, because of the greater specific leaf area of clover. Field studies, again conducted by Davidson et al. (1986), confirmed that white clover and ryegrass had similar relative growth rates in the presence of N fertilizer, while white clover grew faster than ryegrass when no N was applied. This situation is likely to arise in low-N conditions when much of the N supply is being derived from N$_2$-fixation and grass cannot acquire enough N for maximum growth.

To sum up: a critical stage in the grass–clover relationship occurs during canopy development, when differences in the capacity to expand leaf area at lower temperatures/higher N levels change the share of radiation received by the clover.

Nitrogen applied before the canopy closes not only stimulates the expansion of grass leaves: it also increases tiller numbers relative to clover growing points, and the impact of this carries through to later regrowths. Table 4.1 shows the effect of two applications of 50 kg N ha^{-1} during the course of a period of 8 weeks of regrowth (compared with unfertilized controls) was to shift the ratio between numbers of growing points and numbers of tillers entering the next cycle of regrowth in favour of the latter.

An additional effect of N fertilizer is that it reduces transmission through grass leaves, particularly in the red region of the spectrum (Fig. 4.1). Light reflected from N-fertilized grass on to the more horizontally orientated clover leaves below is also deficient in red light and may provoke shade avoidance reactions (such as increases in petiole length and decreases in branch production). Neighbour avoidance reactions in response to changes in the ratio of red to far-red light have been observed by Ballaré et al. (1990) in Sinapis alba and Datura ferox and by Novoplansky et al. (1990) in Portulaca. Conversely, observed differences in the reactions of clover stolons growing into different grass species (Turkington et al., 1991) may relate to the capacity of these species to transmit more red light or to reflect less infra-red light.

Changes in the relative numbers of grass tillers and clover growing points brought about by previous alterations in mineral N supply have the potential to carry through to further regrowths so that the effect of N on the grass–clover relationship is potentially long-lasting (even when much of the N has been removed by subsequent defoliation).

Another situation in which clover is placed at a potential disadvantage arises when day length decreases and temperatures fall. Clover petioles shorten in response to reduced day length (Eagles and Othman, 1986; Juntilla et al., 1990): maximum petiole lengths in young spaced plants of white clover occur in the May to September period (Davies and Jones, 1992). Similar responses in ryegrass are constrained by the length of the sheaths through which the new leaves must emerge (Davies

Table 4.1. Effects of N (2 × 50 kg ha^{-1}) and cutting (two or four cuts in 8 weeks) on ryegrass tiller numbers and clover growing points in mixed swards (Alison Davies, unpublished data).

	+ N	0 N	SE ±	Two cuts	Four cuts	SE ±
Tillers m^{-2}	6148	5701	227.1[a]	6723	5412	227.1[a]
Growing points m^{-2}	1314	1604	74.5[a]	1425	1494	74.5NS

[a]Main effects significant at $P < 0.05$.
N × cutting interactions NS.

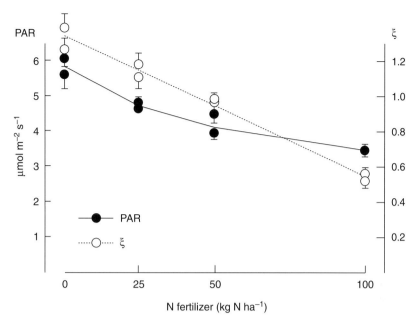

Fig. 4.1. Light transmission and changes in the red/far-red ratio (ξ) of light passing through ryegrass leaves grown at different N levels. PAR, photosynthetically active radiation.

et al., 1983), so that grass leaves and sheaths can remain long if the swards are not defoliated in autumn. The result is that clover may come to occupy a relatively low position in the canopy, where it is unable to compete effectively for light. A relatively mild winter preceded by a final defoliation in September was found to result in the almost complete disappearance of the white clover by the following spring (Davies, 1998; Fig. 4.2): earlier studies in colder winters indicated much smaller differences between autumn managements. Work by Woledge *et al.* (1990) has similarly indicated a greater decline in clover percentage in the warmest of three winters and has shown that the low rate of photosynthesis per unit leaf area relative to that of ryegrass at this time of year is linked to the relatively low position of clover in the canopy, which resulted in greater shading by the ryegrass. Clover height fell from 8.6 cm in November to 2.4 cm in February, while ryegrass fell from 8.0 to 5.6 cm and not at all in the mild winter of 1987/88. It is therefore evident that (in the absence of defoliation) differing height responses of grass and clover to winter temperatures can result in clover suffering severe competition for light.

Since more grass leaf than clover leaf is present in the upper layers of the sward in spring, it is not surprising to find that spring defoliation can be beneficial in terms of clover content (Davies and Evans,

1990; Table 4.2). The removal of a greater proportion of leaf from the taller-growing grass allowed the shorter and relatively less severely defoliated clover to increase numbers of growing points relative to tiller numbers during the period before the sward closed and the production of both tillers and growing points ceased. N application in this experiment stimulated grass growth and (except in the case of the February defoliation, when rain and snow may have resulted in the loss of the applied N by leaching) reduced or prevented the benefit of spring defoliation to clover. Dry weights of clover stolon in June in fertilized plots were, on average, doubled by defoliation in spring; differences in stolon weights in unfertilized plots were not significant.

A similar outcome was reported by Laidlaw *et al.* (1992), who showed that out-of-season grazing by sheep (especially when a March grazing was included) could increase clover content and numbers of growing points later in the season.

It can be concluded that the risk to clover of increased competition from ryegrass is greater in any circumstances that promote increases in ryegrass growth (including leaf extension and the production of new tillers) without having a similar effect on white clover and the number of growing points. It has been shown that these included not only N supply but also the differing effects of seasonal change,

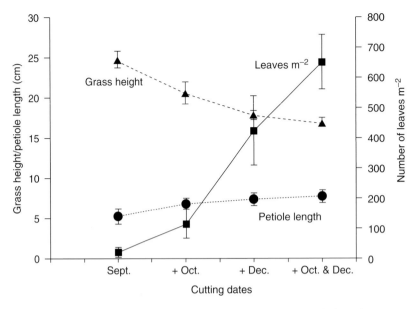

Fig. 4.2. Effect of time of autumn defoliation on grass height, petiole length and white clover content (leaves m^{-2}) in a ryegrass–clover sward. (Redrawn from Davies, 1996.)

probably day-length-mediated, on the location of grass and clover leaves within the sward, and on the interactions between season, N and defoliation.

Leaf Area Distribution – a Modelling Approach

It is evident from the previous section that the way in which the leaf area of clover and grass is distributed in the sward strongly influences their competitive relationship, and a summary of the best available evidence on the seasonal changes in leaf area distribution of ryegrass and white clover is presented in Fig. 4.3. The interaction of such changes with defoliation, defoliation regimes and the choice of varieties can be

aided by mathematical models, such as that recently developed in Wageningen (Nassiri Mahalatti, 1998). In the model the leaf area distributions of grass and clover varieties were effectively represented by triangular leaf area density functions having different heights of maximum leaf area density. A variable dispersion factor for each species was combined with a fixed species-dependent light extinction coefficient to simulate departures from random dispersion. Calculations showed that in the absence of N fertilizer clover captured a significantly higher proportion of photosynthetically active radiation (PAR) than its share of the LAI would indicate and that this was related to the pattern of its leaf area distribution within the canopy. Experiments showed that the addition of N fertilizer greatly increased the propor-

Table 4.2. Clover content in relation to date of cut in spring and to N status (data derived from Tables 2 and 4 in Davies and Evans, 1990).

Date of cut in spring	Mean % clover at time of cut	Mean % clover after 3 weeks' regrowth (DM basis)		Stolon weights in June (g m^{-2})[a]	
		+ N	0N	+ N	0N
Feb.	13.3 ± 0.80	43.3 ± 8.27	38.6 ± 8.27	144	152
March	15.8 ± 2.31	24.4 ± 4.40	36.6 ± 4.40	124	181
April	12.0 ± 4.40	17.2 ± 3.17	31.7 ± 3.17	117	198
Feb. + April	28.4 ± 4.40	16.8 ± 3.17	51.7 ± 3.17	89	199

[a]SE ± 16.8; uncut controls 65 (+ N) and 175 (0N) g m^{-2}.

tion of the grass leaf area in the upper layers of the canopy and that, in such circumstances, a clover variety with long petioles (Olwen) was better able to maintain dry matter production than one with short petioles (S184). In swards that are defoliated frequently the risk of shading in a pasture variety such as S184 is offset by a decreased risk of defoliation, and this point will be dealt with in the next section; for the moment, it is sufficient to conclude that the way in which the leaf area of a legume is distributed in the sward and the factors which alter that distribution have a major impact on its chances of survival, and on the grass–clover relationship as a whole.

Morphological Changes in Grass and Clover in Relation to Season and Management and their Impact on the Relationship between Grass and Clover

The aim of this section is to examine data relating to the growth and survival of clover stolons in grazed swards and to highlight the vulnerability of clover in circumstances in which extended periods of close grazing restrict the development and main-

tenance of an effective stolon network. A series of experimental observations, made in New Zealand in the 1980s, allow useful conclusions to be drawn about the effects of seasonal variation (in temperature, light and moisture supply) and management on stolon growth and death, and the consequences in terms of root and branch production and survival. Moreover, comparisons of New Zealand and UK findings enable some general principles to be developed about the limits of sustainability of an effective clover presence in relation to different management practices. The managements compared in the New Zealand studies, which were representative of those operating in practice, were set-stocking with sheep (SSS) and rotational grazing using either sheep or cattle (RGS or RGC). The effects of transferring sheep from a set-stocking regime (until drafting) to a rotational regime were also investigated (SSS–RGS). Most of the observations on grass were made on the ryegrass component.

Seasonal changes observed in clover growth may help to indicate potential stress periods. The maximum amount of surface stolon was observed in the period extending from just after midsummer through to late autumn (Hay et al., 1983). Stolon extension rates were lowest in winter (Chapman,

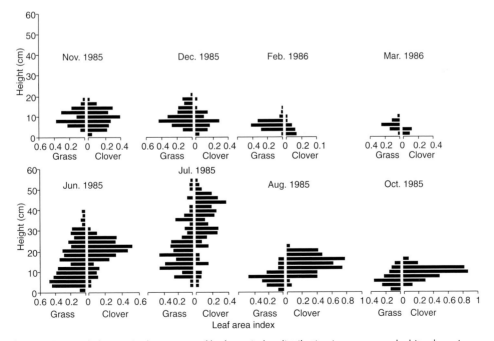

Fig. 4.3. Seasonal changes in the patterns of leaf area index distribution in ryegrass and white clover in mixed swards. (Redrawn from Woledge et al., 1989, 1992.)

1983). Weight per unit length of stolon decreased from 0.57 g m^{-1} in the autumn to 0.37 g m^{-1} in spring, and this was coupled with an increasing amount of stolon burial, associated with treading and worm-casting in wet soil. There were obvious signs of stolon decay in spring, when stolon weight reached a minimum (Hay *et al.*, 1983).

Chapman (1983) observed that patterns of root and branch development showed seasonal differences. Branches developed on 27% and roots on 18% of nodes in summer; values were less than half of this for branches in winter and very few nodes developed roots. These studies seem to indicate that the cooler half of the year is stressful for clover and show that the potential exists for clover to lose ground relative to ryegrass over this period. Critical comparisons with the situation in the grass companion are, however, lacking, largely because of the labour involved in incorporating the number of measurements needed.

Rotational grazing by cattle has been found to favour clover, whereas the constant presence of sheep on pasture results in a low clover content. Calculations of maximum elongation rates made by Chapman (1983) showed that stolons had the potential to explore distances of 29 (SSS), 40 (RGS) and 71 cm (RGC) within the sward in a year; mean elongation rates suggested that the actual extent of stolon movement in these three grazing systems would have been about 5.4, 7.8 and 13.2 cm, respectively. Stolon dry weights for the treatments SSS, RGS and SSS–RGS averaged out at 268, 292 and 323 g m^{-2} and numbers of growing points at 3186, 2640 and 2927 m^{-2} (Hay *et al.*, 1983), while tiller numbers were twice as high in SSS as in RGS. In contrast, stolon dry weights in cattle-grazed swards (RGC) were three times as much as in SSS and a greater proportion of the stolons in the SSS treatment was buried. The percentages of clover in the RGC and SSS treatments were 19 and 8.5%, respectively (Hay, 1983). Rotational grazing by cattle thus clearly

favoured clover, especially relative to the effect of SSS.

When the rotational grazing intervals exceeded 21 days, the interval between defoliations was greater in RGS than that observed in the SSS treatment in all species (Chapman *et al.*, 1984; Table 4.3), but it is important to stress that, although defoliation intervals varied (and more leaves were grazed in the RGC treatment in the colder months), the total amount of leaf removed was approximately the same in all three managements. These observations led Chapman (1986) to conclude that 'despite high stocking rates none of the managements are likely to have restricted assimilate supply through excessive leaf removal as leaves were able to export assimilate for an estimated mean period of 15–17 d[ays] before being removed'. This conclusion, which was to influence the direction of Chapman's future work, needs to be borne in mind when making comparisons between the extent of leaf removal in New Zealand managements and UK continuous sheep-grazing systems.

The radiocarbon studies that were subsequently conducted by Chapman *et al.* (1990) showed that very young leaves are sinks for photosynthate. Rates of photosynthesis equalled rates of respiration when leaflets were clearly separated and each leaflet was approximately 10% unfolded. This stage is equivalent to 0.6 on the Carlson scale (Carlson, 1966).

Defoliation increased the percentages of ^{14}C moving to the parent stolon apex from younger nodes. As the developing leaf proceeded to full expansion, carbon was again exported, principally to the apex and stolon, while most ^{14}C moving to the roots went to older, nodulated roots (Chapman *et al.*, 1991). Stolon and root production depend on the maintenance of sufficient mature leaf tissue to be able to sustain this export.

Studies in continuously grazed UK swards present a completely different picture. Here the use of sward height guidelines (Hodgson *et al.*, 1986), which have been applied to ryegrass-based swards,

Table 4.3. Effect of pasture management on defoliation intervals and mean numbers of leaves over the four seasons, per stolon growing point (data from Chapman, 1986).

	Managements		
	SSS	RGS	RGC
Mean defoliation intervals (days)	23.48	32.49	35.89
Mean number of leaves per stolon	2.49	2.78	3.05

places very considerable pressure on the clover component. Although differences in the proportion of dry weight utilized in the form of grass leaf and clover leaf plus petiole in continuously grazed swards, stocked to maintain sward height at 4–5 cm, were minimal (both averaged out at 50% over 3 years of observations), the proportion of leaf lamina produced which was eaten was greater in clover (66%), indicating greater percentage utilization of photosynthetic tissue (Davies, 1990). Such a difference would contribute to reductions in clover content in the longer term.

Observations made by Jones and Davies (1988) showed that the leaf complement (i.e. the number of leaves per growing point) was low in swards grazed to a height of 4–6 cm by sheep (1.14 leaves per growing point, a leaf being defined as one in which all midribs were exposed when seen from above). Measurements in boxed swards maintained at different leaf complements showed that stolon growth in particular was curtailed when all leaves at or beyond Carlson stage 0.6 (see p. 70) were removed twice a week. Stolon growing points in UK swards grazed continuously to 4–6 cm carry, at any one time, a mean of only one leaf with the midrib visible from above, and an appreciable population (17%) have leaves at an earlier stage of development or no leaves (Jones and Davies, 1988). This contrasts markedly with the more lenient New Zealand managements, in which stolons were found to support 2.49 SSS, 2.78 RGS and 3.05 RGC leaves per stolon (see Table 4.3). The UK sward height management guidelines clearly represent a much more severe defoliation regime than those imposed by the New Zealand feed-budgeting systems (Milligan and Smith, 1984; Shoati et al., 1987) and have a correspondingly greater effect on clover survival.

The difficulty of maintaining a good stolon network over the course of time is reflected in the phenomenon of clover 'crashes', i.e. major reductions in clover content, which typically occur after 3–4 years in the late summer or autumn (Fothergill et al., 1996), though they have been observed earlier in the life of the sward (Davies and Jones, 1988). The predisposing factors seem to be a high clover content and a high overall level of productivity, and the management response has been to increase stock numbers to maintain the prescriptive sward height. The impact of this on the clover component of the sward differs from that on the grass. In clover, the reduction in the mean leaf complement and the constant demand on stored stolon metabolites result in the breakup and fragmentation of complex stolon systems into much smaller and less complex plant units (Fothergill et al., 1997). These units, deprived of their maternal support system, may be defoliated too frequently to sustain sufficient leaf area for survival. Leaf death is also likely to be increased by the greater risk of treading in heavily grazed swards (see pp. 73–74).

At least two factors place grass at less risk. First, treading in ryegrass has the effect of breaking up leaf sheaths and facilitating the growth of short tillers (Davies et al., 1983). Secondly, the more fibrous grass leaves are better able to resist damage from treading and urine scorch than clover leaves (see p. 73). Other factors related to increased grazing pressure may well be identified by closer comparative examination of changes in the amount and character of grass and clover residues.

Indications of the need for remedial action to restore clover content include a marked absence of complex stolon networks (M. Fothergill, personal communication) and a heavy preponderance of short internodes, but more research will be needed before effective practical guidelines can be confidently drawn up to avoid or amend the situation. Staggered sowings, with half the clover seed sown one year and half the next, show some promise (M. Fothergill, personal communication) in helping to avoid synchronization of stolon network breakdown, but are scarcely a practical solution; avoidance of resowing more than one area in a single year, however, seems a practical policy.

Utilization of clover to the extent observed in continuous grazing systems in the UK would rarely occur in less intensive systems in which stocking rates were lower (as in New Zealand systems) and/or grazing animals were free to range over a wide area. The key to increasing clover content is the introduction of a procedure likely to increase clover residues and leaf area at the expense of grass residues. A possible strategy involves the interpolation of a silage cut, which removes relatively little clover stolon but appreciable amounts of grass pseudostem. It also results in the continued production by clover of the large leaves developed in a silage crop.

Table 4.4 shows the results from a box experiment designed to investigate the effect of changing the height of defoliation of the matrix ryegrass plants on the size of petioles and leaves produced by the stolons of clover growing within the matrix. Five

Table 4.4. Effects of a change in cutting height of grass on petiole and midrib length in associated white clover regrowth, 6 days after grass-cutting heights were changed in half the treatments. Cuts were made twice a week.

	Treatments								
	LLO	LLD	LSO	LSD	SSO	SSD	SLO	SLD	SE ±
Petiole length (cm)	10.56	8.82	6.41	4.80	4.63	3.67	7.83	5.16	0.470
Midrib length (cm)	1.47	1.34	1.64	1.33	1.19	0.92	1.23	0.97	0.0572

LL, grass cut five times to 10 cm; LS, grass cut four times to 10 cm, then once to 2.5 cm; SS, grass cut five times to 2.5 cm; SL, grass cut four times to 2.5 cm, then once to 10 cm; O, clover leaves left intact; D, clover leaves removed.

cuts were made at the rate of two per week. The results illustrate the capacity for petiole length and leaf size in white clover to respond independently to defoliation treatment. Petiole length responded rapidly to change in grass-cutting height, decreasing with a change from long to short cut, and increasing with a change from short to long (LL > LS; SS < SL). That is, petiole length changed in relation to current grass-cutting height. This effect occurred whether clover leaves were left intact or removed. In contrast, midrib length of the clover leaflets (a good indication of leaflet area) was related to the earlier grass-cutting height, being greater where early cutting height had been higher (LL > SL; LS > SS), and it remained about the same after the change in cutting height. Both petiole and midrib lengths were reduced if the clover leaves themselves were removed at the same time as the grass was cut.

Similar considerations apply when comparisons are made between clover leaves produced in continuously grazed swards and in swards allowed to proceed to a silage cut. Two weeks after the silage cut, the area per leaf in clover (averaged over four varieties) was found to be 3.76 times greater than the area per leaf in the continuously grazed control swards, i.e. the leaflet diameter had doubled, as had the weight of stolons per unit area (A. Davies, unpublished data). Sward heights did not appear to differ greatly. This uncoupling of leaf area, which is related to leaf complement (see above), from petiole length, which is related to light attenuation and light quality (see pp. 66–67), could, in the right circumstances, be a means of increasing the clover content of swards.

Although the introduction of an early conservation cut in mid-May into a continuous grazing regime has proved ineffective in increasing clover content, some limited success with a late conserva-

tion cut in late June was obtained by Barthram and Grant (1995). A 7-week rest period, commencing on 7 May, resulted in a dramatic increase in clover percentage during that period, especially when grown with Aurora (an early ryegrass variety). The sheep were returned a few days after the cut in early July, but by September clover percentages were once again similar to the continuously grazed controls. Later silage cuts (imposed in the second harvest year after a rest period in early July–late August and incorporating a range of varieties) have met with greater success in terms of clover presence in grid squares (Gooding *et al.*, 1996) and stolon weight per unit length (Gooding and Frame, 1997). Early cuts in late May–early June, which were followed in the same way by a silage cut, were either of no benefit or markedly detrimental. This may possibly be related to the enhanced leaf extension rates characteristic of ryegrass in spring, which have been lost by mid-season (Davies *et al.*, 1989), but observations of the location of clover leaves in the canopy in the different treatments and of changes in populations of tillers and growing points would help to resolve the issue.

The cyclical fluctuations in clover content of UK swards (although a source of annoyance to farmers) are to some extent built into the underlying nature of the grass–clover relationship, at least in the shorter term. They are linked with the transfer of N from clover to grass, the increase in grass growth supported by this transfer and the relatively high proportion of clover leaf area removed in continuous grazing systems based on sward height. There are some indications that swards may, in the longer term, settle down to a more stable equilibrium (Turkington *et al.*, 1991), in which perturbations are less synchronized (see Schwinning and Parsons, 1996b).

The Impact of the Grazing Animal on the Grass–Clover Relationship

The grazing animal may affect the relationship between grass and clover in three ways:

1. By a grazing intensity such that the remaining herbage includes more of the leaf area of one species than the other.
2. By the deposition of dung and urine resulting in: (a) the creation of an uneven pattern of N distribution in the soil; (b) subsequent avoidance of recently contaminated areas.
3. By actively selecting clover-rich areas.

Although defoliation by animals not infrequently removes more clover than grass (see p. 71), it would be wrong to conclude that this is necessarily the result of active selection. Work by Milne *et al.* (1982) showed that sheep grazed clover and grass leaf and stem to the same height and that the increase in clover content of their diet compared with that of the sward could largely (> 80%) be accounted for by the difference in the pattern of distribution of the grass and clover components of the canopy. It may be that modification of this pattern by selection of appropriate grass and clover pair combinations could help to improve clover persistency.

Studies on the effect of different stocking rates in sheep-grazed swards illustrate the effects of overstocking on clover content (Curll and Wilkins, 1982). Herbage on offer declined when stocking rates exceeded 35 sheep ha^{-1} and stolon removal at 55 sheep ha^{-1} accounted for two-thirds of the decline in stolon material. Clover percentage declined at the higher stocking rates (45 and 55 sheep ha^{-1}) and on N-fertilized plots at all four stocking rates (Curll *et al.*, 1985), and it was shown that the herbage ingested by the sheep had a higher clover content than the herbage on offer. Parallel work (Curll and Wilkins, 1983), in which grazethrough cages were used to assess the separate effects of defoliation (D), treading (T) and the return of excreta (E), showed that at a high stocking rate (50 sheep ha^{-1}) clover content declined from 58 to 31% (D), 25% (D and T) and 18% (D, T and E). At the lower stocking rate of 25 sheep ha^{-1}, clover content rose to 75% (D), 71% (D and T) or declined to 45% (D, T and E). Treading and the deposition of excreta account for about 60% of the total animal impact on the white clover component of the sward. Heavy treading by sheep was found by Edmond (1964) to reduce clover more than ryegrass, but not necessarily more than timothy, *Poa* species or *Agrostis* species.

The effect of urine deposition can be particularly damaging, especially in dry conditions. Sheep and cow urine can kill subterranean and white clover; perennial ryegrass was found to be more resistant (Doak, 1954). The photograph (Fig. 4.4) illustrates the effects of sheep urine applied at a volume per unit area similar to that observed in the field on the grass–clover balance of glasshouse swards 1 week after application. No such effects were observed if the urine was injected into the soil (Christophe Dassie, unpublished data), when the outcome was similar to that of an equal amount of N supplied in the form of fertilizer.

Cattle dung pats increase sward heterogeneity. The average diameter of a dung pat is about 25 cm and observations have shown that it takes about 70 days to disappear in a high-rainfall area (Bastiman and Van Dijk, 1975). Dung pats can kill up to 75% of grass tillers and 95% of clover nodes within 25 days (MacDiarmid and Watkin, 1971). The resulting bare patches may last for 6–12 months and provide fresh colonization sites for clover (Weeda, 1967), which has been found to dominate old dung-pat sites for an average of 12–18 months.

Active area selection for clover patches in a sward by animals has been demonstrated in two experiments (both necessarily involving discrete areas sown to grass and to clover or to a grass-clover mixture). In grass swards maintained at 7 cm by grazing, in which mixed grass and clover patches of different sizes and at different spacings had been established, Armstrong *et al.* (1993) observed that the percentage of clover in the diet of weaned

Fig. 4.4. Effects of sheep urine deposition (right) on white clover in mixed ryegrass–clover communities.

lambs was always higher than the percentage in the plots. This was especially true on treatments with a low overall clover presence (< 5%), but represented four times the expected amount even when clover patches occupied 25% of the area. It should not, however, be forgotten that even this situation would be representative of a poor clover presence in practice. Animal preference is also modified by previous experience. Newman *et al.* (1992) showed that, when ewes were offered cut turfs of ryegrass and clover in seed trays, their preference depended on what they had been grazing previously; if it was ryegrass, they preferred clover and chose an intake which was 66–74% clover. If the reverse was the case, they took 0–22% clover (preferring ryegrass). Further studies by Parsons *et al.* (1994) on ewes grazing 0.25 ha swards containing 20, 50 or 80% clover after having previously grazed all grass, all clover or 50/50 grass/clover swards confirmed an initial preference for the opposite species to the one they had been grazing, but showed that this preference was short-lived, and the animals reverted to their earlier diet. Sheep on swards with different clover percentages had a greater proportion of clover in their diet than the proportion of clover in the sward, but this is not necessarily in conflict with the observation of Milne *et al.* (1982) (see p. 73) and may represent a choice of horizon rather than a deliberate search strategy. To sum up: sheep choose a height at which to graze such that they ingest more clover leaf than grass leaf. At the same time clover-rich areas are increasingly likely to be sought out by sheep if clover presence in the sward is low and its distribution is patchy. Both choices will tend to be disadvantageous for clover.

Mathematical Models

Mathematical models have been developed to aid understanding of the contribution of various processes in bringing about changes in clover content over the course of time. The model of clonal growth developed by Cain *et al.* (1995) incorporates an inverse relationship between stolon extension rates and neighbourhood density (based on observations of numbers of stolons with a proximal fully expanded leaf lying within neighbourhoods of 10 cm radius and on average rates of stolon dieback). Persistence in this model is critically related to patch size; stolon growth increased with density at low densities and decreased at high densities. The resulting pattern of patch dissipation over periods of 5–20 years was not, however, supported by field data, since observation showed that clover patches lasted for 1–3 years. Other variables, such as herbivory and N feedback, will need to be incorporated in future versions. A pasture model proposed by Thornley *et al.* (1995) has been developed by Schwinning and Parsons (1996a) to analyse the grass–clover relationships in grazed pastures. The key element of this data-based model is an N-based competitive trade-off between grass and clover; herbivory removes N from the pasture at large, but returns it locally and irregularly, creating a series of relatively short-lived (30–50 days) high-N patches, which become increasingly out of phase with each other. At the same time, widespread depletion occurs in the other areas, although some N returns to the soil as a result of tissue abscission and decomposition. The model assumes that abscised tissues have the same C and N content as equivalent living tissue. Herbivory is related to the relative abundance of grass and clover (and the fraction of clover preferred in the diet). Time delays in the feedback mechanisms that control population densities of grass and clover increase the tendency for these populations to oscillate.

The situation has been further explored in a spatial model by Schwinning and Parsons (1996b), which examines the consequences of patch size oscillations in a grass–clover mixture. The model sward comprises cells that are grass-dominant or clover-dominant or contain grass only (at high or low N). Grass-dominant cells and grass/high-N cells 'age' due to N depletion; clover cells 'age' by N enrichment. Pure grass/low-N cells may be invaded by clover. The probability of urine application is incorporated and increases the period of grass dominance in the affected cell. Initial conditions and the magnitude of disturbances in soil N content dictate the time taken for the system to converge on a stable state, i.e. one in which there is a patchy distribution of clover at different stages in the cycle of depletion and accumulation of N. Clumped distributions, in which many cells are in the same phase, are created by reseeding or can occur as a result of major events, such as high winter mortality of clover or the application of N fertilizer in spring. This phase synchronization can result in field-scale oscillations of the kind observed in practice by Fothergill *et al.* (1996).

Nitrogen Fixation in the Grass–Clover Relationship

The relationship between grass and clover is not, of course, based simply on their respective capacities to exploit and explore above- and below-ground resources, because the grass component stands to benefit from the N fixed by *Rhizobium* bacteria in the clover root nodules. These bacteria, which are present in soils as free-living organisms, attach themselves to the roots, causing the formation of infection threads along which the bacteria are carried to the cortical cells, where they become enclosed in the endoplasmic reticulum of their host. In this form they are known as bacteroids (Sprent and Sprent, 1990). The host provides the energy and the hydrogen used by the *Rhizobium* to fix atmospheric N_2 in the form of ammonia. Relative energy requirements for N_2-fixation and for utilization of combined N (N uptake) vary within and between species, the majority of experiments having indicated that fixation is less energy-efficient (Schubert, 1982). Comparative studies of growth rates in subterranean clover (Silsbury, 1977, 1979) and of respiratory costs in white clover (Ryle *et al.*, 1979; Haystead *et al.*, 1980) seem to support this conclusion. Estimates of the costs vary, however, and there are numerous instances in which no differences can be demonstrated between N-fixing plants and those supplied with mineral N (Schubert, 1982). This rather confused situation arises partly because *Rhizobium* strains vary in host compatibility as well as in N-fixing efficiency.

Host–*Rhizobium* interactions accounted for 74% of the variation observed in fixation in *Vicia faba* (Mytton *et al.*, 1977). Strains of *Rhizobium* vary widely in compatibility and performance (Fig. 4.5), and they interact with clover cultivars. Some clover cultivars perform well with a wide range of *Rhizobium* strains, whereas the growth of others may be restricted (Mytton, 1996; Fig. 4.6). Moreover, although the introduction of more energy-efficient strains is clearly a desirable objective, it is not easily achieved, since indigenous soil bacteria compete with introduced *Rhizobia* for nodule sites (Mytton and Skøt, 1993). It should not be forgotten that *Rhizobium* bacteria reproduce in the soil, not in the legume, and that they follow an evolutionary trajectory that may have little to do with the fate of their leguminous hosts (by which they have, in effect, been hijacked). The grass–clover relationship will be subject to the availability of fixed N in the same way as it is to fertilizer N, though amounts released and the timing of their release will obviously differ (and may affect the outcome).

Fig. 4.5. Effects of different rhizobium strains on the growth of white clover. Left to right – introduced (ineffective) strain, native strain, introduced (effective) strain, introduced (very effective) strain and native strain plus added N. (From *IGAP/Welsh Plant Breeding Station Report* (1987), Vol. 2, p. 18 (Lance Mytton).)

Fig. 4.6. The response of clovers to increasing amounts of N when inoculated with 'restricting' (upper) and 'non-restricting' (lower) rhizobia (Mytton, 1996). 'Restricting' rhizobia are able to prevent clover from making efficient use of soil N.

White clover can up- and down-regulate N_2-fixation within 2–4 days in relation to the availability of mineral N (Davidson and Robson, 1986a). The conclusion that nitrate affects fixation by decreasing the O_2 flux to the bacteroids was reached by Minchin et al. (1986). Their observations showed that an increase in the oxygen diffusion resistance of white clover nodules when plants were transferred to a nutrient solution containing nitrate was associated with reductions in N-linked respiration (and N_2-fixation). N fertilizer applications are normally high enough to result in inhibitory concentrations of nitrate in the soil solution and very high N also affects rhizobial infection (Streeter, 1988). Defoliation reduces nodule dry weight per plant (Chu and Robertson, 1974) and actual losses of nodules were observed within 3 days of defoliation. N_2-fixation ceases within a few hours of defoliation, but nodules can recover after a single defoliation (Gordon et al., 1990). Too frequent defoliation results in the loss of nodules. The possibility that a N shortage might arise in this way and might contribute to the weakening and loss of clover plants from the sward cannot be dismissed.

There is no conclusive evidence to support the idea that low-temperature inhibition of N_2-fixation limits clover growth in spring. The proportion of N derived from fixation in temperature-controlled experiments seems to be related to demand and falls when low temperature reduces that demand (Kessler et al., 1990). Results obtained by Hatch and Macduff (1991) in flowing nutrient culture suggest that N_2-fixation is less sensitive to low root temperature than the process of N uptake. Later results confirm that the down-regulation of N_2-fixation when mineral N is supplied is actually partially inhibited by low root temperatures (Svenning and Macduff, 1996). The observation that low root-zone temperatures inhibit the early stages of nodulation in soybean (Lynch and Smith, 1993) suggests that the problem might be one of restoring activity rather than of reduced activity at low temperatures. Field observations (M. Fothergill, personal communication) indicate that nodules are lost and the proportion of white inactivated nodules increases in winter, while replacement and reactivation are observed in spring. The extent to which growth of white clover in spring is limited by death and inactivation of nodules in winters of differing severity and the impact of this on the grass–clover relationship deserve further study.

Influence of Pests and Diseases

Competition between the grass and legume components of a mixed sward can be influenced by their differing susceptibilities to attack from the pests and diseases present in the environment in which they grow. Thus, although both white clover and ryegrass are susceptible to attack from root-knot and cyst nematodes, clover is susceptible to damage by *Meloidogyne hapla* rather than by *Meloidogyne naasi* (which attacks ryegrass) and to *Heterodora trifolii* rather than to *Heterodora avenae*, *Heterodora mani* or *Heterodora bifenestra*, which also attack ryegrasses. The stem nematode *Ditylenchus dipsaci* is confined to legumes and can cause severe damage to white clover, red clover and lucerne (Cook and Yeates, 1993). Nematodes and slugs (*Deroceras* spp.) are widely distributed; reports from the UK and New Zealand indicate that slugs cause considerable damage to white clover in wet conditions (typically the September/October period in the UK) and least damage in cold and/or dry weather (Murray, 1991). Mob stocking or heavy rolling are recommended as effective control measures.

The principal ryegrass pests and diseases seem to vary with geographical location more than those of clover. Typical UK ryegrass pests include leatherjackets (*Tipula* spp., especially *T. paludosa*), frit fly larvae (*Oscinella* spp.) and fungal diseases, such as crown rust (*Puccinia coronata*), leaf spot (*Drechsleria* spp.), *Rhynchosporium*, *Pythium* and *Erysiphe* (now *Blumeria*) *graminis*, while *Fusarium culmorum* is present both in the UK and New Zealand (Clements, 1994). Pasture grasses in New Zealand are affected by *Listronatus bonariensis* (Argentine stem weevil), *Wiscana* spp. (porina) and *Heteronychus* spp. (black beetle) (Clements and Cook, 1996).

The accidental transfer of pests and diseases from one geographical region to another has caused serious problems in New Zealand in the case of Argentine stem weevil, first recorded in 1927 (Prestidge et al., 1991), and of *Sitona* weevil (*S. lepidus*) in white clover. Resistance to Argentine stem weevil in ryegrass is increased by the presence of *Acremonium* (now *Neotyphodium*), an endophytic fungus, but this endophyte has been found to have adverse effects on sheep performance. Progress is, however, being made in finding endophyte-free ryegrasses, which tolerate or resist stem weevil attack in other ways (Prestidge et al., 1991). The *Sitona* weevil (*S. lepidus*) was first recorded on white clover in May 1994 (Barratt *et al.*, 1996) and is a cause of concern to sheep farmers in particular. *Sitona* larvae feed inside clover root nodules or on root hairs (Murray, 1991) and have been shown to be capable of reducing N_2-fixation to about half that of uninfected controls (Murray *et al.*, 1996). Significant increases have also been observed in the N content of the companion ryegrass as a result of N losses from the damaged clover roots (Murray and Hatch, 1994), with obvious implications in terms of increasing grass competition. White clover varieties with greater resistance to the weevils *Sitona flavescens* and *Sitona lineatus* have been identified by Murray (1996).

Vulnerability to pests and diseases is often greater in establishing swards than it is in older swards (from which some of the most disease-prone genotypes will probably have disappeared). Most interventionist control measures are, in practice, concentrated on ensuring successful pasture establishment, and comparative measurements of the effects of remedial treatments on target and companion species are rare (Cook *et al.* (1992) is a notable exception). Preventive control includes the selection of new varieties with improved natural genetic resistance to specific pests.

Conclusions

Plant species that coexist are able to exploit the natural variation within a non-uniform environment more fully than a single species. Perennial ryegrass is a high N user; white clover fixes N, which returns to the soil either as plant debris or in the excreta from the animals at pasture.

Because of the practical importance of grass–clover associations, they have been studied more extensively than most other species relationships, but they were scarcely the easiest choice as a model system. A number of lessons can, however, be learnt from the studies which have been carried out. Attention in earlier management studies to the effect of seasonal changes in the relative locations of leaves of the two species in the canopy might have been helpful, and the absence of even simple measurements of the relative heights of the two components in different systems and at different times of year has been (and still, to some extent, is) a particularly critical omission. Grass, if not defoliated well into the autumn, can overtop and virtually eliminate clover; strategic spring defoliation may allow it to increase.

In systems involving less frequent defoliation, such as rotational grazing, clover persists, moving from local high-N sites to proliferate in sites where light penetrates more freely. To do this it must, however, be able to accumulate sufficient leaf area to power exploration. New Zealand 'feed-budgeting' systems (Milligan and Smith, 1984; Shoati *et al.*, 1987) permit this, but in productive UK swards continuously grazed to sward height guidelines clover is more heavily utilized and the resulting low leaf complements render plants prone to fragmentation and loss. Short internodes and the breakup of stolon networks are indicative of diminished survival prospects. The situation may even be aggravated by high late-summer clover contents, which lead to high growth rates and correspondingly high stocking numbers. These, in turn, increase the risk of damage from urine scorch and trampling. The interpolation of a silage cut later in the season removes most of the grass leaf and some pseudostem (together with a quantity of N) and benefits clover, in which most of the stolon is below cutting level. Should this fail to restore clover content, a return to rotational grazing is virtually certain to do so (Fig. 4.7).

The unreliability of clover in UK systems from year to year and, in particular, the phenomenon of periodic clover 'crashes' has limited the acceptance of grass–clover systems as an alternative to grass–N systems. There are no easy fixes, but the taking of a silage cut in the later part of the summer is probably the best way of averting the problem, and graziers would be wise to check that clover plants are not becoming excessively fragmented and that the sward has not become dominated by plants with

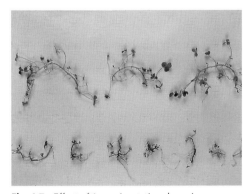

Fig. 4.7. Effect of 1 year's rotational grazing on clover stolons previously subjected to 3 years of continuous grazing (above). Control (4 years of continuous grazing) is shown below.

very short internodes. In addition, what happens out of season is as important as what happens during the growing season and there should be no time at which clover leaves are not visible and receiving adequate light to ensure clover survival.

This chapter has highlighted a number of factors of importance in grass–clover swards that are likely to be of importance in grass–legume association in general. Nitrogen status is one of the most important driving forces, and increases in soil N (whether from fertilizer applications or urine deposition) lead to increases in grass growth and tiller production. Corresponding reductions in the proportion and quality of light passing through or reflected from the grass partner on to the lower-growing clover create an environment less suitable for the development of branches from axillary buds. Increases in stocking rate to exploit a N-promoted increase in herbage production can compound the problem for clover by reducing the number of leaves per growing point to a level of photosynthetic unsustainability, at which the production of new growing points is severely curtailed and stolon dieback increases.

Insufficient attention has been given in the past to changes in the relative locations of grass and clover leaf areas in the canopy in relation to N, time of year and genotypes involved. Simple height measurements could help to identify situations and times of year in which the survival of white clover (or of other pasture legumes) might be at particular risk. Differences in the relative proportions of grass and clover leaf area and dry matter removed by defoliation at different times of year and in different variety combinations result in changes in the character and mass of defoliation residues which have an impact on sustainability.

The introduction of élite rhizobial strains with the capacity to fix N more effectively than existing or native strains would encourage less use of bag N and the N would feed through to the grass component less globally. Existing grass–clover models (see p. 74) could be used to assess the impact on system stability.

The same competitive influences operate both in agricultural swards and in the long-established and relatively stable associations which we would regard as natural, in which N-fixing and soil N-dependent species are accommodated at patch level. The differences between the two situations are of scale rather than of kind; areas of high and low pasture utilization and high and low N availability occur – and change their location over time – in just the same way in unfertilized swards grazed by rabbits and in agricultural swards.

Studies in pasture ecology (particularly in the last 20–30 years) have gone a long way towards providing reasons for the sustainability – and commercial success – of New Zealand grazing systems. The indications are that ongoing studies will continue to furnish information that will not only help to improve the sustainability (both ecological and economic) of grass–clover swards in farming practice under alternative systems of stock management but may also provide a useful blueprint for the study of grass–legume associations in general.

Acknowledgements

I am grateful to the Institute of Grassland and Environmental Research for the use of facilities, and particularly to members of the Library and Publications Section for their advice and help. My special thanks go to Mick Fothergill for numerous useful discussions, and I would also like to acknowledge the helpful advice received from Rosemary Collins, Roger Cook, Frank Minchin and Lance Mytton. Malcolm Dye prepared Figs 4.1, 4.2 and 4.3 and the photographs were taken by Ian Sant.

References

Armstrong, R.H., Robertson, E., Lamb, C.S., Gordon, I.J. and Elston, D.A. (1993) Diet selection by lambs in ryegrass–white clover swards differing in the horizontal distribution of clover. In: *Proceedings of the XVIIth International Grassland Congress, New Zealand*, Palmerston North, New Zealand, pp. 715–716.

Ballaré, C.L., Scopel, A.L. and Sanchez, R.O. (1990) Far-red radiation reflected from adjacent legumes: an early signal of competition in plant canopies. *Science* 247, 329–331.

Barratt, B.I.P., Barker, G.M. and Addison, P.J. (1996) *Sitona lepidus* Gyllenhal (Coleoptera: Curculionidae), a potential clover pest new to New Zealand. *New Zealand Entomologist* 19, 23–30.

Barthram, G.T. and Grant, S.A. (1995) Interactions between variety and the timing of conservation cuts on species balance in *Lolium perenne–Trifolium repens* swards. *Grass and Forage Science* 50, 98–105.

Bastiman, B. and Van Dijk, J.P.F. (1975) Muck breakdown and pasture rejection in an intensive paddock system for dairy cows. *Experimental Husbandry* 28, 7–17.

Beddows, A.R. (1967) Biological flora of the British Isles No. 107 *Lolium perenne* L. *Journal of Ecology* 55, 567–587.

Cain, M.L., Pacala, S.W., Silander, J.A. Jr and Fortin, M.J. (1995) Neighbourhood models of clonal growth in white clover, *Trifolium repens. The American Naturalist* 145, 888–917.

Carlson, G.E. (1966) Growth of clover leaves – developmental morphology and parameters at ten stages. *Crop Science* 6, 293–294.

Chapman, D.F. (1983) Growth and demography of *Trifolium repens* stolons in grazed hill pastures. *Journal of Applied Ecology* 20, 597–608.

Chapman, D.F. (1986) Development, removal and death of white clover leaves under 3 grazing managements in hill country. *New Zealand Journal of Agricultural Research* 29, 39–47.

Chapman, D.F., Clark, D.A., Land, C.A. and Dymock, N. (1984) Defoliation of *Lolium perenne* and *Agrostis* spp. tillers and *Trifolium repens* stolons in set-stocked and rotationally grazed hill pastures. *New Zealand Journal of Agricultural Research* 27, 289–301.

Chapman, D.F., Robson, M.T. and Snaydon, R.W. (1990) The carbon economy of developing leaves of white clover (*Trifolium repens* L.). *Annals of Botany* 66, 623–628.

Chapman, D.F., Robson, M.T. and Snaydon, R.W. (1991) Quantitative carbon distribution in clonal plants of white clover (*Trifolium repens*): source–sink relationships during undisturbed growth. *Journal of Agricultural Science, Cambridge* 116, 229–238.

Chu, A.C.P. and Robertson, A.G. (1974) The effect of shading and defoliation on nodulation and nitrogen fixation by white clover. *Plant and Soil* 41, 509–519.

Clark, H., Newton, P.C.D. and Barber, J. (1999) Physiological and morphological responses to elevated CO_2 and soil moisture deficit of temperate pasture species growing in an established plant community. *Journal of Experimental Botany* 50, 233–242.

Clements, R.O. (1994) *The Importance of Pests and Diseases to Agricultural Grassland in England and Wales.* A review commissioned by the Ministry of Agriculture, Fisheries and Food (MS 14/09), London, 142 pp.

Clements, R.O. and Cook, R. (1996) Pest damage to established grass in the UK. *Agricultural Zoology Reviews* 7, 157–179.

Cocks, P.S. (1973) The influence of temperature and density on the growth of communities of subterranean clover (*Trifolium subterraneum* L.) cv Mount Barker. *Australian Journal of Agricultural Research* 24, 479–495.

Collins, R.P. (in press) The effects of drought stress and winter stress on the persistence of white clover. In: Fisher, G. and Frankow-Lindberg, B. (eds) *Grasslands of Europe – Utilization and Development, Proceedings of FAO/CIHEAM Lowland Grasslands Sub-Network Research Conference 1998, La Coruna, Spain.*

Cook, R. and Yeates, G.W. (1993) Nematode pests of grassland and forage crops. In: Evans, K., Trudgill, D.L. and Webster, J.M. (eds) *Plant Parasitic Nematodes in Temperate Agriculture.* CAB International, Wallingford, UK, pp. 305–350.

Cook, R., Evans, D.R., Williams, T.A. and Mizen, K.A. (1992) The effect of stem nematodes on establishment and early yields of white clover. *Annals of Applied Biology* 120, 83–94.

Curll, M.L. and Wilkins, R.J. (1982) Frequency and severity of defoliation of grass and clover by sheep at different stocking rates. *Grass and Forage Science* 37, 291–297.

Curll, M.L. and Wilkins, R.J. (1983) The comparative effects of defoliation, treading and excreta on a *Lolium perenne–Trifolium repens* pasture grazed by sheep. *Journal of Agricultural Science, Cambridge* 100, 451–460.

Curll, M.L., Wilkins, R.J., Snaydon, R.W. and Shanmugalingham, V.S. (1985) The effects of stocking rate and nitrogen fertilizer on a perennial ryegrass–white clover sward. 1. Sward and sheep performance. *Grass and Forage Science* 40, 129–140.

Davidson, B.R. and Davidson, H.F. (1993) *Legumes: the Australian Experience.* John Wiley & Sons, Research Studies Press, Taunton, UK.

Davidson, I.A. and Robson, M.J. (1984) The effect of temperature and nitrogen supply on the physiology of grass/clover swards. In: Thompson, D.J. (ed.) *Forage Legumes.* Occasional Symposium No. 16, British Grassland Society, Maidenhead, pp. 56–60.

Davidson, I.A. and Robson, M.J. (1986a) Effect of contrasting patterns of nitrate application on the nitrate uptake, N_2-fixation, nodulation and growth of white clover. *Annals of Botany* 57, 331–338.

Davidson, I.A. and Robson, M.J. (1986b) Effect of temperature and nitrogen supply on the growth of perennial ryegrass and white clover. 2. A comparison of monocultures and mixed swards. *Annals of Botany* 57, 709–719.

Davidson, I.A., Robson, M.J. and Drennan, D.S.H. (1986) Effect of temperature and nitrogen supply on the growth of perennial ryegrass and white clover. 1. Carbon and nitrogen economies of mixed swards at low temperature. *Annals of Botany* 57, 697–708.

Davies, A. (1971) Changes in growth rate and morphology of perennial ryegrass swards at high and low nitrogen levels. *Journal of Agricultural Science, Cambridge* 77, 123–134.

Davies, A. (1990) Utilisation of grass and clover in a continuously grazed *Lolium perenne/Trifolium repens* sward. In: *Proceedings of the First Congress of the European Society of Agronomy, Session 1.* European Society of Agronomy, Colmar, France, p. 66.

Davies, A. (1993) Tissue turnover in the sward. In: Davies, A., Baker, R.D., Grant, S.A. and Laidlaw, A.S. (eds) *Sward Measurement Handbook* 2nd edn. British Grassland Society, Reading, UK, pp. 183–215.

Davies, A. (1996) Long-term forage and pasture investigations in the UK. *Canadian Journal of Plant Science* 76, 573–579.

Davies, A. (1998) Pasture research in the UK: present knowledge and future prospects. *Canadian Journal of Plant Science* 78, 211–216.

Davies, A. and Evans, M.E. (1982) The pattern of growth in swards of two contrasting varieties of white clover in winter and spring. *Grass and Forage Science* 37, 199–207.

Davies, A. and Evans, M.E. (1990) Effects of spring defoliation and fertilizer nitrogen on the growth of white clover in ryegrass/clover swards. *Grass and Forage Science* 45, 345–356.

Davies, A. and Jones, D.R. (1988) Changes in the grass/clover balance in continuously grazed swards in relation to sward production components. In: *Proceedings of the 12th General Meeting of the European Grassland Federation, Dublin, Ireland.* Irish Grassland Association, Tuam, Co. Galway, UK, pp. 297–301.

Davies, A. and Jones, D.R. (1992) The production of leaves and stolon branches on established white clover cuttings in relation to temperature and soil moisture in the field. *Annals of Botany* 69, 515–521.

Davies, A., Evans, M.E. and Exley, J.K. (1983) Regrowth of perennial ryegrass as affected by simulated leaf sheaths. *Journal of Agricultural Science, Cambridge* 101, 131–137.

Davies, A., Evans, M.E. and Pollock, C.J. (1989) Influence of date of tiller origin on leaf extension rates in perennial and Italian ryegrass at 15°C in relation to flowering propensity and carbohydrate status. *Annals of Botany* 63, 377–384.

Dear, B.S., Cocks, P.S., Wolfe, E.C. and Collins, D.P. (1998) Established perennial grasses reduce the growth of emerging subterranean clover seedlings through competition for water, light and nutrients. *Australian Journal of Agricultural Research* 49, 41–51.

Dennis, W.D. and Woledge, J. (1985) The effect of nitrogenous fertilizer on the photosynthesis and growth of white clover/perennial ryegrass swards. *Annals of Botany* 55, 171–178.

Djekoun, A. and Planchon, C. (1991) Water status effect on dinitrogen fixation and photosynthesis in soybean. *Agronomy Journal* 83, 316–322.

Doak, B.W. (1954) The presence of root-inhibiting substances in cow urine and the cause of urine burn. *Journal of Agricultural Science, Cambridge* 44, 133–139.

Duke, J.A. (1981) *Handbook of Legumes of World Economic Importance.* Plenum Press, New York, USA.

Eagles, C.F. (1994) Temperature, photoperiod and dehardening of forage grasses and legumes. In: *COST 814 Workshop on Crop Adaptation to Cool, Wet Climates*, European Commission, Brussels, pp. 75–82.

Eagles, C.F. and Othman, O.B. (1986) Effect of temperature, irradiance and photoperiod on morphological characters of seedlings of contrasting white clover populations. *Annals of Applied Biology* 108, 629–638.

Eagles, C.F., Thomas, H., Volaire, F. and Howarth, C.J. (1997) Stress physiology and crop improvement. In: *Proceedings of the XVIIIth International Grassland Congress, Winnipeg, Canada.*

Edmond, D.B. (1964) Some effects of sheep treading on the growth of 10 pasture species. *New Zealand Journal of Agricultural Research* 7, 1–16.

Evans, D.R. and Williams, T.A. (1987) The effect of cutting and grazing managements on dry matter yield of white clover varieties (*Trifolium repens*) when grown with S23 perennial ryegrass. *Grass and Forage Science* 42, 153–159.

Fothergill, M., Davies, D.A., Morgan, C.T. and Jones, J.R. (1996) White clover crashes. In: Younie, D. (ed.) *Legumes in Sustainable Agriculture.* Occasional Symposium No. 30, British Grassland Society, Reading, UK, pp. 172–176.

Fothergill, M., Davies, D.A. and Daniel, E.J. (1997) Morphological dynamics and seedling recruitment in young swards of three contrasting cultivars of white clover (*Trifolium repens* L.) under continuous stocking with sheep. *Journal of Agricultural Science, Cambridge* 128, 163–172.

Fukai, S. and Silsbury, J.H. (1976) Growth responses of subterranean clover communities to temperature. I. Dry matter production and plant morphogenesis. *Australian Journal of Plant Physiology* 3, 527–543.

Fukai, S. and Silsbury, J.H. (1977) Responses of subterranean clover communities to temperature. III Effects of temperature on canopy photosynthesis. *Australian Journal of Plant Physiology* 4, 273–282.

Gooding, R.F. and Frame, J. (1997) Effects of continuous sheep stocking and strategic rest periods on the sward characteristics of binary perennial grass/white clover associations. *Grass and Forage Science* 52, 350–359.

Gooding, R.F., Frame, J. and Thomas, C. (1996) Effects of sward type and rest periods from sheep grazing on white clover presence in perennial ryegrass/white clover associations. *Grass and Forage Science* 51, 180–189.

Gordon, A.J., Kessler, W. and Minchin, F.R. (1990) Defoliation-induced stress in nodules of white clover. I. Changes in physiological parameters and protein synthesis. *Journal of Experimental Botany* 41, 1245–1253.

Gordon, A.J., Minchin, F.R., Skøt, L. and James, C.L. (1997) Stress-induced declines in soybean N_2 fixation as related to nodule sucrose synthase activity. *Plant Physiology* 114, 937–946.

Harris, W., Rhodes, I. and Mee, S.S. (1983) Observations on environmental and genotypic influences on the overwintering of white clover. *Journal of Applied Ecology* 20, 609–624.

Hart, A.L. (1987) Physiology. In: Baker, M.J. and Williams, W.M. (eds) *White Clover.* CAB International, Wallingford, UK, pp. 125–151.

Hatch, D.J. and Macduff, J.H. (1991) Concurrent rates of N_2 fixation, nitrate and ammonium uptake by white clover in response to growth at different root temperatures. *Annals of Botany* 67, 265–274.

Hay, M.J.M. (1983) Seasonal variation in the distribution of white clover (*Trifolium repens* L.) stolons among 3 horizontal strata in 2 grazed swards. *New Zealand Journal of Agricultural Research* 26, 29–34.

Hay, M.J.M., Brock, J.L. and Fletcher, R.H. (1983) Effect of sheep grazing management on distribution of white clover stolons among 3 horizontal strata in ryegrass/white clover swards. *New Zealand Journal of Experimental Agriculture* 11, 215–218.

Haystead, A., King, J., Lamb, W.I.C. and Mariott, C. (1980) Growth and carbon economy of nodulated white clover in the presence and absence of combined nitrogen. *Grass and Forage Science* 35, 123–128.

Hodgson, J., Mackie, C.K. and Parker, J.W.G. (1986) Sward surface height for efficient grazing. *Grass Farmer* 24, 5–10.

Johns, G.G. and Lazenby, A. (1973) Defoliation, leaf area index and water use of four temperate pasture species under irrigated and dryland conditions. *Australian Journal of Agricultural Research* 24, 783–795.

Jones, D.R. and Davies, A. (1988) The effect of simulated continuous grazing on development and senescence of white clover. *Grass and Forage Science* 43, 421–425.

Juntilla, O., Svenning, M.M. and Solheim, B. (1990) Effects of temperature and photoperiod on vegetative growth of white clover (*Trifolium repens*) ecotypes. *Physiologia Plantarum* 79, 427–434.

Kays, S. and Harper, J.L. (1974) The regulation of plant and tiller density in a grass sward. *Journal of Ecology* 62, 97–105.

Kessler, W., Boller, B.C. and Nösberger, J. (1990) Distinct influence of root and shoot temperature on nitrogen fixation by white clover. *Annals of Botany* 65, 341–346.

Laidlaw, A.S., Teuber, N.G. and Withers, J.A. (1992) Out-of-season management of grass/clover swards to manipulate clover content. *Grass and Forage Science* 47, 220–229.

Lynch, D.H. and Smith, D.L. (1993) Soybean (*Glycine max*) nodulation and N_2 fixation as affected by exposure to a low root-zone temperature. *Physiologia Plantarum* 88, 212–220.

MacDiarmid, B.N. and Watkin, A.R. (1971) The cattle dung patch. 1. Effect of dung patches on yield and botanical composition of surrounding and underlying pasture. *Journal of the British Grassland Society* 26, 239–245.

Milligan, K.E. and Smith, M.E. (1984) Pasture allocation in practice. *New Zealand Journal of Agricultural Science, Pasture the Export Earner* 18, 153–156.

Milne, J.A., Hodgson, J., Thompson, R., Souter, W.G. and Barthram, G.T. (1982) The diet ingested by sheep grazing swards differing in white clover and perennial ryegrass content. *Grass and Forage Science* 37, 209–218.

Minchin, F.R., Minguez, M.I., Sheehy, J.E., Witty, J.F. and Skøt, L. (1986) Relationships between nitrate and oxygen supply in symbiotic nitrogen fixation by white clover. *Journal of Experimental Botany* 37(181), 1103–1113.

Mitchell, K.J. and Lucanus, R. (1962) Growth of pasture species under controlled environments. III. Growth at various levels of constant temperature with 8 and 16 hours of uniform light per day. *New Zealand Journal of Agricultural Research* 5, 135–144.

Murray, P.J. (1991) Pests of white clover. In: *Strategies for Weed, Disease and Pest Control in Grassland: Practical Implications of Recent Developments and Future Trends, Proceedings of the British Grassland Society Conference, Gloucester, 27 February 1991.* British Grassland Society, Hurley, UK, pp. 8.1–8.7.

Murray, P.J. (1996) Evaluation of a range of varieties of white clover for resistance to feeding by weevils of the genus *Sitona. Plant Varieties and Seeds* 9, 9–14.

Murray, P.J. and Hatch, D.J. (1994) *Sitona* weevils (Coleoptera, Curculionidae) as agents for rapid transfer of nitrogen from white clover (*Trifolium repens* L.) to perennial ryegrass (*Lolium perenne* L.). *Annals of Applied Biology* 125, 29–33.

Murray, P.J., Hatch, D.J. and Cliquet, J.B. (1996) Impact of root herbivory on the growth and nitrogen and carbon contents of white clover (*Trifolium repens*) seedlings. *Canadian Journal of Botany* 74, 1591–1595.

Mytton, L.R. (1996) Nitrogen from clover. In: *Annual Report.* Institute of Grassland and Environmental Research, Aberystwyth, pp. 51–52.

Mytton, L.R. and Skøt, L. (1993) Breeding for improved symbiotic nitrogen fixation. In: Hayward, M.D., Bosemark, N.O. and Romagoza, I. (eds) *Plant Breeding: Principles and Prospects*. Chapman and Hall, London, UK, pp. 451–473.

Mytton, L.R., El-Sherbeeny, M. and Lawes, D.A. (1977) Symbiotic variability in *Vicia faba*. 3. Genetic effects of host plant, *Rhizobium* strain and of host × strain interaction. *Euphytica* 26, 785–791.

Nassiri Mahalatti, M. (1998) Modelling interactions in grass–clover mixtures. PhD thesis, Wageningen Agricultural University, The Netherlands.

Newman, J.A., Parsons, A.J. and Harvey, A. (1992) Not all sheep prefer clover: diet selection revisited. *Journal of Agricultural Science, Cambridge* 119, 275–283.

Novoplansky, A., Cohen, D. and Sachs, T. (1990) How *Portulaca* seedlings avoid their neighbours. *Oecologia* 82, 490–493.

Parsons, A.J., Newman, J.A., Penning, P.D., Harvey, A. and Orr, R.J. (1994) Diet preference of sheep – effects of recent diet, physiological state and species abundance. *Journal of Animal Ecology* 63, 465–478.

Prestidge, R.A., Barker, G.M. and Pottinger, R.P. (1991) Towards sustainable controls of pasture pests: progress on control of Argentine stem weevil (*Listronotus bonariensis* (Kuschel)). *Proceedings of the New Zealand Grassland Association* 53, 25–31.

Rhodes, I., Collins, R.P. and Evans, D.R. (1994) Breeding white clover for tolerance to low temperature and grazing stress. *Euphytica* 77, 239–242.

Ryle, G.J.A., Powell, C.E. and Gordon, A.J. (1979) The respiratory costs of nitrogen fixation in soybean, cowpea, and white clover. II. Comparisons of the cost of nitrogen fixation and the utilization of combined nitrogen. *Journal of Experimental Botany* 30, 145–153.

Schubert, K.R. (1982) *The Energetics of Biological Nitrogen Fixation*. Workshop Summary, American Society of Plant Physiologists, Rookville, Maryland, USA.

Schwinning, S. and Parsons, A.J. (1996a) Analysis of the coexistence mechanisms for grasses and legumes in grazing systems. *Journal of Ecology* 84, 799–813.

Schwinning, S. and Parsons, A.J. (1996b) A spatially explicit population model of stoloniferous N-fixing legumes in mixed pasture with grass. *Journal of Ecology* 84, 815–826.

Shoati, G.W., Hay, M.J.M. and Giles, K.H. (1987) Managing pastures for grazing animals. In: Neal, A.M. (ed.) *Livestock Farming on Pasture*. Occasional Publication No. 18, New Zealand Society of Animal Production, Hamilton, New Zealand. Chapter 5.

Silsbury, J.H. (1977) Energy requirement for symbiotic nitrogen fixation. *Nature* 267, 149–150.

Silsbury, J.H. (1979) Growth, maintenance and nitrogen fixation of nodulated plants of subterranean clover (*Trifolium subterraneum* L.). *Australian Journal of Plant Physiology* 6, 165–176.

Smith, R.C.G., Biddiscombe, E.F. and Stern, W.R. (1972) Evaluation of five Mediterranean annual pasture species during early growth. *Australian Journal of Agricultural Research* 23, 703–716.

Sprent, J.I. and Sprent, P. (1990) *Nitrogen Fixing Organisms*. Chapman and Hall, London, UK.

Streeter, J. (1988) Inhibition of legume nodule formation and N_2 fixation by nitrate. *CRC Critical Reviews in Plant Science* 7, 1–23.

Svenning, M.M. and Macduff, J.H. (1996) Low root temperature retardation of the mineral nitrogen induced decline in N_2 fixation by a northern ecotype of white clover. *Annals of Botany* 77, 615–622

Terrell, E.A. (1968) Taxonomic and evolutionary relationships. In: Thomson, D.J. (ed.) *US Department of Agriculture Technical Bulletin 1392*, USDA, Washington, DC, p. 4.

Thomas, H. (1984) Effects of drought on growth and competitive ability of perennial ryegrass and white clover. *Journal of Applied Ecology* 21, 591–603.

Thomashow, M.F. (1998) Role of cold-responsive genes in plant freezing tolerance. *Plant Physiology* 118, 1–7.

Thornley, J.H.M., Bergelson, J. and Parsons, A.J. (1995) Complex dynamics in a carbon–nitrogen model of a grass–legume pasture. *Annals of Botany* 75, 79–94.

Turkington, R., Sackville Hamilton, N.R. and Gliddon, C. (1991) Within-population variation in localized and integrated responses of *Trifolium repens* to biotically patchy environments. *Oecologia* 86, 183–192.

Weeda, W.C. (1967) The effect of cattle dung patches on pasture growth, botanical composition and pasture utilisation. *New Zealand Journal of Agricultural Research* 10, 150–159.

Whitlock, R. (1983) *The English Farm*. J.M. Dent and Sons, London, UK.

Woledge, J. and Calleja, S.A. (1983) The growth and photosynthesis of seedling plants of white clover at low temperature. *Annals of Botany* 52, 239–245.

Woledge, J. and Dennis, W.D. (1982) The effect of temperature on the photosynthesis of ryegrass and white clover leaves. *Annals of Botany* 50, 25–35.

Woledge, J. and Parsons, A.J. (1986) the effect of temperature on the photosynthesis of ryegrass canopies. *Annals of Botany* 57, 487–497.

Woledge, J., Davidson, I.A. and Tewson, V. (1989) Photosynthesis during winter in ryegrass/white clover mixtures in the field. *New Phytologist* 113, 275–281.

Woledge, J., Tewson, V. and Davidson I.A. (1990) Growth of grass/clover mixtures during winter. *Grass and Forage Science* 45, 191–202.

Woledge, J., Reyneri, A., Tewson, V. and Parsons, A.J. (1992) The effect of cutting on the proportions of perennial ryegrass and white clover in mixtures. *Grass and Forage Science* 47, 169–179.

5 Plant Competition in Pastures – Implications for Management

D.R. Kemp[1] and W.McG. King[2]

[1]*CRC for Weed Management Systems and Pasture Development Group, University of Sydney, Faculty of Rural Management, Orange, Australia;* [2]*CRC for Weed Management Systems and Pasture Development Group, NSW Agriculture, Orange Agricultural Institute, Orange, Australia*

Introduction

Pastures comprise a mixture of competing species that need to be managed for optimal and sustainable production. In sown pastures, the initial seed mixture can vary from one to several species, with one or more cultivars of each species. Invariably some compromises are made, arising from the need to balance competition and production potential among the sown species and to contain costs by using a minimal amount of seed. After sowing, volunteer species almost inevitably establish within the sward. These species may already be present in the soil seed bank or may invade the community from elsewhere. The resultant pastures are invariably a mixture of species with diverse histories.

Competition in sown pastures occurs from the time of sowing. The sown species compete among themselves and with volunteer species. Subsequent management and environmental conditions then have a major influence on the competitiveness of each species. Often within a relatively short time after sowing, the composition and structure of the pasture are significantly different from that intended (Aarssen and Turkington, 1985; Marriott *et al.*, 1997) and continue to change while it remains a pasture. Worldwide, a broad range of species are sown for pastures. Mostly these are sown for domestic livestock production, but there is an increasing interest in sowing mixtures of pasture species for land management, for nature conservation and for the maintenance of wildlife. In each case, there is an expectation that the resultant pasture will be dominated by the species sown. Whether or not this expectation is met depends, at least in part, upon the outcome of competition among plant species, which in turn is influenced by how the pasture is managed within the overall constraints set by environmental factors. Many sown pastures shift to a 'naturalised' state, where there are more volunteer than sown species. The volunteer species may still be useful at times for animal production (Reed, 1974).

The purpose of this chapter is to discuss, with particular reference to south-eastern Australia, competitive interactions in pastures and how these are modified by management practices. This will include discussion of the general principles of competition within and between sown species and between those sown species and volunteers. Understanding the interconnected nature of the components of pasture plant communities within the conceptual framework of competition provides an ecological context within which to consider the effects of management strategies. The ultimate goal is to contribute to the achievement of truly sustainable pasture-based production systems that will provide an income for livestock producers without degrading the resource base. Emphasis will be on the general principles involved rather than extensive

© CAB *International* 2001. *Competition and Succession in Pastures*
(eds P.G. Tow and A. Lazenby)

detail on the responses of individual species or pasture types and on competition after establishment.

Competition in Pastures

Features that are of agronomic value are rarely ones that confer survival value.

(Donald, 1963)

Species differ in the resources they require to grow, develop and reproduce. If each species required a completely different set of resources from every other species, the only 'resource' they would compete for would be physical space. The common need of species for light, water and nutrients, however, means that competition for several factors is inevitable, even though the amount of resource needed by each species will vary. As resource supply fluctuates, then, so does plant growth, and the competitive interactions between species will also change. Pastures are frequently a mixture of several sown and volunteer species and hence the interactions may become very complex. The competitive relationships between plant species are also strongly influenced by the actions of grazing animals (Brown, 1976). Without careful management, it is usually expected that more grazing-tolerant species will come to dominate the pasture (Lodge and Orchard, 2000). Unfortunately, the more grazing-tolerant species are often those of low nutritive value for livestock.

There are other qualities of pastures that further complicate the consideration of competition. The variable, but frequently high, number of species involved has already been mentioned. They are also harvested at intervals – usually by animals and sometimes by machines. These harvests occur in all seasons and are rarely optimized to enhance the performance of desirable plant species. Apart from consuming forage (usually selectively), animals exert other influences on the pasture. Trampling and deposition of excreta, for example, will affect how those species subsequently grow and develop (Harper, 1977).

The morphology and biochemistry of the interacting species themselves will influence the level of competitive 'interference' that occurs. Individual plants of different species may utilize space differently and will then have reduced interference. Elongated stems that elevate leaves above the canopy (many clovers, for instance) will change competition from that expected by proximity alone. Species that use the C_3 photosynthetic pathway ('temperate' species) have very different sea-sonal growth patterns and, therefore, resource demands from C_4 species ('tropical' species). Legumes fix atmospheric nitrogen, N_2, which may in fact benefit other species. Soil microbiota and invertebrates also influence competitive relationships between pasture plants (Wardle and Barker, 1997; Watkinson and Freckleton, 1997), but they are beyond the scope of this chapter.

Decisions made by farm managers will have profound effects on the pasture through controlling (to some degree) soil fertility, moisture availability and the timing, intensity and selectivity of grazing animals. The interrelations between species in pasture plant communities (Tainton et al., 1996) mean that changes in pasture composition will occur through the direct and indirect effects of any given management. An understanding of competition in pastures provides an ecological framework within which to develop and assess strategies to address management objectives. Studies over the medium to long term and under a range of environmental conditions are required to provide the knowledge to develop principles that can assist pasture managers (Kemp and Dowling, 2000; Kemp et al., 2000).

Pastures vs. crops

Many of the principles underlying plant competition have been developed for crops (e.g. Cousens and Mortimer, 1995). They provide simple systems, often of only the crop and a weed, which interact for a brief period on the single product measured (usually grain yield). Management of competition within a crop aims to maximize the product produced. Pastures, unlike crops, are a compromise and management practices often seek to optimize, rather than maximize, the productivity of the more desirable species. Maximization of the productivity of a pasture ecosystem would require more inputs and controls than are feasible in practice. The ideas developed for the management of competition in crops can rarely be applied directly and simply to pastures, as the interactions are more complex. In crops, weeds can be managed with a single intervention, whereas in pastures that is rarely possible, and crops rarely have to contend with grazing animals.

Many pasture systems have been and still are treated as crops. This approach was heavily promoted in the UK and other temperate zones from the 1950s and is relevant where the 'pasture' is predominantly one species, e.g. *Lolium perenne* under high levels of N or *Medicago sativa* being grown for hay. The promotion of pastures as crops was done to focus more attention on management of the sward to improve productivity and profitability. As the terms of trade for agriculture have declined in many parts of the world, however, and environmental pressures have forced a reduction of inputs to pasture systems, it has not always been possible to maintain such an intensive system. The expanding programmes of 'extensification' in Europe (Marriott and Gordon, 1999) are part of a growing interest in studying and managing pastures more as ecosystems where overall inputs are reduced. The managers of pasture systems now probably learn as much from the studies on natural ecosystems (especially grasslands) by ecologists as from crop agronomists.

A Framework to Assess Competition

Competition within pastures can involve a range of responses and interactions. Plants compete in ways that influence population dynamics (including germination, establishment, recruitment, fecundity and mortality rates), growth rate and phenology. These effects can be measured and are discussed more exhaustively elsewhere in this book. This section aims to develop a framework within which the interactions between plant species and management can be analysed for practical use in pastures.

Early studies of competition often sought to identify what plants were competing for and then to record the impact on biomass. Many experiments attempted to do this by isolating responses to single variables. This can oversimplify reality, however, and has not always advanced our understanding beyond noting that plants can compete for light, water and nutrients and then identifying those species that are more or less competitive for that resource. In practice, the more limiting resources are going to change with time and in space, and so knowledge of the limiting resource at one particular point is unlikely to predict the outcomes of competition in the field.

The analysis of plant competition now focuses more on defining the net effect of resource capture – often measured in terms of growth rates (as an index of the rate of capture of resources) or biomass, with less emphasis on speculating as to which resources are involved.

Analysis of interacting species

Pastures are complex ecosystems and each individual plant will interact with other individuals of the same and different species simultaneously. In addition, with many individuals of many species, higher-order, indirect interactions (which may be characterized by the phrase 'my enemy's enemy is my friend') become more important (Connell, 1990). A tool is then needed to describe simply how all those species interact and then how management may influence that interaction.

In the context of a pasture plant community, the competitive ability of plant species plays an important role in plant diversity, abundance and distribution (Goldberg and Barton, 1992). In addition, the relative competitive abilities of the species that comprise a pasture will help determine which species will increase and which will decrease in abundance as a consequence, for instance, of fertilizer addition or increased grazing (Bakelaar and Odum, 1978; Noy-Meir *et al.*, 1989; Wilson and Shay, 1990). Consideration of competitive interactions, therefore, provides a framework within which to consider pasture management strategies. The measurement of competitive ability, however, is not a trivial task. There are three broad approaches: inference from other measured characters, direct experiments (with, perhaps, extrapolation to community level with models) and inference from field studies.

Inference of competitive ability from plant characters

One of plant ecology's holy grails has been the quest to better account for the ecology and biogeography of species by a search for the principal factors that determine their biology and competitiveness. Grime (1979), Tilman (1988) and, most recently, Westoby (1998) have sought to integrate autecological information into a coherent theory of plant 'strategy' from which the outcomes of competition might be predicted.

Grime's 'CSR' theory maintains that there are three basic plant strategies: competitor, stress-tolerator and ruderal. Defined as the ability to dominate a plant community, competitiveness is not determined by a single character; rather, it is determined by a 'syndrome' of characters, which includes maximum relative growth rate (RGR_{max}), net assimilation rate (NAR) and leaf area ratio (LAR). It has recently been shown (Hodgson *et al.*, 1999) that it is possible to predict the competitiveness score of a given species from 'soft' data as well, such as canopy height, lateral spread, specific leaf area and so on. After an extensive array of experiments, many British plant species have had their 'CSR' scores determined (Grime *et al.*, 1988).

Tilman (1982), in contrast, contended that it is the ability of a species to continue to extract resources down to low concentrations that is a more important predictor of competitive success. This character (denoted R^*) has been experimentally determined for a few species (Tilman and Wedin, 1991) with respect to soil N, where the R^* value of each species did predict the outcome of competition in pairwise mixtures.

The difference between the two schools of thought really comes down to a question of interpretation: the two 'schools' measure competition differently and they are not mutually exclusive (Grubb, 1985; Grace, 1991). Each theory may also apply to differing circumstances. Grime's theory makes useful predictions for productive environments, while Tilman's theory may be more suitable for the understanding of systems with low soil nutrient levels and over a longer time frame.

More recently, Westoby (1998) used data from only a limited number of traits (leaf area index, plant height, seed weight) to classify plant species. The simplicity of such an approach is appealing, but it is difficult to see how so few characters can represent the diversity of the 'strategies' of plant species found in pastures and the usefulness of this approach remains to be determined.

Ultimately, there is continuing uncertainty that the prediction of the outcomes of competition can be made without directly measuring the interaction(s) between competing species. The simple message that emerges from strategy theory – that, in the absence of grazing, those species with the highest 'competitiveness' score or lowest R^* value will dominate the sward – does provide an overarching ecological principle. Within a pasture context, it is not very interesting, however, since it is the grazing interaction that we are interested in. Most importantly, competitive ability will be related to different plant traits in different environments and will also vary over time (Bullock, 1996). In a pasture, for instance, the ability to dominate the sward may be determined more by having low palatability than by any other factor (Briske, 1996).

Modelling of plant competition

Mathematical models of plant competition generally fall into two categories: mechanistic and empirical. Mechanistic models account for the reduction in the growth of a plant due to competition by relating it to the following:

- The density of other plants. If a plant of a given size depends on a certain 'zone of influence' for growth and survival, then growth will be reduced if the available space is smaller than that required for maximum plant size.
- The relative size of other plants. Due to asymmetric competition for light, larger-than-average plants have a competitive advantage over smaller ones (Aikman and Watkinson, 1980).

Used together with functions describing, for example, plant growth, resource capture and fecundity, and scaled up for multiple species, these models can become complex (Thornley, 1998). Despite the enormous value of using such a model to examine the effects of climate change on a pasture (Thornley and Cannell, 1997), the number of parameters describing specific plant attributes required to run such models for even relatively simple pastures is prohibitive for many field studies.

Empirical models are generally much simpler. For instance, Freckleton and Watkinson (1997) argue that the relationship between competing plants is best described by the general yield-loss function:

$$y = ax^{-b} \tag{5.1}$$

where y is the natural logarithm of the biomasses of the target plant, x is the density of neighbour plants, and a and b are constants. This function has some advantages. For instance, it is a formula that may be easily extended to include multispecies communities.

Modelling may be considered to treat competition from a 'plant's-eye view' and is useful from an ecological perspective. It avoids any *a priori* consideration of plant strategy and thus avoids proposing any new hypothesis about how plants interact – it merely measures the response. This approach defines the net relative effect of a plant's ability to capture resources in the presence or absence of others. The principle involved, that of measuring relative change between species, is an appealing one. While the number of parameters involved in species-rich pastures would become unwieldy, a subset of the most important species would yield much information. Using the derived constants, the impact of competition at an agronomic scale could then be estimated.

The model represented by the above formula has been used to analyse simple mixtures and variations of this type of model have been developed to explore the interactions of three species (Silvertown et al., 1994). It would also be possible to use multiple, linked equations of this type to examine spatially explicit processes in a typical, variable, paddock. With increasing complexity, however, the higher-order, indirect interactions, which are not accounted for, become more numerous.

Such models use parameters derived from competition experiments (usually pot-based) and then aim to predict the longer-term outcomes of competition between those species. To be useful in a management context, parameters would need to be derived from experiments designed to explore specific management treatments. Even in a relatively simple uniform pasture, a large number of parameters would be needed. This complexity would be of little value to guide on-farm management decisions. They cannot be readily used to simply transfer data from experiments to farm managers.

In addition, modelling the biomass of pasture species looks at only one aspect of production from the system. It is the most common factor examined, but there are others that are also important to a farm manager. Alternative measures include the yield of nutrients, e.g. minerals or digestible organic matter, and the number and size of propagules. These measures are all important when the major interest is in the continuing productivity of the whole pasture ecosystem for animals. Competition studies have rarely considered these components.

Inference of competitive ability from field data

It is possible to infer competitive relationships between species by studying field data (Wilson and Gitay, 1995). Interpretation of changes in pasture composition over time or with respect to changes in management may be made using multivariate statistical techniques. The main trends of any changes can be extracted and plotted to portray the relationships between pasture species and to show the impact of management practices. In this way, the importance of environmental factors, such as soil fertility, can be graphically represented. Such techniques are well established in ecology, but under-

utilized in agriculture. It is, though, often difficult to compare data sets from different studies within a common framework, due to changes in scale and direction of derived factors. It is also difficult to translate the results from multivariate analyses into an advisory framework.

Pasture composition matrix

An alternative approach (Kemp, 1996) has used a pasture species composition matrix, which incorporates data from all the species present in a pasture. The aim of this approach is to provide a simple framework where data from field experiments could be readily plotted, the trends interpreted and then advisory messages portrayed for use by farmers and their advisers.

Many pastures contain a range of species. These can vary from a few in newly sown mixtures to 20 or so in longer-established swards (Nicholas et al., 1999), to over 100 in naturalized communities of introduced and volunteer species (W.McG. King and D.R. Kemp, unpublished). It would be impossible to display all the interactions between these species in a readily understood and simple framework. It is, though, possible to group species on a functional basis, where differences between species within a functional group are likely to be less than between groups. The grouping of species can be based upon physiological and morphological information. In an agricultural context, some groups can be defined using other criteria, such as economic value or weediness.

The pasture species composition matrix (Kemp et al., 1999b) can be formulated in different ways. One common approach (Fig. 5.1) is to use the four functional groups that are found in many perennial pastures, i.e. desirable and less-desirable grasses (such as perennial and annual grasses) and desirable and less desirable broad-leaved species (such as legumes and thistles). In permanent pasture systems, perennial grasses are more desirable than annuals, as they often provide more forage throughout the year and have deeper root systems, which help manage water-tables and salinity and soil acidity problems. The assumptions behind the subjective choice of these four groups have some basis in agronomy, where the less desirable species may provide poorer-quality forage or have negative animal health effects, and in nature, where ordination analysis of pasture survey data also identified the same group interactions (Fig. 5.2).

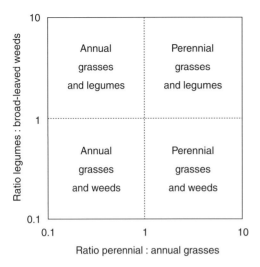

Fig. 5.1. The pasture species composition matrix. This example groups species into four common functional types. The ratio of desirable to less desirable grasses is plotted against the ratio of desirable to less desirable broad-leaved species. Log scales are used on each axis and the ratios are constrained within the limits of 0.1 to 10. In the absence of other information a ratio of 1 : 1 is used on each axis to divide the pasture composition into four 'states' as shown. The most desirable state is the one dominated by perennial grasses and legumes.

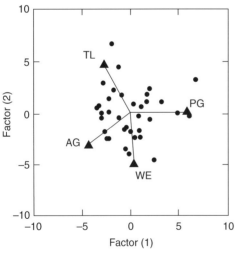

Fig. 5.2. Biplot of a principal components analysis of the functional plant groups (PG, perennial grasses; AG, annual grasses; TL, total legumes; WE, broad-leaved weeds) identified in a survey of 'improved' pastures in central New South Wales (Kemp and Dowling, 1991). Filled circles are points from individual paddocks.

These four groups are often present, but the aim in pasture management is to optimize the mixture to achieve dominance by perennial grasses and legumes. Plotting the ratio of the biomass of the two grass groups against the ratio of the biomass for the two broad-leaved species groups forms the matrix. The data are best plotted on a log scale within a range of 0.1 to 10, which adequately covers the range in dominance from one group to the other. The ratio of components incorporates the principle in Equation 5.1 presented earlier, of measuring the relative change between competing species (or groups). In this case the ratio is not modified by a proportionality constant on one species or the other, as the aim was to keep computations simple and in many cases data are not readily available to derive those constants.

The pasture matrix adds to the ways of analysing plant interactions by also defining 'states' within the pasture. Both desirable and less desirable states are defined. The goal in pasture management is often to achieve a state where perennial grasses and legumes dominate, rather than annual grasses and broad-leaved weeds. These ideas are derived from the 'state and transition' model developed for rangeland ecosystems (Westoby et al., 1989). That model was developed as an advisory tool and was based on the concept that ecosystems often maintain more or less stable states and then periodically move through a transition to another state – often in response to management practices. The pathways of change to and from different states normally differ. The problems with this approach are that experimental data cannot be directly applied to the model and that, in many pasture systems, gradual change is observed, rather than a step change. In temperate pasture systems, there is often a gradual replacement of one species by another (Kemp and Dowling, 1991; Dowling et al., 1996). This evidence supports the general view that the outcomes of competition are not always fixed, but can continually vary, depending upon seasonal, edaphic and managerial conditions.

States within the pasture matrix can be defined in various ways depending upon the goals for that system and the available knowledge on species interactions. In the absence of specific information, a 1 : 1 ratio is a reasonable first guess for a pasture state boundary. In this chapter, the interactions between species and functional groups in response to management will be explored in part using the pasture matrix.

Pasture Dynamics, Competition and Management

Pastures typically comprise a range of species and exist in an environment where there is considerable microvariation in resources. Each species requires different levels of resources and may exist in a niche where it is better able to survive than its competitors. A consequence of these conditions is that spatial variation invariably occurs in the distribution of species across a paddock. All niches are not fully occupied at all times. This then creates opportunities for other species to colonize those sites and, from there, they can then invade other areas. The dynamics of pastures is then initially determined by the species sown, the natural variation in resources/niches and the opportunities for species to colonize and spread within the community.

A problem with many pasture mixtures is that the range of species sown is limited and those species are obviously not fully able to exploit all the available resources. Typically, only a grass and a legume are sown in mixtures. In most cases, weeds are evident in the early life of a pasture, as they can exploit the unused resources. Agronomists have failed to take a sufficiently ecological approach to pasture mixtures and instead rely on the use of management practices to try and maintain a desirable pasture composition (Hoveland, 1999). There is a golden opportunity to improve the longer-term stability and productivity of pasture mixtures by a better understanding of how species exploit niches and what additional species should be included in those mixtures (see also Clark, Chapter 6; and Harris, Chapter 8, this volume). The role of biodiversity in pasture stability and productivity is only starting to be investigated (Tilman *et al.*, 1996; Wardle and Nicholson, 1996).

Soil fertility strongly affects the dynamics of pastures. If the soil fertility is high – especially soil N levels – then grasses or some broad-leaved species will dominate. Often, though, the soil N levels are low and legumes are then able to form a major component of the pasture. In time they fix sufficient N to enable grasses to become more competitive and they replace the legumes at those sites. Subsequently, the grasses deplete the available soil N, become less competitive and the legumes re-establish on those microsites. These dynamics are reasonably well understood and have been observed in practice (Turkington and Harper, 1979) and successfully modelled (Schwinning and Parsons,

1996). In an interesting twist on analysing competitive behaviour, Schwinning and Parsons were able to model this system by considering ryegrass as a predator and clover as prey – a working analogy that helps stimulate the debate as to how plants interact.

Pasture systems in practice are more complicated than outlined above, as legumes are able to exploit soil N if it is available and only fix N when it is not. The balance between grass and legume can then depend upon issues such as competition for light or water (Donald, 1963). In moist environments, a higher legume content can be sustained by maintaining a short sward and preventing the grass from overshadowing the legume (Jones, 1933).

Phenological development influences the pattern of competition among species. During a plant's life cycle its form changes and so does its ability to compete with other species. Many species become more erect when they are producing reproductive organs, and this can elevate them above their neighbours. Growth rates typically increase in grasses during reproductive development above that for vegetative plants growing under the same environmental conditions (Kemp *et al.*, 1989). The timing of events is also important.

Changes in pasture composition

Annual cycles in pastures

The competitive interactions within a pasture during the year normally cause major shifts in the proportion of species and of the 'states' within the pasture (Fig. 5.3). The goal in pasture management may be to achieve a perennial grass–legume-dominant sward, but seasonal conditions may mean that this only occurs for part of the year. Any analysis of management and competition effects needs to be evaluated against the normal seasonal changes that occur. These changes arise from the pasture component species' phenological development, modified by competitive interactions and local environment.

In south-eastern Australia, 'perennial' pastures typically comprise a base of perennial grasses, together with a variable number of annual grasses and legumes and some perennial broad-leaved species (Kemp and Dowling, 1991). These pastures may be considered to start their annual growth in autumn, when the period of reliable rainfall starts. In autumn, perennial grasses (e.g. *Phalaris aquatica*, *Dactylis glomerata*, *L. perenne*) regenerate from

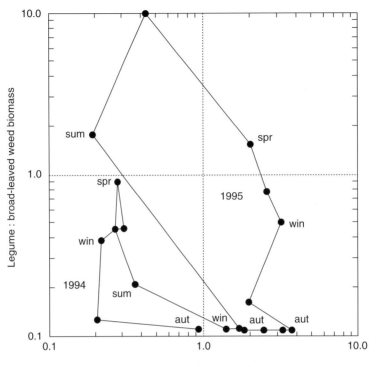

Fig. 5.3. Seasonal changes in composition in a pasture at Newbridge, NSW, Australia. Measurements were taken every 6 weeks over 2 years, from autumn 1994 to autumn 1996. The year 1994 was drier than 1995. The labels indicate the start of each season of the year. Data are for a treatment that consisted of continuously grazing for 9 months and then resting over summer.

crowns, or have retained some green leaf over summer, and initiate further leaf growth from existing tillers. During this period, seedlings of winter-annual species (including the weedy grasses *Vulpia bromoides* and *Bromus hordeaceus*, and the agronomically important *Trifolium subterraneum*) and some perennials germinate and have to compete with established plants. For this reason, the swards at this time can be dominated more by perennial than annual grasses and by broad-leaved weeds (see Fig. 5.3). In winter, effective rainfall is higher, but low temperatures restrict growth, which often constrains legumes more than grasses. As the season progresses from winter to spring (see Fig. 5.3) there is an increasing proportion of legumes (especially *T. subterraneum*) evident. Through spring, many species flower by producing stems to elevate seed heads above the canopy, potential pasture growth rates reach a peak and competition within and between species can be intense. As summer proceeds, the annual species finish producing seeds and

die, while growth of perennial plants is often constrained by declining soil water contents. The legumes often decline faster than the broad-leaf weeds as summer progresses (see Fig. 5.3). The death of plants creates spaces which some species (especially weedy species, such as *Echium plantagineum*) are able to take advantage of and colonize, as they can readily germinate after intermittent summer rainfall events. The plants that establish are then a substantial size when rainfalls that are more reliable return in autumn. The example of these seasonal trends shown in Fig. 5.3 also illustrates how these patterns vary between dry and wetter years. The pasture will often contain less perennial grass biomass in dry years, because of the increased pressure from grazing animals on these more palatable components.

The annual cycle outlined above used to be considered as a successional process, where one or more species or groups of species came to dominate the sward. However, as shown in Fig. 5.3, these processes are more cyclical and continuously dynamic and may not lead to the dominance of any functional group or groups.

Longer-term changes in pastures

Even taking into account the cyclical changes commonly shown within a year, pasture plant communities may also exhibit directional changes. With reduced grazing pressure, for instance, pastures may become increasingly dominated by taller species (Noy-Meir et al., 1989). The relatively stable nature of many temperate pastures (Wilson et al., 1996) suggests that this process of succession is effectively halted by grazing. In this way, pastures may be considered to be in a state of 'arrested successional development' (see Gitay and Wilson, 1995). Ecological theory suggests that, as succession progresses from bare soil after a disturbance, for instance, through a 'pioneer' phase to a more mature community, species of higher competitive ability will come to dominate the sward. Grazing prevents this process from progressing and results in a pasture sward containing species with a mix of competitive abilities. The concept of succession, however, should not be taken to imply that these changes are predictable, linear and potentially reversible. Interactions with local environment, seasonality (as above), stochastic events (Austin and Williams, 1988), inertia (Milchunas and Lauenroth, 1995) and other factors mean that pastures will be highly variable over time and space. For the same reasons, management options, such as crash-grazing (short-duration, high-intensity stocking), will have very different effects on the pasture depending on the timing. This is discussed further below.

Even longer-term trends in pasture composition have been observed at a time-scale of years or decades (Watt, 1981; Dodd et al., 1994; Dunnett et al., 1998) or even longer (e.g. global climate change (see Davis, 1986)). While these emphasize the dynamic nature of pasture communities and the flexible definition of 'stability' in these systems, they are of limited significance to farm managers.

Management Impacts on Competition

Management can have major impacts on the competitive interactions between species and on the composition of the pasture. In sown pastures, the initial management decisions as to what species are sown and how the site is prepared obviously have a critical impact. Emphasis in this section, though, will be on the post-establishment phase and on management practices that aim to optimize the composition of desirable species.

Management practices can affect competition between plant species at any time of the year, but the effect will vary depending upon season and the growth processes that are going on at that time. The evaluation of management impacts can be done from a short- or long-term perspective. In perennial pasture ecosystems, the impact on the long-term persistence of species can be more important. In that case, it is necessary to determine the more reliable reference point for comparison between years. For the example of south-eastern Australia outlined above (see 'Annual cycles'), the better reference points for year-to-year comparisons are the measurements taken in late winter/early spring. This is within the period of more reliable rainfall and at a time of the year where most of the species that occur in the pasture can be identified.

Fertilizer

Few, if any, soils have no nutrient limitations and in most cases nutrient levels will directly influence the productivity of the pasture (Jones et al., 1984). Within the sward, species differ in their abilities to extract and utilize nutrients, which then results in differences in growth. In the absence of competition, virtually every plant species will show some increase in growth rate with raised soil fertility. In a mixed sward, however, fertilizer applications will differentially favour those species whose growth had been most adversely affected by low soil fertility (Shipley and Keddy, 1988). Typically these are inherently highly competitive species with high maximum growth rates, and these profound changes in competitive relationships between species will result in substantial shifts in pasture composition. Often there is a reduction in species richness (Huber, 1994; Eek and Zobel, 1997). The most common use of fertilizers is to alter the grass : legume balance. Phosphate fertilizers tend to favour legumes (Henkin et al., 1996), until soil N levels are high, while applications of N generally lead to grass dominance (Elisseou et al., 1995). In soils where micronutrients are deficient dramatic changes occur following their application. For instance, molybdenum is essential for effective N_2-fixation by legumes and, when applied to deficient soils, legume growth can increase by many orders of magnitude (Lourenco *et al.*, 1989). Lime applications also have the effect of increasing the availability of molybdenum.

Grazing

Pastures by definition support grazing animals. The influence of the grazing animal is critical to the composition and productivity of the pasture. Grazing behaviour and foraging strategy by livestock (Watkin and Clements, 1978) influence what they eat and when, which can have a severe impact on some species relative to others. Controlling total animal intake and the ability of animals to select what they eat by manipulating stocking rate is one of the most potent tools available to farmers in managing the composition of a pasture.

All pasture plants can be considered sensitive to grazing at some times of their life cycles. The more critical times are when establishing from seed, for example, or when regenerating after a period of dormancy or drought and when flowering (Kemp, 1991; Wilson and Hodgkinson, 1991). Damage can be inflicted by grazing, treading, lying on plants, dropping dung and urine and other aspects of animal behaviour. To use grazing as a tactic for pasture management, the general rules are (Kemp, 1993) to rest desirable species at sensitive stages in their life cycles and to pressure the less desirable species at their weak points, for instance during flowering. At other times, controlling the ability of animals to select what they eat so that all species are consumed more or less equally, can lessen any impact on the relative competitive abilities of desirable species. Total pasture biomass must always be considered. Too much biomass results in taller than optimum swards (see Matthew *et al.*, 1995) which will limit legume growth, while too little biomass will reduce livestock foraging efficiency and provide gaps for weed invasion and the exposed soil will be subject to erosion. Indirect effects and feedbacks also occur. Grazing results in the reduction in the incidence of some pests, for example, which in turn will influence the competitive relationships between species (Michael *et al.*, 1999).

Impact of a simple grazing tactic

The impact of a simple grazing practice on changing pasture composition over time is illustrated in Fig. 5.4. These data are from a *D. glomerata*-based

pasture that had been overgrazed and invaded by weed species for some years prior to the establishment of this experiment (Dowling *et al.*, 1996; Kemp *et al.*, 1996). The two treatments shown are the continuously grazed control and a treatment where plots were continuously grazed for 9 months and then rested for 3 months over summer. The data are from measurements in early spring, i.e. some 6 months after the summer rest treatment had been applied each year. Under continuous grazing by sheep, the pasture remained for 6 years in a state where the legume (mostly *T. subterraneum*) content was satisfactory, but annual grasses dominated over the perennials. Broadleaf weeds were more of a problem in drier years, e.g. 1993. The trends in composition under continuous grazing are typical of 'improved' pastures in the region (Kemp and Dowling, 1991).

The summer rest treatment allowed some species to set seed and new plants to establish, depending upon seasonal conditions. Both *D. glomerata* and *Cirsium vulgare* increased over the years from the summer rest treatment. It seemed that either one species or the other, but not both, increased as a proportion of the sward in any one year. The trend in composition was a result of the interaction between management and seasonal conditions. *Cirsium vulgare* tended to increase in drier years, e.g. from 1992 to 1993 (see Fig. 5.4). The changes in *D. glomerata* content (in years 1991/92 and 1994/95) were largely due to an increase in plant numbers, rather than an increase in the size of existing plants.

Year-round grazing practices

Some grazing practices are employed throughout the year and have a continuing impact on pasture composition. *Cichorium intybus* (chicory) is a perennial forage species (Rumball, 1986) that is highly productive and competitive (Alemseged, 1999) but does not persist under continuous grazing, due to its high palatability. The other species that persist in the pasture ecosystem with *C. intybus* depend also upon grazing practices.

Figure 5.5 shows the results of an experiment designed to investigate the impact of different grazing practices on a mixed *C. intybus* pasture (Kemp *et al.*, 1999a). The data are plotted on a triangular diagram, as the system was largely dominated by three species: *C. intybus*, *E. plantagineum* (an annual broad-leaved weed) and *Lolium rigidum* (annual rye-

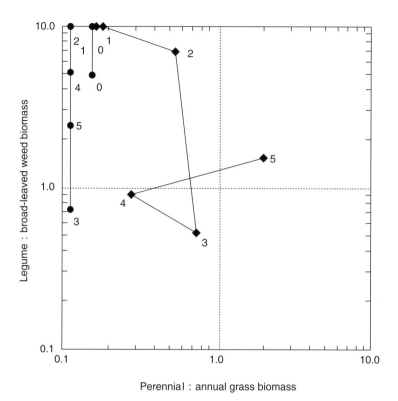

Fig. 5.4. Annual changes in composition of two pasture treatments in a grazing experiment at Newbridge, NSW, Australia. Circles are the unfertilized continuously grazed control and diamonds are from plots rested over summer and continuously grazed the rest of the year. Numbers refer to the years 1990 to 1995.

grass). *Trifolium repens* and *T. subterraneum* were also present, but at low levels, and in this case they fluctuated more in relation to climatic conditions than in relation to the grazing treatments or competitive influences.

Under continuous grazing, *C. intybus* declined and almost disappeared from the pasture after 3 years and the sward was dominated by *E. plantagineum* (Fig. 5.5). Where plots were grazed for 1 week and rested for 5, *C. intybus* remained a major component, along with *L. rigidum*. The 3 weeks' grazing, 3 weeks' rest treatment was intermediate in behaviour. In this study, it was evident that the amount of *C. intybus* was largely related to the frequency of grazing (as found also by Li *et al.*, 1997) and did not seem to decline in response to any invading weeds. The shift between *E. plantagineum* and *L. rigidum* was also a function of the frequency of grazing. Where short-duration, high-density grazing treatments were used, animals could not effectively select what to eat. As a consequence all species were consumed more or less equally and *L. rigidum* remained

a component of the sward. Where graze periods were longer and less intense, animals had more time to select what they would eat. *Lolium rigidum* was consumed in preference to *E. plantagineum* and hence the latter species became dominant.

Grazing practices targeted by phenological development

Grazing tactics can be targeted at specific stages in a plant's life cycle. These are usually at the sensitive stages mentioned earlier. For weed species such as the winter-annual grasses *Vulpia myuros* and *V. bromoides*, a useful target point is when flowering heads are starting to emerge. Reducing the amount of seed set can potentially reduce the competition from these species in subsequent years. In addition, a deferment of grazing in autumn may limit the availability of microsites for germination and establishment (Dowling *et al.*, 1996). Similar tactics have also been suggested for slender thistle (Bendall, 1973).

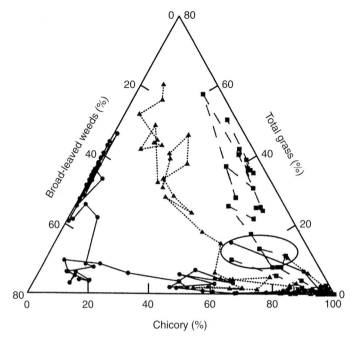

Fig. 5.5. Sequential changes in composition of a chicory pasture over 5 years in response to three grazing management treatments: continuous grazing (circles and solid lines), grazed 3 weeks and then rested 3 weeks (triangles and dotted line), and grazed 1 week, rested 5 weeks (squares and dashed line). Measurements were taken at approximately 6-week intervals. The ellipse identifies the starting position for each treatment. (Data from Kemp *et al.*, 1999a.)

In a large project designed to investigate a range of alternative management practices for the control of *Vulpia* species in a *P. aquatica* pasture, Dowling *et al.* (1998) demonstrated that many treatments resulted in statistically significant changes in pasture composition. Most treatments were, though, of little value in optimizing the proportion of *Phalaris* to *Vulpia* (Fig. 5.6a). Many plots remained dominated by *Vulpia* and *T. subterraneum*. Among those treatments where extra grazing pressure was employed during *Vulpia* tiller elongation (Fig. 5.6b), however, a slightly higher proportion of *Phalaris* was achieved. It is also interesting to note that treatments where grazing was completely excluded also resulted in more *Phalaris*.

The data in Fig. 5.6 are plotted on the pasture species composition matrix using a ratio of 4 *Phalaris* : 1 *Vulpia* (as biomass) to define a desirable target grass composition. This was done to provide a more realistic boundary condition for this grassland type. The 4 : 1 boundary was based upon the likely practical minimum value for *Vulpia* content (10% of the total biomass) that can be achieved in practice, which would mean a *Phalaris*

content of around 40%. A *Phalaris* content below that value would indicate a pasture in decline and consideration would be given to resowing. At the other end of the scale, the 4 : 1 ratio would mean a maximum *Phalaris* content of 80%. At this level *Vulpia* would not exceed 20%, which is within the tolerable range. Of course, this 80% *Phalaris* : 20% *Vulpia* split could only occur in the absence of any other species. In reality, the maximum *Vulpia* content at a 4 : 1 ratio is likely to be around 15%, which is close to the level of control required. No one treatment consistently achieved this ratio, however. The best management strategy for this pasture would probably include grazing management (see Fig. 5.6a,b) and simazine (see Fig. 5.6d) in a system sensitive to seasonal conditions.

Herbicides

The occurrence of weeds in the pasture is often a reflection on past practices, where inappropriate management has resulted in the desirable species becoming less competitive. Those management

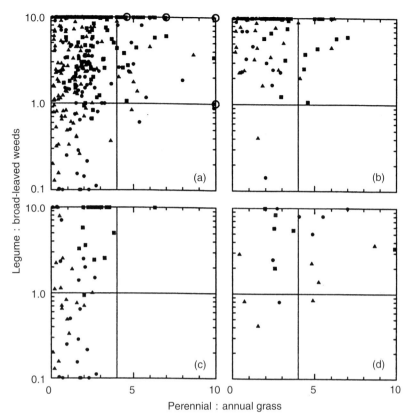

Fig. 5.6. The pasture composition in early spring in 1995 (■), 1996 (▲) and 1997 (●) for (a) all 120 treatments from five experiments investigating management of *Vulpia* species in a *Phalaris* pasture, points within circles are from ungrazed plots; (b) treatments that used extra grazing pressure during tiller elongation; (c) glyphosate; and (d) simazine treatments. Pastures with a perennial : annual grass ratio > 4 and a legume : broad-leaved weed ratio > 1 were considered the more desirable state. (From Dowling *et al.*, 1998.)

practices could be reversed in some way in order to improve the pasture, but in many cases farmers use herbicides first to reduce the weed population in programmes to restore the productivity of pastures.

In the simplest case the use of a selective herbicide can remove an unwanted species. To be successful, other management practices then need to be implemented to enable the more desirable components to colonize the gaps created and strengthen the pasture against further weed invasion. Unfortunately, even the more selective herbicides often cause some collateral damage to desirable species. This can result in reducing the growth rates of the useful species and their ability to compete with weeds.

The influence of herbicides on the competitive interactions among species in a mixed pasture is often complex. In southern Australia the annual

grasses *V. myuros* and *V. bromoides* are major problems in pasture systems. There has been considerable research into herbicides for the management of these species, which has tended to record only their efficacy on the *Vulpia* species, ignoring any wider consideration of how the pasture as a whole has changed.

The large grass control programme referred to earlier (Dowling *et al.*, 1998) explored a range of herbicide and other tactics for the management of *Vulpia* species within a previously sown *P. aquatica* perennial grass pasture. The data from treatments using herbicides are shown in Fig. 5.6.

Simazine is a herbicide recommended for control of *Vulpia* species and in each case there was a significant reduction in the *Vulpia* content. However, the *Phalaris* content often remained low or declined, due to other practices, so that few treatments actually shifted the grass composition into the desirable range. Glyphosate is a more general herbicide that is also recommended for *Vulpia* and did reduce the incidence of this species. However, the collateral damage to other species is clearly evident. Glyphosate reduced the legume and *Phalaris* contents, creating a poorer-quality pasture than had initially existed! By considering the relative interactions among all species, a better appraisal of the full impact of a herbicide is achieved.

Herbicides directly affect the growth of some species and hence their competitiveness and can also have small, but important, effects on other species. They may not kill a plant, but can limit seed set, as done with 'spray-topping' tactics, or reduce growth rates. This means that, when herbicides are used to remove competitive weeds, it is always important to add other tactics that enhance the desirable components and prevent weeds from re-establishing.

Other practices

Fire is a tool that has frequently been used in semi-arid areas to manipulate pasture composition (Wright and Bailey, 1982; Trollope, 1984). It has been used to a lesser extent in higher-rainfall zones and, when it has, this has been more in native grass communities. Very little information is available on the impact of fire in sown pasture systems. However, it is expected that, in some regions, more use could be made of fire for pasture management, as it is potentially a lower-cost technique.

Some general principles are involved in the use of fire. The damage done by fire is directly related to the intensity of the fire, temperatures and soil water content (Wright and Bailey, 1982). Damage to grasses can be minimized by burns when the soil is cool and wet. At such times broad-leaved weed species can be severely affected.

Physical 'renovation' of pastures is used to damage or remove unwanted species. In *M. sativa* stands, annual grasses often establish in the inter-row spaces. The incidence of these grasses can be reduced by a light tine cultivation. Cultivation practices in established swards are now rare, though, as farmers seek more subtle and cost-effective management tools. It can also be difficult to target weed species unless the desirable species are sown in rows.

Consideration of Competition in Pasture Management

The consideration of competition within pastures is an everyday component of pasture management, though many farmers do not see it in that context. More often they focus on the composition of their pastures, i.e. the outcomes of prior competition. Unfortunately, many farmers consider the composition of their pastures only when they are unsatisfied with the performance of their livestock. That can be too late and at a point where some aggressive and more expensive intervention is required to restore the composition to a more desirable state. The challenge for managers is to continually optimize that composition for the particular system involved within the normal seasonal trends and working with an understanding of the competitive interactions between species.

Changes in pasture composition in response to the competitive interactions between species are occurring continuously. Good pasture management, then, requires the continual monitoring of those interactions so that more regular (and simpler) changes in management tactics can be made to keep the composition of the pasture within desirable boundary limits (Michalk and Kemp, 1994). Competition needs to be managed even within the more complex pasture ecosystems. Technologies to do this are being developed, but their effectiveness relies on a good understanding of the underlying mechanisms that affect species interactions. Unfortunately, the emphasis in much agronomic work is on the selection of individual cultivars in isolation, simply for yielding ability in monospecific swards (Hoveland, 1999). As a result, mixed pastures are often very difficult to manage. Recommendations to select at least some species in the presence of the species they will be sown with have been made for many years (Atwood and Garber, 1942), but rarely adopted. An alternative approach, which has not yet been exploited, would be to select communities of species or heterogeneous species.

The management of competition needs to be based upon a clear understanding of the role and growth cycles of the species, or species groups within a pasture. From an agricultural perspective, species can be grouped into more or less desirable categories within functional types and then management planned accordingly. The use of a framework such as the pasture matrix outlined in this chapter can help determine if the pasture is in a realistically desirable state.

Acknowledgements

The ideas presented in this chapter owe much to the continuing discussions among our colleagues, especially Drs D. Michalk and P. Dowling and others within the Sustainable Grazing Systems Programme for southern Australia and the Weeds Cooperative Research Centre. Technical support from G. Millar, J. Tarleton, S. Betts and D. Pickering over the years has been excellent. Our research has been funded by Meat and Livestock Australia, the Cooperative Research Centre for Weed Management Systems, the Woolmark Company and the Land and Water Research and Development Corporation.

References

Aarssen, L.W. and Turkington, R. (1985) Vegetation dynamics and neighbour associations in pasture-community evolution. *Journal of Ecology* 73, 585–603.

Aikman, D.P. and Watkinson, A.R. (1980) A model for growth and self-thinning in even-aged monocultures of plants. *Annals of Botany* 45, 419–427.

Alemseged, Y. (1999) Competitiveness, productivity and management of chicory (*Chicorium intybus* L.) for pasture. PhD thesis, University of New South Wales, Sydney, Australia.

Atwood, S.S. and Garber, R.J. (1942) The evaluation of individual plants of white clover for yielding ability in association with bluegrass. *Journal of the American Society of Agronomy* 34, 1–6.

Austin, M.P. and Williams, O.B. (1988) Influence of climate and community composition on the population demography of pasture species in semi-arid Australia. *Vegetatio* 77, 43–49.

Bakelaar, R.G. and Odum, E.P. (1978) Community and population level responses to fertilization in an old-field ecosystem. *Ecology* 59, 660–665.

Bendall, G.M. (1973) The control of slender thistle, *Carduus pycnocephalus* L. and *C. tenuiflorus* Curt. (Compositae), in pasture by grazing management. *Australian Journal of Agricultural Research* 24, 831–837.

Briske, D.D. (1996) Strategies of plant survival in grazed systems: a functional interpretation. In: Hodgson, J. and Illius, A.W. (eds) *The Ecology and Management of Grazing Systems*. CAB International, Wallingford, UK, pp. 37–67.

Brown, T.H. (1976) Effect of deferred autumn grazing and stocking rate of sheep on pasture production in a Mediterranean-type climate. *Australian Journal of Experimental Agriculture and Animal Husbandry* 16, 181–188.

Bullock, J.M. (1996) Plant competition and population dynamics. In: Hodgson, J. and Illius, A.W. (eds) *The Ecology and Management of Grazing Systems*. CAB International, Wallingford, UK, pp. 69–100.

Connell, J.H. (1990) Apparent versus 'real' competition in plants. In: Grace, J.B. and Tilman, D. (eds) *Perspectives on Plant Competition*. Academic Press, San Diego, USA, pp. 9–26.

Cousens, R. and Mortimer, M. (1995) *Dynamics of Weed Populations*. Cambridge University Press, Melbourne, Australia.

Davis, M.B. (1986) Climatic instability, time lags, and community disequilibrium. In: Diamond, J. and Case, T.J. (eds) *Community Ecology*. Harper and Row, New York, USA, pp. 269–284.

Dodd, M.E., Silvertown, J., McConway, K., Potts, J. and Crawley, M. (1994) Stability in the plant communities of the Park Grass Experiment: the relationships between species richness, soil pH and biomass availability. *Philosophical Transactions of the Royal Society of London, Series B* 346, 185–193.

Donald, C.M. (1963) Competition among crop and pasture plants. *Advances in Agronomy* 15, 1–118.

Dowling, P.M., Kemp, D.R., Michalk, D.L., Klein, T.A. and Millar, G.D. (1996) Perennial grass response to seasonal rests in naturalised pastures of central New South Wales. *Rangeland Journal* 18, 309–326.

Dowling, P.M., Kemp, D.R., Michalk, D.L. and Millar, G.D. (1998) Changes in pasture management needed for improved control of weeds in pastures. In: Michalk, D.L. and Pratley, J.E. (eds) *Proceedings of the 9th Australian Agronomy Conference*. Australian Society of Agronomy, Wagga Wagga, Australia, pp. 559–562.

Dunnett, N.P., Willis, A.J., Hunt, R. and Grime, J.P. (1998) A 38-year study of relations between weather and vegetation dynamics in road verges near Bibury, Gloucestershire. *Journal of Ecology* 86, 610–623.

Eek, L. and Zobel, K. (1997) Effects of additional illumination and fertilization on seasonal changes in fine-scale grassland community structure. *Journal of Vegetation Science* 8, 225–234.

Elisseou, G.C., Veresoglou, D.S. and Mamalos, A.P. (1995) Vegetation productivity and diversity of acid grasslands in northern Greece as influenced by winter rainfall and limiting nutrients. *Acta Oecologia* 16, 687–702.

Freckleton, R.P. and Watkinson, A.R. (1997) Measuring plant neighbour effects. *Functional Ecology* 11, 532–534.

Gitay, H. and Wilson, J.B. (1995) Post-fire changes in community structure of tall tussock grasslands: a test of alternative models of succession. *Journal of Ecology* 83, 775–782.

Goldberg, D.E. and Barton, A.M. (1992) Patterns and consequences of interspecific competition in natural communities: a review of field experiments with plants. *American Naturalist* 139, 771–801.

Grace, J.B. (1991) A clarification of the debate between Grime and Tilman. *Functional Ecology* 5, 583–587.

Grime, J.P. (1979) *Plant Strategies and Vegetation Processes*. John Wiley & Sons, Chichester, UK.

Grime, J.P., Hodgson, J.G. and Hunt, R. (1988) *Comparative Plant Ecology. A Functional Approach to Common British Species*. Unwin Hyman, London, UK.

Grubb, P.J. (1985) Plant populations and vegetation in relation to habitat, disturbance and competition: problems of generalization. In: White, J. (ed.) *The Population Structure of Vegetation*. Dr W. Junk Publishers, Dordrecht, The Netherlands, pp. 595–621.

Harper, J.L. (1977) *Population Biology of Plants*. Academic Press, London, UK.

Henkin, Z., Noy Meir, I., Kafkafi, U., Gutman, M. and Seligman, N. (1996) Phosphate fertilization primes production of rangeland on brown rendzina soils in the Galilee, Israel. *Agriculture, Ecosystems and Environment* 59, 43–53.

Hodgson, J.G., Wilson, P.J., Hunt, R., Grime, J.P. and Thompson, K. (1999) Allocating c-s-r plant functional types: a soft approach to a hard problem. *Oikos* 85, 282–294.

Hoveland, C.S. (1999) Problems in establishment and maintenance of mixed swards. In: Buchanan-Smith, J.G., Bailey, L.D. and McCaughey, P. (eds) *Proceedings of the XVIII International Grassland Congress*. Association Management Centre, Calgary, Canada, Session 22, pp. 411–416.

Huber, R. (1994) Changes in plant species richness in a calcareous grassland following changes in environmental conditions. *Folia Geobotanica et Phytotaxonomica* 29, 469–482.

Jones, H.R., Maling, I.R. and Curnow, B.C. (1984) Prediction of the responsiveness to phosphorus of annual non-irrigated pasture in northern Victoria. *Australian Journal of Experimental Agriculture and Animal Husbandry* 24, 579–585.

Jones, M.G. (1933) Grassland management and its influence on the sward. *Journal of the Royal Agricultural Society of England* 94, 21–41.

Kemp, D.R. (1991) Defining the boundaries and manipulating the system. In: Michalk, D.L. (ed.) *Proceedings of the 6th Annual Conference of the Grassland Society of New South Wales*. NSW Agriculture and Fisheries, Orange, Australia, pp. 24–30.

Kemp, D.R. (1993) Managing pastures by grazing. In: Michalk, D.L. (ed.) *Proceedings of the 8th Annual Conference of the Grassland Society of New South Wales*. NSW Agriculture, Orange, Australia, pp. 32–38.

Kemp, D.R. (1996) Weed management directions in pasture systems. In: Shepherd, R.C.H. (ed.) *Proceedings of the 11th Australian Weed Conference*. Weed Science Society of Victoria, Frankston, Australia, pp. 253–263.

Kemp, D.R. and Dowling, P.M. (1991) Species distribution within improved pastures over central NSW in relation to rainfall and altitude. *Australian Journal of Agricultural Research* 42, 647–659.

Kemp, D.R. and Dowling, P.M. (2000) Towards sustainable temperate perennial pastures. *Australian Journal of Experimental Agriculture* 40, 125–132.

Kemp, D.R., Eagles, C.F. and Humphreys, M.O. (1989) Leaf growth and apex development of perennial ryegrass during winter and spring. *Annals of Botany* 63, 349–355.

Kemp, D.R., Dowling, P.M. and Michalk, D.L. (1996) Managing the composition of native and naturalised pastures with grazing. *New Zealand Journal of Agricultural Research* 39, 569–578.

Kemp, D.R., King, W.M., Michalk, D.L. and Alemseged, Y. (1999a) Weed-proofing pastures: how can we go about it? In: Bishop, A.C., Boersma, M. and Barnes, C.D. (eds) *Proceedings of the 12th Australian Weeds Conference*. Tasmanian Weed Society, Devonport, Australia, pp. 138–143.

Kemp, D.R., Michalk, D.L., Dowling, P.M. and Klein, T.A. (1999b) Analysis of pasture management practices within a pasture composition matrix model. In: Buchanan-Smith, J.G., Bailey, L.D. and McCaughey, P. (eds) *Proceedings of the XVIII International Grassland Congress*. Association Management Centre, Calgary, Canada, Session 22, pp. 105–106.

Kemp, D.R., Michalk, D.L. and Virgona, J.M. (2000) Towards more sustainable pastures: lessons learnt. *Australian Journal of Experimental Agriculture* 40, 343–356.

Li, G.D., Kemp, P.D. and Hodgson, J.G. (1997) Herbage production and persistence of puna chicory (*Cichorium intybus* L.) under grazing management over 4 years. *New Zealand Journal of Agricultural Research* 40, 51–56.

Lodge, G.M. and Orchard, B.A. (2000) Effects of grazing management on Sirosa phalaris herbage mass and persistence in a predominantly summer rainfall environment. *Australian Journal of Experimental Agriculture* 40, 155–170.

Lourenco, M.E.V., De Carvalho, R.J.M. and Da Silva, M.d.L.A.P. (1989) Effects of fertilization and liming on the improvement of native pastures. In: *Proceedings of the XVI International Grassland Congress*. Association Française pour la Production Fourragère, Versailles, France, pp. 57–58.

Marriott, C.A. and Gordon, I.J. (1999) Extensification of sheep grazing systems: effects on soil nutrients, species composition and animal production. In: Buchanan-Smith, J.G., Bailey, L.D. and McCaughey, P. (eds) *Proceedings of the XVIII International Grassland Congress*. Association Management Centre, Calgary, Canada, Session 15, pp. 3–4.

Marriott, C.A., Fisher, J.M., Hood, K.J. and Smith, M.A. (1997) Persistence and colonization of gaps in sown swards of grass and clover under different sward managements. *Grass and Forage Science* 52, 156–166.

Matthew, C., Lemaire, G., Sackville Hamilton, N.R. and Hernandez Garay, A. (1995) A modified self-thinning equation to describe size/density relationships for defoliated swards. *Annals of Botany* 76, 579–587.

Michael, P.J., Grimm, M., Hyder, M. and Doyle, P.T. (1999) Grazing affects pest and beneficial invertebrates in Australian pastures. In: Buchanan-Smith, J.G., Bailey, L.D. and McCaughey, P. (eds) *Proceedings of the XVIII International Grassland Congress*. Association Management Centre, Calgary, Canada, Session 29, pp. 21–22.

Michalk, D.L. and Kemp, D.R. (1994) Pasture management, sustainability and ecosystems theory: where to from here? In: Kemp, D.R. and Michalk, D.L. (eds) *Pasture Management: Technology for the 21st Century*. CSIRO, Melbourne, Australia, pp. 155–169.

Milchunas, D.G. and Lauenroth, W.K. (1995) Inertia in plant community structure: state changes after cessation of nutrient-enrichment stress. *Ecological Applications* 5, 452–458.

Nicholas, P.K., Kemp, P.D., Barker, D.J., Brock, J.L. and Grant, D.A. (1999) Production, stability and biodiversity of North Island New Zealand hill pastures. In: Buchanan-Smith, J.G., Bailey, L.D. and McCaughey, P. (eds) *Proceedings of the XVIII International Grassland Congress*. Association Management Centre, Calgary, Canada, Session 21, pp. 9–10.

Noy-Meir, I., Gutman, M. and Kaplan, Y. (1989) Responses of Mediterranean grassland plants to grazing and protection. *Journal of Ecology* 77, 290–310.

Reed, K.F.M. (1974) The productivity of pastures sown with *Phalaris tuberosa* or *Lolium perenne*. 1. Pasture growth and composition. *Australian Journal of Experimental Agriculture and Animal Husbandry* 14, 640–648.

Rumball, W. (1986) 'Grasslands Puna' chicory (*Cichorium intybus* L.). *New Zealand Journal of Experimental Agriculture* 14, 105–107.

Schwinning, S. and Parsons, A.J. (1996) A spatially explicit population model of stoloniferous N-fixing legumes in mixed pasture with grass. *Journal of Ecology* 84, 815–826.

Shipley, B. and Keddy, P.A. (1988) The relationship between relative growth rate and sensitivity to nutrient stress in twenty-eight species of emergent macrophytes. *Journal of Ecology* 76, 1101–1110.

Silvertown, J., Lines, C.E.M. and Dale, P. (1994) Spatial competition between grasses – rates of mutual invasion between four species and the interaction with grazing. *Journal of Ecology* 82, 31–38.

Tainton, N.M., Morris, C.D. and Mardy, M.B. (1996) Complexity and stability in grazing systems. In: Hodgson, J. and Illius, A.W. (eds) *The Ecology and Management of Grazing Systems*. CAB International, Wallingford, UK, pp. 275–299.

Thornley, J.H.M. (1998) *Grassland Dynamics: an Ecosystem Simulation Model*. CAB International, Wallingford, UK.

Thornley, J.H.M. and Cannell, M.G.R. (1997) Temperate grassland responses to climate change: an analysis using the Hurley Pasture Model. *Annals of Botany* 80, 205–221.

Tilman, D. (1982) *Resource Competition and Community Structure*. Princeton University Press, Princeton, New Jersey, USA.

Tilman, D. (1988) *Plant Strategies and the Dynamics and Structure of Plant Communities*. Princeton University Press, Princeton, New Jersey, USA.

Tilman, D. and Wedin, D. (1991) Plant traits and resource reduction for five grasses growing on a nitrogen gradient. *Ecology* 72, 685–700.

Tilman, D., Wedin, D. and Knops, J. (1996) Productivity and sustainability influenced by biodiversity in grassland ecosystems. *Nature* 379, 718–720.

Trollope, W.S.W. (1984) Fire in savanna. In: Booysen, P.de V. and Tainton, N.M. (eds) *Ecological Effects of Fire in South African Ecosystems*. Springer-Verlag, Berlin, Germany, pp. 149–175.

Turkington, R. and Harper, J.L. (1979) The growth, distribution and neighbour relationships of *Trifolium repens* in a permanent pasture. 1. Ordination, pattern and contact. *Journal of Ecology* 67, 201–218.

Wardle, D.A. and Barker, G.M. (1997) Competition and herbivory in establishing grassland communities – implications for plant biomass, species diversity and soil microbial activity. *Oikos* 80, 470–480.

Wardle, D.A. and Nicholson, K.S. (1996) Synergistic effects of grassland plant species on soil microbial biomass and activity: implications for ecosystem-level effects of enriched plant diversity. *Functional Ecology* 10, 410–416.

Watkin, B.R. and Clements, R.J. (1978) The effects of grazing animals on pastures. In: Wilson, J.R. (ed.) *Plant Relations in Pastures*. CSIRO, Melbourne, Australia, pp. 273–298.

Watkinson, A.R. and Freckleton, R.P. (1997) Quantifying the impact of arbuscular mycorrhiza on plant competition. *Journal of Ecology* 85, 541–545.

Watt, A.S. (1981) A comparison of grazed and ungrazed grassland A in East Anglian Breckland. *Journal of Ecology* 69, 499–508.

Westoby, M. (1998) A leaf–height–seed (lhs) plant ecology strategy scheme. *Plant and Soil* 199, 213–227.

Westoby, M., Walker, B. and Noy-Meir, I. (1989) Opportunistic management for rangelands not at equilibrium. *Journal of Range Management* 42, 266–274.

Wilson, A.D. and Hodgkinson, K.C. (1991) The response of grasses to grazing and its implications for the management of native grasslands. In: Dowling, P.M. and Garden, D.L. (eds) *Native Grass Workshop Proceedings*. Australian Wool Corporation, Melbourne, Australia, pp. 47–57.

Wilson, J.B. and Gitay, H. (1995) Limitations to species coexistence: evidence for competition from field observations, using a patch model. *Journal of Vegetation Science* 6, 369–376.

Wilson, J.B., Lines, C.E.M. and Silvertown, J. (1996) Grassland community structure under different grazing regimes, with a method for examining species association when local richness is constrained. *Folia Geobotanica et Phytotaxonomica* 31, 197–206.

Wilson, S.D. and Shay, J.M. (1990) Competition, fire, and nutrients in a mixed-grass prairie. *Ecology* 71, 1959–1967.

Wright, H.A. and Bailey, A.W. (1982) *Fire Ecology, United States and Southern Canada*. John Wiley & Sons, New York, USA.

6   Diversity and Stability in Humid Temperate Pastures

E. Ann Clark

Department of Plant Agriculture, University of Guelph, Guelph, Ontario, Canada

The thesis that diversity conveys stability has long occupied the creative energies of theoretical and applied ecologists. However, the motivation for exploring this topic in the current chapter is pragmatic; the outcome is directly relevant to commercial graziers. Producers need to know if mixture complexity affects not simply short-term productivity and nutritional quality, but also pasture stability, reliability and, ultimately, longevity. Extending the productive lifespan of the sward bears directly on pasture profitability, particularly when the costs of reseeding are high relative to the value of the increased pasture output.

The hypothesis to be explored in the present chapter is that plant species diversity conveys stability – in time and space – to pasture performance. Due to environmental heterogeneity in both space and time, it is hypothesized that pasture stability will increase as plant species diversity increases up to an optimum that is greater than would be predicted from short-term, small-plot clipping trials. It is theorized that the simple mixtures commonly recommended today predispose swards to yield fluctuation, weed encroachment, a shorter lifespan and reduced profitability. The needs of commercial graziers would be better served by landscape-based evaluation protocols for both species and mixtures.

Definition of Terms and Context

Addressing the hypothesis that diversity conveys stability requires clarification of both terminology and context. Diversity or complexity will refer to the number of plant species in a pasture sward, such that one may refer to a 'complex' mixture as one with three or more species. Conversely, a 'simple' sward will be taken to mean from one or two species.[1] Although space does not permit further amplification, it is recognized that plant species diversity is supported by associated genetic diversity at other trophic levels, all of which combine to affect sward performance (e.g. Sohlenius *et al.*, 1987; Hulme, 1996; Bever *et al.*, 1997; Kaye and Hart, 1997).

Stability, as used by Pearson and Ison (1987), refers to the 'amount of variation experienced by a community of both plants and grazing animals around their dynamic equilibrium following disturbance'. In keeping with this definition, stability in the present chapter will be measured as the variance around the mean of indicated data sets, whether in space (over sites) or in time (over years).

The context of the enquiry will determine the specific parameter used to assess stability. Sustaining a reliable and acceptable level of herbage yield, of animal output or of economic return over years may be of primary importance to those who make their living from the pasture sward. Stability of species composition might be of interest in production systems that capitalize on the strengths of a single species, whether perennial ryegrass (*Lolium perenne*) in the UK or lucerne (*Medicago sativa*) in the prairies of North America. When production depends predominantly on a single species, encroachment by weeds or other inferior species is viewed as destabilizing species composition and reducing sward productivity and, ultimately, stand life. Conversely, those managing land for nature conservation may strive for the opposite – a stable but high-diversity sward (Smith *et al.*, 1997). It follows, then, that the parameter chosen to assess 'stability', and hence the evaluation of stability itself, is context-dependent and, further, that stability in one parameter need not connote stability in another parameter.

The context in which stability is assessed must also be defined in spatial and temporal terms, because the scale of consideration powerfully influences the outcome of the assessment. Stability of yield or species composition in small research plots measured for 2 or 3 production years may not be predictive of stability at the level of interest to a commercial producer, namely, at the field or landscape level over the productive lifespan of the sward.

This chapter will consider first the historical trends in recommended mixture complexity in humid, temperate-zone pastures, with primary emphasis on north-eastern North America. Complications introduced into the study of complexity and stability will be considered, with emphasis on the use of herbage yield as a predictor of animal performance and the relationship between yield and yield stability. Predominant reliance on short-duration small-plot trials will be shown to have compromised the predictive value of mixture research to landscape-level commercial pasture. Sources of environmental heterogeneity and species-specific adaptations to environmental factors will be reviewed, to explain the widely reported trend towards diversification in sown pastures. It will be concluded that a key rationale for more complex species mixtures, namely, to match environmental heterogeneity with genetic diversity, and hence to stabilize and prolong productive stand life, cannot be tested without a landscape orientation to species and mixture recommendations.

Historical Trends in Mixture Recommendations

Simple mixtures have been widely recommended in Ontario and elsewhere in the humid temperate zone since at least the middle of the 20th century. However, it was not always so. Morrison (1979) referenced studies published in 1890 and in 1907

showing that the famous 'fattening' pastures of the Midlands of the UK were a complex mixture with a high proportion of perennial ryegrass, white clover, Kentucky bluegrass (*Poa pratensis*) and other desirable species, together with a low proportion of undesirable species, such as the bentgrasses (*Agrostis* spp.), thistles (*Cirsium* spp.) and buttercups (*Ranunculus* spp.). The original emphasis on complex mixtures early in this century in Ontario was an attempt to re-create the complexity of these productive fattening pastures.

Blaser *et al.* (1952) reviewed historical trends in mixture recommendations, ranging from the 'reasonably complicated' mixture of Stapledon and Davies (1928), with six grass and three legume species, to Willard (1951) who proposed a mixture of 'at least one grass and one legume', although more might provide good insurance. Blaser *et al.* (1952) noted that the trend had been towards simpler mixtures of a few, well-adapted species, because they were easier to manage and more practical.

In reviewing a range of studies conducted in Ontario between 1930 and 1960, Clark and Poincelot (1996) reported that complex mixtures of from six to 12 species generally produced higher yields and persisted better than simpler mixtures of two to three species. To encompass the range of growing conditions in the various regions of the province, the number of recommended pasture mixtures increased from six in 1930 (Ontario Department of Agriculture, 1940, 1943) to 20 by 1960 (Ontario Department of Agriculture, 1954; Parks, 1955). However, by the mid-1990s, most jurisdictions recommended predominantly simple mixtures for pasture (Table 6.1). For example, extension documents provide the following guidance to producers in Ontario and Alberta:

> Ontario: 'Simple mixtures are easier to manage for high yields and should always be used when pasture management is intensive' (Robinson *et al.*, 1990).
> Alberta: 'mixtures do not need to be complex. The use of one legume and one grass, if well adapted to the environment and intended use, will frequently give maximum yield' (Alberta Agriculture, 1981).

The practical outcome of this perspective is that 60–100% of the mixtures recommended to producers in contemporary Alberta (humid and irrigated zone), Ontario and New York consist of one or two species (see Table 6.1).

Table 6.1. Complexity of recommended mixtures in Ontario, Alberta and New York state (adapted from OMAFRA, 1997; Alberta Agriculture, 1981; and Cornell University, undated, respectively).

	Total number of recommended pasture mixtures	Percentage of recommended mixtures with:			
		One species	Two species	Three species	Four species
Ontario	10	0	60	10	30
Alberta[a]	12	17	50	33	0
New York	6	33	67	0	0

[a]For the black and grey luvisolic soil zones and under irrigation.

Assessing Stability: Problems and Prospects

The parameter used most often to compare simple and complex mixtures is herbage yield, typically derived from small-plot studies of 2 or 3 years' duration (Clark *et al.*, 1996). Much less frequently encountered in the literature is evidence of yield stability, particularly over an economically meaningful stand life or across a field-scale landscape. Even less prominent is evidence of the stability of livestock performance or economic return in time or space, and yet these are surely the parameters of greatest interest to the end-user – the commercial grazier.

It will be argued below that the focus on herbage yield rather than animal performance and on yield rather than on stability of yield in time and space may have led researchers to recommendations on mixture complexity which are at variance with the needs of commercial graziers.

Herbage yield and animal performance

Herbage yield, typically under a clipping regime, is the parameter of choice for much breeding and management research in the region. It is apparently presumed that higher biomass yield under clipping will translate into higher animal gain under grazing. This inference is particularly dubious for grazing, where livestock performance depends more completely on voluntary intake. Species- and cultivar-specific differences in preference (Shewmaker *et al.*, 1997) might be expected to affect intake more in grazing than in feeding regimes.

While herbage yields account for much of the evidence referenced in this chapter, it is worth acknowledging the weakness of available evidence to validate the presumption that yield translates

reliably into gain by grazing animals. Indeed, several studies have drawn this premise into question. In Australia, Robinson and Dowling (1985) significantly increased both herbage on offer and sown species contribution to the sward (e.g. *Phalaris aquatica*, tall fescue (*Festuca arundinacea*), cocksfoot (*Dactylis glomerata*) and perennial ryegrass) in a natural pasture through various combinations of reseeding and application of superphosphate and herbicides. However, in the subsequent 4 years of controlled grazing by sheep at three stocking rates, they were unable to detect an improvement in either live-weight or wool production due to improved yield or enhanced pasture species composition.

In a series of replicated, multiyear grazing trials in Minnesota, Gordon Marten and colleagues compared the productive capability of grazed lucerne and bird's-foot trefoil (*Lotus corniculatus*), in monoculture and in combination with other species (Marten and Jordan, 1979; Marten *et al.*, 1987, 1990). As summarized by Clark *et al.* (1993), although bird's-foot trefoil yielded less and sustained 6–8% fewer animal grazing days than lucerne, average daily gain was 6–24% higher on bird's-foot trefoil, and liveweight gain per hectare varied from 5% lower to 18% higher on bird's-foot trefoil than on lucerne. Despite a higher yield potential, resulting from decades of intensive breeding and management research, lucerne produced no higher and perhaps even somewhat lesser levels of animal performance than bird's-foot trefoil. Clearly, factors other than herbage yield accounted for animal performance in these two species.

Similar trends can be found within a single species. Munro *et al.* (1992) compared lamb production from four cultivars of perennial ryegrass, sown alone (with 215 kg N ha^{-1}) or with white clover (with 88 kg N ha^{-1}), over a 2-year interval in the UK (Table 6.2). Cultivars differed in performance in

Table 6.2. Inconsistent ranking of perennial ryegrass cultivars when evaluated for lamb production, for hay herbage and for grazed herbage (adapted from Munro *et al.*, 1992).

Perennial ryegrass cultivar	Cultivar maturity	Lamb production, kg ha^{-1}		Hay herbage yield, t ha^{-1}	Grazed herbage yield, t ha^{-1}	
		Grass only	Grass/clover		Grass only	Grass/clover
Aurora	Very early	908[a]	1055[a]	9.6[b]	11.0[ab]	9.8[a]
Frances	Early	764[b]	953[ab]	10.8[a]	11.9[a]	10.2[a]
Talbot	Intermediate	837[ab]	896[bc]	10.5[ab]	10.7[b]	9.8[a]
Melle	Late	866[ab]	821[c]	9.9[ab]	10.7[b]	8.5[b]
Effect of cultivar		**		*	*	
Effect of sward type		**			**	

*Significantly different ($P<0.05$); **Significantly different ($P<0.01$). Within individual columns, figures with the same letter are not significantly different.

three assessment regimes – lamb gain, hay yield and grazed herbage yield. However, the rank order of cultivar performance differed under the three assessment regimes. For example, the very early cv. Aurora yielded less dry matter than cv. Frances under haying (9.6 vs. 10.8 t ha^{-1}), but not under grazing. However, when evaluated in terms of animal gain in the grass-only treatment, Aurora produced 19% more than Frances, despite similar grazed herbage yields (11 vs. 11.9 t ha^{-1}). Similar, but statistically insignificant patterns were observed in the grass–clover treatments. In another contrast, cv. Talbot yielded more grazed herbage in the grass–clover treatment than did cv. Melle (9.8 vs. 8.5 t ha^{-1}), but produced similar levels of lamb production (896 vs. 821 kg ha^{-1}) (Munro *et al.*, 1992).

Animal performance appeared to reflect not simply dry matter yield or herbage on offer, but other, little understood factors such as preference to grazing stock (Shewmaker *et al.*, 1997). Further, from a commercial perspective, the magnitude of the unexplained cultivar differences in animal performance was large. Consider, for example, that Aurora yielded similarly to Melle under hay management (9.6 vs. 9.9 t ha^{-1}) but produced 30% more lamb in a grass/clover sward (1055 vs. 821 kg ha^{-1}) (Munro *et al.*, 1992).

These examples suggest that pasture performance assessments based on yield or herbage on offer – often the only parameters measured – may not extrapolate predictably to commercial performance, let alone performance stability, on grazed pastures.

Yield vs. yield stability

One of the apparent justifications for concentrating contemporary production research on simple mix-

tures, often profiling the strengths of a single species, such as perennial ryegrass or lucerne, was the presumption that one or a few highly bred, management-responsive species would necessarily yield more than any mixture including lesser yielding species. However, in the UK, Frame (1990) demonstrated that species such as creeping red fescue (*Festuca rubra*) and Kentucky bluegrass produced comparable yields to perennial ryegrass when grown with a single application of 50 kg N ha^{-1}. Subsequent work by Frame (1991) demonstrated that perennial ryegrass only outyielded the best of a range of ten secondary grasses at N rates above 240 kg N ha^{-1}. Studies by Marten and colleagues, summarized by Clark *et al.* (1993) have already been referenced to challenge the perceived superiority of lucerne for animal gains within the humid temperate region.

Stability of herbage yield, particularly on temporal or spatial scales of interest to commercial graziers, is seldom encountered in refereed literature. Evidence bearing on both the presumed yield superiority of simple mixtures and on the relationship between yield and yield stability is drawn from two studies below.

Study 1

The Ontario Forage Crops Committee (OFCC) conducted a pair of replicated small-plot field trials in the 1940s and 1950s, from which it was concluded that simple rather than complex mixtures would henceforth be recommended in the province (OFCC, 1962, 1963). In one trial, 15 mixtures consisting of from one to seven species were sown at each of five imperfectly drained locations in the province. A parallel trial involved ten mixtures containing from two to seven species, sown at five well-drained sites. Both trials were subjected to a three-cut harvest regime, with data recorded for 3 production years.

HERBAGE YIELD. Under well-drained conditions, 3-year mean yield ranged from 7.4 to 7.7 t ha^{-1} and was unaffected by mixture complexity (Fig. 6.1). Conversely, among mixtures suitable for imperfectly drained conditions, yield ranged from 5.0 to 6.2 t ha^{-1}, with mixtures containing four or five species producing about 10% more than either the two- or three-species mixtures or the combination of six- or seven-species mixtures (see Fig. 6.1). Within this small range of mixture complexity, the relationship between diversity and yield was optimal, centring on the four- or five-species range, analogous to the humpbacked relationship between species density and biomass yield in grazed and ungrazed meadows in Finland (Grace and Jutila, 1999).

STABILITY. Using the coefficient of variation (CV) for yield among sites as a measure of stability, CV declined modestly with increasing mixture complexity in both trials, in both the first and second years, although the opposite occurred in the third year, when CV was generally higher (Fig. 6.2). Three-year mean CV declined with mixture complexity under imperfectly drained conditions, but was unrelated to complexity under well-drained conditions. Within-site heterogeneity may have

been greater at the imperfectly drained sites, perhaps revealing the greater capacity of a more complex mixture to occupy diverse niches and sustain yield.

Thus, in this study, complex mixtures were at least as productive and stable as simple mixtures on well-drained sites, and may have had a modest advantage in both yield and stability on imperfectly drained soils. It should also be noted that the relative merit of simple vs. complex mixtures for both yield and stability was not generalizable, but varied with the particular species and growing conditions of the test.

Study 2

Hopkins *et al.* (1990) compared the yield of permanent and sown grasslands at five rates of nitrogen (N) fertilizer (from 0 to 900 kg N ha^{-1}) at 16 diverse sites in the UK. Plots were harvested for yield at 4- or 8-week intervals, with data reported for 3 production years. Sown plots were planted either to perennial ryegrass and white clover (for the 0 N control) or to perennial ryegrass (e.g. simple mixtures). Species composition in the permanent swards was complex and representative of that in swards with a history of low management inputs, e.g. less than 30% perennial ryegrass and a range of other wild or sown species.

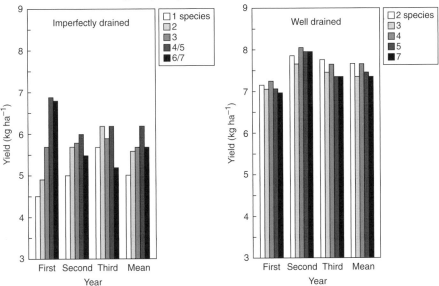

Fig. 6.1. Mixture complexity and yield in Ontario (adapted from OFCC, 1962, 1963). (Key is number of species per mixture.)

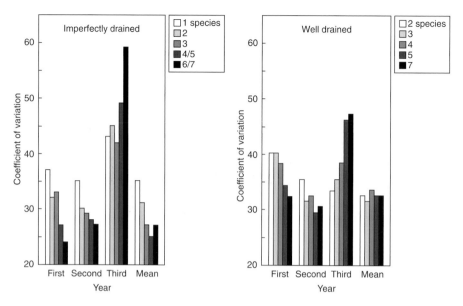

Fig. 6.2. Mixture complexity and coefficient of variation for yield in Ontario (adapted from OFCC, 1962, 1963). (Key is number of species per mixture.)

HERBAGE YIELD. Yields from sown swards were superior to those from permanent pasture at the same level of N in the first production year, after which the advantage of the sown swards was evident only at 450 and 900 kg N ha^{-1} (Hopkins et al., 1990). At lesser N rates (0, 150 and 300 kg ha^{-1}), permanent swards yielded as well as or better than the sown plots in years 2 and 3. Although yield potential differed greatly among sites, similar patterns of relative production by permanent and sown swards were generally observed.

As demonstrated by Smith and Allcock (1985) and others, Hopkins et al. (1990) found that swards consisting of wild and other sown species can yield as well as perennial ryegrass, at the same level of management. Similarly, in a 3-year study from the UK reported by Wilkins (1986), cattle gains from a permanent sward were 93% those in a resown perennial ryegrass sward when both swards received 400 kg N ha^{-1} (Garwood, unpublished, cited in Wilkins, 1986). Thus, as previously reported by Peel (1979), Hodgson (1990), and Hopkins and Hopkins (1993), species composition, and, specifically, the contribution of perennial ryegrass, was shown to be more a dependent variable responding to fertilizer, drainage and grazing management rather than an independent effector of pasture performance.

STABILITY. The CV for yield at 300 kg N ha^{-1} year^{-1} over 16 sites and 3 years (n = 48) was identical for the sown and permanent swards harvested at 4-week intervals (CV = 21 and 21, respectively), while for swards harvested at 8-week intervals, the CV was somewhat lower for permanent than for sown swards (CV = 16 vs. 20, respectively, calculated from Hopkins et al., 1990) (Fig. 6.3).

The relative yield advantage of complex (permanent) vs. simple (sown) swards varied among production years, while the relative stability of complex and simple swards varied with harvest interval. As in the OFCC study above, results were not generalizable over levels of mixture complexity, but were specific to a given contrast.

While suggestive, the above yield-based studies present far from compelling evidence that complex mixtures elicit greater stability than simple mixtures. They none the less challenge the justification that simple mixtures are necessarily higher-yielding, particularly in the longer term. Remarkably few studies appear to have been published with a sufficient range of sites, years and/or mixtures to effectively test this thesis. Perhaps stronger inferences can be drawn from the ecological literature.

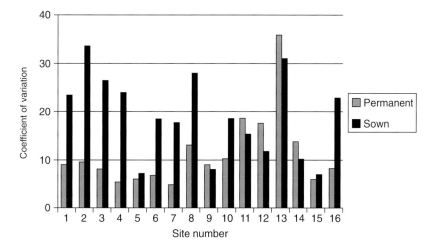

Fig. 6.3. Variability over years (3) in yield of permanent and sown swards at 16 sites in the UK (adapted from Hopkins *et al.*, 1990).

Towards a Landscape Approach

Historically and even to the present day, mixture comparisons are usually generalized from small-plot trials to commercial pastures. The preference for simple mixtures in contemporary recommendations appears to have been derived largely from controlled, small-plot studies. However, the last 20 years have seen a virtual explosion in awareness of within-field environmental heterogeneity (Caldwell and Pearcy, 1994; Dale, 1999), and the role of biotic and abiotic factors in creating patchiness, biodiversity and megacommunities within a single field (Turkington and Harper, 1979; Aarssen and Turkington, 1985; Palmer and Dixon, 1990; Silvertown *et al.*, 1992; Schwinning and Parsons, 1996a,b; Turkington and Jolliffe, 1996; Dale, 1999).

Landscape-level diversity and hence the influence of diversity on stability of commercial sward performance are necessarily ill-predicted by small-plot studies. Realization of in-field variability is the foundation of the modern-day emphasis on 'site-specific' or 'precision' agriculture (various references in Jaynes and Colvin, 1997). Recognizing the natural variability that exists even in well-managed arable land, satellite-guided, tractor-mounted tracking systems monitor yield and soil nutrient status and compensate by varying fertilizer rate. By the same token, the spatial scale of small plots cannot encompass the range of environmental variability that occurs within a pasture field.

The implications of this scaling problem are central to the present analysis of diversity and stability. It will be argued below that one of the chief advantages of a more complex mixture – and one that would be impossible to discern in a typical small-plot study – is the provision of a sufficient range of genetic variation to more fully occupy the diverse niches that exist in a commercial pasture. Better niche occupancy by sown species would be expected to translate into more consistent productivity, less weed encroachment and a longer stand life, outcomes that could not be revealed in a short-term, small-plot study.

By the same logic, short-term studies of the sort that predominate in the contemporary research-funding environment (Clark *et al.*, 1996) can hope to capture but a small part of the year-to-year variability in weather in which the mixtures will be expected to perform commercially. Short-term studies are particularly ill-suited to evaluating the merit of mixture complexity in perennial swards, swards that reflect not just the current management and environment, but also those of the past.

Thus, the paucity of clear evidence bearing on the effect of mixture complexity on stability of either sward or livestock performance may be the legacy of inappropriate scaling to address the question, owing to the prominence of short-term, small-plot studies for mixture assessment.

Environmental Heterogeneity: Sources and Selection

In order for a complex mixture to be better able than a simple mixture to occupy the niche diversity created in time and space and to sustain stand life, at least two features must be present. First, the growing environment must be sufficiently heterogeneous – in time and/or space – to exceed the adaptation of a simple mixture and, secondly, species-specific differences in adaptation must be sufficiently large to manifest themselves in sward compositional changes in time and space – which is, after all, the basis of ecological succession (Connell and Slatyer, 1977).

Sources of niche diversity in a pasture sward are many. According to Grime (1994), variability in space and time or 'patch and pulse phenomena' are characteristic of resource supply in nature. Writing in the same volume, Bell and Lechowicz (1994) consider environmental heterogeneity to be ubiquitous, the norm rather than the exception. Rosenzweig (1995) argued persuasively that 'the greater the habitat variety, the greater the species diversity', and demonstrated the generalizability of a positive log–log relationship between number of species and size of the sampled area for a surprising range of species. Heterogeneity in resource availability has direct implications for sown mixture complexity.

In the parlance of Grime (1987), different habitats select for plant species with different life strategies. Sites characterized by low resource availability (e.g. nutrients, light, water and other abiotic factors) and high disturbance, such as intensive grazing, are termed 'disturbed'. Disturbed sites have been found to favour ruderal species with a rapid growth rate able to capitalize on brief resource pulses. Conversely, sites with high resource availability and low disturbance select for competitive species, which are able to compete aggressively by virtue of size, both above and below ground.

For example, in less productive grasslands, most nutrients are immobilized in plant or microbial tissues and pulses of nutrient availability are likely to be brief and unpredictable. When pulse duration was less than 10 h, species adapted to infertile soils, such as sheep's fescue (*Festuca ovina*) exhibited a higher relative growth rate and higher specific N absorption rate than species adapted to fertile soils, such as *Arrhenatherum elatius*. This ranking was reversed for pulse durations longer than 10 h

(Crick and Grime, 1987; Campbell and Grime, 1989).

Both natural edaphic and managerial factors can create a mosaic of diverse habitats within the same managed pasture. For example, soil nutrients and/or water may be deficient during part or all of the growing season on eroded or south-facing slopes, while being less limiting at more favoured sites in the same field. Patchy distribution of urine and faeces would create the same variation on a smaller scale. Similarly, paddocks that are withheld from grazing to conserve the spring excess would favour taller and more competitive species than paddocks subjected to season-long grazing. Non-uniform in-paddock grazing, as occurs at the lower stocking rates, could create the same patchwork of high and low disturbance on a smaller scale. Thus, on both a macro- and a microscale basis, diverse habitats can occur within the same field, favouring different species mixtures in different places.

Sheath and Boom (1985) related variation in species composition in hill-country pastures in New Zealand to soil moisture and nutrient variations associated with sheep camping/tracking areas and topographic position. They found, for example, that brown top (*Agrostis tenuis*) content increased, while perennial ryegrass decreased, on steeper, drier land. Thus, the topographic and edaphic attributes of the pasture site interacted with grazing management to create a mosaic of distinct communities across the pasture landscape. Each community was uniquely adapted to its own 'niche', a reality that would have been invisible in a small-plot mixture comparison.

Grazing can itself create heterogeneity, not owing simply to grazing preferences but also to the redistribution of soil nutrients. In a comparison conducted in dune grasslands in the UK, grazing by sheep increased soil heterogeneity in the concentration of most soil ions in the summer, and of nitrate and phosphate in the autumn (Gibson, 1986; Fig. 6.4). In summer, tiller density of Kentucky bluegrass was positively correlated with ammonium ion concentration in the ungrazed but not in the grazed sward, while that of creeping red fescue was positively correlated with ammonium, phosphate and sodium ion concentration in the ungrazed sward, but only with phosphate in the grazed sward.

Steinauer and Collins (1995) found that urine patches affected the balance between C_3 and C_4 species at prairie sites in Kansas and Nebraska, although the effect varied depending on the history of burning and litter accumulation. Parish and

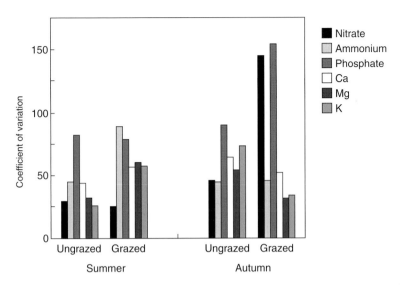

Fig. 6.4. Spatial and temporal heterogeneity in soil ion supply under grazed and ungrazed grassland (adapted from Gibson, 1986).

Turkington (1990) documented the effect of mole-hills and dung pats on plant species abundance and distribution in British Columbia. Species differed in response to disturbance, with white clover abundance increasing in the year of the disturbance, while perennial ryegrass abundance responded to disturbance from the previous season. In areas of high disturbance, *Holcus lanatus* declined, while other species, such as cocksfoot, couch grass (*Agropyron repens*) and dandelion (*Taraxacum officinale*) increased. Thus, grazing serves to redistribute and concentrate soil nutrients in discrete patches, presumably via urination and defecation, creating the potential for different species associations from those that might occur under ungrazed conditions.

The amount and distribution of resource-enriched patches can alter species distribution and sward composition, because species differ in their ability to capitalize on sites of resource enrichment. Sterling *et al.* (1984) documented the influence of microtopography on species distribution across pastures abandoned for varying lengths of time. Effects on species distribution were already evident after 2 years, but were strongest in pastures abandoned for 7 years, with nitrophilous, nutrient-demanding and drought-intolerant species concentrating in surface depressions. Grime (1994) considered that plant communities were structured with both dominant and subordinate species, to furnish the capa-

bility to produce in both stable and perturbed conditions. Thus, the variety of plant species present within a given sward may be viewed as a functional, community response to an unpredictably varying environment. Seen in this way, the utility of sward diversity can be channelled intentionally to support more consistent and sustained productivity in commercial pastures.

Jackson *et al.* (1990) demonstrated that sagebrush (*Artemesia tridentata*) and two perennial tussock-grass species (*Agropyron desertorum* and *Agropyron spicatum*) exhibited differences in phenotypic plasticity in rate of P uptake when exposed to P-enriched soil patches. While root proliferation in an enriched patch was evident in *A. desertorum* within 1 day, no root proliferation was exhibited by *A. spicatum* within 14 days. Rate of P uptake by sagebrush increased by as much as 80% within days of treatment, and was particularly noteworthy on soils very low in available P. Jackson and Caldwell (1993) noted that species have been shown to differ not simply in root proliferation but in changes in nutrient uptake kinetics and changes in the frequency of mycorrhizal infection, all of which could contribute to species distribution dynamics in pasture swards.

In a related paper, Jackson and Caldwell (1996) reported a measured threefold and 12-fold range in soil P and nitrate concentrations, respectively, around individual plants of *A. desertorum.* Modelling studies revealed that root plasticity in response to soil heterogeneity was much more important for uptake of nitrate than for P, primarily because of the greater documented range in nitrate concentration in the soil. Species-specific differences in response to nutrient-rich patches were considered important in competitive acquisition of nutrients. Papers cited in Jackson and Caldwell (1996) discussed the factors to be considered in addressing the potential agronomic importance of plant plasticity in responding to soil heterogeneity.

The role of temporal variation in growing conditions on the dynamics of perennial sward composition, independent of sward age, is perhaps less well documented. Halvorson *et al.* (1997) tabulated a range of soil quality parameters known to vary in time frames termed highly dynamic (< 1 year), dynamic (1–10 years) and relatively static (10–1000 years). Highly dynamic factors, which could advantage or disadvantage sward components and change species composition during the course of a single year, included porosity, infiltration rate, compaction, temperature, water-holding capacity and soluble nutrient concentration. Particular emphasis was placed on the issue of scale, with the caution that extrapolating from fine-scale to landscape-scale phenomena – in either spatial or temporal terms – assumes a 'scale-independent uniformitarianism of patterns and processes (which we know to be false)' (Wiens, 1989).

Micheli *et al.* (1999) noted that biotic and abiotic factors can interact to influence variability in species composition, which can in turn stabilize aggregate plant community parameters. Silvertown *et al.* (1994) showed that, over the 90-year Park Grass Experiment in the UK, species composition was much more variable than herbage yield. The primary driving force for variation in yield was rainfall, but the impact of rainfall on yield was moderated by changes in species composition. Species composition changed under the influence of species-specific differences in tolerance to water stress. The net effect was to stabilize year-to-year variation in yield by increasing variation in species composition.

The Homogeneity of Managed Grasslands

It may be argued that managed grasslands are, by definition, more uniform and hence less diverse than the natural environment, thus diminishing the merit of an argument for biotic diversity. This point may have merit for simple mixtures sown as short-term leys under a high level of management. However, the degree of variablity in arable cropland is currently considered sufficient to warrant redress through precision agriculture technologies (Jaynes and Colvin, 1997). None the less, for longer-term pastures managed more extensively, the premise of sufficient homogeneity to justify simple mixtures can be challenged on the following grounds:

1. Even well-managed high-value land exhibits sufficient environmental heterogeneity as to select for a complex mosaic of diverse plant communities over time. The 20-ha Northfork research pasture near Elora, Ontario, consists of CLI Class 1^2 tile-drained land that was originally sown uniformly to a complex mixture. Most of the nine sown species are still present, but distribution is no longer uniform on either a microscale basis within an individual paddock (Fig. 6.5) or on a macroscale basis among fields within the pasture (Fig. 6.6).

2. Simple mixtures do not stay simple for very long. Diversification, both within and among species, commonly occurs as pastures age. In the UK, on-farm surveys published in the early 1970s and mid-1980s documented a decline in sown species contribution with age, although the latter survey showed a lesser rate of decline (Fig. 6.7; Morrison and Idle, 1972; Hopkins *et al.*, 1985). After 5 years, sown species accounted for an estimated 55 and 75% of the sward in the older and newer surveys, respectively, while after 10 years, comparable figures were 40 and 65%, respectively.

Aarssen and Turkington (1985) compared pastures originally sown in 1939, 1958 and 1977 to approximately the same complex mixture of seven species. Overall, species diversity increased to 28 sown or naturalized species, but, while diversity did not vary among the three pastures, the relative importance of individual species varied with age. Frequency of stable, non-random species associations (positive or negative) increased from nil in the youngest pasture to six in the oldest. In the oldest pasture, negative associations were observed between cocksfoot and *H. lanatus* and between perennial ryegrass and *Poa compressa*, while white clover was positively associated with each of *H. lanatus*, perennial ryegrass, couch grass and *P. compressa*. Thus, the trend with time was for progressive spatial re-sorting to achieve more stable positive or negative species associations.

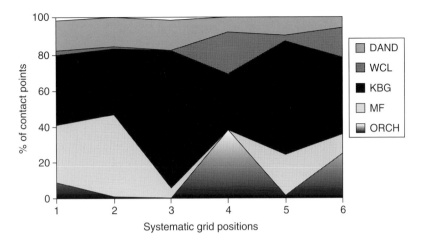

Fig. 6.5. Microscale species sorting in six 1 m² grids across a 0.15 ha paddock within a 9-year-old pasture at Elora, Ontario (from E.A. Clark, unpublished). DAND, dandelion; WCL, white clover; KBG, Kentucky bluegrass; MF, meadow foxtail; ORCH, orchardgrass.

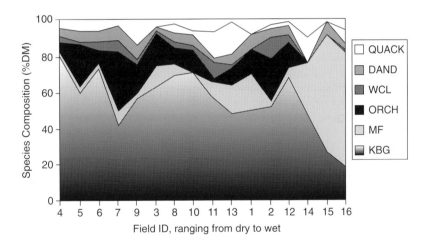

Fig. 6.6. Macroscale species sorting among 16 1.25 ha fields in a 20 ha, 12-year-old pasture at Elora, Ontario (from E.A. Clark, unpublished). DM, dry matter. QUACK, quackgrass; DAND, dandelion; WCL, white clover; ORCH, orchardgrass; MF = meadow foxtail; KBG, Kentucky bluegrass.

Turkington and Mehrhoff (1990) cited a range of papers showing that permanent pastures are characterized by a rich diversity of up to about 50 species and, further, that a high level of genetic diversity is also found within individual species. Encroachment by unsown species suggests an adaptation to prevailing conditions which exceeds that found in the components of a sown simple mixture. Conversely, when the sown mixture is complex, as in the above example for the Northfork pasture, the contribution of unsown species (weeds) to herbage yield remains low, even after 12 years of intensive grazing (see Fig. 6.6).

3. High levels of resource input can retard or even reverse diversification, resulting in higher yields from simpler mixtures. This should not be taken to mean, however, that the higher yields arose from the simpler mixtures. For example, Tilman (1987)

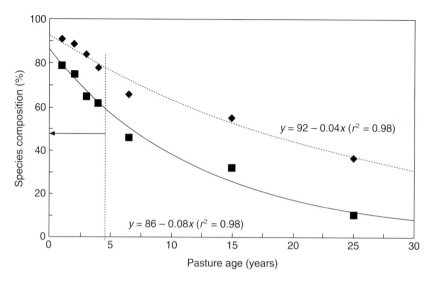

Fig. 6.7. Sown species persistence (visual estimates) on well-drained soils in the UK (adapted from Morrison and Idle, 1972; Hopkins *et al.*, 1985). Vertical and horizontal lines refer to per cent sown species remaining at 5 years of stand life.

demonstrated that homogenizing the growing environment through applied N acted to reduce species diversity while increasing productivity. However, Goldberg and Miller (1990) clarified that it was the nutrients rather than the decreased diversity which enhanced productivity. Augmenting the supply of another resource, such as water, also increased productivity but without compromising diversity.

Therefore, it may be concluded that, even on well-managed, high-value land, sufficient environmental heterogeneity exists to warrant consideration of more complex mixtures. The trend towards simplification at very high levels of resource input reflects the effect of the resources *per se*, rather than an inherent advantage of simplified swards. This is an important distinction to make, given the current trend towards more extensive, less resource-intensive approaches to livestock production (Haggar and Peel, 1993).

Concluding Comments

Recommendations for humid temperate pasture seedings in North America have tended towards simple, one- or two-species mixtures in recent decades. Recommended mixtures often centre on a single dominant species, such as lucerne in North America or perennial ryegrass in Western Europe

and New Zealand. It was assumed, apparently without validation, that such mixtures were higher-yielding and more stable in commercial pastures. This perception, however, appears to have been based on studies ill-suited to support generalization to landscape-scale applications.

Virtually all available evidence bearing on the question of diversity and stability in the region was founded upon herbage yield rather than animal performance, and on yield without explicit consideration of stability in units of space as well as time of relevance to commercial graziers. Flawed foundations appear to have weakened the link between research-based seeding recommendations and commercial graziers. The preference for simple mixtures in the region may also have been an artefact of the historic reliance on short-duration small-plot studies to assess mixture performance. More recent awareness of landscape-level environmental heterogeneity challenges these historical precedents, and supports the need for a new approach to comparing and recommending mixtures.

A substantive database to directly address the issue of diversity and stability does not appear to be available, particularly involving units and scaling of interest to commercial graziers. None the less, evidence is available to infer greater stability in mixtures that are more complex than those currently recommended in the region. Evidence to support this finding includes the following:

1. The pasture environment is heterogeneous, in both time and space.
2. Species differ in adaptation to parameters known to vary in the pasture environment.
3. Pasture diversification in time and space is the norm in sown pastures, providing clear evidence of the instability of a simple mixture in terms of species composition.
4. Diversification arises from encroachment by unsown species, from spatial sorting and both positive and negative associations or communities among species and from genetic changes within sown species.
5. Changes in species composition buffer against environmental variability, serving to stabilize herbage yield.

Sward diversification or the decline in sown species composition typically evokes at least two responses from researchers. The first is to rationalize frequent reseeding to maintain sown species composition, on the premise that ingressing species will depress yield or animal performance. Evidence has been presented to challenge these perceptions. The second, and perhaps less common, response is to urge reconsideration of the relative value of ingressing vs. sown species under comparable levels of management (various references cited in Peel *et al.*, 1985; Smith and Allcock, 1985; Hopkins, 1986; Frame, 1990, 1991; Hopkins *et al.*, 1990).

At a third and perhaps more fundamental level, however, researchers might wish to consider why unsown species are able to invade commercial swards so routinely. The ready dilution of sown species composition in contemporary swards – and hence the inducement for frequent reseeding, short amortization intervals and reduced profitability – may well be a direct result of the inability of a simple sown mixture to sustain niche occupancy in time and space. The needs of commercial graziers may be better served by research that explicitly acknowleges environmental heterogeneity in time and space, and harnesses the adaptive benefits of more complex mixtures to stabilize performance, prolong stand life and, ultimately, enhance profitability.

Notes

1. A wider range in species composition between simple and complex mixtures would have made a more compelling test of the hypothesis; however, it proved difficult to find literature making wider contrasts over time or space.
2. CLI = Canada Lands Inventory classification system, where Class 1 is the category with highest yield potential.

References

Aarssen, L.W. and Turkington, R. (1985) Vegetation dynamics and neighbour associations in pasture-community evolution. *Journal of Ecology* 73, 585–603.

Alberta Agriculture (1981) *Alberta Forage Manual*. Agdex 120/20-4, Alberta Agriculture, Edmonton.

Bell, G. and Lechowicz, M.J. (1994) Spatial heterogeneity at small scales and how plants respond to it. In: Caldwell, M.M. and Parcy, R.W. (eds) *Exploitation of Environmental Heterogeneity by Plants*. Academic Press, San Diego, USA, pp. 391–414.

Bever, J.D., Westover, K.M. and Antonovics, J. (1997) Incorporating the soil community into plant population dynamics: the utility of the feedback approach. *Journal of Ecology* 85, 561–573.

Blaser, R.E., Skrdla, W.H. and Taylor, T.H. (1952) Ecological and physiological factors in compounding forage seed mixtures. *Advances in Agronomy* 4, 179–219.

Caldwell, M.M. and Pearcy, R.W. (eds) (1994) *Exploitation of Environmental Heterogeneity by Plants*. Academic Press, San Diego, USA.

Campbell, B.D. and Grime, J.P. (1989) A new method of exposing developing root systems to controlled patchiness in mineral nutrient supply. *Annals of Botany (London) (new series)* 63, 395–400.

Clark, E.A. and Poincelot, R.P. (eds) (1996) *The Contribution of Managed Grasslands to Sustainable Agriculture in the Great Lakes Basin*. Food Products Press, New York, USA.

Clark, E.A., Buchanan-Smith, J.G. and Weise, S.F. (1993) Intensively managed pasture in the Great Lakes Basin: a future-oriented review. *Canadian Journal of Animal Science* 73, 725–747.

Clark, E.A., Christie, B.R. and Weise, S.F. (1996) The structure and function of agricultural research. *Canadian Journal of Plant Science* 76, 603–610.

Connell, J.H. and Slatyer, R.O. (1977) Mechanisms of succession in natural communities and their role in community stability and organization. *American Naturalist* 111, 1119–1144.

Cornell University (undated) *Cornell Field Crops Handbook.* Cooperative Extension, New York State College of Human Ecology, and New York State College of Agriculture and Life Sciences, Ithaca, New York, USA.

Crick, J.C. and Grime, J.P. (1987) Morphological plasticity and mineral nutrient capture in two herbaceous species of contrasted ecology. *New Phytology* 107, 403–414.

Dale, M.R.T. (1999) *Spatial Pattern Analysis in Plant Ecology.* Cambridge University Press, Cambridge, UK.

Frame, J. (1990) Herbage productivity of a range of grass species in association with white clover. *Grass and Forage Science* 45, 57–64.

Frame, J. (1991) Herbage production and quality of a range of secondary grass species at five rates of fertilizer nitrogen application. *Grass and Forage Science* 46, 139–151.

Gibson, D.J. (1986) Spatial and temporal heterogeneity in soil nutrient supply measured using *in situ* ion-exchange resin bags. *Plant and Soil* 96, 445–450.

Goldberg, D.E. and Miller, T.E. (1990) Effects of different resource additions on species diversity in an annual plant community. *Ecology* 71, 213–225.

Grace, J.B. and Jutila, H. (1999) The relationship between species density and community biomass in grazed and ungrazed coastal meadows. *Oikos* 85, 398–408.

Grime, J.P. (1987) The C–S–R model of primary plant strategies: origins, implications, and tests. In: Gottlieb, L.D. and Jain, S.K. (eds) *Evolutionary Plant Biology.* Chapman and Hall, London, UK, pp. 371–394.

Grime, J.P. (1994) The role of plasticity in exploiting environmental heterogeneity. In: Caldwell, M.M. and Parcy, R.W. (eds) *Exploitation of Environmental Heterogeneity by Plants.* Academic Press, San Diego, USA, pp. 1–19.

Haggar, R.J. and Peel, S. (1993) *Grassland Management and Nature Conservation.* Occasional Symposium No. 28, British Grassland Society, Cambrian Printers, Aberystwyth, UK.

Halvorson, J.J., Smith, J.L. and Papendick, R.I. (1997) Issues of scale for evaluating soil quality. *Journal of Soil Water Conservation* 52(1), 26–30.

Hodgson, J. (1990) *Grazing Management. Science into Practice.* Longman Scientific and Technical, Longman Group, UK.

Hopkins, A. (1986) Botanical composition of permanent grassland in England and Wales in relation to soil, environment and management factors. *Grass and Forage Science* 41, 237–246.

Hopkins, A. and Hopkins, J.J. (1993) UK grasslands now: agricultural production and nature conservation. In: Haggar, R.J. and Peel, S. (eds) *Grassland Management and Nature Conservation.* Occasional Symposium No. 28, British Grassland Society, Cambrian Printers, Aberystwyth, UK, pp. 10–19.

Hopkins, A., Matkin, E.A., Ellis, J.A. and Peel, S. (1985) South-west England grassland survey 1983. 1. Age structure and sward composition of permanent and arable grassland and their relation to manageability, fertilizer nitrogen and other management features. *Grass and Forage Science* 40, 349–359.

Hopkins, A., Gilbey, J., Dibb, C., Bowling, P.J. and Murray, P.J. (1990) Response of permanent and reseeded grassland to fertilizer nitrogen. 1. Herbage production and herbage quality. *Grass and Forage Science* 45, 43–55.

Hulme, P.E. (1996) Herbivores and the performance of grassland plants: a comparison of arthropod, mollusc and rodent herbivory. *Journal of Ecology* 84, 43–51.

Jackson, R.B. and Caldwell, M.M. (1993) The scale of nutrient heterogeneity around individual plants and its quantification with geostatistics. *Ecology* 74, 612–614.

Jackson, R.B. and Caldwell, M.M. (1996) Integrating resource heterogeneity and plant plasticity: modelling nitrate and phosphate uptake in a patchy soil environment. *Journal of Ecology* 84, 891–903.

Jackson, R.B., Manwaring, J.H. and Caldwell, M.M. (1990) Rapid physiological adjustment of roots to localized soil enrichment. *Nature* 344, 58–59.

Jaynes, D.B. and Colvin, T.S. (1997) Spatiotemporal variability of corn and soybean yield. *Agronomy Journal* 89, 30–37.

Kaye, J.P. and Hart, S.C. (1997) Competition for nitrogen between plants and soil microorganisms. *TREE* 12, 139–143

Marten, G.C. and Jordan, R.M. (1979) Substitution value of birdsfoot trefoil for alfalfa-grass in pasture systems. *Agronomy Journal* 71, 55–62.

Marten, G.C., Ehle, F.R. and Ristau, E.A. (1987) Performance and photosensitization of cattle related to forage quality of four legumes. *Crop Science* 27, 138–145.

Marten, G.C., Jordan, R.M. and Ristau, E.A. (1990) Performance and adverse response of sheep during grazing of four legumes. *Crop Science* 30, 860–866.

Micheli, F., Cottingham, K.L., Bascompte, J., Bjornstad, O.N., Eckert, G.L., Fisher, J.M., Keitt, T.H., Kendall, B.E., Klug, J.L. and Rusak, J.A. (1999) The dual nature of community variability. *Oikos* 85, 161–169.

Morrison, J. (1979) Botanical change in agricultural grassland in Britain. In: Charles, A.H. and Haggar, R.J. (eds) *Changes in Sward Composition and Productivity.* Proceedings, Occasional Symposium No. 10, British Grassland Society, Hurley, UK.

Morrison, J. and Idle, A.A. (1972) *A Pilot Survey of Grassland in S.E. England.* Technical Report No. 10, Grassland Research Institute, Hurley, UK, cited in Morrison (1979).

Munro, J.M.M., Davies, D.A., Evans, W.B. and Scurlock, R.V. (1992) Animal production evaluation of herbage varieties. 1. Comparison of Aurora with Frances, Talbot and Melle perennial ryegrasses when grown alone and with clover. *Grass and Forage Science* 47, 259–273.

OFCC (Ontario Forage Crops Committee) (1962) *Ontario Forage Crop Investigations*. Ontario Forage Crops Committee, University of Guelph, Guelph, Ontario, Canada.

OFCC (Ontario Forage Crops Committee) (1963) *Ontario Forage Crop Investigations*. Ontario Forage Crops Committee, University of Guelph, Guelph, Ontario, Canada.

OMAFRA (Ontario Ministry of Agriculture, Food, and Rural Affairs) (1997) *Field Crop Recommendations 1997–1998*. Publication 296, Queen's Printer for Ontario, Toronto, Canada.

Ontario Department of Agriculture (1940) *Pasture is Paramount for Milk and Meat Production in Ontario*. Circular No. 28, Statistics and Publications Branch.

Ontario Department of Agriculture (1943) *Good Seed Mixtures for Hay and Pasture in Ontario*. Circular No. 64, Statistics and Publications Branch.

Ontario Department of Agriculture (1954) *Hay and Pasture Mixtures in Ontario*. Circular No. 239, 13 pp.

Palmer, M.W. and Dixon, P.M. (1990) Small-scale environmental heterogeneity and the analysis of species distributions along gradients. *Journal of Vegetation Science* 1, 57–65.

Parish, R. and Turkington, R. (1990) The influence of dung pats and molehills on pasture composition. *Canadian Journal of Botany* 68, 1698–1705.

Parks, D.L. (1955) *Successful Crop Production in Eastern Canada*. McClelland and Stewart.

Pearson, C.J. and Ison, R.L. (1987) *Agronomy of Grassland Systems*. Cambridge University Press, Cambridge, UK.

Peel, S. (1979) The effect of botanical composition on the utilized metabolizable energy output of permanent grassland farms. In: Charles, A.H. and Haggar, R.J. (eds) *Changes in Sward Composition and Productivity*. Proceedings, Occasional Symposium No. 10, British Grassland Society, Hurley, UK.

Peel, S., Matkin, E.A., Ellis, J.A. and Hopkins, A. (1985) Southwest England grassland survey 1983. 2. Trends in land use, reseeding and sward composition 1970–83. *Grass and Forage Science* 40, 467–472.

Robinson, G.G. and Dowling, P.M. (1985) The effect of proportion of sown grasses on pasture and animal production from fertilised pastures on the northern tablelands of New South Wales. *Australian Range and Journal* 7, 88–92.

Robinson, S., Clare, S. and Leahy, M. (1990) *Pasture Production*. Publication 19, Ontario Ministry of Agriculture and Food, Queen's Printer for Ontario, Toronto, Canada.

Rosenzweig, M.L. (1995) *Species Diversity in Space and Time*. Cambridge University Press, Cambridge, UK.

Schwinning, S. and Parsons, A.J. (1996a) Analysis of the coexistence of mechanisms for grasses and legumes in grazing systems. *Journal of Ecology* 84, 799–813.

Schwinning, S. and Parsons, A.J. (1996b) A spatially explicit population model of stoloniferous N-fixing legumes in mixed pasture with grass. *Journal of Ecology* 84, 815–826.

Sheath, G.W. and Boom, R.C. (1985) Effects of November–April grazing pressure on hill country pastures. 3. Interrelationship with soil and pasture variation. *New Zealand Journal of Experimental Agriculture* 13, 341–349.

Shewmaker, G.E., Mayland, H.F. and Hansen, S.B. (1997) Cattle grazing preference among eight endophyte-free tall fescue cultivars. *Agronomy Journal* 89(4), 695–701.

Silvertown, J., Holtier, S., Johnson, J. and Dale, P. (1992) Cellular automaton models of interspecific competition for space – the effect of pattern on process. *Journal of Ecology* 80, 527–534.

Silvertown, J., Dodd, M.E., McConway, K., Potts, J. and Crawley, M. (1994) Rainfall, biomass variation, and community composition in the Park Grass Experiment. *Ecology* 75, 2430–2437.

Smith, A. and Allcock, P.J. (1985) The influence of species diversity on sward yield and quality. *Journal of Applied Ecology* 22, 185–198.

Smith, H., McCallum, K. and Macdonald, D.W. (1997) Experimental comparison of the nature, conservation value, productivity and ease of management of a conventional and a more species-rich grass ley. *Journal of Applied Ecology* 34, 53–64.

Sohlenius, B., Bostrom, S. and Sandor, A. (1987) Long-term dynamics of nematode communities in arable soil under four cropping systems. *Journal of Applied Ecology* 24, 131–144.

Stapledon, R.G. and Davies, W. (1928) *Welsh Plant Breeding Station Bulletin Series H* 8, 150–162 (cited in Blaser *et al.*, 1952).

Steinauer, E.M. and Collins, S.L. (1995) Effects of urine deposition on small-scale patch structure in prairie vegetation. *Ecology* 76, 1195–1205.

Sterling, A., Peco, B., Casado, M.A., Galiano, E.F. and Pineda, F.D. (1984) Influence of microtopography on floristic variation in the ecological succession in grassland. *Oikos* 42, 334–342.

Tilman, D. (1987) Secondary succession and the pattern of plant dominance along experimental nitrogen gradients. *Ecological Monographs* 57, 189–214.

Turkington, R. and Harper, J.L. (1979) The growth, distribution and neighbour relationships of *Trifolium repens* in a permanent pasture. II. Inter- and intra-specific contact. *Journal of Ecology* 67, 219–230.

Turkington, R. and Jolliffe, P.A. (1996) Interference in *Trifolium repens–Lolium perenne* mixtures: short- and long-term relationships. *Journal of Ecology* 84, 563–571.

Turkington, R. and Mehrhoff, L.A. (1990) The role of competition in structuring pasture communities. In: Grace, J.B. and Tilman, D. (eds) *Perspectives on Plant Competition*. Academic Press, San Diego, USA, pp. 307–340.

Wiens, J.A. (1989) Spatial scaling in ecology. *Functional Ecology* 3, 385–397.

Wilkins, R.J. (1986) Permanent and sown grasslands: productivity and support energy use. In: *European Grassland Federation General Meeting, Setubal*.

Willard, C.J. (1951) *Forages*. Iowa State College Press, Iowa, USA, pp. 431–447 (cited in Blaser *et al.*, 1952).

7   The Population Dynamics of Pastures, with Particular Reference to Southern Australia

E.C. Wolfe[1] and B.S. Dear[2]

[1] School of Agriculture, Charles Sturt University, Wagga Wagga, New South Wales, Australia; [2] NSW Agriculture, Wagga Wagga, New South Wales, Australia

Introduction

In Australia, as in other continents, agricultural development has resulted in substantial changes to grassland ecosystems. Australian agriculture has passed through several distinct phases since white settlement in 1788 (Table 7.1; Shaw, 1990; Barr and Cary, 1992). In the 19th century, an initial exploration phase was supplanted by exploitation of the Australian landscape for grazing and crop production, and then followed periods of consolidation, amelioration and restoration during the 20th century. Not surprisingly, in response to this development, profound changes occurred in the botanical composition of Australian grasslands (Moore, 1970). Among the original native species and among those species that were either accidentally or deliberately introduced, there have been notable failures and survivors.

In this chapter, a brief historical assessment will be given of the impact of development on grasslands in the agricultural areas of southern Australia, which are characterized by Mediterranean-type (south-western, southern) and temperate (south-eastern) climates; these grasslands have a minimum mean annual rainfall of about 300 mm. Examples of the botanical changes that occurred in response to grazing, plant introduction and fertilization on the tablelands (predominantly non-arable, grazing) and slopes/plains (arable, farming) will be explored in relation to three conceptual models that have been used, worldwide, to describe and explain the interplay of climatic, edaphic and biotic factors on

the dynamics of plant communities. Then, in a series of case-studies, pasture species/varieties that have been notably successful in Australian agriculture will be related to these models, and evaluated in terms of the processes and principles that have underpinned their competitive success. In a concluding section, past experience and experimentation will be considered in relation to the future integration of conservation and agriculture in Australian rural lands.

The Impact of Agricultural Development on the Pastures of Southern Australia

Early in the 19th century, the occupation of rural Australia by British immigrants proceeded slowly but, by 1860, driven in part by the discovery of gold in 1851, land settlement encompassed the south-eastern quarter of Australia from north of Adelaide to beyond Brisbane, as well as part of western Australia (Shaw, 1990). In 1862, New South Wales (NSW) and Victoria were both supporting about 6 million sheep, with Queensland only a little behind. There were many difficulties during this exploration phase (Shaw, 1990; Barr and Cary, 1992), but the availability of suitable grasslands was not one of them (Barr and Cary, 1992).

According to Shaw (1990), the period between 1850 and 1890 marked the heyday of the pastoralists, who developed huge grazing properties with

© CAB *International* 2001. *Competition and Succession in Pastures*
(eds P.G. Tow and A. Lazenby)

Table 7.1. Agricultural development in Australia, 1820–2000.[a]

Phase	Year	Livestock numbers ($\times 10^6$) Sheep	Cattle	Wheat area (ha $\times 10^6$)
Exploration	1820	0.3	< 0.1	0.01
phase	1842	6	< 0.1	0.06
	1851	17	0.2	0.1
Exploitation	1861	21	3.8	0.3
phase	1871	40	4.3	0.5
	1881	65	8.0	1.2
	1891	106	11.1	1.3
Consolidation	1901	72	8.5	2.1
phase	1911	97	11.8	3.0
	1921	86	14.4	3.9
	1931	111	12.3	6.0
	1941	125	13.6	4.9
	1950/51	116	15.2	4.2
Amelioration	1960/61	153	17.3	6.0
phase	1970/71	172	24.4	6.5
Restoration	1980/81	131	25.2	11.3
phase	1990/91	162	23.6	9.2
	1996/97	120	26.8	10.9

[a]The statistics for 1820–1950/51 were taken from Shaw (1990) and the more recent values from ABARE (Knopke *et al.*, 1995) and the Australian Bureau of Statistics.

axes, dams, artesian bores, fences and better techniques of sheep breeding and husbandry. By 1891 there were 62 million sheep in NSW, 20 million in Queensland and 13 million in Victoria (see Table 7.1); between 1860 and 1894, the Australian sheep population had risen from about 20 million to 100 million and cattle from 4 million to more than 12 million (Shaw, 1990).

A combination of factors, notably overstocking, the invasion of pastoral lands by rabbits introduced in 1859, a general economic depression and bank collapses in the 1890s, and a severe drought from 1895 to 1902, arrested and then reversed the pastoral boom. Erosion, pasture degradation and a fall in livestock numbers were the consequences, outcomes predicted by P.E. de Strzelecki. In an 1840 report to Governor Gipps on his travels to the Australian Alps and Gippsland, Strzelecki expressed his concern about the exploitative practices of overgrazing and burning (Hancock, 1972). Meanwhile, the farming of wheat, barley and oats had expanded steadily in area (see Table 7.1), first in South Australia (Meinig, 1954) and Victoria (Barr and Cary, 1992) and then, with the development of a

railway network, in NSW. A common problem was the decline of cereal yields on land that had been farmed for several years (Donald, 1967).

However, not all of the 19th century was characterized by exploitation. The rapid expansion of mechanization and transportation from 1870 and the establishment of agricultural colleges and experimental farms during the 1890s were notable developments in terms of their immediate benefit and future impact.

As outlined by Barr and Cary (1992), the southern Australian wheat industry was rescued from decline at the turn of the century by the new techniques of dry farming, purposeful wheat breeding and superphosphate fertilizer. However, crop yields on the poor Australian soils were still low by world standards and bare fallowing both depleted the soil of organic matter and rendered it liable to wind erosion. Green manuring with oats or lucerne was advocated, but it was not until the 1930s that a technique of ley farming[1] with annual pasture legumes was developed and promoted (Puckridge and French, 1983). Subterranean clover (*Trifolium subterraneum* L.), which had been discovered and promoted unsuccessfully by Amos Howard in the 1890s, was the basis of the ley farming system introduced at Rutherglen Research Station in north-eastern Victoria (Barr and Cary, 1992). Ley farming with annual medics (*Medicago* spp.) was in use in the 1930s at the Roseworthy Agricultural College in South Australia (Callaghan, 1935). Fertilized ley pastures, by setting in train a new succession of botanical changes in response to added fertilizer and N_2-fixation, had the potential to alter markedly the nature of grazed pastures in the farming zones. This topic will be dealt with later in this chapter. Because of the depression and then war, there was little change in on-farm practices or outputs between 1930 and 1950 (Donald, 1967; Gruen, 1990), and land degradation continued. However, investments in agricultural research during this consolidation phase produced some notable discoveries that were to underpin the rapid expansion of improved pastures from 1950 to 1970 on both arable and non-arable sites. These discoveries included successful searches (mainly within Australia (Cocks *et al.*, 1980)) for new varieties of subterranean clover that were intermediate between the mid-season variety Mount Barker (found in South Australia, commercialized in 1906) and the early strain Dwalganup (commercialized in western Australia, 1929); a strain (cv. Hannaford) of annual medic (barrel medic, *Medicago truncatula* Gaertn.) that was commercialized in South Australia in 1938; the development as a sown species of a strain of *Phalaris tuberosa* L. (syn. *aquatica*), a perennial grass that was capable of surviving summer drought (Oram and Culvenor, 1994); the realization that deficiencies of phosphorus and sulphur were widespread in Australian soils, many of which also needed one or more of the minor (trace) elements copper, zinc, molybdenum, manganese, iron and boron (Williams and Andrew, 1970); elucidation of the legume – *Rhizobium* symbiosis (Williams and Andrew, 1970); and the development of a virulent strain of the myxomatosis virus for rabbit control (described by Barr and Cary, 1992).

The above knowledge, allied with the 1950s boom in wool prices, financial incentives for investment in agriculture (Gruen, 1990) and the advent of aerial agriculture (Campbell, 1992), ushered in an amelioration phase in temperate Australian grasslands. According to the account of this phase by Crofts (1997), the area of sown pastures increased from around 5 million ha in 1950 to in excess of 25 million ha by 1970: the pasture area

fertilized annually more than doubled between 1950 (7 million ha) and 1973, when a superphosphate subsidy was removed. These improved pastures consisted mainly of subterranean clover, annual medics, lucerne (*Medicago sativa* L.) and, in the higher-rainfall areas (> 600 mm) of NSW, Victoria and Tasmania, perennial grasses, such as phalaris, perennial ryegrass (*Lolium perenne* L.) and cocksfoot (*Dactylis glomerata* L.).

Finally, during the 1970s, 1980s and 1990s, a reappraisal of pasture development took place. This reappraisal occurred in response to: (i) a worsening cost-price squeeze (Gruen, 1990); (ii) an apparent widespread decline in the productivity of pasture legumes following the occurrence of new disease, pest and weed problems (Gramshaw *et al.*, 1989); and (iii) evidence of widespread land degradation phenomena, such as eucalyptus dieback, soil acidification and salinization (Goldney and Bauer, 1998). The area of sown pastures in Australia has remained static at around 27 million ha since 1970, with declines from 1970 to 1985 in the area fertilized and the rate of fertilizer applied (Gramshaw *et al.*, 1989). Since then, problems of oversupply in the wool industry and changing community attitudes towards conservation have given rise to initiatives that aim to integrate conservation and agricultural production, with the objective of achieving sustainable production systems. This restoration phase (Goldney and Bauer, 1998) has been marked by: a reduction in the sheep population (see Table 7.1); better documentation of the effects of and solutions to the environmental problems created by agriculture; programmes and legislation to protect areas of native grasslands and woodlands; a shift in emphasis from action at the farm level to the catchment level; and improved partnerships between scientists, farmers and the community.

Pasture Dynamics During the Phases of Agricultural Development

Models of pasture dynamics

There are at least three different conceptual models that have been used as a framework for describing and explaining the interplay of climatic, edaphic and biotic factors on the dynamics of pasture communities and for managing these communities. The first two of these models, the Clementsian theory of succession and the state and transition model, were recently discussed fully by Humphreys (1997); these models are illustrated and explained in a following section. A third theory, the competition–stress–disturbance (CSD) model, was developed and used by Grime (1977) to classify plants according to the combination of characteristics they display in response to three primary ecological factors. These factors are: (i) competition (with the high vegetative competitive ability of some species accounting for their dominance); (ii) stress (with certain plant species adapted to and tolerating unproductive conditions); and (iii) disturbance (with some species adapted to grazing disturbance, and with ruderal species possessing an ability to invade and grow in severely disturbed but potentially productive environments). All three theories – succession, state and transition, and CSD – have some usefulness in interpreting the changes in vegetation that have occurred in natural and improved pastures in southern Australia.

Species changes during the exploration and exploitation phases

Moore (1970) outlined the changes in pasture species that occurred over several decades in typical *Eucalyptus* woodland – grassland communities on the slopes and tablelands of southern NSW in response to clearing, higher grazing pressures (from sheep, cattle and rabbits) and the application of superphosphate fertilizer. The original climax vegetation, dominated by tall, warm-season, perennial tussock grasses, such as kangaroo grass (*Themeda triandra* Forsskal), plains grass (*Stipa aristiglumis* F. Muell.) and poa tussock (*Poa labillarderi* Steud.), was presumably well adjusted to the ebb and flow of the native herbivores (kangaroos, wallabies, bird life) and occasional fires. Once sheep and cattle were introduced to the tablelands and slopes in the 1830s–1840s, accompanied by timber-clearing operations, there began a sequential progression (Fig. 7.1) in the botanical composition of the grasslands, towards a disclimax community (or, more correctly, a number of disclimax communities). Such communities contained an array of grazing-tolerant, cool- and warm-season native grasses, together with various naturalized annual grasses and forbs that had been introduced into Australia in agricultural seeds and feeds (see also Garden and

Bolger, Chapter 11, this volume). This pathway of change in plant communities, from a pre-settlement, climax (stable) vegetative state through an unstable continuum of several disclimax stages, where stability was more or less maintained by 'management', was consistent with the original linear succession model first proposed by F.E. Clements in 1916 (Humphreys, 1997). In terms of Clements's theory, the activities of fire, grazing, clearing and fencing opposed the natural successional tendency towards pristine, climax grassland.

Loss of grazing-susceptible plant species, opening the sward to native and exotic invading species, nutrient redistribution and changes in the seasonal extraction and replenishment of soil water are the processes that presumably influenced the outcome of plant competition and the succession of plant communities in Australian grasslands, over time and space. While the main catalyst for the botanical changes that took place was grazing (a disturbance factor, in terms of Grime's CSD model), the evidence for its specific effects is largely anecdotal. Early reports made during the exploration and exploitation phases (Barr and Cary, 1992) indicated that kangaroo grass was abundant and palatable to livestock. Kangaroo grass was sensitive either to defoliation or treading or both; it did not persist wherever sheep or cattle were grazed (Moore, 1970). Subsequently, it was shown in South Africa (O'Connor, 1996) that the persistence of *T. triandra* depended on lax defoliation, which enhanced both the recruitment and the survival of seedlings.

Several other native perennial grasses were protected in part from grazing by mechanisms such as less palatable herbage (for example, *Bothriochloa ambigua* S.T. Blake, red grass) or spiny seeds (*Stipa* spp., spear or corkscrew grasses). On the northern slopes of NSW, three-awned speargrass (*Aristida ramosa* R.Br.), an unpalatable species, became co-dominant with red grass on extensive areas of lightly grazed grasslands (Williams, 1979). However, it has subsequently been shown that *A. ramosa* is sensitive to defoliation and the balance can be shifted back towards more palatable species, notably wallaby grass (*Danthonia* spp.) and subterranean clover, by heavy grazing with flocks of sheep applied strategically in summer–autumn, coinciding with the flowering and seedling establishment of the speargrass (Lodge and Whalley, 1985).

Johnston (1996) noted that the causal relationship between grazing (defoliation, trampling) and grass species composition has not been seriously

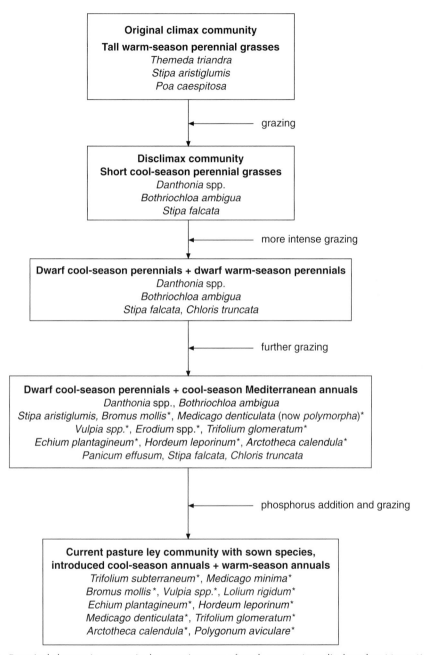

Fig. 7.1. Botanical changes in pasture in the cropping zone of south-eastern Australia, based on Moore (1970) and modified by Dear (1998). Asterisks denote naturalized species, to distinguish them from native species.

challenged. He listed several factors that may have been involved in the replacement of the original native grasses with other grasses; these include changes in plant–soil water and nutrient relations, as well as timber clearing and reduced fire frequency. According to Johnston (1996), the persistence to the present day of certain productive and nutritious C_3 native grasses, notably weeping meadow grass

(*Microlaena stipoides* (Labill.) R.Br.) and the wallaby grasses, may be due, at least in part, to their preference for shade and/or the complementarity of their growth cycle and phenology to the original C_4 grasses, which are taller and better adapted to dry habitats and hot seasons.

Other important components of the degraded grasslands included many grasses, forbs and legumes that became naturalized after their entry to Australia. Sometimes this naturalization followed their accidental introduction in agricultural produce from various destinations along the sea routes from Europe and the Mediterranean region to Australia (for example, *Vulpia* spp., silver grass, *Hordeum leporinum* (Link), barley grass, and *Arctotheca calendula* (L.), Levyns, capeweed). In other cases, plants may have been introduced deliberately as ornamentals (for example, *Oxalis pes-caprae* L., soursob, *Silybum marianum* (L.) Gaertn., variegated thistle, and *Echium plantagineum* (L.), Paterson's curse or 'Salvation Jane' (Michael, 1970)). Most of the strains of subterranean clover found in suburbs of Perth by J.S. Gladstones (1966) entered Australia (accidentally, not deliberately) before 1870, and some annual *Medicago* spp. were naturalized by the early 1900s (Crawford *et al.*, 1989).

Michael (1970) attributed the competitive success of these alien species to their adaptability to disturbed environments and/or the absence of their native pests and competitors. Why they were so successful in competition with the native perennial grasses is open to speculation. A possible reason is that the winter-growing Mediterranean annuals depleted soil water in spring such that the summer-growing perennials were unable to survive the long, dry summers. Another factor may have been selective grazing of the perennials by sheep over summer, thereby reducing the above-ground green material of the perennials and presumably lowering energy reserves in their roots and crowns. In the case of annual legumes, one clear advantage was their capacity to fix N and thrive in the low-nitrogen (N) soils of southern Australia.

Instability of grazed pasture communities over the last half-century

During the last 50 years (the amelioration and restoration phases), the dynamics of pastures became even more complex in response to developments that accompanied or followed the pasture improvement revolution (1950–1970). In addition to the grazing factor, superphosphate application and N_2-fixation by legumes became important agents in determining the direction and pathways of botanical change. Both benefits and problems (Table 7.2) were initiated by pasture improvement, and this period of rapid change was followed by one of reassessment. Consequently, the pathways of botanical change in pasture communities became numerous and complex.

Humphreys (1997) postulates many reasons why the dynamics of grasslands are better described by a 'state and transition' model instead of Clements's succession model. The state and transition model is able to represent a set of multiple pathways and vegetation states occurring in response to several sets of factors. In addition, it is flexible in accommodating the notion of resilience (ability to recover) in an ecological system, whereas Clements's theory is too focused on the concept of a single climax state and on grazing management as the dominant factor that drives succession.

In Fig. 7.2, adapted from Lodge and Whalley (1989) and Garden *et al.* (1996), the state and transition model is used in an attempt to summarize the nature and timing of the main changes that have occurred in grasslands on the tablelands of NSW, particularly during the last half-century. At one extreme of the time–management continuum, native and naturalized pastures were ploughed up, sown with perennial grasses and clovers and fertilized; initially, such pastures were dominated by the sown legumes until soil N levels from rhizobial activity were sufficient to allow the grasses (and weeds) to compete effectively with the legumes. The botanical progression from initial legume dominance towards eventual grass (or other non-legume) dominance was dependent on a minimum annual rate of superphosphate application (to stimulate and favour legume growth) and grazing (to enhance the transfer of fixed N to the associated species). For example, in grass–white clover pastures on the northern tablelands of NSW, the clover-dominant phase was intensified, but the onset of grass dominance was hastened following high annual rates of superphosphate (375 kg ha^{-1}) instead of intermediate (125–188 kg ha^{-1} year^{-1}) rates (Wolfe and Lazenby, 1973). The cessation of superphosphate application to perennial ryegrass–white clover pastures in the same locality resulted in a loss of the sown species and a corresponding increase in the proportion of the native red grass (Cook *et al.*, 1978).

Table 7.2. An updated list of the factors that are, or may be, associated with the decline of pasture legumes in southern Australia.

Medic decline, South Australia (Carter *et al.*, 1982)
   Reduced spraying to control insect pests of pastures (earth mites and lucerne flea)
   Reduced application of superphosphate fertilizer to pasture
   Spread of sitona weevil
   Increased cropping intensity and consequent grazing pressure
   Poor grazing management and fodder conservation practices
   Increased use of herbicides in the cropping phase of the rotation
   Reduced undersowing of medic into cereal crops
   Rapid spread of pasture aphids (spotted alfalfa aphid, blue-green aphid, pea aphid) after introduction in 1977/78
   Increased use of nitrogen fertilizers on crops (less need for medic ley)
   Apathy and despondency concerning value of medics

Medic decline, South Australia (Denton and Bellotti, 1996)
   Drought
   Sulphonylurea herbicides (suspected)
   Higher populations of root lesion nematodes
   Zinc deficiency
   Rhizoctonia solani

Subterranean clover decline, southern slopes, NSW (Hochmann *et al.*, 1990)
   Suboptimal supply of phosphorus
   Root rot associated with *Phytophthora clandestina*
   Soil acidity

White clover decline, northern tablelands, NSW (Hutchinson *et al.*, 1995)
   Climatic stress, particularly the effect of January–March (late summer) rainfall on stolon survival
   Set-stocking at high rates, encouraging the presence of competitive annuals
   Soil nitrogen (N) build-up (of secondary importance)

Annual legumes, Australia (Gramshaw *et al.*, 1989)
   Economic trends (lower returns for beef and wool compared with cropping, removal of fertilizer subsidies, farm cost inflation during the 1970s)
   Insect pests and diseases (lucerne pathogens, clover scorch, root rots of *Trifolium* spp., effects of spotted alfalfa and blue-green aphids on lucerne, medics and other pasture legumes)
   Land degradation (soil acidity, salinization, compaction)
   Changes in crop production techniques (more frequent cropping, longer cropping sequences, partial substitution of N fertilizers and pulses for pasture legumes)

High-rainfall zone and annual zone pastures, Australia (Wilson and Simpson, 1993)
   No regular and standardized surveys of the state of pastures, but *ad hoc* surveys have indicated suboptimal clover content and the significant presence of unsown annual grasses, particularly in legume leys in the cropping zone
   Quantitative evidence is lacking on both the extent and the nature of change in botanical composition and productivity
   Factors that are claimed to be associated with low legume content, such as stocking rate, climatic variation, pasture age and management, are listed and reviewed. The list is similar to that of Carter *et al.* (1982) at the top of this table

At the other extreme of the management continuum, large areas of native pastures on the NSW tablelands were aerially sown with subterranean clover and treated with superphosphate from time to time. Introduced perennial grasses were usually not sown from the air; if they were, their establishment was frequently unsuccessful (Wolfe, 1968) unless the guidelines developed by Campbell (1992) were used. These guidelines were to sow the grasses when rainfall is likely to be effective, use herbicide to minimize competition from the existing vegetation and minimize seed theft by ants. In contrast to the difficulties associated with grass establishment, the establishment of aerially sown

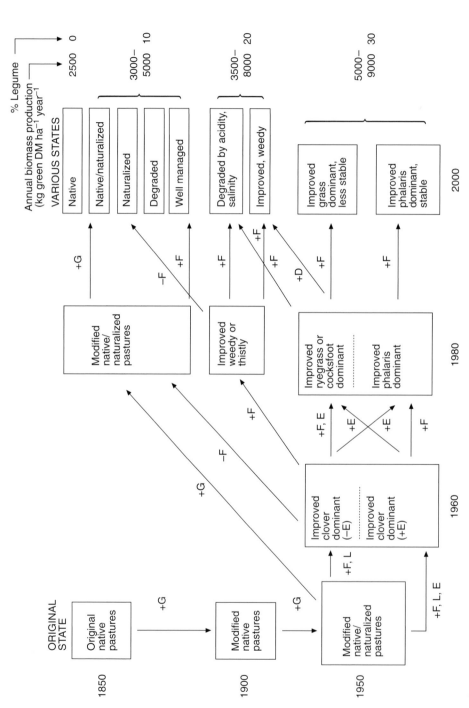

Fig. 7.2. The application of the state and transition model to the generalized pathways of botanical change that have occurred in permanent pastures, south eastern Australia. The boxes are 'states', and arrows represent the general direction of transition due to the main agents of change (G, grazing; F, fertilizer; L, legumes; E, exotic grasses; D, drought). Depending on the circumstances, a transition may or may not be reversible.

legumes (subterranean clover and/or white clover) was generally satisfactory. Further, so long as phosphate applications were maintained, legumes became dominant (Willoughby, 1954) until the pastures were invaded by N-loving species, such as annual grasses and forbs (Rossiter, 1964) or thistles (Michael, 1970; see also Fig. 7.2). Many such species were more aggressive for moisture, light and nutrients and/or were better able to exploit the soil conditions. Thus, many of the native perennial grasses were suppressed or eliminated by competition or by high grazing pressure.

There were several intermediate scenarios between these extremes. For example, less palatable tussocky grass species, such as poa tussock (*P. labillarderi* Steud.), were present in low numbers on the southern tablelands of NSW. They increased in frequency and size over time in response to the N fixed by the introduced legumes (Fisher, 1972); then, heavy grazing pressure on the palatable species in the gaps between the tussocks exacerbated tussock dominance, response to fertilizer declined and the swards deteriorated into unproductive, N-deficient grasslands. Much the same sequence occurred on parts of the north-western slopes of NSW, where three-awned spear-grass became dominant (Williams, 1979), and on the central tablelands, where the aggressive and indigestible serrated tussock (*Nassella trichotoma* (Nees) Arech.) colonized tracts of non-arable country (Campbell, 1998). This latter species, introduced from South America, became a serious weed of the grasslands of South Africa and New Zealand, as well as Australia (McLaren *et al.*, 1998).

On many soil types, the prolific growth of pasture legumes that was evident in the early years of fertilized pastures in most areas of southern Australia did not continue. There were several reasons for this. From the 1970s, the general malaise of 'pasture legume decline' (Carter *et al.*, 1982) was evident, together with progressively developing incidence of specific problems, such as soil acidity (Cregan and Scott, 1998), and the occurrence of certain pests and diseases of pastures (Panetta *et al.*, 1993). Some of the possible biological, nutritional, economic and social factors involved in pasture decline are listed in Table 7.2.

Many of the environmental phenomena leading to change were episodic rather than progressive. Examples of episodic events include severe droughts, which occurred in most decades (leading to the loss of drought-susceptible, introduced perennials, such as perennial ryegrass, cocksfoot and white clover); the apparently accidental introduction of three pasture aphids (the spotted alfalfa aphid (*Therioaphis trifolii* (Morrell) f. *maculator*), the blue-green aphid (*Acyrthosiphon kondoi* Shinji) and the pea aphid (*Acyrthosiphon pisum* Harris)) into Australia in the late 1970s, seriously reducing the productivity and persistence of lucerne and annual medic pastures (Panetta *et al.*, 1993); and the development of new races of plant pathogens, such as *Phytophthora clandestina* races that attack subterranean clover (Dear *et al.*, 1993b).

In summary, both progressive and episodic occurrences produced in Australian grasslands a range of pasture states (see Fig. 7.2). The potential and actual changes, including those that occurred before the amelioration phase, are represented adequately by the state and transition model, which conforms fully with the ecological principle of succession (progressive change) but which avoids the inadequacies (linearity, inflexibility) that are embodied in the concept of a linear succession towards a monoclimax (Humphreys, 1997).

Ley pastures in the cropping belt

Ley pastures in Australian croplands (Puckridge and French, 1983) represent a different situation from that of the permanent pastures used for grazing sheep and cattle on the coast and tablelands of south-eastern Australia. In such leys, the opportunity is available to resow pasture legumes, and the time frame of the pasture phase between crop cycles is short (typically 1–5 years).

A useful conceptual representation of the population dynamics of the species in ley pastures is Grime's (1977) triangular CSD model (Fig. 7.3) of the competition, stress and disturbance factors that influence the distribution and abundance of plant species. As noted by McIvor (1993), pasture and grassland species are located near the centre of the CSD triangle; competition (C), stress (S) and disturbance (D) are all important but none is overwhelming. The annual legumes used in leys are ruderal (R) species that tolerate disturbance. Ruderal species (C-R, S-R or C-S-R plants with short life cycles, high reproductive effort or other strategies for dealing with disturbed sites) take advantage of disturbance to colonize; they are favoured in productive croplands, which are frequently disturbed by crop phases (cultivation, herbicide application) and by grazing during the pasture phase.

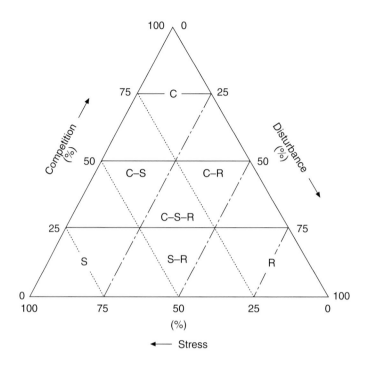

Fig. 7.3. Adapted from Grime's (1977) competition–stress–disturbance model representing the factors of competition, stress and disturbance on plants, and the location of plants that possess primary strategies (C, competitors; S, stress-tolerators; R, ruderal species that are adapted to disturbance) and secondary strategies (e.g. C–S, C–S–R) of adaptation to these influences.

At the other extremes of Grime's model, stress-tolerators (S – low growth rates and low reproductive effort) are favoured on less disturbed, less productive locations (e.g. acid soils). In contrast, competitors (C – high growth rate, low reproductive effort) are favoured in undisturbed, productive situations (such as abandoned cropland or lightly grazed pastures). In permanent pastures or long leys in Australian agriculture, pasture species that exemplify success are tolerant of stress (for example, drought) and disturbance (grazing) and/or, in favourable locations, are efficient competitors for light and nutrients. Examples of such plants are phalaris, lucerne and, in locations with high annual rainfall (> 750 mm), perennial ryegrass.

However, in any community, the CSD balance varies both spatially and temporally (McIvor, 1993). This is particularly so in pasture leys where the annual species are examples of C–S–R plants, which are adapted variously to competition (for moisture, light and nutrients), stress (acid soils, seasonal water-logging) and disturbance (ruderal species are capable

of profuse seeding and several have dormancy or other seed conservation mechanisms and/or they are adapted in some way to grazing). A list of some common components of ley pastures and an overview of their CSD tolerance are given in Table 7.3. The population dynamics of subterranean clover and its companion species will be discussed in a later section.

Problems have been reported in the ley pasture system in Australia. They are due in part to landowners reacting to higher economic returns from crops than from livestock (Reeves and Ewing, 1993) and to the apparent decline in the legume content of ley pastures (Carter *et al.*, 1982; Hochmann *et al.*, 1990; Gramshaw *et al.*, 1989). Some of the reports of legume decline and the factors that may be or are associated with it are listed in Table 7.2. Solutions have been found in liming (Cregan and Scott, 1998), improved agronomy (Hochmann *et al.*, 1990) and the selection and release of tolerant cultivars of annual legumes (Collins and Stern, 1987). A promising technique – winter cleaning with herbicides (a combination of

Table 7.3. Characteristics of some common temperate pasture species in relation to their tolerance of competition (C), stress (S) and disturbance (D).

Species		Characteristics
Subterranean clover *Trifolium subterraneum*	C	May be outcompeted by annual grasses for P and for light; subclover effective in suppressing some weeds in their rosette stage (for example, St John's wort)
	S	Tolerant of Al and Mn toxicities in acid soils; *yanninicum* subspecies tolerates waterlogging; *brachycalycinum* subspecies is adapted to neutral–alkaline soils
	D	Outstanding tolerance of grazing – seedlings unattractive to livestock, plus defoliation in winter stimulates inflorescence production, burr burial and seed yield; prolific seed production, and effective seed conservation mechanisms (embryo dormancy, hard seed)
Annual medics *Medicago* spp.	S	More effective than subclover in tolerating water stress and Zn deficiency, but sensitive to acid soils
	D	High levels of hard seed guarantee persistence in semi-arid localities
Serradella *Ornithopus compressus*	S	Very tolerant of Al toxicity in acid soils, deep rooting allows it to access Zn at depth in sandy soils
	D	Prefers sandy soils, coverage of seed by drifting sand is important in the breakdown of hard seeds
Balansa clover *T. michelianum*	S	Tolerant of waterlogging
	D	High levels of hard seed
Persian clover *T. resupinatum*	S	Tolerant of waterlogging, heavy-textured soils
	D	Prolific seeder, high levels of hard seed
Lucerne *M. sativa*	C	Deep-rooted, crown habit – competes well for water
	S	Some cultivars susceptible to insects, root and crown rots
	D	Adapted to rotational grazing
Perennial ryegrass *Lolium perenne*	C	Competes well for light with white clover but shallow-rooted habit means susceptibility to Australian droughts
	D	Free-tillering, recovers well from grazing; will re-establish well from seed
Phalaris *Phalaris aquatica*	C	Allelopathic, most persistent of the introduced perennial grasses due to deep rooting habit
	S	Susceptible to soil acidity
	D	Strong tuber, resists overgrazing
Annual ryegrass *Lolium rigidum*	C	Competes well for light and nutrients with clover, crops
	S	Very tolerant of Al and Mn toxicities in acid soils
	D	Prolific seeding, with dormancy mechanisms
Barley grass *Hordeum leporinum*	C	Effective scavenger for soil P
	D	Prolific seeder, seeds unattractive to livestock
Silvergrass *Vulpia* spp.	C	Allelopathic
	D	Herbage unattractive to livestock; prolific seeder
Capeweed *Arctotheca calendula*	D	Mature plants unattractive to livestock; prolific seeding and complex seed dormancy mechanisms
Paterson's curse *Echium plantagineum*	C	Tall when mature, shading out competitors
	D	Not eaten by cattle and horses at any stage, or by sheep when mature
Saffron thistle *Carthamus lanatus*	D	Prolific seeding; herbage not attractive to livestock
Skeleton weed *Chondrilla juncea*	S	Deep-rooting and therefore drought-tolerant; susceptible to biocontrol agents
	D	Prolific seeding

simazine and paraquat) – has been developed to reduce weed content and boost the legume content of ley pastures (Thorn, 1992). Further improvements in the effectiveness of legumes in ley pastures are likely through the increased use of lucerne as a component of leys (Peoples *et al.*, 1998) and the adoption of pasture monitoring protocols to help farm managers identify key indicators of pasture legume production and to guide management accordingly (Crosby *et al.*, 1993; Paul, 1999).

However, there are additional threats, such as the evolution or introduction of new weed ecotypes or new disease races and the development of herbicide resistance in weeds, that are likely to cause problems in legume leys. The failure of Group A and Group B selective herbicides to control annual ryegrass (*Lolium rigidum* Gaudin), a common component of wheat-belt pastures before the advent of these selective herbicides, is one issue that is now causing particular concern (Powles *et al.*, 1997). Another concern with herbicides is the deleterious effect of some residual herbicides on pasture legumes, particularly on alkaline soils, where the degradation of these herbicides is slow (Gillett and Holloway, 1996).

Summary

During the last two centuries, there have been profound shifts in the botanical composition of temperate pastures in Australia, and the pace of change has increased in recent decades. Because of the complexity of the interactions between pasture species and environmental phenomena, no one model is available to explain the dynamics of pastures. However, the state and transition model is a useful and flexible way of representing and understanding pasture dynamics. The focus of the CSD model is on the short-term response of individual plant species to these three factors and thus it complements the plant community focus of the state and transition model. The use of these two models is recommended as a way of understanding both the ecology and agronomy of grasslands/pastures. They are used in the next section in case-studies that illustrate the success of some pasture species in the Australian environment.

Successful Australian Pasture Species

Pasture components – exotic or native?

A recent estimate (Hill, 1996) of the potential areas of adaptation in Australia for the main sown pasture species – all of which were originally introduced into Australia – is given in Table 7.4. This vast potential may be compared with another recent estimate, that of the relative importance of pasture types in NSW (total area of sown pastures 5–6 million ha; Australia 25–30 million ha) (Table 7.5),

derived from an Australian temperate pastures database (Pearson *et al.*, 1997). The current distribution of introduced pasture species is broadly depicted in Fig. 7.4 (after Moore, 1970; Hill, 1996). The main features of this map are inland aridity barriers to the occurrence of all species, together with other boundaries (climatic, soil pH) that are applicable to the distribution of annual legumes (Donald, 1970). However, despite the emphasis in Australian agronomy on exotic pasture species, native pastures still cover a much larger area than that of introduced plants (see Table 7.5). While pasture legumes must be based on the locally adapted variants of exotic germ-plasm, there is a strong case for the use of management options that utilize adapted native grasses and for the inclusion of a wider range of perennial grasses, particularly C_4 species, for pastoral use (Johnston *et al.*, 1999). Johnston *et al.* (1999) questioned the amount of research and development effort that has gone into replacing indigenous grasses with exotic introductions, many of which fail to persist over the frequent droughts that characterize the Australian climate. They reviewed the evidence for the persistence, productivity and nutritive value of several species of native grasses in grazed pastures, and argued for strategies that utilize adapted, palatable grasses, such as wallaby grass, in low-input situations, or even C_4 native grasses in areas that are prone to hydrological imbalance. Importantly, Johnston *et al.* (1999) acknowledged the folly of a 'one-or-the-other' philosophy, compared with an approach that achieves complementarity between a low-input, conservative approach to pasture management and the high-input, exotic approach to pasture improvement, which has produced notable gains in the productivity of Australian agriculture.

In the winter-dominant rainfall areas of southern NSW and northern Victoria, the presence of a persistent perennial grass in permanent, grazed pastures appears essential for several reasons. Annual pastures in such areas have a potential midwinter imbalance in the supply (relatively high) and demand (low) for water and nitrate in the soil profile (Johnston *et al.*, 1999). A persistent perennial component increases the pre-winter soil water deficit (for example, 135 mm for annuals, 210 mm for perennials (Whitfield, 1998)) and thereby helps overcome this potential imbalance. Minimizing the flows of water beyond the plant roots counters dryland salinity (Passioura and Ridley, 1998) and reduces the rate of soil acidification (Ridley *et al.*,

Table 7.4. Estimated potential areas of adaptation for nine temperate annual and perennial pasture species for Australia (Hill, 1996).

Pasture species	Area (million ha)	% of freehold + leasehold land
South-eastern Australia		
Subterranean clover	60.8	67.8
Balansa clover	23.7	26.5
Persian clover	30.7	34.3
Barrel medic	27.0	30.1
Serradella	37.4	41.7
White clover	20.7	23.0
Lucerne	86.4	96.4
Perennial ryegrass	19.4	21.6
Phalaris	33.9	37.8
Total, south-eastern	89.6	100.0
South-western Australia		
Subterranean clover	20.8	94.0
Balansa clover	4.9	22.3
Persian clover	5.8	26.0
Serradella	7.2	35.6
Barrel medic	17.1	77.3
White clover	0.2	1.0
Lucerne	9.3	42.1
Perennial ryegrass	< 0.2	< 1.0
Phalaris	2.7	12.0
Total, south-western	22.1	100.0

Table 7.5. The relative importance (on an area basis) of pasture species, genera and types in NSW, 1995 (from Pearson *et al.*, 1997).

Description	Percentage by area
Unimproved native pasture	62.5
Improved (fertilized) native pasture	16.8
Subterranean clover	14.8
White clover	6.0
Lucerne	4.5
Phalaris	3.4
Cocksfoot	2.2
Annual medics	2.1
Perennial ryegrass	1.5
Tall fescue	1.1
Annual ryegrasses	0.4
Serradella	0.2

1999). Such differences are meaningful to the achievement of hydrological balance, particularly between the 600 and 800 mm average annual rainfall isohyets in northern Victoria and southern NSW. Soil degradation is common in these areas due to a combination of factors (recharge, salt loads in the soil profile, poorly buffered soils) (Passioura and Ridley, 1998).

The better water use of palatable perennial grasses enhances livestock production compared with pastures based on annuals or poorly persistent perennials (Axelsen and Morley, 1968). Furthermore,

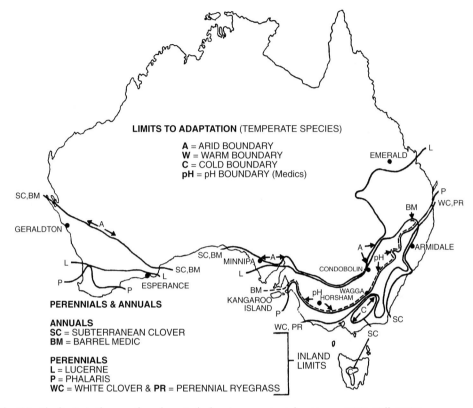

Fig. 7.4. The limits to the growth and survival of pasture species (after Moore, 1970; Hill, 1996).

a persistent perennial grass component in pasture prevents or reduces the incidence of undesirable weeds, such as thistles (Michael, 1970), and the ingress of unpalatable native and introduced tussock grasses, as discussed earlier.

Case-studies

Some case-studies of successfully introduced pasture plants are reviewed below. Successful pasture species in Australia are those which, in our environment, are stable (resistant to change), resilient (able to recover) and productive. Case-studies of a number of introduced pasture plants that have been successful in the Australian environment are relevant to the future ideal of sustainable production. They demonstrate some principles that apply to the success or failure of desirable pasture species and highlight lessons that should have been learnt. The dynamics of phalaris swards and subterranean clover pastures are treated in most detail, reflecting their success and greater importance in Australian agriculture than is the case elsewhere.

Phalaris – a persistent grass in Australia

The 'Australian' commercial cultivar of phalaris is, according to experiments (e.g. Hill, 1985) and experience (Watson, 1993), the most persistent of the temperate perennial grasses that have been introduced and sown in Australia. In southern NSW and Victoria, varieties of phalaris are capable of reliable persistence down to a median annual rainfall of 500 mm, whereas perennial ryegrass will persist only where the rainfall is at least 700 mm. Between these isohyets, where phalaris and, to a lesser extent, cocksfoot are the only reliable perennial grasses, there is a considerable area of land (see Fig. 7.4) at risk due to land degradation. Even in wetter tableland localities where more than 700–800 mm rainfall is received annually, phalaris has survived occasional droughts that have killed cocksfoot, tall fescue (*Festuca arundinacea* Schreb) and perennial ryegrass (FitzGerald *et al.*, 1995).

A deep-rooted and prostrate habit, dense tillering, partial summer dormancy and the presence of a large underground rhizome are features associated with the outstanding persistence of the original 'Australian' strain of phalaris (Oram and Culvenor, 1994). The persistence of 'Australian' phalaris and the importance of stocking rate was illustrated by Hutchinson (1992), who compared over 28 years the presence of phalaris with annual grasses in pastures grazed by sheep on the northern tablelands of NSW (Fig. 7.5). The stability and resilience of phalaris following grazing and drought, at least at low (ten sheep ha^{-1}) and medium (20 sheep ha^{-1} in earlier years, then 15 sheep ha^{-1}) stocking rates (see Fig. 7.5), contrasts with the death of white clover (Hutchinson *et al.*, 1995; Hutchinson and King, 1999), perennial ryegrass, tall fescue and cocksfoot in improved pastures on the northern tablelands in droughts such as those that occurred in 1965, 1980–1982 and 1994 (FitzGerald *et al.*, 1995). Similar observations on the susceptibility of perennial grasses to high stocking rates and drought have been made in long-term grazing experiments and on farm paddocks on the central tablelands (Kemp and Dowling, 1991) and the southern tablelands (Axelsen and Morley, 1968) of NSW. The ecological importance of a relatively stable perennial grass component and the high cost of re-establishing this component if it is lost make it important that persistence, rather than 'productivity', be the first priority in selecting grass cultivars.

Many farmers are reluctant to grow phalaris (Barr and Cary, 1992) because of perceivable but avoidable management problems (palatability, dominance, animal disorders) and other beliefs (establishment difficulty) associated with growing

the plant. There are also some qualifications that need to be made to the use of phalaris to create relatively stable improved pastures.

First, the newer, winter-active cultivars of phalaris released during the 1960s and subsequently (cvs Sirocco, Sirolan, Sirosa) have been found to be up to 50% less persistent, measured as basal cover, than the 'Australian' phalaris cultivar (Culvenor and Oram, 1992). According to Oram and Culvenor (1994), 'Australian' phalaris has a more prostrate, densely tillered habit than the newer, winter-active cultivars (Sirosa, Sirolan), and it has more capacity for rhizomatous spread; for these reasons it is suited to heavy grazing. Selection in grazed swards for basal cover, which has a higher heritability than the spreading ability of spaced plants (Oram and Culvenor, 1994), is a potential way of improving the persistence of future cultivars of phalaris.

Secondly, as reported in Oram and Culvenor (1994), phalaris was found to be intolerant of high soil aluminium (Al) levels, which, together with a high availability of soil manganese, may adversely affect plants in highly acid soils (pH$_{Ca}$ < 5.2) (Cregan and Scott, 1998). While a build-up in soil Al levels may be constrained by the ability of well-established, deep-rooted phalaris to capture leached nitrate, attempts to establish phalaris on soils that are already acid may be thwarted by the sensitivity to Al of the root growth of the establishing phalaris seedlings, particularly if phalaris is competing with a companion weed species that is tolerant of high acidity. The effect of differential tolerance to Al on competitive relationships in perennial grass–annual

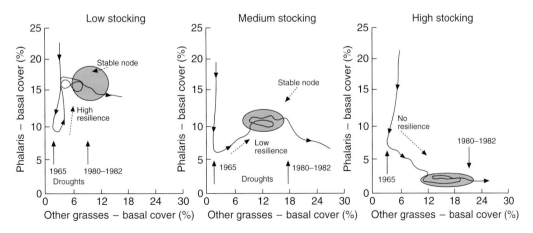

Fig. 7.5. The stability and resilience of phalaris from 1964 to 1991 (inclusive) on the northern tablelands of NSW (after Hutchinson, 1992).

ryegrass binary mixtures 24–36 weeks after sowing was shown in a recent pot experiment (Rubzen, 1996; Table 7.6). The grasses used were phalaris (cv. Sirosa, sensitive to Al) and cocksfoot (cv. Porto, highly tolerant), and the ryegrass ecotypes were either collected from an acid soil site where Al was present at potentially toxic levels (A-ryegrass) or from an alkaline soil site that contained no free aluminium (B-ryegrass). In the alkaline soil pots, both ryegrasses suppressed phalaris or cocksfoot, but in acid soil the competitive outcomes were determined by the level of tolerance of each of the grasses to aluminium: the A-ryegrass ecotype suppressed the sensitive phalaris but A-ryegrass was suppressed by the highly Al-tolerant cocksfoot cultivar. These results indicate that the difficulty of establishing a sensitive plant type on acid soils may be exacerbated by the presence of an adapted competitor. Sources of aluminium tolerance for phalaris are currently being used in a breeding programme to produce a tolerant cultivar (Oram and Culvenor, 1994).

In summary, phalaris is the most persistent of the perennial species introduced into Australian agriculture. However, due to the constant disturbance of grazing, successional change in perennial pastures is consistently directed towards annualization (Hutchinson and King, 1999). According to Hutchinson and King (1999), the loss of sown and palatable perennial grasses can be explained 'as a balance between their strong competitive ability under conservative stocking *versus* their vulnerability to increased levels of disturbance from the combination of high stocking rates and grazing preference'. Stresses, such as drought and poor fertilizer management (Cook *et al.*, 1978; Hutchinson and King, 1999), exacerbate this vulnerability.

Lucerne, and competition between the components of lucerne mixtures

In southern Australia, lucerne (*M. sativa* L.) has an area of potential adaptation that is greater than that for any other pasture species (Hill, 1996; see also Fig. 7.4). In practice, the wider adoption of lucerne has been constrained by its susceptibility to acid soils, the need for a rotational grazing regime to ensure its survival (Leach, 1978; Lodge, 1991), the incidence of insect pests and disease (Lodge, 1991) and the attitudes of farmers and graziers, who perceived lucerne to be a special-purpose pasture rather than an all-rounder. Hence, lucerne was usually sown in 'hay paddocks' as the only component of a pasture. However, there is advocacy of lucerne as a component of permanent or semi-permanent grass–clover pastures and with subterranean clover or medic in ley pastures, as a means of enhancing livestock production (Wolfe *et al.*, 1980), N_2-fixation (Peoples *et al.*, 1998) and/or water extraction from the soil profile

Table 7.6. Relative changes in the yield[a] of the components[b] of 50 : 50 binary mixtures (replacement series design) compared with the yield of each component in monoculture (Rubzen, 1996).

Soil treatment in pot	Binary mixtures	
Acid soil (pH 3.7, 2.3 µg ml Al)	A-ryegrass/phalaris +87% > −82%	B-ryegrass/phalaris −50% < +39%
	A-ryegrass/cocksfoot −37% < +40%	B-ryegrass/cocksfoot −85% < +86%
Alkaline soil (pH 8.2, 0.01 µg ml Al)	A-ryegrass/phalaris +73% > −70%	B-ryegrass/phalaris +69% > −65%
	A-ryegrass/cocksfoot +68% > −77%	B-ryegrass/cocksfoot +72% > −67%

[a] Regrowth measured from the second to the third harvest, i.e. from 24 to 36 weeks after sowing.
[b] Grasses were cocksfoot (cv. Porto): ED Al_{50} value[c] = 272 µM, very highly tolerant
phalaris (cv. Sirosa): ED Al_{50} value = 28 µM, sensitive
Ryegrasses were A-ryegrass: ED Al_{50} value = 188 µM, highly tolerant
B-ryegrass: ED Al_{50} value = 21 µM, highly susceptible
[c] ED Al_{50} value is the equivalent dose (concentration) of aluminium in solution that would produce a 50% toxic response (in this case, a 50% reduction in root extension compared with the nil-effect control).

(Crawford and Macfarlane, 1995; Dear, 1998). This potential warrants a short consideration of the dynamics of lucerne mixtures.

In permanent pastures, lucerne is not a particularly compatible component when associated with a perennial grass. Either the grass competes strongly for moisture with lucerne during summer and autumn (Wolfe and Southwood, 1980), or the rotational grazing management that is essential for the plant's survival (McKinney, 1974; Leach, 1978) does not suit the growth/survival of the companion grass.

In ley pastures, too, there is incompatibility between the components of the recommended lucerne–annual legume mixtures. Annual legumes, such as subterranean clover, are usually grown with lucerne to provide winter feed when lucerne growth is poor. These lucerne mixtures are frequently unstable (Leach, 1978) and, in lower-rainfall environments (<500 mm annual rainfall), gaps develop between the lucerne plants. The gaps are either bare or, when soil N levels increase, occupied by opportunistic annual grasses, such as *Vulpia* spp. Surface soil erosion is encouraged, along with high levels of dust contamination in the wool. Wolfe and Southwood (1980) found that lucerne grew best with Geraldton, the least productive of three subterranean clovers used, and that seed yield of subterranean clover could be increased by increasing the lucerne row spacing. More recently, it has been shown that lucerne can reduce subterranean clover persistence in two ways: reducing the seed set of the subclover and reducing the establishment of clover seedlings in autumn. Lucerne dries out the soil surface rapidly during clover seedling germination and establishment, particularly when there is an early autumn break and temperatures and evaporation rates are high (Dear and Cocks, 1997). This results in a smaller proportion of the seed pool successfully establishing (Dear, 1998).

Hence, while lucerne is a valuable pasture species, its use has been somewhat constrained, due in part to its need for rotational grazing and to its incompatibility with other pasture components. The improvement of relationships between lucerne and companion species is a worthwhile topic for future research.

The success of subterranean clover and other components of ley pastures

CHARACTERISTICS OF SUBTERRANEAN CLOVER. In Mediterranean-type climates, subterranean clover is a species that is well adapted to continuous grazing

with livestock (Collins, 1978). During summer, the pool of hard seed (seed with an impermeable testa) is partially protected from grazing by burr (pod) burial, but losses of up to 50% of the seed bank over summer have been recorded (Dear and Jenkins, 1992). During winter, when grazing pressure is high (supply of green herbage < demand for green feed), the continued close grazing of the prostrate herbage stimulates seed production in spring (Collins, 1978). During spring, low grazing pressure on green herbage (supply > demand) and burr burial protect the developing inflorescences. The only time when subterranean clover is vulnerable to grazing is at the seedling stage; fortunately, livestock avoid grazing the young seedlings.

In spite of the adaptation of subterranean clover to grazing, pastures commonly contain a number of other vigorous sown and invading species, with the sown clover component comprising 30% or less of the pasture biomass. The nature of competition between subterranean clover and other species and between strains of subterranean clover is important from several perspectives. These include the selection of cultivars that are more vigorous and successful in environments where clover decline is occurring and the need to replace oestrogenic strains (high-formononetin strains, which cause ewe infertility) with low-formononetin cultivars. An account of the key population characteristics that determine the success of the species provides an ideal introduction to a consideration of the relationships of subterranean clover with its competitors.

Rossiter (1966) reviewed the characteristics that seemed to be associated with the long-term success of strains of subterranean clover, namely their survival/productivity over a number of seasons or crop/pasture cycles. He listed the seed-producing capacity of a strain as an important determinant of success or failure. However, it is our experience with testing hundreds of subterranean clover strains and crossbreds over several years in NSW that, rather than inherent differences between strains in their potential seed production, success differs only according to how well the fit is between their inflorescence development (triggered by the temperature and light regime (Archer *et al.*, 1987)) and the soil moisture profile in spring. For maximum seed production of subclover (similar to annual medic, about 200 g m^{-2} (Wolfe, 1985)), moisture conditions must be non-limiting during the flowering and seed-development phases (Collins, 1981; Blumenthal and Ison, 1993), a total of about 70 days. This ideal is rarely achieved in the Australian environment, and a target seed production of 60 g m^{-2} is sufficient for a high-quality subclover pasture (Dear *et al.*, 1993a). A practical guideline for optimizing the balance between spring herbage production (later-maturing cultivars produce more spring herbage) and the amount of seed is to choose a cultivar that flowers and sets seed just before soil moisture usually runs out in spring.

The strong relationship between rainfall and time to maturity of subterranean clover observed in environments of southern Australia does not always hold in the plant's countries of origin. Piano *et al.* (1993) found that, although mean populations of *T. subterraneum* subsp. *subterraneum* collected in Sicily increased with increasing rainfall, in *T. subterraneum* subsp. *brachycalycinum* the effect of altitude was greater and that of rainfall less marked. There is also evidence that seed-production characteristics of subterranean clover strains of similar maturity can vary. Piano and Pecetti (1997) found that, despite Geraldton and Seaton Park commencing flowering at a similar date, Geraldton had a shorter flowering period and generally a higher seed yield. Small-seeded strains of subterranean clover have also been shown to reach a mature seed weight faster than larger seed lines (Pecetti and Piano, 1994). This may contribute, at least in part, to the success and dominant position of small-seeded strains of subterranean clover, such as cvs Goulburn, Denmark and Leura, recent releases from the Australian breeding programme where high seed production is considered a high priority.

Rossiter (1966) also identified conservation of seed, particularly through hard-seededness, as a crucial trait for success. The importance of hard-seededness has been borne out consistently in wheat-belt environments in western Australia (Smith *et al.*, 1996) and NSW (Wolfe, 1985; Dear *et al.*, 1993a). In western Australia, where the incidence of false breaks (early rains separated from the main rainfall, resulting in germination of the seed and subsequent death of seedlings) is low, hard-seededness is important to enable subterranean clover seed reserves to carry through the 1–2-year cropping phase. In NSW, the cropping phase is 4–6 years long and most pastures are resown, rather than regenerating from hard seed. However, in NSW, the seasonal 'break' is much less well defined than in the more typical Mediterranean climate of south-western Australia (Cornish, 1985); false breaks may reduce the number of seedlings that survive to well below the target of about 1500 plants m^{-2} that is needed to optimize the early dry-matter production of subclover. Hence, in most environments, moderate to high levels of residual hard seededness (20–60%) are sought in subclover cultivars, even in mid-season strains (Dear *et al.*, 1993a).

Increased levels of hard-seededness can also be used to guard against a failure to set adequate levels of seed due to herbicide damage, insect attack or drought (Dear and Sandral, 1997; Fig. 7.6). Indeed, near the arid boundary for annual legumes, such as at Condobolin in central NSW (Cornish, 1985), annual medics are preferred to subterranean clover – not only because they are more tolerant of water stress during seed development (Wolfe, 1985), but also because cultivars with very high levels of residual hard-seededness are available (Cocks *et al.*, 1980). On the other hand, in cool, moist environments, such as northern NSW (Lodge *et al.*, 1990) and Tasmania (Evans and Hall, 1995), only a low level of residual hard-seededness (< 20%) is needed for persistence of subterranean clover; the cool climate and summer growth modify the soil temperature profile, slow the rate of hard-seed breakdown during summer and enhance the survival of seedlings in autumn.

Smith *et al.* (1996) showed that, among legume species that have a similar hard-seed level, the pattern of breakdown from hard to soft (permeable) seed differed, with some breaking down more rapidly in summer and early autumn, while others had a delayed pattern. The latter pattern resists false breaks, but it may decrease competitive ability with rapidly germinating weed species. Thus, in a summer-dominant rainfall environment, Lodge (1996) showed that the seedling recruitment of subterranean clover and barrel medic was highest when they germinated in midsummer, whereas with other legumes it was highest with an autumn germination.

Other notable characteristics of subterranean clover are those that confer resistance or tolerance to various limiting factors, including diseases, such as clover scorch (Collins and Stern, 1987) and root rots (Dear *et al.*, 1993b), insect pests (Gramshaw *et al.*, 1989), soil nutrient deficiencies/toxicities and waterlogging (Reed *et al.*, 1985). These characteristics are important in ensuring the adaptability and competitive ability of particular cultivars and strains of subclover in stressful (*sensu* Grime, 1977) situations.

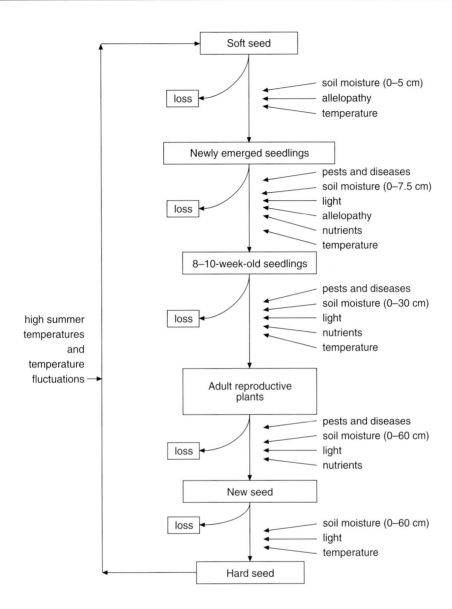

Fig. 7.6. Stages in the life cycle of subterranean clover and pathways by which the presence of perennial pasture plants or their residues may influence the outcome.

As noted by Rossiter (1966, 1977), there are attributes of subterranean clover, other than those mentioned above, that are not well understood in terms of their effects on competition between strains within a season. Little attention has been given to the vegetative stage properties listed by Rossiter (1977); properties such as seedling establishment capability, leaf size and petiole length may bear on the success or failure of one strain growing with another strain or with other species. For example, when the defoliation of a mixture of subterranean clover strains was infrequent, a later-maturing, longer-petiole strain (Clare) was able to overtop early-maturing, shorter strains (Seaton Park, Daliak) (Hill and Gleeson, 1991). This result agreed with an earlier study by Rossiter and Pack

(1972), who found that increased stocking rates during vegetative growth improved the competitive performance of cv. Geraldton, a short-stature strain, by restricting the overtopping effect of the taller Woogenellup cultivar.

Cocks *et al.* (1982) found that the high for-mononetin content of some subterranean clover strains was an important characteristic associated with their long-term success in mixed populations, since sheep prefer to eat cultivars with a low content of formononetin (Rossiter, 1974). More recently, Cocks (1992) found that levels of genistein, another of the three phyto-oestrogens in subterranean clover, tended to increase with time in divergent genotypes of subterranean clover in the field. Genistein is thought to be one of the phytoalexins, substances produced by plants to activate enzyme defences in response to infection by fungi and bacteria (Bell, 1981, reported in Cocks, 1992).

In summary, the ability of subterranean clover to survive and compete in plant communities is determined by a number of known factors, of which time to reach maturity and hard-seededness are paramount. However, the influence of some other characteristics of the species on competitiveness is not precisely known.

COMPETITION BETWEEN SUBTERRANEAN CLOVER AND OTHER ANNUAL SPECIES. There have been a number of long-term studies undertaken on the dynamics of annual pastures in the Mediterranean climate of southern Australia (Rossiter, 1966; Cocks, 1994a). These studies are relevant, at least in an ecological sense if not strictly an agronomic one, to species frequency and productivity in short-term subterranean clover leys that regenerate or are sown after a cropping phase. The conditions at the end of the cropping phase, such as depleted levels of soil N and low levels of disease organisms, favour the early growth of clover. Thereafter, the success of subterranean clover during the pasture phase is determined by factors such as the size of the populations of other annual grasses and herbs, seasonal conditions, phosphate supply and defoliation/grazing (Rossiter, 1966).

In southern Australia, the main competitors with subterranean clover in annual ley pastures are barley grass, annual ryegrass, silvergrass, brome grasses (*Bromus* spp.), capeweed and erodium/storksbill (*Erodium* spp.) (Rossiter, 1966). Low annual legume contents in mixed pastures can be traced to differences between non-legumes and legumes in a number of characters associated with

establishment. For example, McWilliam *et al.* (1970) found differences between legumes and grasses in the water absorption of seeds, rates of germination, and early root elongation; in their experience, annual ryegrass was superior to annual legumes in its ability to germinate under moisture stress. The evidence for subterranean clover, reviewed by Rossiter (1966), indicated that, once germinated, subterranean clover seedlings were more susceptible to moisture stress than common associate species, such as barley grass, capeweed, erodium and Paterson's Curse. According to Rossiter (1966), non-legumes were robust in years in which 'false breaks' occurred in mid-autumn, while the contribution of subterranean clover declined as a result of insufficient numbers of surviving plants. However, if the early break was followed by moist conditions, subterranean clover was favoured; late rains favoured a range of species including subclover (Rossiter, 1966).

In addition to population density, phosphate supply and defoliation regimes also affect the botanical composition of annual pastures in southern Australia (Rossiter, 1966). In southern NSW, a series of investigations on ley pastures grazed by sheep (Ayres *et al.*, 1977) revealed that subterranean clover dominated heavily grazed leys at moderate levels of phosphate supply and low levels of soil N, but barley grass quickly became dominant if superphosphate was top-dressed at moderate to high rates (> 150 kg ha^{-1} annually), particularly if stocking rates were low. This and other work on defoliation (Collins, 1978) and management (FitzGerald, 1976) of subterranean clover led on to recommendations for clover leys in southern NSW (Southwood and Wolfe, 1978). These recommendations, which are still current (Dear and Sandral, 1997) were: choose a persistent cultivar of subterranean clover; establish at 3–10 kg ha^{-1} under a light seeding rate (< 20 kg ha^{-1}) of a cover crop (barley); fertilize the cover crop with additional superphosphate for the pasture phase (a total of 180 kg ha^{-1}, drilled); avoid or minimize top-dressed phosphate during the pasture phase by monitoring the trend in soil phosphate levels; graze leys continuously with at least moderate stocking rates of sheep during the vegetative phase of growth; and avoid excessive defoliation (grazing, hay cuts) during spring.

In recent years, the population balance has probably swung towards subterranean clover due to the availability and use of clover cultivars that are more persistent than the ones they replace (Dear and Sandral, 1997) and to the application of herbicides to reduce the presence of pasture species that are weeds during the cropping phase (Powles *et al.*, 1997). One species that declined was annual ryegrass, which, from the 1970s, was the target of a range of selective herbicides, but it has now developed resistance to most of these chemicals (Powles *et al.*, 1997). Another species that declined was barley grass, which was a common associate of subterranean clover in well-fertilized pastures (Ayres *et al.*, 1977); lower rates of superphosphate applied to leys, higher stocking rates and herbicides were the probable factors that led to a lower frequency of this grass.

Conversely, silvergrass (*Vulpia* spp.) has apparently increased in frequency in ley pastures, presumably due to less competition with ryegrass and higher stocking rates. According to Leys *et al.* (1991) and Dowling *et al.* (1997), the incidence of vulpia has increased substantially in the last decade, primarily through its tolerance of selective herbicides, which have removed one of its competitors, annual ryegrass, and the adoption of reduced tillage practices, which leave the seed bank intact near the soil surface. Other factors that have favoured an increasing incidence of vulpia and attributes that contribute to its competitiveness are listed in Table 7.7.

The herbicide simazine is one of the widely used options to control vulpia but there is concern that the current use of simazine on triazine-tolerant canola varieties and on lupin crops and its possible use to winter-clean pastures will hasten the development of simazine-tolerant vulpia biotypes. The development of non-chemical methods for controlling vulpia is now a priority, but progress may be slow. The findings of Leys *et al.* (1993) and Dowling *et al.* (1997) indicated that competition from other species was by itself insufficient to suppress vulpia (Table 7.8). Although increasing the density of clover reduced seed set by vulpia in the first year, there was no difference between the density treatments by the third year; the inclusion of a competitive grass (annual ryegrass, itself a weed) was the most effective strategy in the experiments of Leys *et al.* (1993). Dowling *et al.* (1997) concluded that the key components of an integrated control strategy for vulpia were a persistent clover strain, the inclusion of a competitive grass, forage conservation and other seed capture techniques and grazing management.

SUBTERRANEAN CLOVER IN MIXTURES WITH PERENNIAL SPECIES. Most studies of the ecology of subterranean clover have involved monocultures and few have studied its ability to coexist with perennial species. A summary of the stages and processes by which perennials may have a potential impact on the annual legume life cycle is presented in Fig. 7.6.

There are two main environments where subterranean clover is grown with perennials. The first is in predominantly permanent pastures in the higher rainfall (> 500 mm annual rainfall), elevated tableland environments of NSW, Victoria and Tasmania, where subterranean clover occurs with phalaris or cocksfoot. The second is in parts of the cropping zone where subterranean clover has been sown in

Table 7.7. Factors favouring dominance of *Vulpia* spp. in pasture swards, southern NSW (based on Leys *et al.*, 1993; Dowling *et al.*, 1997).

Management and edaphic factors
    Reduced use of disc plough
    Widespread use of selective herbicides to control annual ryegrass
    Reduced vigour of annual legumes due to pathogens, lower fertilizer use
    Heavy grazing
    Allelopathic effects of plant residues on annual legumes
    Transfer of seed in hay
    Lower soil pH
    Occurrence of droughts

Characteristics of *Vulpia* contributing to competitiveness
    High seed production capacity
    Low palatability
    Delayed early growth
    Multiple germinations, seed dormancy

Table 7.8. Effect of subterranean clover density on *Vulpia* invasion in pastures at Wagga Wagga, southern NSW (Leys *et al.*, 1993).

| | *Vulpia* density and seed production m^{-2} | | | | |
| | 1990 | | 1991 | | 1992 |
Pasture density/composition	Plants	Seeds	Plants	Seeds	Plants
Low subclover density	720	1,071,300	22,610	432,000	9,200
Medium subclover density	860	536,200	13,370	399,700	7,700
High subclover density	550	239,300	8,890	386,100	9,720
Medium subclover density + annual ryegrass	640	179,400	4,290	173,300	2,530

combination with lucerne as part of 4–6-year ley phase between cropping phases.

The ability of subterranean clover to form stable mixtures in swards with perennial grasses has been poor (Dear, 1998). Typically, the annual legume content declines over time, leading to grass-dominant, N-deficient, less productive swards. There are a number of possible reasons for the inability of subterranean clover to respond to the decreasing grass vigour as N availability declines. The hard-seed breakdown of the subterranean clover seed bank may be depressed by the smaller diurnal temperature fluctuations at the soil surface, due in part to the increased cover of residual herbage. Another view is that allelopathic chemicals released by the grass residues inhibit the germination and survival of subterranean clover seedlings (Leigh *et al.*, 1995). More recently, it has been shown that the seedling survival and early growth of subterranean clover can be greatly reduced by competition for moisture with perennials; perennials such as lucerne or phalaris decrease the favourable period of soil moisture following a rainfall event in late summer or early autumn (Dear and Cocks, 1997). This deficiency leads to a more rapid desiccation and smaller size of clover seedlings and increased seedling mortality. Furthermore, a combination of shading by the perennials and lower available nitrate levels in the presence of perennial grass decreased clover seedling growth rates (Dear *et al.*, 1998).

In spring, also, competition between subterranean clover and perennials (phalaris, lucerne) was found to be important – the quantity of seed set by subterranean clover was inversely related to the amount of light intercepted by the canopy of perennial plants (Dear, 1998). Seed production by subterranean clover is known to be sensitive to shading (Collins *et al.*, 1978). In Dear's (1998) experiments, the presence of perennials did not apparently increase the level of moisture stress experienced by the companion annual legumes in spring.

An extreme case of the shading effect may account for the strong competitive advantage of white clover over subterranean clover. The combination of these two species is most likely to occur in environments such as the northern tablelands of NSW, where at least 60% of the total annual rainfall (> 700 mm) occurs in summer, outside the growing period of the annual. The studies of Smith and Crespo (1979) and Hill and Gleeson (1990) pointed to the ability of white clover to elongate its petioles in spring and shade the companion subterranean clover, resulting in large reductions in seed set of the annual. Another important factor could be the large losses in subterranean clover seed reserves that can occur when the environment remains moist after seed set (Archer, 1990).

SUMMARY. In subterranean clover pastures, recent work has filled in a number of gaps in our understanding of competition between and within species. Higher stocking rates, lower rates of superphosphate applied to ley pastures, shorter leys, the release of improved cultivars and winter cleaning are factors that, if anything, are making it easier for farmers to grow legume-dominant subterranean clover pastures. However, the advent of herbicide-resistant weeds and the increasing use of perennial species are issues that could intensify competition between subterranean clover and non-legumes.

Species dynamics in annual
Medicago pastures

The available reviews of medic pastures (Cocks *et al.*, 1980; Crawford *et al.*, 1989) place less emphasis on the competitive ability of medics than on the ability of individual species to grow, produce seed and persist. This emphasis is understandable, since in Australia medics are traditionally grown in short-term pasture/crop leys, where the survival and regeneration of the medic after the cropping phase is of paramount importance (Puckridge and French, 1983). Another possible constraint to long-term studies of species relationships in medic pastures is the medic decline syndrome (see Table 7.2), which was first reported in the 1980s (Carter *et al.*, 1982) after a devastating occurrence of the spotted alfalfa aphid and the blue-green aphid in the late 1970s. This decline has continued in pastures in South Australia, where low soil phosphorus and zinc levels and *Pratylenchus* nematodes have also been implicated (Denton and Bellotti, 1996). The extensive use of residual sulphonylurea herbicides is another factor considered to affect the persistence of annual *Medicago* spp. grown in rotation with crops. In recent years, medic breeding and evaluation in Australia have been dominated by the need to develop aphid-tolerant varieties; the programme has resulted in the release of new barrel medics (*M. truncatula*), cvs Parraggio, Sephi, Caliph and Mogul, and strand medics (*Medicago littoralis*), cvs Harbinger AR and Herald.

The vulnerability of medic pods to ingestion by sheep and goats is a critical factor determining the amount of the germinable seed pool present in autumn–winter, thereby influencing the ability of the medic to maintain high plant populations in longer term pastures. Dear and Jenkins (1992) found that sheep could remove 200–600 kg ha^{-1} medic seed in 7 days when grazed at a high stocking rate. The greater proportion (up to 95%) of the seed is digested, with the recovery of seed in sheep pellets being inversely proportional to the size of seed (Russi *et al.*, 1992).

Shallow seed burial has been shown to enhance the rate of softening in the first year (Cocks, 1993), as well as preventing ingestion of the pod by stock. Hence most successful farming systems involving medics require frequent cropping and shallow seed incorporation. Deep ploughing, as commonly practised in western Asia and North Africa, results in poor regeneration (Cocks, 1994b). Another benefit of the shallow burial of seed is the improvement in seed–soil contact, which assists in uptake of water when rainfall is light (Cocks, 1993). Seeds in surface pods often fail to germinate with light falls of rain and, although this can be an advantage in that it prevents premature germination and the depletion of the seed bank, it can shorten the growing season and reduce dry-matter yield in good seasons (Cocks, 1993).

The benefit of sowing mixtures of annual medics in Australian farming systems has not been widely reported. Latta and Carter (1996) concluded that there were benefits in combining an early-maturing small-seeded cultivar, such as Harbinger AR, with a mid-season, larger, soft-seeded cultivar (Paraggio), to maximize seed production and regeneration. The relationships between annual medics and competing species should be included in any future monitoring programme to evaluate the long-term contribution of these newer varieties to agriculture.

Increasing the diversity and stability of annual pasture swards

The legume improvement programmes in Australia are changing their emphasis after 50 years of mainly concentrating on developing new cultivars of subterranean clover and annual medics. They are being refocused towards new, previously uncultivated, annual legume species. This development in part reflects the significant progress that has been made in subterranean clover towards the objectives of lowering oestrogens, increasing hard-seededness and improving resistance to leaf and root diseases by breeding programmes (Nichols *et al.*, 1994). But it also reflects a need to find species adapted to niches not suited to subterranean clover and to introduce new, adaptive mechanisms not contained within subterranean clover or annual *Medicago* spp. These include greater adaptation to waterlogging, different patterns of hard-seed breakdown, small seeds that can pass through the rumen undigested, aerial seeding for ease of harvesting and greater insect tolerance. A good example of how other species may fill gaps in pastures is the colonization by *T. clusii* Godr. & Gren. of poorly drained niches within subterranean clover pastures (Cocks, 1994a). Cocks (1994a) has suggested that the lack of species diversity in Australian pastures may be slowly resolving itself through a combination of hybridization, mutation and the spread of species by grazing animals. This process can be accelerated by carefully targeting selection and introduction programmes.

Equally importantly, the change in emphasis is also an attempt to increase the diversity of species in pasture swards. In response to temporal and spatial variation, diversity may produce small but important improvements in ecosystem stability and resilience, at least in terms of biomass, if not in botanical composition (Tilman, 1996). Diversity is also a step away from the annual legume monocultures, which have dominated Australian pastoral systems, monocultures that have been vulnerable to diseases, such as clover scorch (*Kabatiella caulivora*) or root rots in subterranean clover, or pests, such as the spotted alfalfa aphid.

The use of more diverse mixtures of legume species opens up a whole new area of plant competition, which has received very limited study. While many of the factors identified earlier in competition between genotypes of the same species may be important, other factors will be involved. For example, a study by Dear and Coombes (1992) found that when *T. subterraneum*, *Medicago murex* and *Trifolium michelianum* Savi were grown in binary mixtures, seed production by *T. michelianum* was, relative to the monoculture yields, more sensitive to competition than either of the other species, setting only 59–81 kg ha^{-1} in mixtures, compared with 287–514 kg ha^{-1} by the other two species. Whether this was due to the crash grazing system employed or the shading of one species by another could not be determined. Another finding was the need for heavy grazing over summer for *T. michelianum* to regenerate satisfactorily the following autumn (Dear and Coombes, 1992). In contrast, the seed reserves of larger-seeded species, such as *M. murex*, contained in a large pod on the soil surface, can be severely depleted through consumption by livestock over summer, resulting in poor recruitment the following year.

These findings reinforce the challenge of choosing a grazing strategy that maintains the diversity, profitability and sustainability of pasture mixtures. In extensive areas where grazing management is less well controlled, it may be more important to select species combinations that can survive under a particular grazing regime rather than attempting to force species together with divergent grazing needs, despite the desirability of other agronomic characteristics. These combinations have important implications for breeding and selection programmes and how new species are evaluated. Evaluating species in monocultures in small plots may be inappropriate, and it may be preferable to evaluate combinations of species under realistic grazing regimes to discover their dynamics, productivity, stability and resilience. Such an approach is in line with proposals by Cocks (1992), who concluded that, due to the complexity of the possible interactions, the evaluation of species mixtures should be conducted in farmers paddocks under practical management conditions.

Conclusions

This review has highlighted the main developments that have occurred and how they have shaped the dynamics of species relationships in southern Australian pastures. The main phases were a century and a half (1800–1950) of exploitation of Australian native grasslands (a phase that is continuing, albeit at a slower rate), a short and intensive period (1950–1970) of pasture improvement with fertilizers and exotic species and a period of re-evaluation during the last 30 years of the 20th century. This last period has provided an opportunity to consolidate the knowledge of scientists and the experience of farmer/graziers and to reflect on successful and unsuccessful aspects of pasture management in relation to the future of permanent and ley pastures in Australian agriculture.

In this review, a number of gaps in knowledge were indicated. First, there was a lack of knowledge on the effects of 'management' (subdivision, grazing and fertilization) on plant communities, covering a representative range of native, naturalized and exotic species. Fortunately, work is under way to enhance the understanding of agriculturists, ecologists and pastoralists on the separate effects of defoliation, trampling and nutrient addition/depletion on plant–soil–water relations in native and improved grasslands. A feature of this work is a greater emphasis on the impact of species on soil erosion and on the hydrology of catchments (Johnston *et al.*, 1999).

Secondly, there is a need for a more theoretical approach to grassland dynamics, in which the observed changes in productivity and botanical composition are related to useful frameworks that might help integrate knowledge and explain phenomena. In this respect, the application of the state and transition model (Humphreys, 1997) to plant succession in permanent pastures and of the CSD model (Grime, 1977) to ley pastures is recommended. Both models eschew the dogma that is associated with the traditional Clementsian (linear) model of succession, and they are flexible enough to encompass not only breadth of knowledge but also the targeting of a particular problem or issue.

Thirdly, the review has pointed to a lack of quantitative evidence available for the decline in legume and total pasture productivity that has reportedly (Carter *et al.*, 1982) occurred in pastures in southern Australia. A recent survey (Australian Bureau of Agricultural and Resource Economics (ABARE), 1999, personal communication) revealed that farmers in the cropping zone of Australia were predominantly satisfied with both the productivity and quality of their pastures. The separate views of researchers and farmer/graziers need further exploration and, as suggested by Wilson and Simpson (1993), there is scope for regular and standardized surveys of the state of pastures. Strategic long-term studies of pasture relationships, similar to those undertaken by Hutchinson and colleagues (Hutchinson, 1992; Hutchinson *et al.*, 1995; Hutchinson and King, 1999) on permanent pastures on the northern tablelands of NSW, are warranted. The lack of such studies for some pasture types, such as those based on annual medics, must be overcome.

In Australian croplands, developments such as the changing spectrum of crop weeds and the advent of herbicide resistance are likely to force agronomists and managers to place a greater value on monitoring, understanding and manipulating the population dynamics of pasture/weed species, during both the cropping and pasture phases of rotations. In grazing lands, too, the scene is changing, due to enhanced community perceptions of the sustainability of catchments and the aesthetic value of landscapes. Developments such as new pests and diseases or the loss of key herbicides due to community or corporate imperatives will ensure the need for theoretical and applied studies of plant competition and pasture dynamics.

Note

1. In Australia, there is a trend towards the use of 'ley' to denote a 1-year pasture and 'phase' to denote several years of pasture, but we prefer the use of the terms 'ley' and 'ley farming' to embrace pastures of one to several years' duration in the crop/pasture rotation.

References

Archer, K.A. (1990) The effects of moisture supply and defoliation during flowering on seed production and hardseededness of *Trifolium subterraneum* L. *Australian Journal of Experimental Agriculture* 30, 515–522.

Archer, K.A., Wolfe, E.C. and Cullis, B.R. (1987) Flowering time of cultivars of subterranean clover in New South Wales. *Australian Journal of Experimental Agriculture* 27, 791–797.

Axelsen, A. and Morley, F. (1968) Evaluation of eight pastures by animal production. *Proceedings of the Australian Society of Animal Production* 7, 92–98.

Ayres, J.F., McFarlane, J.D., Gilmour, A.R. and McManus, W.R. (1977) Superphosphate requirements of clover-ley farming. 1. The effects of topdressing on productivity in the ley phase. *Australian Journal of Agricultural Research* 28, 269–285.

Barr, N.F. and Cary, J.W. (1992) *Greening a Brown Land: an Australian Search for Sustainable Land Use.* MacMillan, Melbourne, Australia.

Blumenthal, M.J. and Ison, R.L. (1993) Water use and productivity in sub. clover and murex medic swards. II. Seed production. *Australian Journal of Agricultural Research* 44, 109–119.

Callaghan, A.R. (1935) Activities at Roseworthy Agricultural College 1934–35. *Journal of the Department of Agriculture of South Australia* 39, 310–318.

Campbell, M.H. (1992) Extending the frontiers of aerially sown pastures in temperate Australia: a review. *Australian Journal of Experimental Agriculture* 32, 137–148.

Campbell, M.H. (1998) Biological and ecological impact of serrated tussock (*Nassella trichotoma* (Nees) Arech.) on pastures in Australia. *Plant Protection Quarterly* 13, 80–86.

Carter, E.D., Wolfe, E.C. and Francis, C.M. (1982) Problems of maintaining pastures in the cereal–livestock areas of southern Australia. In: *Proceedings of the 2nd Australian Agronomy Conference, Wagga Wagga*, pp. 68–82.

Cocks, P.S. (1992) Evolution in sown populations of subterranean clover (*Trifolium subterraneum* L.) in South Australia. *Australian Journal of Agricultural Research* 43, 1583–1595.

Cocks, P.S. (1993) Seed and seedling dynamics over four consecutive years from a single seed set of six annual medics (*Medicago* spp.) in north Syria. *Experimental Agriculture* 29, 461–472.

Cocks, P.S. (1994a) Colonisation of a South Australian grassland by invading Mediterranean annuals and perennial pasture species. *Australian Journal of Agricultural Research* 45, 1063–1076.

Cocks, P.S. (1994b) Effect of tillage system on the spontaneous regeneration of two annual medics (*Medicago* spp.) after wheat in north Syria. *Experimental Agriculture* 30, 237–248.

Cocks, P.S., Mathison, M.J. and Crawford, E.J. (1980) From wild plants to pasture cultivars: annual medics and subterranean clover in Australia. In: Summerfield, R.J. and Bunting, A.H. (eds) *Advances in Legume Science*. Royal Botanic Gardens, Kew, UK, pp. 569–596.

Cocks, P.S., Craig, A.D. and Kenyon, R.V. (1982) Evolution of subterranean clover in South Australia. 2. Change in genetic composition of a mixed population after 19 years' grazing on a commercial farm. *Australian Journal of Agricultural Research* 33, 679–695.

Collins, W.J. (1978) The effect of defoliation on inflorescence production, seed yield and hardseededness in swards of subterranean clover. *Australian Journal of Agricultural Research* 29, 789–801.

Collins, W.J. (1981) The effects of length of growing season with and without defoliation on seed yield and hardseededness in swards of subterranean clover. *Australian Journal of Agricultural Research* 32, 783–792.

Collins, W.J. and Stern, W.R. (1987) The national subterranean clover improvement program – progress and directions. In: Wheeler, J.L., Pearson, C.J. and Robards, G.E. (eds) *Temperate Pastures, Their Production, Use and Management*. Australian Wool Corporation/CSIRO, Melbourne, Australia, pp. 276–278.

Collins, W.J., Rossiter, R.C. and Monreal, R.A. (1978) The influence of shading on the seed yield of subterranean clover. *Australian Journal of Agricultural Research* 29, 1167–1175.

Cook, S.J., Lazenby, A. and Blair, G.J. (1978) Pasture degeneration. 2. The importance of superphosphate, nitrogen and grazing management. *Australian Journal of Agricultural Research* 29, 19–29.

Cornish, P.S. (1985) Adaptation of annual *Medicago* to a non-Mediterranean climate. In: Hochmann, Z. (ed.) *The Ecology and Agronomy of Annual Medics*. Technical Bulletin 32, Department of Agriculture, Sydney, New South Wales, Australia, pp. 13–16.

Crawford, E.J., Lake, A.W.H. and Boyce, K.G. (1989) Breeding annual *Medicago* species for semiarid conditions in southern Australia. *Advances in Agronomy* 42, 399–437.

Crawford, M.C. and Macfarlane, M.R. (1995) Lucerne reduces soil moisture and increases livestock production in an area of high groundwater recharge potential. *Australian Journal of Experimental Agriculture* 35, 171–180.

Cregan, P. and Scott, B. (1998) Soil acidification – an agricultural and environmental problem. In: Pratley, J. and Robertson, A. (eds) *Agriculture and the Environmental Imperative*. CSIRO, Melbourne, Australia, pp. 98–128.

Crofts, F.C. (1997) Australian pasture production: the last 50 years. In: Lovett, J.V. and Scott, J.M. (eds) *Pasture Production and Management*. Inkata Press, Melbourne, Australia, pp. 1–16.

Crosby, J.R., Bellotti, W.D., Kerby, J.S. and Harrison, R. (1993) Farmer attitudes towards pastures in the cereal–livestock zone of South Australia. In: *Proceedings of the 7th Australian Agronomy Conference, Adelaide*, pp. 321–324.

Culvenor, R.A. and Oram, R.N. (1992) The persistence of phalaris under grazing. In: *Proceedings of the 6th Australian Agronomy Conference, Armidale*, p. 551.

Dear, B.S. (1998) Ecology of subterranean clover growing in association with perennial pasture species. PhD thesis, University of Western Australia, Perth, Australia.

Dear, B.S. and Cocks, P.S. (1997) Effect of perennial pasture species on surface soil moisture and early growth and survival of subterranean clover (*Trifolium subterraneum* L.) seedlings. *Australian Journal of Agricultural Research* 48, 683–693.

Dear, B.S. and Coombes, N. (1992) Population dynamics of *Trifolium subterraneum*, *Medicago murex* and *Trifolium balansae* grown in monocultures and mixtures. In: *Proceedings of the 6th Australian Agronomy Conference, Armidale*, pp. 284–287.

Dear, B.S. and Jenkins, L. (1992) Persistence, productivity, and seed yield of *Medicago murex*, *M. truncatula*, *M. aculeata*, and *T. subterraneum* on an acid red earth soil in the wheat belt of eastern Australia. *Australian Journal of Experimental Agriculture* 32, 319–329.

Dear, B.S. and Sandral, G.A. (1997) *Subterranean Clover in NSW – Identification and Use*, 2nd edn. Agfact P2.5.16, NSW Agriculture, Orange, Australia.

Dear, B.S., Cregan, P.D. and Murray, G.M. (1993a) Comparison of the performance of subterranean clover cultivars in southern New South Wales. 1. Persistence, productivity, and seed yields. *Australian Journal of Experimental Agriculture* 33, 581–590.

Dear, B.S., Murray, G.M., Cregan, P.D. and Taylor, P.A. (1993b) Comparison of the performance of subterranean clover cultivars in southern New South Wales. 2. Effects of *Phytophthora clandestina* and bromoxynil on seedling survival, growth, and seed set. *Australian Journal of Experimental Agriculture* 33, 591–596.

Dear, B.S., Cocks, P.S., Wolfe, E.C. and Collins, D.P. (1998) Established perennial grasses reduce the growth of emerging subterranean clover seedlings through competition for water, light and nutrients. *Australian Journal of Agricultural Research* 49, 41–51.

Denton, M.D. and Bellotti, W.D. (1996) Factors involved in the annual medic decline syndrome in the Murray Mallee, South Australia. In: *Proceedings of the Eighth Australian Agronomy Conference, Toowoomba*, pp. 192–195.

Donald, C.M. (1967) Innovation in agriculture. In: Williams, D.B. (ed.) *Agriculture and the Australian Economy*, 1st edn. Sydney University Press, Sydney, Australia, pp. 57–86.

Donald, C.M. (1970) Temperate pasture species. In: Moore, R.M. (ed.) *Australian Grasslands.* ANU Press, Canberra, Australia, pp. 303–320.

Dowling, P.M., Leys, A.R. and Plater, B. (1997) Effect of herbicide and application of superphosphate and subterranean clover seed on regeneration of vulpia in pastures. *Australian Journal of Experimental Agriculture* 37, 431–438.

Evans, P.M. and Hall, E.J. (1995) Seed softening patterns from single seed crops of subterranean clover (*Trifolium subterraneum* L.) in a cool temperate environment. *Australian Journal of Experimental Agriculture* 35, 1117–1121.

Fisher, H.J. (1972) Effect of nitrogen fertiliser on a kangaroo grass (*Themeda australis*) grassland. *Australian Journal of Experimental Agriculture and Animal Husbandry* 14, 526–532.

FitzGerald, R.D. (1976) Effect of stocking rate, lambing time and pasture management on wool and lamb production on annual subterranean clover pastures. *Australian Journal of Agricultural Research* 27, 261–275.

FitzGerald, R.D., Harris, C.A. and Ayres, J.F. (1995) Impact of drought on survival of pasture species. In: *Proceedings of the 10th Annual Conference, Grassland Society of NSW, Armidale*, p. 76.

Garden, D., Jones, C., Friend, D., Mitchell, M. and Fairbrother, P. (1996) Regional research on native grasses and native grass-based pastures. *New Zealand Journal of Agricultural Research* 39, 471–485.

Gillett, D.C. and Holloway, R.E. (1996) The effects of a sulphonyl urea herbicide on annual medic in alkaline soil. In: *Proceedings of the Eighth Australian Agronomy Conference, Toowoomba,* p. 653.

Gladstones, J.S. (1966) Naturalised subterranean clover (*Trifolium subterraneum* L.) in Western Australia: the strains, their distributions, characteristics and possible origins. *Australian Journal of Botany* 14, 329–354.

Goldney, D.C. and Bauer, J.J. (1998) Integrating conservation and agricultural production: fantasy or imperative? In: Pratley, J. and Robertson, A. (eds) *Agriculture and the Environmental Imperative.* CSIRO, Melbourne, Australia, pp. 15–34.

Gramshaw, D., Read, J.W., Collins, W.J. and Carter, E.D. (1989) Sown pastures and persistence: an Australian overview. In: Marten, G.C., Matches, A.G., Barnes, R.F., Brougham, R.W., Clements, R.J. and Sheath, G.W. (eds) *Persistence of Forage Legumes.* American Society of Agronomy/Crop Science Society of America/Soil Science Society of America, Madison, USA, pp. 1–21.

Grime, J.P. (1977) Evidence for the existence of three primary strategies in plants and its relevance to ecological and evolutionary theory. *American Naturalist* 111, 1169–1194.

Gruen, F.H. (1990) Economic development and agriculture since 1945. In: Williams, D.B. (ed.) *Agriculture and the Australian Economy*, 3rd edn. Sydney University Press, Sydney, Australia, pp. 19–26.

Hancock, W.K. (1972) *Discovering Monaro – a Study of Man's Impact on his Environment.* Cambridge University Press, London, UK.

Hill, B.D. (1985) Persistence of temperate perennial grasses in cutting trials on the central slopes of New South Wales. *Australian Journal of Experimental Agriculture* 25, 832–839.

Hill, M.J. (1996) Potential adaptation zones for temperate pasture species as constrained by climate: a knowledge-based logical modelling approach. *Australian Journal of Agricultural Research* 47, 1095–1117.

Hill, M.J. and Gleeson, A.C. (1990) Competition between white clover (*Trifolium repens* L.) and subterranean clover (*Trifolium subterraneum* L.) in binary mixtures in the field. *Grass and Forage Science* 45, 373–382.

Hill, M.J. and Gleeson, A.C. (1991) Competition between Clare and Seaton Park, and Clare and Daliak subterranean clovers in replacement series mixtures in the field. *Australian Journal of Agricultural Research* 42, 161–173.

Hochmann, Z., Osborne, G.J., Taylor, P.A. and Cullis, B. (1990) Factors contributing to reduced productivity of subterranean clover (*Trifolium subterraneum* L.) pastures on acidic soils. *Australian Journal of Agricultural Research* 41, 669–682.

Humphreys, L.R. (1997) *The Evolving Science of Grassland Improvement.* Cambridge University Press, Cambridge, UK.

Hutchinson, K.J. (1992) The grazing resource. In: *Proceedings of the 6th Australian Agronomy Conference, Armidale*, pp. 54–60.

Hutchinson, K.J. and King, K.L. (1999) Sown temperate pasture decline – fact or fiction? In: *Proceedings of the 14th Annual Conference, Grassland Society of NSW, Queanbeyan*, pp. 78–86.

Hutchinson, K.J., King, K.L. and Wilkinson, D.R. (1995) Effects of rainfall, moisture stress, and stocking rate on the persistence of white clover over 30 years. *Australian Journal of Experimental Agriculture* 35, 1039–1047.

Johnston, W.H. (1996) The place of C$_4$ grasses in temperate pastures in Australia. *New Zealand Journal of Agricultural Research* 39, 527–540.

Johnston, W.H., Clifton, C.A., Cole, I.A., Koen, T.B., Mitchell, M.L. and Waterhouse, D.B. (1999) Low input grasses useful in limiting environments (LIGULE). *Australian Journal of Agricultural Research* 50, 29–53.

Kemp, D.R. and Dowling, P.M. (1991) Species distribution within improved pastures over central NSW in relation to rainfall and altitude. *Australian Journal of Agricultural Research* 42, 647–659.

Knopke, P., Furmage, B., Walters, P. and Krieg, A. (1995) Agriculture in the Australian economy. In: Douglas, F. (ed.) *Australian Agriculture*, Morescope Publishing, Melbourne, Australia, pp. 17–34.

Latta, R.A. and Carter, E.D. (1996) Annual medic cultivar mixes in semi-arid farming systems. In: *Proceedings 8th Australian Agronomy Conference, Toowomba*, pp. 361–364.

Leach, G.J. (1978) The ecology of lucerne pastures. In: Wilson, J.R. (ed.) *Plant Relations in Pastures*. CSIRO, Melbourne, Australia, pp. 290–308.

Leigh, J.H., Halsall, D.M. and Holgate, M.D. (1995) The role of allelopathy in legume decline in pastures. 1. Effects of pasture and crop residues on germination and survival of subterranean clover in the field and nursery. *Australian Journal of Agricultural Research* 46, 179–188.

Leys, A.R., Cullis, B.R. and Plater, B. (1991) Effect of spraytopping applications of paraquat and glyphosate on the nutritive value and regeneration of vulpia (*Vulpia bromoides* (L.) S. F. Gray). *Australian Journal of Agricultural Research* 42, 1405–1415.

Leys, A.R., Dowling, P.M. and Plater, B. (1993) The effect of pasture density and composition on vulpia. In: *Proceedings of the 10th Australian Weeds Conference, Brisbane*, pp. 193–197.

Lodge, G.M. (1991) Management practices and other factors contributing to the decline in persistence of grazed lucerne in temperate Australia: a review. *Australian Journal of Experimental Agriculture* 31, 713–724.

Lodge, G.M. (1996) Seedling emergence and survival of annual pasture legumes in northern New South Wales. *Australian Journal of Agricultural Research* 47, 559–574.

Lodge, G.M. and Whalley, R.D.B. (1985) The manipulation of species composition of natural pastures by grazing management on the northern slopes of New South Wales. *Australian Rangeland Journal* 7, 6–16.

Lodge, G.M. and Whalley, R.D.B. (1989) *Native and Naturalised Pastures on the Northern Slopes and Tablelands of New South Wales*. Technical Bulletin 35, New South Wales Agriculture and Fisheries, Sydney, Australia.

Lodge, G.M., Murison, R.D. and Heap, E.W. (1990) The effect of temperature on the hardseed content of some annual legumes grown on the northern slopes of New South Wales. *Australian Journal of Agricultural Research* 41, 941–955.

McIvor, J.G. (1993) Distribution and abundance of plant species in pastures and rangelands. In: Baker, M.J. (ed.) *Grasslands for Our World*. SIR Publishing, Wellington, pp. 100–104.

McKinney, G.T. (1974) Management of lucerne for sheep grazing on the southern tablelands of New South Wales. *Australian Journal of Experimental Agriculture and Animal Husbandry* 14, 726–734.

McLaren, D.A., Stajsic, V. and Gardener, M.R. (1998) The distribution and impact of South/North American stipoid grasses (Poaceae: Stipeae) in Australia. *Plant Protection Quarterly* 13, 62–70.

McWilliam, J.R., Clements, R.J. and Dowling, P.M. (1970) Some factors influencing the germination and early seedling development of pasture plants. *Australian Journal of Agricultural Research* 21, 19–32.

Meinig, D.W. (1954) *On the Margins of the Good Earth*. South Australian Government Printer, Adelaide, Australia.

Michael, P.W. (1970) Weeds of grasslands. In: Moore, R.M. (ed.) *Australian Grasslands*. ANU Press, Canberra, Australia, pp. 349–360.

Moore, R.M. (1970) Southeastern temperate woodlands and grasslands. In: Moore, R.M. (ed.) *Australian Grasslands*. ANU Press, Canberra, Australia, pp. 170–190.

Nichols, P., Collins, W.J., Gillespie, D. and Barbetti, M. (1994) Developing improved varieties of subterranean clover. *Journal of Agriculture, Western Australia* 35, 60–65.

O'Connor, T.G. (1996) Hierarchical control over seedling recruitment of the bunch-grass *Themeda triandra* in a semi-arid savanna. *Journal of Applied Ecology* 33, 1094–1106.

Oram, R.N. and Culvenor, R.A. (1994) Phalaris improvement in Australia. *New Zealand Journal of Agricultural Research* 37, 329–339.

Panetta, F.D., Risdell-Smith, T.J., Barbetti, M.J. and Jones, R.A.C. (1993) The ecology of weeds, invertebrate pests and diseases of Australian sheep pastures. In: Delfosse, E.S. (ed.) *Pests of Pastures: Weed, Invertebrate and Disease Pests of Australian Sheep Pastures*. CSIRO, Melbourne, Australia, pp. 87–114.

Passioura, J.B. and Ridley, A.M. (1998) Managing soil water and nitrogen to minimise land degradation. In: *Proceedings of the 9th Australian Agronomy Conference, Wagga Wagga*, pp. 99–106.

Paul, J.A. (1999) Critical parameters for monitoring ley pastures. PhD thesis, Charles Sturt University, Wagga Wagga, Australia.

Pearson, C.J., Brown, R., Collins, W.J., Archer, K.A., Wood, M.S., Petersen, C. and Bootle, B. (1997) An Australian temperate pastures database. *Australian Journal of Agricultural Research* 48, 453–465.

Pecetti, L. and Piano, E. (1994) Observations on the rapidity of seed and burr growth in subterranean clover. *Journal of Genetics and Breeding* 43, 225–228.

Peoples, M.B., Gault, R.R., Scammell, G.J., Dear, B.S., Virgona, J.M., Sandral, G.A., Paul, J., Wolfe, E.C. and Angus, J.F. (1998) The effect of pasture management on the contributions of fixed N to the N economy of ley-farming. *Australian Journal of Agricultural Research* 49, 459–474.

Piano, E. and Pecetti, L. (1997) Effect of water stress during flowering on seed yield in subterranean clover (*Trifolium subterraneum* L.) cultivars. Current situation and future prospects for forage cultivation in grasslands and pastures, Lodi, Italy. *Revista di Agronomia* 31, 229–232.

Piano, E., Spanu, F. and Pecetti, L. (1993) Structure and variation of subterranean clover populations from Sicily, Italy. *Euphytica* 68, 43–51.

Powles, S.B., Preston, C., Bryan, I.B. and Jutsum, A.B. (1997) Herbicide resistance: impact and management. *Advances in Agronomy* 58, 57–93.

Puckridge, D.W. and French, R.J. (1983) The annual legume pasture in cereal–ley farming systems in southern Australia: a review. *Agriculture, Ecosystems and Environment* 9, 229–267.

Reed, K.F.M., Schroder, P.M., Eales, J.W., McDonald, R.M. and Chin, J.F. (1985) Comparative productivity of *Trifolium subterraneum* and *T. yanninicum* in south-western Victoria. *Australian Journal of Experimental Agriculture* 25, 351–361.

Reeves, T.G. and Ewing, M.A. (1993) Is ley farming in mediterranean zones just a passing phase? In: Baker, M.J. (ed.) *Grasslands for Our World.* SIR Publishing, Wellington, pp. 810–818.

Ridley, A.M., Simpson, R.J. and White, R.E. (1999) Nitrate leaching under phalaris, cocksfoot and annual ryegrass pastures and implications for soil acidification. *Australian Journal of Agricultural Research* 50, 55–63.

Rossiter, R.C. (1964) The effect of phosphate supply on the growth and botanical composition of annual type pasture. *Australian Journal of Agricultural Research* 15, 61–76.

Rossiter, R.C. (1966) Ecology of the Mediterranean annual-type pasture. *Advances in Agronomy* 18, 1–56.

Rossiter, R.C. (1974) The relative success of strains of *T. subterraneum* L. in binary mixtures under field conditions. *Australian Journal of Agricultural Research* 25, 757–766.

Rossiter, R.C. (1977) What determines the success of subterranean clover strains in south-western Australia? In: Anderson, D.J. (ed.) *Exotic Species in Australia – Their Establishment and Success.* Ecological Society of Australia, Sydney, Australia, pp. 76–88.

Rossiter, R.C. and Pack, R.J. (1972) Effect of stocking rate and defoliation on relative seed production in a mixture of two subterranean clover strains. *Journal of the Australian Institute of Agricultural Science* 38, 209–211.

Rubzen, B.B. (1996) The adaptation of plant and weed species to acid soil environments. PhD thesis, Charles Sturt University, Wagga Wagga, Australia.

Russi, L., Cocks, P.S. and Roberts, E.H. (1992) The fate of legume seeds eaten by sheep from a Mediterranean grassland. *Journal of Applied Ecology* 29, 772–728.

Shaw, A.G.L. (1990) Colonial settlement 1788–1945. In: Williams, D.B. (ed.) *Agriculture and the Australian Economy,* 3rd edn. Sydney University Press, Sydney, Australia, pp. 1–18.

Smith, F.P., Cocks, P.S. and Ewing, M.A. (1996) Short-term patterns of seed softening in *Trifolium subterraneum, T. glomeratum,* and *Medicago polymorpha. Australian Journal of Agricultural Research* 47, 775–785.

Smith, R.C.G. and Crespo, M.C. (1979) Effect of competition by white clover on seed production characteristics of subterranean clover. *Australian Journal of Agricultural Research* 30, 597–607.

Southwood, O.R. and Wolfe, E.C. (1978) *Identifying and Using Subterranean Clovers,* 2nd edn. Bulletin P473, New South Wales Division of Plant Industry, NSW Government Printer, Sydney, Australia.

Thorn, C.W. (1992) Management of annual legume pastures for crop and animal production. In: Angus, J.F. (ed.) *Transfer of Biologically Fixed Nitrogen to Wheat.* GRDC, Canberra, Australia, pp. 93–108.

Tilman, D. (1996) Biodiversity: population versus ecosystem stability. *Ecology* 77, 350–363.

Watson, R.W. (1993) *Phalaris.* Agfact P 2.5.1, NSW Agriculture, Orange, Australia.

Whitfield, D.M. (1998) Hydrologic utility of phase farming based on winter rainfall in south eastern Australia. In: *Proceedings of the 9th Australian Agronomy Conference, Wagga Wagga,* pp. 823–826.

Williams, A.R. (1979) A *Survey of Natural Pastures in the North-west Slopes of New South Wales.* Technical Bulletin 22, Department of Agriculture, New South Wales, Australia.

Williams, C.H. and Andrew, C.S. (1970) Mineral nutrition of pastures. In: Moore, R.M. (ed.) *Australian Grasslands.* ANU Press, Canberra, Australia, pp. 321–338.

Willoughby, W.M. (1954) Some factors affecting grass–clover relationships. *Australian Journal of Agricultural Research* 5, 157–180.

Wilson, A.D. and Simpson, R.J. (1993) The pasture resource base: status and issues. In: Kemp, D.R. and Machalk, D.L. (eds) *Pasture Management Technology for the 21st Century.* CSIRO, Melbourne, Australia, pp. 1–25.

Wolfe, E.C. (1968) Cattle bloat in southern New England, NSW, 1961–66. *Proceedings of the Australian Society of Animal Production* 7, 123–128.

Wolfe, E.C. (1985) Subterranean clover and annual medics – boundaries and common ground. In: Hochmann, Z. (ed.) *The Ecology and Agronomy of Annual Medics.* Technical Bulletin 32. Department of Agriculture, Sydney, New South Wales, Australia, pp. 23–28.

Wolfe, E.C. and Lazenby, A. (1973) Grass–white clover relationships during pasture development. 2. Effect of nitrogen fertiliser with superphosphate. *Australian Journal of Experimental Agriculture and Animal Husbandry* 13, 575–580.

Wolfe, E.C. and Southwood, O.R. (1980) Plant productivity and persistence in a mixed pasture containing lucerne at a range of densities with subterranean clover or phalaris. *Australian Journal of Experimental Agriculture and Animal Husbandry* 20, 189–196.

Wolfe, E.C., FitzGerald, R.D., Hall, D.G. and Southwood, O.R. (1980) Beef production from lucerne and subterranean clover pastures. *Australian Journal of Experimental Agriculture and Animal Husbandry* 20, 678–687.

8 Formulation of Pasture Seed Mixtures with Reference to Competition and Succession in Pastures

Warwick Harris

Lincoln Botanical, Lincoln, Canterbury, New Zealand

Introduction

What does a grassland farmer need to know about competition between plants in pasture in order to successfully formulate mixtures of seed and to manage the growth of these into productive pasture after they are sown? This contribution considers this question especially for pastures in humid temperate regions and mostly draws on experience in New Zealand. It aims to illustrate principles by which pasture seed mixtures have been formulated in the past, showing how these formulations have brought into action the processes of inter- and intra-specific competition. From this historical perspective, it then goes on to explore how principles of plant competition can be applied to the formulation of mixtures in current grassland farming. This involves an examination of the consequences of mixing species and cultivars in respect of yield and its seasonal distribution and the dynamics of species and genotype composition.

The need to reiterate principles and practices of mixing pasture species and cultivars is prompted by the large increase in the availability of both herbage species and cultivars that has resulted from plant introduction, breeding, selection and modification in recent decades. In 1998 grassland farmers in New Zealand had a choice of 17 grass, 12 legume and three herb species and hybrids, and this choice was enlarged by diversification of these into at least 113 cultivars (Table 8.1). Thirty years before they had about 17 species and hybrids to choose from

and the cultivar numbers were much less, with single New-Zealand bred cultivars of perennial ryegrass, hybrid ryegrass, Italian ryegrass and white clover (Harris, 1968). This process of herbage cultivar diversification has occurred in other countries (e.g. see Frame *et al.*, 1996).

The greater number of entities available for inclusion in pasture seed mixtures has the potential to confuse farmers, causing them to use inappropriate mixtures and to undo the efforts of breeders, selectors and modifiers of pasture plants by unreasoned mixing of cultivars of the same species. Thus a viewpoint put forward in this chapter is that just as much can be gained in pasture yield, seasonal growth distribution and herbage quality by the scientific formulation of pasture species mixtures as can be achieved by the development of new cultivars.

Proliferation of cultivars has been encouraged by acquisition of ownership of segregates of the genetic variation of species through plant variety rights (see Table 8.1). In 1998, 36% of the certified herbage cultivars available in New Zealand had plant variety rights, whereas in 1968 all the available cultivars were publicly owned (Harris, 1968). Ownership of cultivars provides an incentive to include these in proprietary mixtures in order to increase profit from seed sales rather than for a proven enhancement of the performance of the mixture. With knowledge of the principles applying to the formulation of pasture seed mixtures, farmers will be better equipped both to make up mixtures appropriate to their own forage

© CAB *International* 2001. *Competition and Succession in Pastures*
(eds P.G. Tow and A. Lazenby)

Table 8.1. Pasture grass, legume and herb species, the numbers of their cultivars and the number of the cultivars protected by plant variety rights (PVR) grown in New Zealand in 1998.

Grasses	Cultivars	Protected by PVR	Legumes and herbs	Cultivars	Protected by PVR
01. Perennial ryegrass (*Lolium perenne*)	18	8	18. White clover (*Trifolium repens*)	13	8
02. Hybrid ryegrass (*Lolium × boucheanum*)	10	2	19. Red clover (*Trifolium pratense*)	7	3
03. Italian ryegrass (*Lolium multiflorum*)	17	6	20. Strawberry clover (*Trifolium fragiferum*)	2	0
04. Cocksfoot (*Dactylis glomerata*)	9	3	21. Caucasian clover (*Trifolium ambiguum*)	1	1
05. Timothy (*Phleum pratense*)	1	0	22. Lotus (*Lotus uliginosus*)	2	1
06. Crested dogstail (*Cynosurus cristatus*)	2	1	23. Bird's-foot trefoil *L. corniculatus*	1	1
07. Tall fescue (*Festuca arundinacea*)	5	2	24. Lucerne (*Medicago sativa*)	4	1
08. Phalaris (*Phalaris aquatica*)	1	0	25. Serradella (*Ornithopus sativus*)	2	0
09. Prairie grass (*Bromus willdenowii*)	1	0	26. Sulla (*Hedysarum coronarium*)	1	0
10. Upland brome (*Bromus sitchensis*)	1	0	27. Alsike clover (*Trifolium hybridum*)	1	0
11. Grazing brome (*Bromus stamineus*)	1	1	28. Crown vetch (*Coronilla varia*)	1	0
12. Browntop (*Agrostis capillaris*)	1	0	29. Subterranean clover (*Trifolium subterraneum*)	Several imported	0
13. Yorkshire fog (*Holcus lanatus*)	2	1	30. Chicory (*Cichorum intybus*)	1	0
14. Paspalum (*Paspalum dilatatum*)	1	0	31. Plantain (*Plantago lanceolata*)	2	2
15. Perennial ryecorn (*Secale delmaticum*)	1	0	32. Sheep's burnett (*Sanguisorba minor*)	1	0
16. Pubescent wheat grass (*Agropyron trichophorum*)	1	0			
17. Oat grass (*Arrhenatherum elatius*)	1	0			
Total grasses	73	24	Total legumes and herbs	40	17

production needs and to assess the merits of mixtures offered by seed merchants.

The course of this chapter will be first to examine the composition of pasture seed mixtures that were largely derived from the practical experience of farmers establishing new pastures and renewing old ones. Then principles of plant competition that were derived from this practical experience and supporting experimentation are considered. Finally, special attention is paid to mixing cultivars of the same species in pasture seed mixtures. These considerations are used to provide guidance as to whether the mixing of herbage species and cultivars from the wide choice now available can provide further improvements of pasture yield, quality and stability. How processes of plant competition and succession in pastures can provide these outcomes will also be considered.

The Principles of Environmental Matching, Cover and Versatility

There are three principles that are basic to the formulation of pasture seed mixtures. They were largely derived from the practical experience of farmers and early grassland scientists. Here they are largely illustrated by recommendations for seed mixtures given by Levy (1970). Levy's recommendations were developed in the course of the history of pasture development in New Zealand, much of which involved the sowing of pasture seed on land from which forest had been cleared by felling and burning. However, these principles also apply where primary vegetative cover other than forest is removed and where old sown pastures are removed by cultivation, herbicide application and other means of clearance before resowing.

The principles also apply where new species or cultivars are added to existing pasture, most simply by adding seed into a sward by the process defined as oversowing. More sophisticated methods of insertion of seed into existing pasture involve controlled grazing and mechanized procedures, variously described as undersowing and direct drilling which provide partial cultivation, targeted application of herbicide and fertilizer and precision placement of seed. These sophisticated methods provide microenvironments that enhance the success of seed germination and establishment and reduce the competitive effects of plants from the old sward.

The first principle is the matching of species and cultivars to the environment where they are to be sown. This matching is dependent on the species' response to temperature, water and nutrient availability. Consequently, species included in the mixtures formulated by Levy were those considered to be the best adapted to survive, grow and compete in what were anticipated to be the environmental conditions of the site cleared for pasture development. Essentially, in order to compete and play a role in the successional processes of a pasture in a given environment, pasture species need to be able to successfully establish and grow in that environment.

The second principle is to provide early cover by including species or cultivars in the mixture that can quickly occupy ground that inhibits the establishment of weeds that may regenerate or germinate at the time the pasture seed mixture is sown. The facet of competition that is important in this second principle is that plants that can gain a foothold in a pasture quickly usually have a competitive advantage over plants that establish more slowly.

The third principle is to select a mixture with versatility that allows both for the heterogeneity of environmental microsites or ecological niches within the area being sown and for the uncertainty about the course of succession of the species after sowing. This is covered by the inclusion of more species and cultivars in the pasture seed mixture, thus ensuring that at least a proportion of the sown species will find niches suitable for establishment and growth.

Farmer experience in the outcome of pasture plant competition and succession over time enabled the composition of pasture seed mixtures to be simplified to some extent. Essentially, this came down to using species that would establish and grow well on farms, with preference for those that established quickly to provide competitive suppression of weeds as well as forage for livestock at the earliest possible date. As quickly establishing species are often not persistent, mixtures for long-term pasture were designed to include both rapidly establishing, short-lived species and long-lived perennials or self-regenerating annual species.

Replacement of natural vegetation

In sowing pasture species on an area for the first time, the species and cultivar composition of seed mixtures and management of its establishment will be influenced by the nature of the vegetation on the area to be sown. At one extreme is pasture development on land covered by perennial native or wild naturalized woody or herbaceous vegetation. Another extreme is where seasonal cold or aridity fosters communities of annual species.

Much of the area of existing humid temperate grassland had forest as its climax vegetation. In Europe most grasslands have been cleared of woody vegetation for centuries. In contrast, in Australia and New Zealand, most of the forest and scrub clearance for pasture development has taken place in the last century, and even more recently in some tropical regions.

Where temperate forest is the climax vegetation in New Zealand, the natural course is for succession through coarse herbs and ferns, shrubs and finally trees. As these plants are largely unpalatable to domestic livestock, grassland farmers see this natural succession as reversion. Consequently, in the conversion of woody vegetation to pasture, grassland farmers endeavour to destroy the existing vegetation as comprehensively as possible. This has the dual purpose of removing vegetation that is unsuitable for livestock production and vegetation that competes with the herbage plants established from the sown seed mixture.

Topography and rainfall in New Zealand markedly influenced the efficiency with which forest and scrub plant communities could be destroyed to reduce the competition these imposed on sown pasture mixtures. In drier areas, hotter burns could be achieved to destroy both trees that had been felled and other vegetation left standing. Further, clearance and preparation of a seed-bed by cultivation was more easily achieved on areas of lower slope.

There was varied opinion about what pasture seed should be sown on forest areas felled and burned for pasture establishment. On the one hand, it was believed that only the best of the 'English grasses' should be sown, especially as these initially grew well on the ash of the forest burn. But cost considerations also led to the sowing of much poor-quality seed. The varied nature of the land on which conversion from forest to pasture was carried out determined variation in the composition of seed mixtures sown. The overall trend was towards a high number of components in the mixtures as an insurance against the uncertainty about which species would establish and survive on the newly cleared land. In areas of varied topography, it was recommended that spurs, ridges, steep slopes and unfavourable aspects should be sown with one mixture and flats, lower slopes and favourable slopes with another.

A versatile mixture

Early in the history of pasture development in New Zealand, seed mixtures containing as many as 20 species were sown to establish pastures on areas of forest felled and burned (primary forest burn). From experience of which of these species survived and contributed reliably to pasture production Levy (1970) recommended the less complex seed mixture given in Table 8.2 for primary forest burns.

Other additions to this mixture were 1 kg lotus (*Lotus uliginosus*) for high-rainfall areas, 2 kg subterranean clover (*Trifolium subterraneum*) for dry, sunny

areas and 2 kg ha^{-1} of *Paspalum dilatatum* in the warmer region of the northern North Island of New Zealand. These variations took into account differences between species in their water and temperature responses. Also, Levy (1970) recommended variations in the seed mixtures according to assessment of the soil fertility of the area of primary forest burn to be sown. On infertile soils it was recommended that browntop seed should be increased to 2–3 kg ha^{-1} and that 3 kg ha^{-1} of Chewings fescue (*Festuca rubra*) should be added, whereas on fertile areas these species and danthonia should be replaced with timothy (*Phleum pratense*), Alsike clover (*Trifolium hybridum*) and more red clover.

Also of interest is the fact that Levy recommended that the hybrid ryegrass component of the primary forest burn mixture should be a blend of the two cultivars available in New Zealand in 1970. These were 'Grasslands Manawa' short-rotation ryegrass (*Lolium perenne* × *multiflorum*) and the then recently released 'Grasslands Ariki' long-rotation ryegrass (*L. perenne* × (*L. perenne* × *multiflorum*)). Neither of these are persistent grasses. Their role was to respond to the flush of nutrients provided by the ash of the forest burn and provide quick cover and herbage from the newly sown pasture.

Harris (1970) provides an illustration of the outcome of sowing a versatile mixture into heterogeneous environments. From a seed mixture of perennial and hybrid ryegrass cultivars and other grass and legume species, the occurrence of these components in the established pasture varied between wet and dry and between excreta and inter-excreta micro sites. This environmental sorting through the competitive interaction between species and of genotypes of the same species is considered further in the section on mixing cultivars of the same species.

The key principle underlying the formulation of this versatile mixture is the first one of using species likely to succeed in the environment in which they were sown. However, the mixture is still relatively complex to meet the requirements of the third principle, based on the foresight that the initial flush of fertility following forest clearance would be followed by lower-fertility conditions over all or part of the sown area. In these lower-fertility conditions, browntop and danthonia compete more effectively with ryegrass and cocksfoot and the succession is towards their dominance as the pasture ages.

Table 8.2. Levy's (1970) recommended seed mixture for primary forest burns.

Component	Botanical name	Seed kg ha^{-1}
Perennial ryegrass	*Lolium perenne*	11
Hybrid ryegrass	*Lolium × boucheanum*	13
Cocksfoot	*Dactylis glomerata*	6
Crested dogstail	*Cynosurus cristatus*	3
Browntop	*Agrostis capillaris*	1
Danthonia	*Notodanthonia biannularis*	3
White clover	*Trifolium repens*	2
Red clover	*Trifolium pratense*	2

A mixture providing cover to exclude unwanted species

Ryegrass cultivars that require high soil fertility were not included in the seed mixture recommended for sowing after burning of secondary growth. This frequently occurred on wet hill country where the primary burn and seed sowing did not succeed and unpalatable weeds, fern and scrub took over. The need in this situation was to establish a dense pasture tolerant of low fertility that did not leave spaces for the establishment of the species of secondary regrowth. 'Grasslands Ruanui' was the only perennial ryegrass purposefully selected for New Zealand conditions available at the time. Levy (1970) recommended the seed mixture given in Table 8.3 for secondary growth burns.

Thus the non-persistent high-fertility responsive hybrid ryegrass was excluded from this mix, as its early death could leave gaps for the establishment of weeds. It also allowed dense-tillering, low-fertility-tolerant browntop to more effectively form a dense sward, providing long-term cover that limited the formation of spaces suitable for the establishment of secondary growth species.

The key principle underlying the variation of this mixture from the versatile mixture is the second one of providing cover to keep out unwanted weeds. It demonstrates a variation in the application of this principle in that the use of a rapidly establishing high-fertility ryegrass cultivar could competitively suppress the sown, low-fertility grasses so severely that they could not resist the natural successional process back to woody vegetation.

A mixture matching grassland improvement

With the advent of aerial top-dressing after the Second World War application of phosphatic fertilizer to New Zealand hill country became common practice. The recommended seed mixture oversown on to these aerial-fertilized grasslands excluded the low-fertility grasses browntop and danthonia. Seed of danthonia is not available today and very little browntop is sown. However, both remain as common, well-naturalized species, and their replacement by ryegrass, cocksfoot and other high-fertility-responding grasses has become a goal in pasture improvement. A key to successful oversowing was

Table 8.3. Levy's (1970) recommended seed mixture for secondary growth burns.

Component	Seed kg ha^{-1}
Perennial ryegrass ('Grasslands Ruanui')	11
Crested dogstail	4
Browntop	1
Lotus	0.5
Danthonia	3
White clover	2
Subterranean clover	2 on sunny faces

the use of good strains of white clover. Consequently, 'Grasslands Huia', the only cultivar of white clover purposefully bred for New Zealand conditions available at that time, was a key component of the mixture Levy recommended for oversowing phosphate-fertilized hill country (Table 8.4).

This mixture can be further simplified to a perennial ryegrass, a hybrid ryegrass and an effective legume, which meet all the principles of formulating pasture seed mixtures. Reliable maintenance of high soil fertility by application of phosphorus (P) fertilizer and effective nitrogen supply by nitrogen- (N_2-) fixation by the legume allows confident use of only high-fertility grasses. The hybrid ryegrass establishes quickly to provide early forage, suppresses weeds in the establishment phase and may, with appropriate combinations of climate, soil fertility and grazing management, provide a long-term component, together with perennial ryegrass (Brougham and Harris, 1967; Harris and Brougham, 1968, 1970).

Matching species in mixtures to environments

There is now much more information about herbage species and cultivars, which allows their matching to different environments and livestock systems, than that available to Levy and his contemporaries (Scott et al., 1985). However, in extending pasture development into new areas, or evaluating new species and selections, the most direct and practically useful test is to sow and follow the growth and survival of different species or cultivars in different environments. This allows an ecological sorting of those that are suitable for given situations.

Two contrasting approaches to this kind of testing are provided by experiments in New Zealand in recent times. Scott (1993) subjected a mixture of many species to a range of fertilizer and grazing treatments to obtain an ecological natural selection of those most suitable for the grasslands of the high-country region of the South Island. This approach can be likened to beginning with a versatile mixture, from which particular species survive better in the heterogeneity of environments provided by the different experimental treatments. While this approach provided a rigorous screening involving intense interspecific competition, it was also prejudicial against species that might have established and survived had they been sown in simpler mixtures. Tall fescue (*Festuca arundinacea*) provides an example for this. While this species provides a better alternative to ryegrass as a persistent component of pastures in drier regions of New Zealand it is very severely suppressed by competition from ryegrass if sown in mixture with this species (Brock, 1972).

The contrasting approach, used by Williams et al. (1990), was designed to match lines of subterranean and white clover cultivars to a range of hill-country environments, and involved transplanting plants into resident pasture at eight widely separated and mostly seasonally dry sites. Consequently this approach by-passed the intense selection by competitive exclusion that occurs during pasture establishment. However, it clearly demonstrated that, even if they had established from sowing, all white clover lines were unsuitable for the driest sites (Hoglund, 1990; MacFarlane et al., 1990b; Sheath et al., 1990) and all subterranean clover lines were unsuitable for the coldest and wettest site (Widdup and Turner, 1990). Amongst the white clover lines it was shown that, overall, those of New Zealand origin were better than overseas lines. Also, small- and medium-leaved lines did better, because of

Table 8.4. Levy's (1970) recommended seed mixture for oversowing phosphate-fertilized hill country.

Component	Seed kg ha^{-1}
Perennial ryegrass ('Grasslands Ruanui')	7
Hybrid ryegrass ('Grasslands Ariki')	4
Crested dogstail	2
Subterranean clover (on sunny aspects)	2
Lotus (wet regions and shady faces)	0.5
White clover ('Grasslands Huia')	2

their tolerance of intensive sheep grazing, which occurred at most of the sites (Chapman and Williams, 1990b; MacFarlane *et al.*, 1990a, b; Widdup and Turner, 1990). The main exception was at the northernmost and seasonally wet site at Kaikohe where the differences between the lines in nematode resistance and the response of 'Kopu' white clover to the laxer grazing gave a different order of response (Rumball and Cooper, 1990). 'Tallarook' was the most suitable subterranean clover line where this species did well (Chapman and Williams, 1990a; MacFarlane *et al.*, 1990b; Sheath and MacFarlane, 1990a, b; Sheath *et al.*, 1990), because its flowering time and prostrate habit favoured seed set, except at the North Canterbury site (Hoglund, 1990). At the North Canterbury site, earlier onset of seasonal drought impaired the flowering of 'Tallarook'.

These results show the usefulness of information about both the general and the specific matching of species and cultivars to environments in selecting the components of seed mixtures. Thus, for white clover, there is the general matching of New Zealand-bred lines for use in wetter localities in this country, and more specific matching that points to 'Kopu' being a suitable white clover cultivar in northern regions of the country, providing grazing is lax.

It is emphasized that, before a species can begin to compete and play a role in the successional processes in a pasture, it has to be able to successfully establish and grow in the environment of the pasture and under the management practices applied to the pasture.

The provision of cover

The experience of Levy (1970) – that the non-persistence of annual or hybrid ryegrass included to provide quick cover in hill country could leave gaps allowing the ingress of unwanted species – points to the care needed in including cover species in seed mixtures. Examples of refinements applying to the inclusion of species to provide early cover and aspects of management that need to accompany this practice are illustrated in a series of experiments undertaken by Brougham (1954a, b, c). These experiments were prompted by the agronomic evaluation of the then recently released short-rotation ryegrass (*L. perenne* × *multiflorum*), also known as H1 ryegrass (Corkill, 1945) and later as 'Grasslands Manawa'.

Variation of seed rate

In an autumn sowing, Brougham (1954a) varied the seeding of 'Grasslands Manawa' over seven rates from 0 to 67 kg ha^{-1}, while holding the rates of red clover and white clover at 4.5 and 3.4 kg ha^{-1}, respectively. High seed rates of the ryegrass provided more yield initially but suppressed clover yield. The suppression of clover led to a reduced level of N$_2$-fixation, resulting in a reduction of ryegrass yield later in the establishment year, where it was sown at high rates. It was concluded that the optimum ryegrass seed rate in terms of seed cost, grass–clover balance, weed control and yield was around 20 kg ha^{-1}. This result relates to the classic crop yield-seed density relationship, with the modification that ryegrass can respond to seed density by adjustment of tiller number and size (e.g. see Lazenby and Rogers, 1962). Also, as clovers are of key importance in the N economy of this kind of pasture, adjustment of ryegrass seed rate was needed to minimize clover suppression by shading.

Seed rate and defoliation

Brougham (1954b) investigated how grazing in the year of pasture establishment could modify shading of clover by ryegrass. 'Grasslands Manawa' ryegrass was sown in autumn at rates of 17 and 45 kg ha^{-1}, with a common rate of red and white clover, and the pasture was grazed when it reached a height of 7.5 or 22.5 cm. Later in the year, these grazing heights were reversed for half of the area. The combination of high ryegrass seed rate and grazing at 22.5 cm suppressed clover and caused poor ryegrass growth in spring, as a consequence of lower clover N$_2$-fixation. Grazing at 7.5 cm enabled good clover growth, irrespective of ryegrass seed rate. It was concluded that there was an advantage in sowing the ryegrass at a lower rate because it enabled more latitude in grazing management without an adverse effect on the clover.

Suppression from quick cover

Barley was included in the mixture of 'Grasslands Manawa' ryegrass and clovers as a cover crop for winter green feed in the third experiment (Brougham, 1954c). While this boosted winter yield, it decreased ryegrass and total annual yield, although clover yield was increased in spring. This result highlights the requirement of adjusting the seed rates of the components of mixtures when new entities that differ in their competitive ability are either added to or replace those of a standard mixture. For example, 'Grasslands Nui' ryegrass, when it became available as an alternative to 'Grassland Ruanui', was found to cause greater competitive suppression of white clover (Harris, 1980) and adjustments of seed rates were required to establish a suitable ryegrass–white clover content. Similarly, when 'Grasslands Pitau' became an available alternative to 'Grasslands Huia' as a white clover selected for New Zealand grasslands, it was also found to be more competitive against the grass component of mixtures (Brock, 1971). To some extent, this negated the improved yield and different seasonal growth of these cultivars selected when they were grown without competition.

Practice and theory

This section has indicated that seed mixtures for moist temperate perennial pastures in New Zealand primarily evolved through the practical observations and trial and error of pioneering grassland farmers. These farmers were probably not particularly aware of the processes of plant competition and succession, and it is suggested that most modern-day farmers have little awareness of these processes either. But it is essential to the survival of all grassland farmers that they know which forage species are best matched to the environments of their farms to provide the optimal amounts and seasonal spread of herbage to feed their livestock. Further, their farming experience will make them aware of which species will provide the cover required to keep out unwanted weeds while allowing the best forage species to dominate their pastures. Finally, because of the many environmental variables that influence the performance of pastures, farmers will be inclined to sow more complex versatile mixtures, unless they are very sure of the outcome of sowing a less complex mixture.

To progress beyond practical experience, the question that arises is whether a scientific understanding of plant adaptation and competition and their interaction to change species presence in pasture succession can lead to the formulation of even more effective pasture mixtures. Providing an answer to this question is the intent of the sections that follow.

Pasture Renewal and the Components of Seed Mixtures

The three principles of environmental matching, provision of cover to reduce the content of weed species and versatility for heterogeneous environments continue to be relevant, especially in areas of low-intensity pastoral farming, where topography is variable. In areas of more intensive pastoral farming other considerations relating to the formulation of pasture seed mixtures become more important. In these areas there is greater certainty about which species and cultivars will be the most persistent and productive, there is no risk of secondary woody weed growth, although the herbaceous weed factor remains, and there is less within-field micro site heterogeneity. The reason why a farmer elects to clear and resow an established pasture also has an important bearing on the formulation of pasture seed mixtures for pasture renewal.

Reasons for pasture renewal

Farmers may routinely renew an area of established pasture where it is a field in a crop–pasture rotation. In areas of long-term, permanent pastures, changes in botanical composition that have led to an undesirable content of species which are low-yielding, unpalatable or harmful to stock health can be a reason for pasture renewal. Marked reduction of pasture cover brought about by atypical climatic conditions, particularly drought, can also be a prompt for pasture renewal. This has been an important motivation in New Zealand in recent years, driven by the opinion that overuse of ryegrass in drought-prone areas has aggravated losses of pasture production in dry years. Consequently there has been promotion of the use of the more drought-tolerant species cocksfoot, tall fescue and phalaris (*Phalaris aquatica*) (Rumball, 1983; Milne *et al.*, 1993).

Pasture renewal might also be prompted by a wish to capitalize on the improved qualities of newly released herbage cultivars. These qualities may include greater dry-matter yield, redistribution of yield into seasons when pasture growth is normally below stock requirements, better feed quality, improved pest and disease resistance and removal of herbage factors injurious to stock health. In respect of the last two qualities, understanding of the biology of ryegrass fungal endophytes has had a marked effect on the direction of ryegrass cultivar development and use in New Zealand in recent years (MacFarlane, 1990).

It should also be remembered that pasture renewal provides the opportunity to establish a mixture of species that enhances all the qualities that can be advanced by plant breeding and selection. Also, attention to the composition of seed mixtures and their management after establishment will enhance the chances that the characteristics of new herbage cultivars will be expressed in an established pasture. To achieve these enhancements, it is necessary that equal emphasis be given to effort on pasture plant breeding and to formulation of pasture mixtures.

Seed number and size

Establishing a suitable ratio of grass and clover is a key consideration in pasture systems where the N nutrition is dependent on fixation by the legume. In the early stages of establishment, changing the proportion of grass and clover in the seed mixture can influence this ratio. However, where white clover is the sown legume, its spread by stoloniferous growth will quickly override any effect of different proportions of its seed in the sown mixture (Harris and Thomas, 1973). It is curious that the common seed rate of 20 kg ryegrass and 5 kg clover ha^{-1} that is current in New Zealand provides about the same number of seed of the two species per unit area. Has this seed ratio, arrived at through practice rather than through theory, a fundamental biological significance?

Information about the individual seed weight of pasture species is a good starting-point in the design of pasture mixtures. This information for pasture species used in New Zealand is tabulated by Charlton (1991). Together with information about the viability of seed, an estimate can be made of the potential number of seedlings that can be established from a given weight of seed. Restricting this seed to a given area of land brings into play density-related thinning of seedlings and this is a key determinant of the percentage establishment of viable seed. Other key causes in lowering the percentage of sown viable seeds that establish include seed predators, disease, climatic situations unsuitable for germination and establishment in the period after sowing, and placement in microsites unsuitable for germination and establishment. The last two causes are modified by choice of the season when sowing takes place

and methods employed in the preparation of seedbeds and the placement of seed. The preferred season of pasture sowing in New Zealand is autumn. This is because temperature and moisture levels are satisfactory for establishment at this time, winter cold is not usually injurious to seedlings and the risk of drought-induced death of seedlings that may follow spring sowing is avoided.

As well as the adjustment of the ratio of species in a mixture according to the number of viable seeds, a further adjustment that is required derives from the general response that large seed have more rapid and higher establishment rates (Brown, 1977; Hampton, 1986). As a consequence seedlings from large seed have a competitive advantage over small seed during establishment. This effect is a factor in the commonly observed suppression of cocksfoot and timothy when they are sown in mixture with ryegrass (Cullen, 1964). The 1000-seed weights (g) of these species are ryegrass 2.0, cocksfoot 0.9 and timothy 0.4. If seed cost was not a factor to consider and the objective was to establish a pasture with similar contents of the species early in the life of the pasture, it could be better to sow similar weights of the species. Sowing each species at a rate of 1 kg ha^{-1} would provide 50 diploid ryegrass, 100 cocksfoot and 250 timothy seeds m^{-2}.

Inherent growth rates and plant size

There are also differences between species in establishment rate independent of seed weight. Tall fescue provides a good example of this, for, although its 1000-seed weight of 2.6 g is higher than that of diploid ryegrass, it is noted for its slow establishment, which makes it susceptible to competition from ryegrass (Brock, 1972). In theory, this difference could be compensated for by sowing a slow-establishing species, such as tall fescue, before a fast-growing species, such as ryegrass. The more practical option is to either markedly reduce or eliminate the ryegrass content where the objective is to establish slow-establishing herbage species. In promoting the use of cocksfoot, tall fescue and phalaris in dry eastern regions of New Zealand in recent years, the preference of farmers for ryegrass had to be overcome by practical demonstration of successful establishment of these three drought-tolerant species (Milne et al., 1993).

It is also relevant to consider the potential size a plant of a sown species can attain. Thus, while it is more difficult to establish cocksfoot, tall fescue,

timothy and phalaris in pasture than it is for ryegrass, the first four species can develop taller plants
with larger basal areas. Plant height is an important
determinant of the outcome of competition for
light and, if plants of these four species are established in sufficient numbers, they can shade and
suppress ryegrass, where defoliation is infrequent
and well above the soil surface.

Different tillering rates and tiller survival will
determine the rate of expansion of the basal area of
plants to occupy space from which other plants are
excluded. Also very important is the capacity of a
species to spread vegetatively by stolons or rhizomes.
This capacity has already been noted for white clover.
It is also a key characteristic of browntop, which, particularly in wet hill-country pastures in New Zealand,
where soil fertility is low and grazing is frequent and
hard, can spread vegetatively from low initial contents
to dominate a pasture (Harris and Thomas, 1972;
Harris, 1973a; Fig. 8.1). Provided a sufficient number
of plants of the species given as examples can survive
the initial establishment phase, cutting or grazing
management can be used to alter their content in a
fully established or renewed pasture.

Seeds as the building blocks of ideal pasture mixtures

Pasture renewal can be undertaken with reference
to the history of the pasture that is to be replaced.
Thus there can be more informed choice of the
species to be used and also the opportunity to take
up the most advanced products of pasture plant
breeding. Building a seed mixture in a precise way
requires information about the characteristics of the
seeds of the species to be mixed, the rate at which
these seed germinate and establish, the patterns of
the vegetative spread of the species and the potential
size and form of the component species.

Establishing a pasture using only one species, as
if it were a crop, is a very simple matter compared
with establishing a pasture with a mixture of species
in defined proportions. There would be little point
in going to the trouble of growing mixtures unless
they provided greater quantity and better quality
of herbage, produced in a seasonal pattern that
better matches livestock requirements than that
obtained by growing a series of forage species as
one-species crops.

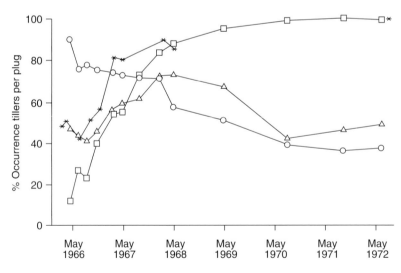

Fig. 8.1. Changes over 6 years in the presence of ryegrass, white clover ('Grasslands Huia') and unsown
browntop (*Agrostis capillaris*) in a sowing on cultivated North Island, New Zealand hill country. The sown
ryegrass consisted of a 50:50 mixture of 'Grasslands Ruanui' perennial ryegrass and 'Grasslands
Manawa' short-rotation ryegrass. Changes in the proportions of these cultivars in the pasture were
followed for 2 years. *, Per cent 'Grasslands Ruanui' in the surviving ryegrass population; □, per cent
presence of browntop; △, per cent presence of white clover; ○, per cent presence of ryegrass. (From
Harris, 1973a.)

Optimizing the Use of Environmental Resources in Different Spaces

Getting more from mixtures than from monocultures

So far, the principles applying to the formulation of pasture seed mixtures that have been considered relate mostly to defining mixtures of species that will establish and survive to provide cover against unwanted species and produce forage in varying environments and for different livestock systems. This leads on to a consideration of whether mixtures of species can provide both more forage and a better seasonal distribution of forage production for livestock needs than species grown in monocultures.

If an environment has a single pool of resources to which all plant species or cultivars have equal access for their growth requirements, then it could be recommended that the species that can produce the most herbage suitable for livestock production in that environment should be the only one sown. Alternatively, if there are different pools of growth resources to which different species or cultivars have different levels of access, then it can be envisaged that a mixture of these would provide more forage. A further possibility is that a series of species grown as monocultures, each best adapted to growth conditions at different times of the year, might produce more forage than if they were grown in a single mixture. These alternatives are considered within the concept of 'resource space' implicit in the theoretical model of plant competition developed by de Wit (1960).

Resources in 'spaces'

Experimental demonstration of this model involves growing species in monocultures and in mixtures in which the ratio of the species is varied but with the number of plants per unit area at establishment held constant. There are three yield outcomes.

First, where species compete for the same 'space' containing a limited resource, if one species is more successful in capturing this resource, it gains in biomass more than the other. This species then becomes dominant, and the biomass of the suppressed species is reduced in proportion. In this outcome one of the species monocultures produces the highest yield, and usually this species is dominant in mixture. This outcome can be described as competition for the same 'space'.

The second yield outcome is where a mixture of two species provides more yield than the average yield of the two species grown in monoculture. This outcome is of particular interest if the mixture yield is higher than those of both monocultures. It arises where, as well as competing for the same 'space', one or both of the species have access to resources in a 'space' or 'spaces' that are not available to the others. While one species may become dominant in this situation, the suppression of the other is not as great as would occur if both species were competing for the same limited resource in the same 'space'. Mixtures of grasses and legumes provide the most obvious example of this situation (de Wit et al., 1966; Harris and Thomas, 1973). Both grass and legume compete for mineral N in the 'soil space' but only the legume has access to N in the 'atmospheric space' through symbiosis with N-fixing rhizobia.

When the yield of a mixture is less than the average of the yields of species grown as monocultures this can be explained by the action of one or both of the species reducing the availability of a limiting resource to both species. The most frequent explanation that is offered for this outcome is that one species secretes a substance that is toxic to the other species. The reality of this phenomenon, known as allelopathy (Newman and Rovira, 1975), has been much debated. A more likely cause of a pasture mixture yielding less than its constituent species grown in monoculture is when the balance of competitive dominance and suppression interferes with the seasonal growth or successional patterns of the species (Harris and Thomas, 1970).

There has been debate about the appropriateness of the terms competition and 'space' as used here to describe the interactions between plants growing closely together in pasture (Hall, 1974). Interference has been suggested as a better term for the mutual influences between plants growing closely together (Harper, 1964). It is true that plants in pasture interfere with the growth of each other in ways that may be positive or negative. Nevertheless, a consequence of this interference is that it influences which species or plants of a species win or lose in the competition for resources required for plant growth. De Wit's concept of 'space' was criticized from the viewpoint that plants did not compete for physical space in the strict sense. In fact, by occupying physical space, plants do deny access of other plants to this space. For example, occupancy of a site by a mat of browntop stolons and rhizomes provides an impediment to the growth and spread of white clover, even where the resources required for growth have been modified to favour white clover (Jackman and Mouat, 1972; Harris, 1974).

The concept of different 'resource spaces' and an understanding of how species vary in their ability to compete for the resources in these different 'spaces', are a key to designing mixtures that yield more than the monoculture yields of their component species. A brief consideration of the key resources that plants compete for, how these resources can be contained in different 'spaces', and examples of how species can differ in their ability to gain access to these 'spaces' follows.

Shoots and the light resource

Competition for the 'light resource' in a pasture principally involves differences in the ability of plants to place their leaves above those of adjacent plants in the pasture. Leaves that are highest in a pasture canopy are the first to receive light, which provides energy for photosynthesis, while leaves lower in the canopy are shaded to varying degrees. If a plant has a high proportion of shaded leaves, its net gain of energy by photosynthesis may well be less than the respiratory energy it requires for maintaining its tissue. Such a plant, by falling below the photosynthetic compensation point, is a loser in the competition for light. If it can maintain itself under shade by the respiration of carbohydrate reserves stored in its tissue, defoliation of the pasture by cutting or grazing may return the plant to light levels that lift it above the compensation point (Harris, 1978; Rhodes and Stern, 1978).

Plant species differ in their ability to raise their leaves above ground, and may be loosely classified as tall- and short-growing species. Tall and short species can enter into different 'light spaces', which are in large part created by the species themselves. Tall species, by placing their leaves highest in the pasture canopy will be exposed to high-intensity light. In most situations, the light intensity at the top of the pasture canopy will be more than is required to meet the energy requirement for photosynthesis, i.e. the leaves in direct sunlight (sun leaves) will be above light saturation. This suits

plagiophile leaves, characteristic of grasses with long, slender, erect leaves, which allow light to penetrate into the pasture canopy. As the intensity of the light and also its quality change as it passes through the canopy, different 'light spaces' are formed. Light in these 'spaces' may be better used by a shade-tolerant species than by tall, sun-loving species. In a shaded 'light space', a horizontal or planophile arrangement of leaves will intercept light more directly, abruptly reducing its penetration to ground level, where its energy will be lost to photosynthesis. Planophile leaves are a characteristic of white clover, and furthermore the phototrophic movements of white clover leaves place the surface of their laminae at right angles to the sun as it moves across the sky.

An old concept of a sound pasture was the presence of desirable 'bottom species' that limited the ingress of weeds into gaps at the base of swards opened out by rank growth of sown tall species. *Poa trivialis* is an example of a species once sown as a bottom grass (Vartha, 1972). As well as reducing the ingress of less desirable species, these bottom species may also function to more effectively utilize the 'light space' near ground level, either immediately after defoliation of a sward that has been allowed to grow tall or more continuously where defoliation is frequent and close.

Roots and water and nutrient resources

Inherent differences between species in the depth their roots can penetrate (Evans, 1977, 1978) can be envisaged as enabling species in mixtures to utilize different 'spaces' containing water and mineral nutrients. This is seen as a useful feature of agroforesty systems, where the root systems of trees can tap into much deeper layers of the soil than surface-rooting herbaceous species. The shallower root systems of hawkweeds (*Hieracium* spp.) compared with those of fescue tussock (*Festuca novae-zelandiae*) was given as a factor in the invasion of New Zealand tussock grassland by *Hieracium*. By their position in the soil profile, it was suggested that hawkweeds had first share of inadequate levels of applied fertilizer and intermittent rainfall (Fan and Harris, 1996).

As well as the physiological differences between grasses and legumes that allow them to tap different nitrogen sources, there may be others that enlarge the 'nutrient resource space' available when species are mixed. This appeared to be the case for a mixture of *Rumex acetosella* and *Trifolium repens*, and the ability of *R. acetosella* to raise the pH of soil in which it grows was indicated as a factor in this response (Harris, 1971). Hall (1974, 1978) speculated that interspecific differences in uptake of P from the labile pool and from alternative sources in the soil might enlarge the 'nutrient resource space' utilized by a mixture. This was based on information that crop species differed markedly in the utilization of the various soil P fractions and that this was most obvious on soils of unusual P behaviour with a high degree of P fixation.

Temperature, time and 'space'

A consequence of different seasonal growth patterns of plants is that the level of demand by the component species of a mixture on the resources of the 'space' that they occupy varies with time. This may provide temporal niche separation so that species do not exclude each other by competition. It also provides the opportunity for seed mixture formulations that spread forage production more evenly throughout the year. By utilizing the 'resource space' more continuously, such mixtures may also provide more yield than if the component species were grown separately.

Different temperature response curves are the main determinant of variation of species' seasonal growth patterns. The optimal temperatures for growth of ryegrass and white clover of 18–21 and 24°C, respectively (Mitchell, 1956), underlie the seasonal shift of dominance of these species. This is an important component of the success of ryegrass–white clover pastures (Harris and Hoglund, 1980; Harris, 1987, 1990; Harris and Rhodes, 1989).

The opportunity for gaining yield by mixing species with different temperature-determined seasonal growth patterns is very much influenced by the climate of the site to be sown (Harris, 1990). For example, seasonal shifts of dominance are limited where high temperatures and adequate water throughout the year, as in the wet tropics, support continuously high growth rates of tropical or C_4 grasses. Alternatively, for climates that regularly have severe winter cold or summer drought, marked restriction of growth to part of the year restricts the expression of different seasonal growth patterns. An experiment on the northern tablelands of New South Wales, Australia, where winters are cool and moist and summers hot and dry, showed that mixtures of ryegrass and paspalum yielded more than the grasses grown in monoculture under irrigated but not under dryland conditions (Harris and Lazenby, 1974).

Warm-temperate or subtropical climates, with even distribution of rainfall throughout the year, provide good opportunities for mixtures of C_3 and C_4 species, which provide higher and more evenly distributed herbage yield. These climatic conditions are characteristic of northern New Zealand, where the subtropical grasses paspalum (*P. dilatatum*) and kikuyu (*Pennisetum clandestinum*) commonly grow in association with ryegrass and white clover. In a controlled climate experiment, monocultures of ryegrass yielded most when temperature was maintained at 14°C day/8°C night, whereas monocultures of paspalum yielded most when temperature was maintained at 24°C day/18°C night. However, when moved from one temperature regime to another to simulate a temperature-regulated seasonal change, the mixtures of these grasses provided more yield for the full period of the experiment than the monocultures. Cutting had a marked effect on how much the yield of mixtures exceeded the monocultures. More frequent defoliation provided a more even balance between the component species, allowing a more effective switch of species dominance when mixtures were moved from one temperature regime to another (Harris *et al.*, 1981a, b).

The need for the very specific matching of species with different temperature-regulated growth periods to particular environments is illustrated by evaluations of tropical grass introductions to northern New Zealand (Davies and Hunt, 1983). Six tropical grasses, each mixed with ryegrass and white clover, were compared with ryegrass–white clover alone at three sites in close geographical proximity. The mixture did not work at one site because frost resulted in non-persistence of the tropical grasses, while at another site the drought-prone sandy soil resulted in non-persistence of ryegrass and white clover. However, at the third site, mixtures including the tropical grasses, and most particularly *Hemarthria altissima*, had higher annual yields than ryegrass–white clover alone. This was because the temperate and tropical components were able to coexist at this third site, the tropical grasses providing the bulk of the growth in summer and the temperate species during the cooler months of the year.

Successful matching of grass and legume species to specific environmental conditions is further illustrated by studies of mixtures of *Digitaria eriantha* and lucerne (*Medicago sativa*) in the subhumid environment of the north-west slopes of New South Wales, Australia (Tow *et al.*, 1997a, b, c). This environment has both summer and winter rainfall, allowing the species to grow when seasonal temperatures were optimal for their growth. As a consequence, the species were able to complement each other's growth and their mixture yielded more than their monocultures over the year. The yield advantage of the mixture was influenced by nitrogen level, the susceptibility of the *Medicago* to flooding and control of the level of dominance of the *Digitaria* in summer.

As indicated by effects of cutting frequency and controlled temperature switches on species dominance (Harris *et al.*, 1981a, b), and by the observation of excessive summer dominance of *Digitaria* (Tow *et al.*, 1997c), pasture defoliation management can have a key role in exploiting temperature regulated 'spaces' in time. Bringing about a switch of summer to winter dominance of different species, along with cleaning out dead-matter residues (Hunt and Brougham, 1967), was the basis for the recommendation of hard 'autumn clean-up' grazing of pasture in New Zealand made by Brougham (1960).

Efficient use of resources in different spaces by species mixtures

It is clear from the foregoing that purposefully constructed mixtures of species can achieve more comprehensive use of resources available in an environment and consequently produce more biomass than monocultures. This property of mixtures is most apparent where the component species of a mixture have physiological differences that allow them to acquire their requirements for biomass production from different sources. Legumes provide the most obvious example of an effective physiological difference, their ability to fix atmospheric N_2 avoiding their need to compete with non-legume components for soil N, as well as increasing the total supply of N to the pasture. This is a phenomenon well utilized by farmers. The first step in its utilization is the inclusion of forage legumes in pasture seed mixtures.

Not so well consciously utilized by farmers is the different seasonal growth periodicity of pasture

species, which is largely regulated by differences of temperature responses of plant species. It is apparent that this phenomenon is most effective in climatic zones between tropical and temperate regions that do not have a marked dry season. Expression of this phenomenon is enhanced by defoliation management that fosters the attainment of dominance of species in the season that most favours their growth. In many pasture regions, utilizing the differences of seasonal growth periodicity of species should be an important consideration in designing seed mixtures. As well as having the potential to provide more annual yield, mixing species of different seasonal growth periodicity will provide a more even spread of herbage production through the year.

Different 'light spaces' exploited by shoots and different 'soil spaces' exploited by roots for water and nutrients other than N are more subtle phenomena, which are not easily utilized by farmers and are deserving of more scientific study. As a suggestion, a mixture of shallow-rooted self-regenerating annual species and deep-rooted perennial species can be envisaged as providing an effective combination for areas of deep soils and markedly seasonal rainfall patterns.

Mixing Cultivars of the Same Species

Degrees of difference and complementation

So far, the mixtures that have been considered have been those for which the components differ genetically to the extent that they can be identified as species or as interspecific hybrids. These components also have well-defined physiological differences, which match their agronomic use to specified ecological and farm management systems. It has also been indicated how marked physiological differences can act to complement the growth of species in mixtures to improve both yield and its seasonal distribution.

Genetic and physiological differences between cultivars of a species are much smaller than those between species. A generally accepted principle is that competition between individuals of the same species is much more intense than that between individuals of different species. This is because they have the same requirements and the same mechanisms of acquiring the light, water and nutrients required for their growth. This is borne out by studies of the competitive interactions between white clover genotypes undertaken by Annicchiarico and Piano (1994, 1997). They showed that competitive–interference effects were greater between the clover genotypes than between clover and ryegrass. Compatibility of the white clover genotypes with the grasses they grew them with was solely determined by the yield of the grass, i.e. the higher the yield of the grass, the greater the competitive suppression of the clover. They concluded that there was no gain of yield by mixing white clover genotypes.

Is mixing cultivars of the same herbage species common?

In 1998 an attempt was made to gather information on the composition of pasture seed mixtures currently being sown in New Zealand. This was designed to update earlier surveys of seeds mixtures (Harris, 1968; Sangkkara et al., 1982), when the number of available species, and especially cultivars, was much less than at present. Seed merchants approached were uncooperative in providing details of specific mixtures sold to farmers. However, an indication of what was being recommended to farmers was provided by a booklet (no date or author) published by a company that is a leading seller of pasture seed in New Zealand. Of the five mixtures in this booklet, four recommended the mixing of cultivars of both ryegrass and white clover.

One of these mixtures included 'Grasslands Tahora', 'Grasslands Pitau' and 'Grasslands Kopu', which, respectively, are small-leaved, medium large-leaved and erect, large-leaved cultivars of white clover, together with three ryegrass cultivars. No explanation was given of the purpose of mixing these white clover cultivars. Mixing them undoes the selection by breeders of the three white clover leaf types, known to be suited for persistence and performance in different grazing systems. This was a proprietary mixture, with the designer of the mixture having ownership rights over five of the six cultivars it contained.

Evans et al. (1995) studied the white clover yields of mixtures of the small-leaved white clover varieties 'S184' or 'Gwenda' with medium-leaved 'Menna' or 'Donna' sown together with a commercial grass mixture. In introducing this study, they referred to studies that showed that the white clover content of pasture can be considerably altered by the choice of the cultivar sown. They also com-

mented that blends of white clover cultivars have long been included in seed mixtures in the UK, even though there was no scientific evidence that supported this practice. Their results showed that the lower first year yield of 'S184' could be overcome by mixing with the slightly larger-leaved 'Menna', but in the longer term clover content was reduced where 'Donna', which has larger leaves than 'Menna', was mixed with the small-leaved cultivars. A likely explanation of this result is that 'Donna' excessively suppressed 'S184' by overtopping and shading it in the first year, and 'S184' was weakened to the extent that it could not fill the gap left by the poor persistence of 'Donna' in later years.

Evans et al. (1995) recommended that care should be taken in mixing white clover cultivars and that this should only be done according to sward management and cultivar persistence. Their study is a rare example of the investigation of the outcomes of mixing white clover cultivars. It emphasizes that unreasoned mixing can have negative results, but that rational mixing of cultivars can produce useful outcomes.

Genotype × environment interaction

To acquire plant variety rights for a herbage cultivar, it is necessary to establish that the selection is distinctive, uniform and stable. As the number of protected herbage cultivars increases, it probably follows that, in order to meet the requirement of distinctiveness, the genetic basis of herbage cultivars becomes narrower. From the narrowing of the genetic basis of a cultivar, it follows that there is a need for more specific definition of the environment and management system in which the cultivar can express those characteristics that make it distinctive and agronomically superior in that system. In another system, the cultivar may grow differently and is likely to be inferior in its agronomic performance to cultivars designed for this other system. This is an expression of the phenomenon of genotype × environment interaction.

Genotype × environment interaction, whereby individual plants or populations (cultivars) of the same or similar genetic composition perform differently in different environments, has long been problematic for both plant breeders (Fejer, 1959; Hill, 1975) and those who undertake the comparative evaluation of herbage cultivars (Lazenby and Rogers, 1964). Even though the destiny of selected herbage plants is to be grown in swards where they will be subjected to both inter- and intraspecific competition of high intensity, their breeders find it both more practical and easier to base their selections on widely spaced plants. As demonstrated by Lazenby and Rogers (1964), when grown at wide spacing, relative differences of ryegrass cultivar yields are exaggerated and sometimes reverse the relative yield of the cultivars grown at densities where inter-plant competition occurs. Also both breeders of herbage cultivars and those who make comparative evaluations of bred herbage cultivars are limited in the number and scope of the environments in which they can do their work.

Of key relevance also is the difference of performance of plants under cutting and grazing (Korte and Harris, 1987). Smaller scale, easier control of standards and simpler measurement techniques make cutting evaluation the preferred option, even though grazing provides the more realistic test of plant selections destined for use in grazed pastures. Further, examiners for plant variety rights focus their attention on characters such as tiller size and flowering date, which are expressed more consistently in different growth conditions but bypass important agronomic characters that relate to yield, quality and persistence, which are strongly modified by environmental conditions.

Genetic shifts in sown mixtures of genotypes

The genotype × environment interactions recorded by Lazenby and Rogers (1964) were derived from unit area yields of ryegrass cultivars grown in monoculture. A further step in investigating genotype × environment interaction is to follow shifts in the genetic structure of populations (cultivars) or individual genotypes when mixed together in swards and subjected to different environments and competitive situations. This was done in a series of experiments in New Zealand. When the cultivars 'Grasslands Manawa' and 'Grasslands Ruanui' were mixed (Brougham and Harris, 1967; Harris and Brougham, 1970; Fig. 8.2), genetic shifts in the grazed pasture of the hybrid cultivar 'Grasslands Manawa' ryegrass were significant (Brougham et al., 1960), and were rapid and in different directions according to grazing management. Further, marked genotype × environment interactions in response to density, cutting height and the presence of white clover were shown by individual ryegrass genotypes

(Harris, 1973b). Harris and Thomas (1970) investigated details of the competitive interactions underlying these genetic shifts and genotype × environment interactions. Competition for light was the main factor determining the genetic shifts, taller genotypes overtopping and shading shorter genotypes with less frequent and higher defoliation. Removal of elevated stem apices with close defoliation also acted to shift the genetic structure towards shorter, more prostrate genotypes where defoliation was frequent and close.

Determining changes in the genetic structure of the mixtures containing 'Manawa' and 'Ruanui' ryegrass was possible through tracking the frequency of seed fluorescence under ultraviolet light and of awning (see Fig. 8.2) These characters taxonomically distinguish *Lolium multiflorum* from *L. perenne*, the parents of 'Manawa'. Different genetic markers are required to track genetic shifts in cultivar populations selected from within the two species. This was done using allozymes to track the frequency of 'Grasslands Nui' ryegrass relative to resident unsown ryegrass in a series of soil fertility and grazing treatments (Sanders et al., 1989). Five and a half years after sowing, the proportion of Nui in the various treatments varied from 5 to 44%. This illustrates the difficulty of introducing a new cultivar to replace a resident ryegrass population, even without the complication of purposefully mixing cultivars.

A similar technique was applied to determine the persistence of 'Grasslands Kopu' white clover mixed with 'Grasslands Nui' ryegrass 10 years after sowing (Prins et al., 1989). The 43% content of 'Grasslands Kopu' determined by this technique showed that this large-leaved cultivar was persistent under the rotational grazing management applied to the sowing. Nevertheless, the resident white clover content was considerable.

The studies reviewed show how quickly desirable agronomic traits of herbage cultivars can be lost as a consequence of competitive interactions between sown and/or resident genotypes of the same species in pasture. They lead to the question of whether cultivars of the same species should ever be mixed. If the answer to this question is yes, then, if there is a wide choice of cultivars for particular species, as is the case for ryegrass and white clover in New Zealand (see Table 8.1), the issue becomes that of which cultivars to mix.

The most likely situation where sowing a mixture of cultivars could be beneficial is where the area

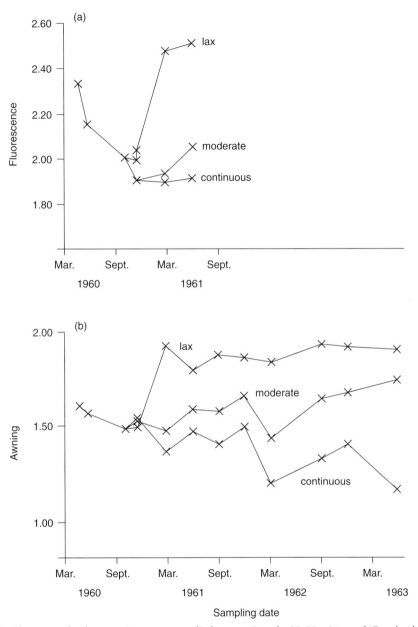

Fig. 8.2. Changes under three grazing systems applied to a sowing of a 50:50 mixture of 'Grassland Ruanui' perennial ryegrass and 'Grasslands Manawa' short-rotation ryegrass of the proportions of ryegrass genotypes, showing (a) root fluorescence under ultraviolet light and (b) awns on seed. Fluorescence and awning are absent in 'Grasslands Ruanui'. Lax grazing, grazed when herbage reached a height of 23–30 cm down to a height of 8–10 cm; moderate grazing, grazed when herbage reached a height of 15–20 cm down to a height of 5–8 cm; continuous grazing, continuously grazed to a height of approximately 3–5 cm. From Brougham and Harris, 1967.

being sown is heterogeneous in time and space, thus providing a diversity of environmental microsites to which different genotypes are more specifically adapted. The nature of this situation is considered with reference to studies of the heterogeneity of microsites present in New Zealand hill-country pastures.

Genotype matching to hill-country pasture microsites

The heterogeneous nature of New Zealand hill-country climate, soil and topography and the way this influences botanical composition and pasture growth have been defined by several studies (Radcliffe, 1968, 1982; Rumball and Esler, 1968; Brougham *et al.*, 1973; Gillingham, 1973, 1982; Grant and Brock, 1974; Lambert and Roberts, 1976, 1978; Gillingham and Bell, 1977). Collections were made to ascertain whether the diversity of hill-country environments might have brought about the natural selection of ecotypes adapted to them (Suckling and Forde, 1978; Forde and Suckling, 1980; Caradus *et al.*, 1990a, b). From these collections new cultivars have been selected for their suitability for hill-country pastures, e.g. 'Grasslands Tahora' white clover, a small-leaved, densely stoloned, low-growing white clover suited for continuously grazed pasture on wet hill country, and 'Grasslands Muster' browntop, selected as a grass for similar conditions.

Caradus *et al.* (1990a) found that dry hill-country populations of white clover were more cyanogenic, larger-leaved and taller and had larger tap-root diameters and different leaf mark characters compared with those from wet hill country. This study also found differences between populations from north and south aspects of dry hill farms but not wet hill farms, the difference relating to characters that confer drought tolerance. However, in wet hill country, populations from pastures rotationally grazed with cattle were taller than those grazed by sheep (Caradus *et al.*, 1990b).

Description of resident ryegrass genotypes from North Island New Zealand hill country showed considerable variation of types within a restricted range, characterized by many, fine-leaved, prostrate tillers (Wedderburn *et al.*, 1989). The variation within this range was not associated with specific microenvironments within the hill-country area studied. 'Grasslands Ruanui' perennial ryegrass fell within the same range of variation as the hill-country genotypes, but more recently selected cultivars of perennial ryegrass, 'Grasslands Nui', 'Ellet' and 'Yatsyn' and the hybrid 'Grasslands Manawa', clearly fell outside this range by having fewer and larger semi-erect tillers (Fig. 8.3). The rate of loss of 'Grasslands Manawa' from a hill-country sowing (Harris, 1973a; see Fig. 8.1) indicates that the segregation of ryegrass genotypes suited to hill country shown by Wedderburn *et al.* (1989) occurs rapidly.

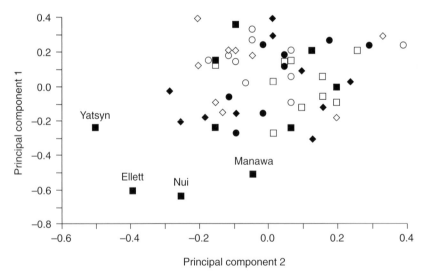

Fig. 8.3. Principal component analysis of the morphological characteristics of resident hill-country ryegrass genotypes and bred ryegrass cultivars. Bred cultivars are marked by ■. The other symbols relate to genotypes from hill-country microsites defined by slope, aspect and stock camping. The bred cultivars 'Yatsyn', 'Ellet', 'Nui' and 'Manawa' were separated from the hill-country ryegrass genotypes by this analysis. The cultivars 'Ruanui', 'Ariki', 'Takapau Persistor' and 'Droughtmaster' and local hill-country populations from Tara Hills and Matawai fell within the range of variation of the hill-country ryegrass genotypes specifically sampled for the study. (From Wedderburn *et al.*, 1989.)

Wedderburn *et al.* (1996) studied the effects of herbicide and oversowing ryegrass and white clover cultivars on the genetic structure of these species in hill country. Their results suggested that, where soil fertility was high and there was not a large resident ryegrass seed pool, oversowing could change the genetic structure of ryegrass populations. Changes in the content and genetic composition of white clover by oversowing a new cultivar of this species were more difficult to achieve. Wedderburn and Pengelly (1991) concluded that a proportion of the numerous and subtly different environmental niches present in hill country could be exploited by the genetic variation present within resident hill-country ryegrass populations. They recommended that management to foster the growth and spread of these resident populations was better than attempting to introduce new cultivars. The management practices most likely to increase the competitive advantage of the resident ryegrass and white clover populations in these hill-country pastures would be rotational grazing with cattle and increased application of phosphate fertilizer (Lambert *et al.*, 1986).

Cultivar design and their suitability for mixing

Genetic variation within cultivars

To further address the questions of which cultivar to sow and whether there is value in mixing cultivars, three categories of herbage cultivars are defined: (i) those with broad genetic bases; (ii) those with narrow genetic bases, either derived from locally adapted populations (ecotypes) or purposefully selected for yield and persistence in specified environments and management systems; and (iii) those selected for specific factors relating to herbage quality, animal health and disease and pest resistance.

The trend towards private ownership of herbage cultivars and control of ownership by plant variety rights, which requires meeting standards of distinctiveness, uniformity and stability, is acting to reduce the availability of cultivars with broad genetic bases. Hill (1975) was of the opinion that the success of the old British perennial ryegrass cultivar 'S. 23' derived from the broad genetic base on which it was founded. The same opinion applies to its New Zealand equivalent, 'Grasslands Ruanui'

perennial ryegrass. Thirty years ago, it was generally considered that this one truly perennial ryegrass met the requirement for *L. perenne* in all New Zealand pastures. Levy (1970) wrote that, within true perennial ryegrass, it was not possible to select widely divergent plant forms. He considered it fortunate that *L. perenne* and *L. multiflorum* were freely interfertile and considered that, by mixing their hybrids and the parent species, pastures of better palatability, spread of production and yield could be achieved.

Levy (1970) was firmly of the opinion that the use of natural ecotypes was 'agronomically unsound' because they were low producers and had extended periods of dormancy. This opinion was linked to his preference for selection of herbage plants under conditions largely non-limiting to their growth, rather than conditions characteristic of where the selection was likely to be used. This philosophy dominated the direction of plant breeding and pasture management in New Zealand for many years. However, with regard to the prime species, perennial ryegrass, practical experience of the use of 'Grasslands Ruanui' showed that, even with its broad genetic base, it did not persist in all New Zealand pastoral situations. Further, it was not economically feasible in many of these situations to develop soil fertility and water availability levels satisfactory for its growth or persistence. Alternatively, it was found that, in some high-production pastures, local ecotypes of perennial ryegrass outperformed 'Grasslands Ruanui' ryegrass. These experiences led to a new direction in plant selection and breeding in New Zealand (Burgess, 1987).

The observations of Wedderburn *et al.* (1989, 1996; see section on genotype matching to hill-country pasture microsites) point to the difficulty of selecting cultivars that surpass 'Grasslands Ruanui' as a perennial ryegrass suitable for low-soil fertility hill-country sowings. This situation has prompted the selection and improvement of alternative species that are more stress-tolerant and persistent but with lower yield potential than ryegrass. 'Grasslands Muster' browntop provides an example of this approach. However, for high-fertility, adequately watered pastures, 'Mangere' an ecotype from the district of that name south of Auckland, and then 'Ellet' and 'Grasslands Nui' selected from this ecotype, came to replace 'Grasslands Ruanui' as New Zealand's premier perennial ryegrass cultivar. In the period 1993–1997, 53% of the certified perennial ryegrass seed produced in New Zealand was 'Grassland Nui' and 'Grasslands Ruanui' provided only 3% of the seed (MAF, 1998).

With this background of the derivation of herbage cultivars in New Zealand, a series of recommendations can be made for including and combining cultivars according to their genetic structures or for sowings in different environmental and farming system situations.

Using cultivars with a broad genetic base

The first recommendation applies to cultivars that have a broad genetic base and thus have the potential for wide ecological adaptation. There remains a place for such cultivars where the paddock to be sown contains a variety of microsites or where variation of environmental conditions in time is not easily predictable. In New Zealand grassland farming, the most important year-to-year variation in environmental conditions is the intensity of summer drought. Buffering the impact of drought on the persistence and herbage production of a species through variation in the response of its consistent genotypes to moisture stress will also be enhanced by variation of the genotypes' response to different intensities of grazing. This follows from the intensity of defoliation by grazing being greater in drought years, when pasture production cannot keep up with stock requirements for herbage.

Broadening the genetic base of a sown species can also be purposefully achieved by mixing cultivars and can be especially effective if it is known that the cultivars chosen for mixing fit into well-defined parts of the range of variation of a species. Good examples of this practice were mixtures that included 'Grasslands Ruanui' perennial ryegrass, 'Grasslands Manawa' hybrid ryegrass and 'Grasslands Paroa' Italian ryegrass (Harris, 1968). Mixing of cultivar entities of this kind needs to take into account the intended life of the pasture, through adjustment of the seed ratios of the components and grazing management that allows the constituent cultivars to express their agronomic characteristics effectively (Brougham, 1954d). Sound information about the breeding and genetic structure of cultivars is needed for mixing to achieve designed broadening of the genetic structure of a sown species. Rather than introducing a mixture of new cultivars, it may be just as effective, as recommended by Wedderburn and Pengelly (1991), to use the genetic diversity present in resi-

dent populations in heterogeneous environments by applying management that fosters the growth and spread of the surviving and presumably adapted genotypes.

Using cultivars in environments for which they have been selected

The second recommendation is that, where a cultivar has been specifically selected for its suitability for defined environmental or management situations, it should be sown only in those situations and not mixed with other cultivars of the same species. This may seem a self-evident recommendation, but nevertheless mixing that is contradictory to this recommendation does occur, e.g. 'Grasslands Tahora', 'Grasslands Pitau' and 'Grasslands Kopu' white clovers in the proprietary mixture discussed previously. The likelihood of inappropriate mixing of cultivars selected for different environments and management systems increases as more cultivars of this kind become available (see Table 8.1). Release of cultivars without adequate information about their genetic structure, environmental tolerances and management requirements will further enhance the likelihood that they will be inappropriately mixed.

If not sown in its intended area of use, a cultivar is likely to be rapidly eliminated by natural selection, as a consequence of competition with better-adapted species. This process will be accelerated if there is also a resident adapted population of the species of the sown cultivar or if it is blended with a cultivar with a broad genetic base. This recommendation is supported by the observed interactions between and survival of resident and sown populations of white clover and ryegrass in hill-country pasture (Wedderburn et al., 1996), 'Grasslands Nui' ryegrass under different management systems (Sanders et al., 1989) and 'Grasslands Kopu' white clover under rotational grazing (Prins et al., 1989).

Using cultivars selected for characters other than those for ecological adaptation

The third recommendation relates to cultivars that have been selected or modified for herbage quality, effects on animal health or resistance to plant diseases and pests. A recent New Zealand example of selection for herbage quality is 'Aries HD' perennial ryegrass, claimed to be the first commercial perennial ryegrass selected specifically for improved digestibility (Bluett *et al.*, 1997).

Control of the presence or absence of different strains of the ryegrass fungal endophyte (*Neotyphodium lolii*) in ryegrass cultivars is regarded as modification rather than selection. Understanding of the biology of this endophyte has played a dominant role in the improvement of ryegrass in recent years (Latch, 1994). Presence of the endophyte can have negative effects on the health and performance of livestock (Fletcher, 1993), but can have a positive effect by ameliorating damage to ryegrass by Argentine stem weevil (*Listronotus bonariensis*) (Rowan *et al.*, 1990). Cultivars of ryegrass with high or low levels of the endophyte have been developed with recommendations for their use in different areas according to the incidence of Argentine stem weevil or negative effects on animal health (MacFarlane, 1990). More recently, strains of the endophyte that retain negative effects on Argentine stem weevil but are without negative effects on stock performance have been isolated. Inoculation of these endophyte strains into existing low-endophyte cultivars of ryegrass will markedly increase the number of cultivars protected by plant variety rights in New Zealand (L.R. Fletcher, Lincoln, New Zealand, 1998, personal communication).

Mixing a herbage cultivar with a specific quality or resistance property with cultivars of the same species or a hybrid without the property would directly act to dilute the advantage of the property. Further, it might activate natural selection against the specifically improved property, e.g. where animals preferentially graze high-digestibility genotypes or avoid toxic plants. Recognition of the importance of maintaining a specific cultivar characteristic in a pasture is illustrated by efforts to avoid contamination of endophyte-free ryegrass-based dairy pastures by volunteer ryegrass with high endophyte levels (van Vught and Thom, 1997). Nevertheless, a current proprietary ryegrass white clover seed mixture for rotational grazing systems recommends the mixing of high-digestibility 'Aries HD' perennial ryegrass, 'Embassy' perennial ryegrass and 'Maverick Gold' hybrid ryegrass. These cultivars are indicated to have average, high and low levels of endophyte, respectively, although a low-endophyte option of 'Embassy' was offered for pastures not attacked by Argentine stem weevil. The company recommending this mixture has proprietary rights over the three ryegrass cultivars it contains.

The recommendation that a cultivar with specific herbage quality or pest and disease resistance property should not be mixed with other cultivars of the same species or a hybrid is made more strongly than the similar recommendation given for cultivars selected for their performance in defined pasture habitats. Confidence to include only one cultivar of a species in a mixture should be increased with more precise matching of selections to habitats and greater control of herbage quality and resistance properties.

Conclusions

When establishing new pastures or renovating old ones, farmers will sow seed of those species that they believe will establish and provide them with the greatest amount of herbage of the best possible quality. They will prefer those species that provide herbage in a pattern of seasonal availability that will best meet the food requirements of their livestock. Sowing these species together in a mixture quickly leads to competitive interactions between them, with weeds and with resident plants of pastures being renovated. The species sown and their competitive interactions will set the course of succession as a pasture ages, but this course can by altered by fertilizer, defoliation and other tools of pasture management.

Farmers are inclined to use complex seed mixtures if they are uncertain about the outcomes of establishing or renovating pastures. Understanding of the competitive interactions between species and how this regulates successional changes of botanical composition and greater control of the pasture environment by more intensive management allow simplification and greater precision in the formulation of seed mixtures.

An understanding of the competitive abilities of the species available for sowing in a mixture can be used to determine the balance of species in a pasture by varying the ratio of seed of the species in the mixture. Defining these seed ratios is assisted by information about the biological characteristics of both the seed of species and their mature plants. Sown species that establish quickly and exert strong competition are useful for their ability to exclude unwanted plants from pasture. This property has to be used carefully as it can also suppress slow-establishing species included in mixtures to provide long-term persistent components of pasture.

Mixtures can achieve more comprehensive use of the resources available in pasture environments if the component species can draw their growth requirements from different 'spaces'. To varying degrees, this allows species to avoid competition with each other. The ability of legumes to obtain their N requirements from the atmosphere or the soil provides the most obvious example for this. It also occurs for mixtures of species that have temperature-regulated responses that attain their optimum levels in different seasons of a year. There is a need for research to gain better understanding of the different 'spaces' in which resources required for plant growth are held. Related to this is the need for improved understanding of the physiological characteristics of species that allow them to obtain their growth requirements from different 'spaces'.

Competition is intensified the closer the genotype of interacting plants. This is because genetically similar plants have the same requirements for growth and identical means of obtaining these requirements. This relationship underlies the need to exercise caution in including more than one cultivar of a species in a pasture seed mixture. Knowledge of the genetic structure of cultivars allows them to be sown in environments where competitive interactions will not act to eliminate them by natural selection and where they will grow to express their desirable agronomic traits selected by pasture plant breeders.

References

Annicchiarico, P. and Piano, E. (1994) Interference effects in white clover genotypes grown as pure stands and binary mixtures with different grass species and varieties. *Theoretical and Applied Genetics* 88, 153–158.

Annicchiarico, P. and Piano, E. (1997) Response of white clover genotypes to intergenotypic and interspecific interference. *Journal of Agricultural Science* 128, 431–437.

Bluett, S.J., Hodgson, J., Kemp, P.D. and Barry, T.N. (1997) Animal evaluation of Aires HD perennial ryegrass selected for high digestibility. *Proceedings of the New Zealand Grassland Association* 59, 245–249.

Brock, J.L. (1971) A comparison of 'Grasslands 4700' and 'Grasslands Huia' white clovers in establishing ryegrass–clover pastures under grazing. *New Zealand Journal of Agricultural Research* 14, 368–378.

Brock, J.L. (1972) Effect of sowing depth and post-sowing compaction on the establishment of tall fescue varieties. *New Zealand Journal of Experimental Agriculture* 1, 11–14.

Brougham, R.W. (1954a) Pasture establishment studies. I. The effect of grass seeding rate on the establishment and development of pasture of short-rotation ryegrass, red and white clovers. *New Zealand Journal of Science and Technology* A35, 518–538.

Brougham, R.W. (1954b) Pasture establishment studies. II. The influence of herbage height at grazing and of grass seeding rate, on the establishment and development of pasture of short-rotation ryegrass, red and white clovers. *New Zealand Journal of Science and Technology* A35, 539–549.

Brougham, R.W. (1954c) Pasture establishment studies. III. Barley as a cover crop in establishing pastures for the production of late autumn and winter green-feed. *New Zealand Journal of Science and Technology* A36, 47–59.

Brougham, R.W. (1954d) Pasture establishment studies. IV. A comparison of mixtures containing short-rotation ryegrass, perennial ryegrass, or both, as the grass component. *New Zealand Journal of Science and Technology* A36, 365–374.

Brougham, R.W. (1960) The effect of frequent hard grazing at different times of the year on the productivity and species yields of a grass–clover pasture. *New Zealand Journal of Agricultural Research* 3, 125–136.

Brougham, R.W. and Harris, W. (1967) Rapidity and extent of changes in genotypic structure induced by grazing in a ryegrass population. *New Zealand Journal of Agricultural Research* 10, 56–65.

Brougham, R.W., Glenday, A.C. and Fejer, S.O. (1960) The effects of frequency and intensity of grazing on the genotypic structure of a ryegrass population. *New Zealand Journal of Agricultural Research* 3, 442–453.

Brougham, R.W., Grant, D.A. and Goodall, V.C. (1973) The occurrence of browntop in the Manawatu. *Proceedings of the New Zealand Grassland Association* 35, 86–94.

Brown, K.R. (1977) Parent seed weight, plant growth, and seeding in 'Grassland Tama' Westerwolds ryegrass. *New Zealand Journal of Experimental Agriculture* 5, 143–146.

Burgess, R. (1987) The development of our pasture plant resources: then and now. *Proceedings of the New Zealand Grassland Association* 48, 89–92.

Caradus, J.R., MacKay, A.C., Charlton, J.F.L. and Chapman, D.F. (1990a) Genecology of white clover (*Trifolium repens* L.) from wet and dry hill country pastures. *New Zealand Journal of Agricultural Research* 33, 377–384.

Caradus, J.R., Chapman, D.F., MacKay, A.C. and Lambert, M.G. (1990b) Morphological variation among white clover

(*Trifolium repens* L.) populations collected from a range of moist hill country habitats. *New Zealand Journal of Agricultural Research* 33, 607–614.

Chapman, D.F. and Williams, W.M. (1990a) Evaluation of clovers in dry hill country 8. Subterranean clover at 'Ballantrae', New Zealand. *New Zealand Journal of Agricultural Research* 33, 569–576.

Chapman, D.F. and Williams, W.M. (1990b) Evaluation of clovers in dry hill country 9. White clover at 'Ballantrae', and in central Wairarapa. *New Zealand Journal of Agricultural Research* 33, 577–584.

Charlton, J.F.L. (1991) Some basic concepts of pasture seed mixtures for New Zealand farms. *Proceedings of the New Zealand Grassland Association* 53, 37–40.

Corkill, L. (1945) Short rotation ryegrass, its breeding and characteristics. *New Zealand Journal of Agriculture* 71, 465–468.

Cullen, N.A. (1964) Species competition in establishing swards: suppression effects of ryegrass on establishment and production of associated grasses and clovers. *New Zealand Journal of Agricultural Research* 7, 678–693.

Davies, L.J. and Hunt, B.J. (1983) Growth of tropical grass introductions in mixed swards with ryegrass and clover under mowing. *New Zealand Journal of Agricultural Research* 26, 415–422.

de Wit, C.T. (1960) On competition. *Verslagen van Landbouwkundige Onderzoekingen* 668, 1–82.

de Wit, C.T., Tow, P.G. and Ennik, G.C. (1966) Competition between grasses and legumes. *Verslagen van Landbouwkundige Onderzoekingen* 687, 1–30.

Evans, D.R., Williams, T.A., Jones, S. and Evans, S.A. (1995) The effect of blending white clover varieties and their contribution to a mixed grass/clover sward under continuous sheep stocking. *Grass and Forage Science* 50, 10–15.

Evans, P.S. (1977) Comparative root morphology of some pasture grasses and clovers. *New Zealand Journal of Agricultural Research* 20, 331–335.

Evans, P.S. (1978) Plant root distribution and water use patterns of some pasture and crop species. *New Zealand Journal of Agricultural Research* 21, 262–265.

Fan, J. and Harris, W. (1996) Effects of soil fertility level and cutting frequency on interference among *Hieracium pilosella*, *H. praealtum*, *Rumex acetosella* and *Festuca novae-zelandiae*. *New Zealand Journal of Agricultural Research* 39, 1–32.

Fejer, S.O. (1959) Intraspecific competition as a factor in ryegrass breeding. *New Zealand Journal of Agricultural Research* 2, 107–123.

Fletcher, L.R. (1993) Grazing ryegrass/endophyte associations and their effect on animal health and performance. In: *Proceedings of the Second International Symposium on Acremonium/Grass Interactions, Plenary Papers.* AgResearch, Grasslands Research Centre, Palmerston North, New Zealand, pp. 89–93.

Forde, M.B. and Suckling, F.E.T. (1980) Genetic resources in high-rainfall pastures of New Zealand II. Description of the ryegrass collection. *New Zealand Journal of Agricultural Research* 23, 179–189.

Frame, J., Baker, R.D. and Henderson, A.R. (1996) Advances in grassland technology over the past fifty years. In: Pollott, G.E. (ed.) *Grasslands into the 21st Century: Challenges and Opportunities.* Occasional Symposium of the British Grassland Society 19, British Grassland Society, Maidenhead, UK, pp. 31–63.

Gillingham, A.G. (1973) Influence of physical factors on pasture growth on hill country. *Proceedings of the New Zealand Grassland Association* 35, 77–85.

Gillingham, A.G. (1982) Topographic and management effects on dung distribution by grazing sheep. *Proceedings of the New Zealand Grassland Association* 43, 161–170.

Gillingham, A.G. and Bell, L.D. (1977) Effect of aspect and cloudiness on grass and soil temperatures at a hill site in Raglan county. *New Zealand Journal of Agricultural Research* 20, 37–44.

Grant, D.A. and Brock, J.L. (1974) A survey of pasture composition in relation to soils and topography on a hill country farm in the southern Ruahine Range, New Zealand. *New Zealand Journal of Experimental Agriculture* 2, 243–250.

Hall, R.L. (1974) Analysis of the nature of interference between plants of different species. I. Concepts and extension of the de Wit analysis to examine effects. *Australian Journal of Agricultural Research* 25, 739–747.

Hall, R.L. (1978) The analysis and significance of competitive and non-competitive interference between species. In: Wilson, J.R. (ed.) *Plant Relations in Pastures*, CSIRO, East Melbourne, Australia, pp. 163–174.

Hampton, J.G. (1986) Effect of seed and seed lot 1000-seed weight on vegetative and reproductive yields of 'Grasslands Moata' tetraploid Italian ryegrass (*Lolium multiflorum*). *New Zealand Journal of Experimental Agriculture* 14, 13–18.

Harper, J.L. (1964) The nature and consequence of interference among plants. In: *Proceedings of the 11th International Conference of Genetics, The Hague, 1963.* Macmillan, New York, USA, pp. 465–481.

Harris, W. (1968) Pasture seeds mixtures, competition and productivity. *Proceedings of the New Zealand Grasslands Association* 30, 142–153.

Harris, W. (1970) The distribution of ryegrass cultivars and other herbage species in respect to environmental heterogeneity within fields. *New Zealand Journal of Agricultural Research* 13, 862–868.

Harris, W. (1971) The effect of fertiliser and lime on the competitive interactions of *Rumex acetosella* L. with *Trifolium repens* L. and *Lolium* spp. *New Zealand Journal of Agricultural Research* 14, 185–207.

Harris, W. (1973a) Why browntop is bent on creeping. *Proceedings of the New Zealand Grassland Association* 35, 101–109.

Harris, W. (1973b) Ryegrass genotype–environment interactions in response to density, cutting height, and competition with white clover. *New Zealand Journal of Agricultural Research* 16, 207–222.

Harris, W. (1974) Competition among pasture plants V. Effects of frequency and height of cutting on competition between *Agrostis tenuis* and *Trifolium repens*. *New Zealand Journal of Agricultural Research* 17, 251–256.

Harris, W. (1978) Defoliation as a determinant of the growth, persistence and composition of pasture. In: Wilson, J.R. (ed.) *Plant Relations in Pastures*. CSIRO, East Melbourne, Australia, pp. 67–85.

Harris, W. (1980) An approach to evaluate a large number of mixtures under grazing. *Proceedings of the 13th International Grassland Congress*. Akademie-Verlag, Berlin, Germany, pp. 389–393.

Harris, W. (1987) Population dynamics and competition. In: Baker, M.J. and Williams, W.M. (eds) *White Clover*. CAB International, Wallingford, UK, pp. 203–297.

Harris, W. (1990) Pasture as an ecosystem. In: Langer, R.H.M. (ed.) *Pastures, their Ecology and Management*. Oxford University Press, Auckland, New Zealand, pp. 75–131.

Harris, W. and Brougham, R.W. (1968) Some factors affecting change in botanical composition in a ryegrass–white clover pasture under continuous grazing. *New Zealand Journal of Agricultural Research* 11, 15–38.

Harris, W. and Brougham, R.W. (1970) The effect of grazing on the persistence of genotypes in a ryegrass population. *New Zealand Journal of Agricultural Research* 13, 263–278.

Harris, W. and Hoglund, J.H. (1980) Influences of seasonal growth periodicity and N-fixation on competitive combining ability of grasses and legumes. In: *Proceedings of the 13th International Grassland Congress*. Akademie-Verlag, Berlin, Germany, pp. 239–243.

Harris, W. and Lazenby, A. (1974) Competitive interaction of grasses with contrasting temperature responses and water stress tolerances. *Australian Journal of Agricultural Research* 25, 227–246.

Harris, W. and Rhodes, I. (1989) Comparison of ryegrass–white clover competitive interactions in New Zealand and Wales. In: *Proceedings of the 16th International Grassland Congress*. Association Francaise pour la Production Fourragere, Nice, France, pp. 617–618.

Harris, W. and Thomas, V.J. (1970) Competition among pasture plants I. Effects of frequency and height of cutting on competition between two ryegrass populations. *New Zealand Journal of Agricultural Research* 13, 833–861.

Harris, W. and Thomas, V.J. (1972) Competition among pasture plants II. Effects of frequency and height of cutting on competition between *Agrostis tenuis* and two ryegrass cultivars. *New Zealand Journal of Agricultural Research* 15, 19–32.

Harris, W. and Thomas, V.J. (1973) Competition among pasture plants III. Effects of frequency and height of cutting on competition between white clover and two ryegrass cultivars. *New Zealand Journal of Agricultural Research* 16, 49–58.

Harris, W., Forde, B.J. and Hardacre, A.K. (1981a) Temperature and cutting effects on the growth and competitive interactions of ryegrass and paspalum I. Dry matter production, tillers numbers, and light interception. *New Zealand Journal of Agricultural Research* 24, 299–307.

Harris, W., Forde, B.J. and Hardacre, A.K. (1981b) Temperature and cutting effects on the growth and competitive interactions of ryegrass and paspalum I. Interspecific competition. *New Zealand Journal of Agricultural Research* 24, 309–320.

Hill, J. (1975) Genotype–environment interactions – a challenge for plant breeding. *Journal of Agricultural Science* 85, 477–493.

Hoglund, J.H. (1990) Evaluation of clovers in dry hill country 10. Subterranean clover in North Canterbury. *New Zealand Journal of Agricultural Research* 33, 585–590.

Hunt, L.A. and Brougham, R.W. (1967) Some changes in the structure of a perennial ryegrass sward frequently but leniently defoliated during the summer. *New Zealand Journal of Agricultural Research* 10, 397–404.

Jackman, R.H. and Mouat, M.C.H. (1972) Competition between grass and clover for phosphate I. Effect of browntop (*Agrostis tenuis* Sibth) on white clover (*Trifolium repens* L.) growth and nitrogen fixation. *New Zealand Journal of Agricultural Research* 15, 653–666.

Korte, C.J. and Harris, W. (1987) Effects of grazing and cutting. In: Snaydon, R.W. (ed.) *Managed Grasslands: Analytical Studies*. Ecosystems of the World 17B, Elsevier, Amsterdam, The Netherlands, pp. 71–79.

Lambert, M.G. and Roberts, E. (1976) Aspect differences in an unimproved hill country pasture I. Climatic differences. *New Zealand Journal of Agricultural Research* 19, 459–467.

Lambert, M.G. and Roberts, E. (1978) Aspect differences in an unimproved hill country pasture II. Edaphic and biotic differences. *New Zealand Journal of Agricultural Research* 21, 255–260.

Lambert, M.G., Clark, D.A., Grant, D.A. and Gray, Y.S. (1986) Influence of fertiliser and grazing management on North Island moist hill country. *New Zealand Journal of Agricultural Research* 29, 23–31.

Latch, G.C.M. (1994) Influence of *Acremonium* endophytes on perennial grass improvement. *New Zealand Journal of Agricultural Research* 37, 311–318.

Lazenby, A. and Rogers, H.H. (1962) Selection criteria in grass breeding. I. *Journal of Agricultural Science* 62, 285–298.

Lazenby, A. and Rogers, H.H. (1964) Selection criteria in plant breeding. II. Effect, on *Lolium perenne*, of differences in population density, variety and available moisture. *Journal of Agricultural Science* 59, 51–62.

Levy, E.B. (1970) *Grasslands of New Zealand*, 3rd edn. Government Printer, Wellington, New Zealand.

MacFarlane, A.W. (1990) Field experience with new pasture cultivars in Canterbury. *Proceedings of the New Zealand Grassland Association* 52, 139–143.

MacFarlane, M.J., Sheath, G.W. and McGowan, A.W. (1990a) Evaluation of clovers in dry hill country 5. White clover at Whatawhata, New Zealand. *New Zealand Journal of Agricultural Research* 33, 549–556.

MacFarlane, M.J., Sheath, G.W. and Tucker, M.A. (1990b) Evaluation of clovers in dry hill country 6. Subterranean and white clovers at Wairakei, New Zealand. *New Zealand Journal of Agricultural Research* 33, 557–564.

MAF (1998) Quantities (kilograms) of seed of each certified class dressed during each of the years 1993–1997. In: *Seed Certification Statistics 1997/98*. Seed Certification Bureau, Ministry of Agriculture and National Seed Laboratory, Palmerston North, New Zealand, pp. 34–72.

Milne, G.D., Moloney, S.D. and Smith, D.R. (1993) Demonstration of dryland species on 90 East Coast North Island farms. *Proceedings of the New Zealand Grassland Association* 55, 39–44.

Mitchell, K.J. (1956) The influence of light and temperature on the growth of pasture species. In: *Proceedings of the 7th International Grassland Congress*. The Congress, Wellington, New Zealand, pp. 58–69.

Newman, E.I. and Rovira, A.D. (1975) Allelopathy among some British grassland species. *Journal of Ecology* 63, 727–737.

Prins, E., Sanders, P.M., Lyons, T.B. and Wewala, G.S. (1989) Use of electrophoretic techniques to identify the proportion of an improved white clover cultivar ('Grassland Kopu') in a mixed sward. *New Zealand Journal of Agricultural Research* 32, 515–520.

Radcliffe, J.E. (1968) Soil conditions on tracked hillside pastures. *New Zealand Journal of Agricultural Research* 11, 359–370.

Radcliffe, J.E. (1982) Effects of aspect and topography on pasture production on hill country. *New Zealand Journal of Agricultural Research* 25, 485–496.

Rhodes, I. and Stern, W.R. (1978) Competition for light. In: Wilson, J.R. (ed.) *Plant Relations in Pastures*. CSIRO, East Melbourne, Australia, pp. 175–189.

Rowan, D.D., Dymock, J.J. and Brimble, M.A. (1990) Effect of fungal metabolite peramine and analogs on feeding and development of Argentine stem weevil (*Listronotus bonariensis*). *Journal of Chemical Ecology* 16, 1683–1695.

Rumball, P.J. and Cooper, B.M. (1990) Evaluation of clovers in dry hill country 2. Subterranean and white clover at Kaikohe, New Zealand. *New Zealand Journal of Agricultural Research* 33, 527–532.

Rumball, P.J. and Esler, A.E. (1968) Pasture pattern on grazed slopes. *New Zealand Journal of Agricultural Research* 11, 575–588.

Rumball, W. (1983) Breeding for dryland farming. *Proceedings of the New Zealand Grassland Association* 44, 56–60.

Sanders, P.M., Barker, D.J. and Wewala, G.S. (1989) Phosphoglucoisomerase-2 allozymes for distinguishing perennial ryegrass cultivars in binary mixtures. *Journal of Agricultural Science* 112, 179–184.

Sangkkara, R., Roberts, E. and Watkin, B.R. (1982) Grass species used and pasture establishment practices in central New Zealand. *New Zealand Journal of Experimental Agriculture* 10, 359–364.

Scott, D. (1993) Multi-species mixtures as a pasture research technique. In: *Proceedings of the XVII International Grasslands Congress*. The Association, Palmerston North, New Zealand, pp. 304–306.

Scott, D., Keoghan, J.M., Cossens, G.G., Maunsell, L.A., Floate, M.J.S., Wills, B.J. and Douglas, G. (1985) Limitations to pasture production and choice of species. In: Burgess, R.E. and Brock, J.L. (eds) *Using Herbage Cultivars*. Grasslands Research and Practice Series 3, New Zealand Grassland Association, Palmerston North, New Zealand, pp. 9–15.

Sheath, G.W. and MacFarlane, M.J. (1990a) Evaluation of clovers in dry hill country 3. Regeneration and production of subterranean clover at Whatawhata, New Zealand. *New Zealand Journal of Agricultural Research* 33, 533–539.

Sheath, G.W. and MacFarlane, M.J. (1990b) Evaluation of clovers in dry hill country 4. Components of subterranean clover regeneration at Whatawhata, New Zealand. *New Zealand Journal of Agricultural Research* 33, 541–547.

Sheath, G.W., MacFarlane, M.J. and Crouchley, G. (1990) Evaluation of clovers in dry hill country 7. Subterranean and white clovers at Porangahau, Hawkes Bay, New Zealand. *New Zealand Journal of Agricultural Research* 33, 565–568.

Suckling, F.E.T. and Forde, M.B. (1978) Genetic resources in high-rainfall pastures of New Zealand I. Collection of ryegrass, browntop and white clover. *New Zealand Journal of Agricultural Research* 21, 499–508.

Tow, P.G., Lovett, J.V. and Lazenby, A. (1997a) Adaptation and complementarity of *Digitaria eriantha* and *Medicago sativa* on a solodic soil in a subhumid environment with summer and winter rainfall. *Australian Journal of Experimental Agriculture* 37, 311–322.

Tow, P.G., Lazenby, A. and Lovett, J.V. (1997b) Effects of environmental factors on the performance of *Digitaria eriantha* and *Medicago sativa* in monoculture and mixture. *Australian Journal of Experimental Agriculture* 37, 323–333.

Tow, P.G., Lazenby, A. and Lovett, J.V. (1997c) Relationship between a tropical grass and lucerne on a solodic soil in a subhumid, summer-winter rainfall environment. *Australian Journal of Experimental Agriculture* 37, 335–342.

van Vught, V.T. and Thom, E.R. (1997) Ryegrass contamination of endophyte-free dairy pastures after spray-drilling in autumn. *Proceedings of the New Zealand Grassland Association* 59, 233–237.

Vartha, E.W. (1972) Effects of *Poa trivialis* L. on growth of perennial ryegrass and white clover. *New Zealand Journal of Agricultural Research* 15, 620–628.

Wedderburn, M.E. and Pengelly, W.J. (1991) Resident ryegrass in hill country pasture. *Proceedings of the New Zealand Grassland Association* 53, 91–95.

Wedderburn, M.E., Pengelly, W.J., Tucker, M.A. and di Menna, M.E. (1989) Description of ryegrass removed from New Zealand North Island hill country. *New Zealand Journal of Agricultural Research* 32, 521–529.

Wedderburn, M.E., Adam, K.D., Greaves, L.A. and Carter, J.L. (1996) Effect of oversown ryegrass (*Lolium perenne*) and white clover (*Trifolium repens*) on the genetic structure of New Zealand hill pastures. *New Zealand Journal of Agricultural Research* 39, 41–52.

Widdup, K.H. and Turner, J.D. (1990) Evaluation of clovers in dry hill country 11. Subterranean and white clover on the Hokonui Hills, Southland, New Zealand. *New Zealand Journal of Agricultural Research* 33, 591–594.

Williams, W.M., Sheath, G.W. and Chapman, D.F. (1990) Evaluation of clovers in dry hill country 1. General objectives and description of sites and plant material. *New Zealand Journal of Agricultural Research* 33, 521–526.

9 Effects of Large Herbivores on Competition and Succession in Natural Savannah Rangelands

Christina Skarpe

Norwegian Institute for Nature Research, Trondheim, Norway

Introduction

Aims of this chapter

The aim of this chapter is to discuss how large herbivores influence competition between plants and thereby may drive or change succession of vegetation in savannah rangelands and thus influence the structure and function of ecosystems. Savannahs are managed to support many different production systems based on consumptive and/or non-consumptive use of a variety of herbivorous mammals. The impact of large herbivores, including all herbivorous mammals, domestic and wild, weighing more than *c.* 5 kg, i.e. about the size of a dikdik, will thus be considered in general terms. Large herbivores in savannah rangelands utilize both herbaceous and woody plants, and impact by herbivory on both these vegetation components will be discussed.

Large herbivores influence the patterns of competition in savannah vegetation by changing the resource uptake of some plants, hence altering resource availability for others. To understand how competition between plants is affected by grazing and browsing animals, the traits and strategies that plants adopt to obtain or maintain competitive power and fitness in a savannah environment with large herbivores will be considered in the section 'adaptations by plants to herbivory'. Thereafter, in the section 'large herbivores and plant communities', ways in which the competitive success of plants with different characteristics may influence vegetation succession and the composition of plant communities will be discussed. The section 'Herbivory and plant competition on an ecosystem scale' will consider how large herbivores interact directly and indirectly with ecosystem processes and thereby change the system within which plants compete for resources. Emphasis will be on African savannahs but examples from other regions will be included.

Savannah vegetation and competition

Savannas can be subdivided into different types according to structural or functional characteristics. Functionally, savannahs may be divided either along a moisture gradient, into, for example, humid, subhumid, semi-arid and arid savannahs, or according to nutrient availability, into rich and poor savannah. On a continental scale, the rainfall gradient and the soil fertility gradient tend to be negatively correlated (Huntley, 1982; Scholes and Walker, 1993), resulting in semi-arid savannahs generally being nutrient-rich and subhumid savannahs tending to be nutrient-poor. Savannahs may also be divided into those with a long and continuous evolutionary history of herbivory by large mammals and those in which herbivory has been discontinued for hundreds or thousands of years prior to the introduction of livestock.

Water availability determines the length of the growing season in most savannahs, and there is generally a strong correlation between rainfall and annual biomass production of grasses (Walter, 1939; Rutherford, 1980). In wet savannahs, with much grass biomass, competition between plants is primarily for light, giving a competitive advantage to tall-growing species. In dry savannahs, biomass is restricted by water limitation, plants often have a low stature and much of the competition is for under-ground resources, such as water and nutrients (Milchunas *et al.*, 1988). There is evidence that grasses and woody species largely compete in different soil layers; grasses are generally shallow-rooted, whereas woody species have access to both shallow and deeper soil layers, but have exclusive access to the latter (Walter, 1954; Walker and Noy-Meir, 1982). Nutrient availability in the environment influences plant growth rate as well as plant chemical composition. While plants in nutrient-rich environments are often adapted to fast growth, plants in nutrient-poor systems tend to be inherently slow-growing. Competition for resources is probably important, at least periodically, in both systems (Grubb, 1992).

The savannah large herbivores

Large herbivores eat many kinds of plant biomass in savannahs, including roots and tubers, flowers, fruits, bark, wood, leaves and shoots (Bergström, 1992). Generally, large herbivores select food items in order to maximize the rate of intake of digestible nutrients and energy, while minimizing intake of harmful substances. Depending on diet and anatomy of mouth and digestive system, large herbivores have been divided into: (i) 'browsers' or 'concentrate selectors', including animals feeding on leaves and twigs from woody plants and on forbs; (ii) 'bulk and roughage eaters' or 'grazers', eating grass and grasslike plants (graminoids); and (iii) 'intermediate feeders' mixing both food categories (Hofmann and Stewart, 1972, Hofmann, 1989). Most animals mix different food categories, at least to some extent, and the feeding classes should be seen as points along a continuum.

As the general physiological requirements are similar for different mammalian herbivores, food preferences may be assumed to be more or less similar. This has also been demonstrated for species with comparable body size and the same digestive system (Owen-Smith and Cooper, 1987). However, the ability to utilize food of poor quality increases with increasing body size, and also differs with digestive system (Owen-Smith, 1988; Hofmann, 1989). Thus, the mixed large herbivore communities of natural savannahs exert a more diverse utilization of the vegetation than a system with one herbivore species (Prins and Olff, 1998), and hence have a more diverse impact on competition between plants.

In terms of metabolic mass, both domestic and indigenous large herbivore communities in savannahs are in most cases dominated by grazers, e.g. cattle, sheep or wildebeest (Cumming, 1982). Most savannahs are used as rangelands for livestock: cattle (Fig. 9.1), sheep, goats (Fig. 9.2), Indian buffaloes, equids and camelids. In addition, the African savannahs include areas with the highest

Fig. 9.1. Masai herd in open savannah, Tanzania.

Fig. 9.2. Domestic mixed feeder – browsing goats in degraded arid savannah, Sudan.

species richness and biomass of wild large herbivores on earth (Sinclair and Arcese, 1995), including antelopes of many species, African buffalo, pigs, zebras, giraffes, rhinoceroses, hippopotamus and elephant. Present-day South American savannahs have comparatively few indigenous large herbivores, including deer and capybara (Ojasti, 1983). In Australia the indigenous herbivores are all marsupials, some of which have shown a marked increase in numbers with the provision of water for livestock (Newsome, 1975, 1983). The Asian savannahs, to a large extent of anthropogenic origin, also have few indigenous large herbivores, but livestock grazing there has in some places a very long history (Misra, 1983, and references therein).

Herbivory and plant–plant interactions

Large herbivores, as a rule, eat only part of the plants they feed upon and, except for small annuals and seedlings, direct mortality of plants caused by large herbivores is low (Crawley, 1983). The effect of large herbivores on plant survival and fitness is largely mediated by changes in interactions between plants, including competition, apparent competition (i.e. when plants have adverse effects on one another via the effects both have on a third organism, e.g. a herbivore), facilitation and mutualism, involving, for example, mycorrhizas and nitrogen-fixing bacteria (Allen and Allen, 1990; Louda et al., 1990; Tainton and Hardy, 1999). Many effects of herbivory on vegetation have been described, and emphasis has been on different aspects, such as grazing resources for livestock, plant biomass production or species diversity. Correspondingly, different models have been employed to explain the effects, as described below.

The increaser–decreaser–invader concept (Sampson, 1919) is based on grazing-induced changes in species composition. In 1975, Stoddart et al. described herbivore impact in Clementian terms as a regressive succession away from a climax situation, with changes after cessation of herbivory as the corresponding progressive succession back to the climax. The predation hypothesis of Paine (1966, 1971) has been used to describe herbivore impact in terms of predation. It predicts that predators, by breaking the dominance of the most successful competitor in the local prey community, will release resources for other prey species, thus increasing species diversity. The intermediate disturbance

hypothesis (Grime, 1973; Connell, 1978) suggests a bell-shaped response of species diversity along an axis of disturbance, with low diversity/high dominance caused at one end by competitive exclusion and at the other by excessive disturbance. Huston (1979, 1985) describes the development of vegetation as a result of the dynamic balance between the rate of competitive exclusion and the reduction of populations through disturbance. Milchunas et al. (1988) stress the importance of gradients from similar to divergent selection pressures for competition and herbivory tolerance for vegetation changes following herbivory. They suggest that plant adaptations for competition and for herbivory tolerance diverge with increasing primary production. Further, adaptations tend to change from intolerance to tolerance of grazing with an increasingly long evolutionary history of grazing (Milchunas et al., 1988).

The hypotheses of Paine (1966, 1971), Grime (1973) and Huston (1979, 1985) deal generally with predation and disturbance, respectively, rather than explicitly with large herbivore impact. Some of these hypotheses are based on two assumptions, namely that: (i) herbivory is a disturbance; and (ii) herbivory leads to reduced competition between plants. The two assumptions are closely related both to each other and to plant competition, as disturbance on a community or ecosystem level can be understood as an event leading to directed change in vegetation, i.e. a succession (Skarpe, 1992). Further, changes in plant species composition during a succession are largely caused by changes in competitive hierarchies.

Whether or not herbivory is a disturbance is a matter both of the history of the vegetation concerned (Milchunas et al., 1988; Milchunas and Lauenroth, 1993) and of the spatial and temporal scale of observation (Skarpe, 1992). The introduction of herbivory to a vegetation developed without grazing/browsing will change the competitive hierarchy between plant species, leading to a succession, and would hence be a disturbance. On the other hand, in a vegetation developed with herbivory, the cessation of grazing or browsing would be a disturbance and lead to vegetation changes. The concept of disturbance is scale-dependent, and patches of vegetation may undergo directed change at some period of time, while the vegetation on a larger scale only shows non-directed dynamics (Skarpe, 1992). On the scale of plant individuals, the defoliation of single tillers in a continuously grazed sward takes place as discrete events (O'Connor, 1992), triggering a response in the plant, and may temporarily change its competitive ability. Such defoliation may thus be a disturbance for the plant, but not for the sward as a whole.

Herbivory does not necessarily lead to relaxed competition between plants. Only herbivory of such a high intensity that the total vegetation is reduced to an extent where limiting resources are not fully utilized by plants would lead to a reduction in competition. Changes in herbivory may lead to changes in the kind of competition, e.g. a shift from canopy competition for light to root competition for nutrients (Milchunas *et al.*, 1988). Competition always takes place within constraints set by biotic and abiotic environmental factors (Tilman, 1977) and, in order to maintain competitiveness, plants adapt to the combined effects of these factors in the most resource-economic way. This implies that different plant traits give competitive advantage in environments with different combinations of impact factors, of which herbivory may be one.

While most vegetation ecologists regard competition as one important ecological factor, few good field experiments have been carried out showing competition in interaction with environmental constraints, e.g. herbivory. In addition, there is still considerable debate on theories and definitions related to competition (Tilman, 1977, 1987; Grime, 1983; Keddy, 1989; Grace, 1990, 1991). It is, therefore, often difficult to state the role of competition and other forms of plant–plant interactions in individual cases of herbivory-induced change or difference in vegetation. The suggestion by Tilman (from Grace, 1991) to measure competitive success as the ability to dominate in the habitat will be adopted in the following sections.

Adaptations by Plants to Herbivory

What should a clever plant do?

In 1982 Norman Owen-Smith and Peter Novellie published a paper with the title 'What should a clever ungulate eat?' As plants are not just passive prey, but respond in various ways to being eaten, the corresponding question 'What should a clever plant do to compete successfully for resources in an environment with hungry herbivores?' is equally relevant.

What plants do and can do in relation to herbivory is a result of adaptation of the species during evolution and of the triggers and opportunities provided by the environment to express these adaptive traits.

Large herbivores remove biomass, often photosynthesizing tissue, and nutrients therein from the plant. They thereby deprive the plant both of its nutrient capital and its means of production. This implies changes in the plant's ability to acquire resources, both absolute and relative to neighbours, and hence influences competitive interactions (Dirzo and Harper, 1982; Fowler and Rausher, 1985). Plants respond to herbivory by changing resource allocation, and will either maximize nutrient uptake and compensate for lost resources or minimize the loss of resources (Grime, 1983; Berendse and Elberse, 1990; Grubb, 1992). The competitive success of the different strategies depends on, among other factors, the relationship between resource availability in the environment and the herbivory pressure (Berendse, 1985; Berendse *et al.*, 1992; Grubb, 1992). This theme is developed further in the following paragraphs on plant compensation for eaten biomass and on plant resistance to herbivory.

Compensation for eaten biomass

Herbivory may either promote or reduce growth of plants or plant parts (Noy-Meir, 1993). External factors known to influence the degree of compensatory growth include timing of herbivory relative to the phenological development of the plant (Laycock, 1967; Wolfson, 1999), intensity and frequency of herbivory (Mueggler, 1972; Harper, 1977; Wallace *et al.*, 1985; Olson and Richards, 1988), plant-available nutrients (Bryant *et al.*, 1983; McNaughton and Chapin, 1985), plant-available water (Cox and McEvoy, 1983), age and kind of tissue eaten (Milthorpe and Davidson, 1966; Crawley, 1983), history of herbivory in the system (Olson and Richards, 1988) and events following the herbivory, e.g. much or little rainfall, fire or repeated grazing/browsing (Walker, 1987). The degree of compensation is the sum of positive and negative effects of herbivory on the productivity of the plant under the given conditions (Noy-Meir, 1993). Herbivory, as a rule, reduces the leaf area, implying loss in total photosynthesis and hence production; on the other hand, light levels for remaining leaves may increase, and so may photosynthesis rate per leaf area or mass (Gifford and Marshall, 1973;

Senock *et al.*, 1991). Nutrients stored in shoots or leaves are lost with herbivory, but at the same time competition between remaining plant modules for resources and often the ratio between net and gross production (i.e. between production of biomass and that plus 'maintenance costs' for the plant) increases, as total biomass is reduced (Jefferies *et al.*, 1994). Active apical meristems are often removed, but dormant meristems may be activated for vegetative and eventually generative growth (Murphy and Briske, 1992; Järemo *et al.*, 1996).

Compensatory growth can result in exact compensation, when the production of the grazed plant is the same as in an ungrazed control, or in under- or overcompensation, when a grazed plant produces less or more, respectively, than the control (Noy-Meir, 1993). Compensatory growth has mostly been measured as mass of shoots and leaves, sometimes as generative output and rarely as total plant biomass production.

Genetic traits, which are often connected with ability for compensatory growth, include high potential resource uptake, high potential growth rate, availability of dormant, easily activated buds/meristems and nutrient stores, and meristems that are inaccessible to the herbivores concerned (Noy-Meir, 1993). Where nutrient availability is high relative to herbivore pressure, rapidly compensating plants may be competitively advantaged and come to dominate the vegetation, as will be exemplified in the following. The higher the herbivory pressure, the more the plant is deprived of nutrients, active and dormant buds and photosynthesizing tissue, thus reducing its uptake of nutrients, production and competitive ability. Where nutrient availability is low relative to the intensity of herbivory, plants with strategies other than fast growth are likely to gain competitive dominance. Thus, herbivory pressure and nutrient availability are both important for the net outcome of plant compensatory growth (McNaughton and Chapin, 1985; Maschinski and Whitham, 1989) and competitive success (Dirzo and Harper, 1982; Crawley, 1983). McNaughton *et al.* (1983), in a controlled experiment, found that the sedge *Kyllinga nervosa* Steud, a very common plant in part of the Serengeti, overcompensated for above-ground biomass removed at moderate defoliation and high N availability, undercompensated at severe defoliation and just compensated under other conditions. Coughenour *et al.* (1985) showed that *Themeda triandra* Forsk., also from the Serengeti, undercompensated for

experimentally cut biomass under all conditions, but less so with N fertilization.

Grasses are generally tolerant to grazing, and may show vigorous compensatory growth. Except when flowering, most species have meristems positioned low, under the main grazing level, allowing rapid regrowth of tillers after grazing (Vickery, 1984). At flowering, the apical meristem in most grasses is elevated to a level where it is vulnerable to grazing (Wolfson, 1999). Removal of the apical meristem often stimulates tillering from activated axillary buds, but this is not always the case (Olson and Richards, 1988; Murphy and Briske, 1992). Generally, low-growing grasses are better competitors under grazing than tall ones, as a reduced proportion of their biomass is grazed. In grazed vegetation competition for light, favouring tall species, is less important than in ungrazed swards (Milchunas *et al.*, 1988). Grazing-tolerant grasses often respond to herbivory by attaining low stature and high shoot density while maintaining a large leaf area, and by physiological and morphological processes that enhance compensatory growth following herbivory (McNaughton *et al.*, 1983). In this way grasses maximize resource capital and uptake and hence competitive ability. Oesterheld and McNaughton (1988) found that three clones of the grass *Themeda triandra* from the Serengeti differed in height inversely to the grazing pressure at the sites where they were collected. Field observations of *Panicum repens* in a grassland in Zimbabwe showed the grass to have higher shoot density, shorter shoots and shorter internodes at the sites with highest grazing pressure (estimated from faecal counts). There was no difference in leaf area index with grazing pressure (Skarpe, 1997). Rhizomatous grasses with an ability for vegetative reproduction have an advantage over those with obligate reproduction from seed in environments where flowers have a high probability of being eaten. O'Connor (1991) found that the proportion of a rhizomatous genotype of *Digitaria eriantha* increased consistently with heavy grazing.

Browse from trees and shrubs (from here on referred to as trees) is an important food resource, seasonally or all the year, for many species of livestock and wild large herbivores in savannah (Cumming, 1982). In a review of browsing in African savannah, Bergström (1992) distinguishes three modes of browsing on trees: the pruning of shoots with or without leaves and the picking or stripping of leaves. Twig browsing often leads to vigorous compensatory growth in savannah trees. A number of studies of natural browsing or experimental clipping of acacias show high degrees of compensation (Pellew, 1984; Dangerfield and Madukanele, 1996). Twig browsing often results in bigger, but fewer, annual shoots (du Toit *et al.*, 1990). Simulated browsing (clipping) and elephant impact (stumping) on *Combretum apiculatum* in a savannah in Botswana, resulted in the clipped trees producing fewer shoots than the controls, but of about the same total biomass. In contrast, the stumped trees produced the biggest individual shoots, but only reached 15% of the total shoot biomass on control trees (Bergström *et al.*, 2000). Browsing of leaves may lead to different responses from those to twig pruning, at least in temperate environments (Danell *et al.*, 1994). Compensation for defoliation in savannah trees varies with time and intensity of browsing. Most studies have found rapid regrowth of leaves after defoliation early in the growing season, but not if the impact was late in the season (Guy *et al.*, 1979 (cited in Bergström, 1992); Cissè, 1980; Teage, 1985). This seems resource-effective, if the cost of refoliation is weighed against the expected photosynthetic lifetime of the leaf. Many trees refoliate after high but not after low intensity of defoliation (Cissé, 1980).

It is known that trees and herbaceous vegetation in savannah may inhibit each other and that excessive grazing of the field layer may lead to competitive advantage of trees (Walker and Noy-Meir, 1982; van Vegten, 1983; Stuart-Hill and Tainton, 1989; Skarpe, 1990). However, little is known about the influence of browsing of trees on competitive interactions between trees or between trees and herbaceous vegetation (Bergström and Danell, 1987; Owen-Smith, 1988). Stuart-Hill and Tainton (1989) found that simulated browsing (clipping) of *Acacia karroo* led to decreased production of a surrounding grass sward, implying that the clipping increased the competitiveness of the tree. Browsing megaherbivores, primarily African elephant, have a strong influence on tree growth and survival and hence on tree competitiveness with both other trees and the herbaceous layer (Owen-Smith, 1988).

Plant resistance to herbivory

While compensatory growth and competition for resources involve similar adaptations in plants (Järemo *et al.*, 1996), resistance to herbivory has a

cost in resources withdrawn from growth and reproduction. Plants can avoid herbivory by large mammals either by having a major proportion of the resources unavailable for the herbivore, e.g. in lignified stems or underground organs (Fig. 9.3), or by developing defences of different kinds. In the first case, a cost for the plant is the reduced proportion of photosynthetic tissue; in the latter, it is allocation of resources to structural or chemical defence. Therefore, 'the production of defences is only favoured by natural selection' – i.e. gives a competitive advantage – 'when the cost of production is less than the benefit of enhanced protection from herbivores', according to the optimal defence theory (Coley *et al.*, 1985). A number of hypotheses have developed this optimal defence theory further, in efforts to predict more precisely under which conditions plants would evolve defence mechanisms of different types, in order to maximize competitive ability and fitness.

The plant apparency hypothesis (Feeny, 1976) predicts that plants that are apparent in space (i.e. have high cover) or in time (i.e. have palatable parts that persist during periods when most of the vegetation is unpalatable), and hence risk a high rate of herbivory, would invest in defences, particularly in chemical defences. Similar ideas were expressed by Rhoades and Cates (1976). The resource availability hypothesis (Coley, 1983), further developed by Coley *et al.* (1985), proposes that plants growing at resource-poor sites are generally obligatory slow-growing and not well able to compensate for lost nutrients/biomass; they would be expected to invest more in defence than would plants growing in more

Fig. 9.3. To have a large proportion of biomass underground may be an adaptation to herbivory, drought and/or fire: *Tylosema esculenta* in the Kalahari, Botswana.

resource rich environments. The carbon-nutrient hypothesis of Bryant *et al.* (1983; 1991) is based on the metabolic costs of different plant defences. Grubb (1992) formulated the multidimensional scarcity–accessibility hypothesis, basically stating that resources that are scarce (for the animal or for the plant), attractive and available must be defended. Such situations include plants in patches of nutrient-rich vegetation in a generally resource poor-landscape, nutrient-rich plants in a permanently or seasonally nutrient-poor vegetation, and nutrient-rich plant parts, such as single meristems in some species, which are scarce and difficult for the plant to replace.

Defence mechanisms can be: (i) structural, e.g. spines of different shapes and origins; or (ii) chemical, with different chemical composition and effects on the animals (Bergström, 1992). More indirect plant defences include, for example, ant-symbiosis systems (Fig. 9.4) (Janzen, 1979; Sabelis *et al.*, 1997). Plant defence mechanisms are inherited by the species, but the degree of expression can be modified by the environment, and both chemical and structural defence can be induced by herbivory (Milewski *et al.*, 1991). Georgiadis and McNaughton (1988) found that the grass *Cynodon plectostachyus* developed high levels of the potent poison cyanide in response to defoliation. Gowda (1997) found that pruning of young *Acacia tortilis* resulted in increased growth of spines. That physical defence reduces bite size and/or feeding rate and/or total biomass eaten from the plant has been

found in studies of both wild browsers in South Africa (Cooper and Owen-Smith, 1986) and domestic goats in Tanzania (Gowda, 1996).

What clever plants do – some conclusions

In most vegetation, competition is an important determinant of plant survival and fitness. To gain competitive superiority, plants must maximize resources acquired relative to competitors under the prevailing environmental conditions, including, for example, droughts, frost, fires or herbivory. To do this, plants have, during evolution, attained specific characteristics, or the ability to express such characteristics, to minimize any negative impact by environmental factors on resource economy.

A number of plant features or strategies are believed to have evolved, at least partly, as means to cope with loss of biomass to large herbivores. In situations where the potential for resource uptake is high relative to losses to herbivores, plants with adaptations for rapid compensatory growth may gain competitive dominance. Where potential resource uptake is low relative to the risk of losses to grazing and browsing animals, competition may instead favour plants that minimize loss of resources. This includes slow-growing plants in resource-poor environments and plants or plant parts that constitute a scarce resource for large herbivores, e.g. plants in nutrient-rich patches in a generally nutrient-poor vegetation and particularly nutrient rich-plants or plant parts. Strategies to minimize loss of resources include having a large proportion unavailable to herbivores or developing structural or chemical defence against such animals.

Changes in herbivory regime may change competitive hierarchies between plants with different adaptive traits or strategies. This may lead to vegetation succession and changes in plant communities, which are discussed in the following section.

Large Herbivores and Plant Communities

The importance of net primary production and evolutionary history of herbivory

As seen above, different traits allow plants to survive and compete successfully in environments with different resource availability and rate of herbivory. Thus, a change in grazing or browsing regime may

Fig. 9.4. Morphological and indirect defences: thorns and ant galls in *Acacia drepanolobium*, Tanzania.

modify competitive relationships between plants and lead to vegetation succession. A vegetation developed with large herbivores will change if herbivory is discontinued, and an ungrazed vegetation will change if herbivory is introduced.

Milchunas and Lauenroth (1993) compared grazed and ungrazed vegetation in a worldwide data set from 236 sites. For dry 'grasslands' and 'shrublands', including savannahs, they found changes in community physiognomy, i.e. general features of the vegetation, species composition, species diversity and potential for invasion by alien species to be largely a function of: (i) net primary productivity or biomass production; (ii) evolutionary history of herbivory; and (iii) level of current herbivory. These three factors, in decreasing importance, explained more than 50% of the variance in species composition.

The importance of biomass production for community response to herbivory is consistent with the hypothesis for, on the one hand, convergence of adaptive traits for herbivore tolerance and competition for below-ground resources in dry savannah with low productivity, and, on the other hand, the contrast between adaptive traits for herbivore tolerance and for competition for light in subhumid, highly productive savannahs (Milchunas et al., 1988). In the first case, for example, low stature and stoloniferous growth are suitable adaptations both for underground competition, which is important in dry savannahs, and for dealing with herbivory. In the latter case, tall growth is an advantage in competing for light, which is important in subhumid savannahs, but a disadvantage if herbivore pressure is high. This would imply less difference between grazed and ungrazed vegetation in semi-arid than in subhumid environments, as found by Milchunas and Lauenroth (1993).

In vegetation with a long evolutionary history of grazing most or all plant species are adapted to herbivory, whereas in vegetation without such a history most plant species will not be adapted to herbivory. A substantial change in species composition is thus likely to occur if grazing/browsing is introduced. Consequently, Milchunas and Lauenroth (1993) found less difference between ungrazed and grazed vegetation with a long evolutionary history of grazing than in vegetation without such a history.

O'Connor (1991), working in dry South African savannah with a long history of herbivory, found changes in species composition of small tussock grasses to be more related to differences in rainfall between years than to differences in grazing between sites. Only a strongly stoloniferous genotype of D. eriantha consistently increased with grazing pressure (O'Connor, 1991). In some neotropical savannahs, which for a considerable time have evolved without much herbivory, bunch grasses with low tolerance for defoliation often dominate (Sarmiento, 1992). Rusch and Oesterheld (1997) found that grazing in the subhumid flooding pampa of Argentina led to decreased biomass production and invasion of exotic forbs. Sarmiento (1992) refers to a Venezuelan savannah managed as a mowed lawn for 15 years. After this time, none of the original perennial grasses persisted, but were replaced with Panicum maximum and Hyparrhenia rufa (aliens of African origin, adapted to grazing) and with annuals and two indigenous perennials, Paspalum virgatum and Axonopus compressus, usually found as pioneers in disturbed sites.

Herbivory intensity

In the global studies by Milchunas et al. (1988), actual grazing pressure was less important than productivity and grazing history for variation in composition between grazed and ungrazed vegetation. On a smaller geographical scale, given a certain rainfall regime and evolutionary history of herbivory, the intensity of grazing/browsing is important in determining competitive hierarchies between plant species, and hence vegetation composition. Pronounced vegetation changes can be seen along gradients in grazing pressure, e.g. with distance from watering points (Lange, 1969; Skarpe, 2000). Some heavily grazed vegetation in the Kalahari resembles vegetation in drier areas of the same system in terms of composition of species and growth forms (Skarpe, 1986). This can be interpreted in the light of the model by Milchunas et al. (1988) – that plants with traits making them successful competitors in an arid environment are also well adapted to herbivory.

The increase of Karoo species, mainly dwarf shrubs, in South African savannah (Fig. 9.5), is described by Acocks (1975) as a response to heavy livestock grazing. The dwarf shrubs employ a different strategy from the grasses, in that they minimize the loss of nutrients (see Berendse, 1985), rather than maximizing the uptake, hence gaining competitive advantage with intensive herbivory in a relatively nutrient-poor environment. Since the mid-1980s,

Fig. 9.5. Differently managed semi-arid rangeland: heavily grazed with mainly dwarf shrubs on one side of the fence, moderately grazed and grass dominated on the other, South Africa.

there has been an increase in medium tall grasses at the expense of dwarf shrubs and prostrate grasses around some Kalahari pans (natural temporary water-holes surrounded by slightly saline soils) used by wildlife (C. Skarpe, unpublished). The change coincides in time with a sharp decline in the wildebeest population in 1983, until then the main bulk grazer in the area. If there is a cause-and-effect relation behind the coincidence in time, it might be explained as the reverse of the succession recorded by Acocks (1975).

Savannah vegetation, particularly arid savannahs, are dynamic, shifting between dominance of different plant species and life-forms as a result of stochastic rainfall, occasional fires and small and large herbivores (Frost *et al.*, 1986; Walker, 1987). Indigenous large herbivores in such environments are highly mobile, and leave the area when food gets scarce and low in moisture content (Noy-Meir, 1979/80). Domestic stock is often supplemented with water and tends to use the arid savannahs more permanently. This may result in a reduction in regeneration of perennial palatable species, which are outcompeted, probably as seedlings, by ephemerals and/or unpalatable species (Acocks, 1975; Werger, 1977). Le Houérou (1989) hypothesises that this process long ago caused a change in much of the Sahel, from vegetation dominated by perennial grasses to the present predominantly annual vegetation.

Some conclusions on herbivores and plant communities

Milchunas and Lauenroth (1993) showed that, on a global scale, net primary production, evolutionary history of herbivory and current level of herbivory, in decreasing order, explained the major part of plant community changes with grazing/browsing. The pattern was explained as differences in plant adaptation to herbivory causing differences in competitive ability and hence in, for example, species composition and potential for invasion by alien species. However, on a local scale, variation in potential net primary production and evolutionary history is small, and intensity of herbivory becomes an important factor for plant community changes (Frost *et al.*, 1986; Skarpe, 1986; Walker, 1987).

Herbivory and Plant Competition on Ecosystem Scale

Large herbivores setting the scene for plant competition

This chapter began with a consideration of the adaptive traits or strategies that give plants a competitive advantage with herbivory under different conditions of resource availability. There is now considerable evidence that large herbivores can influence resource availability for plants both directly and indirectly, and hence set the scene for direct interaction between large herbivores and plant competitive patterns (Botkin *et al.*, 1981; Pastor *et al.*, 1993; Ritchie *et al.*, 1998). For example, large herbivores have been shown to influence nutrient cycling, plant biomass production and spatial patterns in vegetation, fire regime and rate and direction of successional processess, as well as switching of systems between alternative states (Hobbs, 1996). All these phenomena influence and are influenced by competition between plants adapted for different environments and herbivory, as shown below.

Large herbivores and nutrient cycling

The cycling rate of nutrients is often more important for nutrient availability to plants than the total nutrient capital in the system (Vitousek, 1982; Aber and Melillo, 1991). Large herbivores influence nutrient cycling in a number of ways, e.g. by changing the quality and quantity of above- and below-ground plant litter and by depositing dung, urine and carcasses. There is evidence for a positive relationship between plant palatability for large herbivores and plant litter decomposition rate (Horner *et al.*, 1988). Fast-growing young plant parts, including compensatory growth after herbivory, have often been shown to be palatable for large herbivores (du Toit *et al.*, 1990; Price, 1991), and the litter decomposes and mineralizes rapidly (Horner *et al.*, 1988; Seagle *et al.*, 1992). A contributing factor to the rapid turnover rate of nutrients in productive grazed or browsed areas is often the conversion of a large proportion of the plant biomass to dung and urine. This often – but not always – increases mineralization rate (Day and Detling, 1990). Thus, under some conditions, herbivory may enhance nutrient cycling and contribute to higher nutrient availability (Botkin *et al.*, 1981; Hobbs, 1996; Mazancourt *et al.*, 1998). In such situations, fast-growing, rapidly compensating plants may gain competitive dominance (Roux, 1969; McNaughton, 1985; Pastor and Naiman, 1992; Seagle *et al.*, 1992). Animals will repeatedly feed in such patches (du Toit *et al.*, 1990; Lundberg and Danell, 1990) and may thereby maintain the competitive hierarchies between plants and hence the patchy structure of vegetation. McNaughton (1988) describes such preferentially grazed patches in the Serengeti, deviating in nutrient availability from the surroundings. Lock (1972) found a similar patchy vegetation structure induced by hippopotami in Uganda.

The maintenance of the competitive advantage for fast-growing palatable plant species probably depends on fine-tuned interactions between herbivory, vegetation and soil processes. Many situations have been described (Hobbs, 1996), when herbivory depresses the competitiveness of the nutrient-rich selectively eaten plants, which are then outcompeted by nutrient-poor, slow-growing, often chemically defended species. Litter produced by such plants tends to decompose and mineralize slowly, reducing the rate at which nutrients become available to plants (Horner *et al.*, 1988; Bryant *et al.*, 1991; Hobbie, 1992; Pastor *et al.*, 1993). Ritchie *et al.* (1998) found that exclusion of large herbivores in a North American oak savannah led to increased N cycling in the soil and more plants with relatively high N content. In cases where herbivory reduces nutrient cycling and net primary production, interactions between plants would favour slow-growing, nutrient-poor and/or chemically defended plant species. Under such conditions, animals would avoid feeding in a previously grazed/browsed patch (Hobbs, 1996), resulting in decreasing spatial heterogeneity in vegetation composition and ecosystem processes.

Large herbivores and fires in savannah

Fire, as well as large herbivores, is an important determinant of savannah structure and function, and most savannah plant species are more or less adapted to fire.

Savannah fires are generally spread by dry grass fuel and, in situations where grazing reduces the standing biomass of grasses beyond the point where it can carry a fire, the frequency and intensity of fires are much reduced (McNaughton, 1992). In such situations, plant species that have relatively poor fire tolerance may gain a competitive advantage. This applies, for example, to seedlings and saplings of some woody species, and a change in fire regime may influence species composition, density and spatial distribution of woody vegetation (Menaut *et al.*, 1990; Skarpe, 1991; Bond and van Wilgen, 1996).

Switches between wooded and open savannah

By direct and indirect effects on ecosystem processes, large herbivores may cause rapid successions or switches between different states of a system – for example, between densely wooded savannah and open grassland (Noy-Meir, 1982; Hobbs, 1996).

A change of wooded savannahs to more open grasslands has been described, particularly from rangelands used by game in Africa. It has often been attributed to excessive killing of trees by elephants, but, as pointed out by Pellew (1983) and Belsky (1984), poor recruitment of saplings into the tree layer due to fires and browsers may be more important than increased mortality of mature trees. Woody seedlings and saplings resprouting after browsing or fire may suffer from competition with the field-layer species for both light and soil resources. Once established, woody species may outshade high light-intensity-demanding grass species, and may compete for soil resources in a deeper soil layer than the grasses (Walter, 1954; Walker and Noy-Meir, 1982).

The encroachment of woody vegetation into open savannah (Fig. 9.6), has been described from savannah grazing lands all over the world (Walter, 1954; van Vegten, 1983; Hacker, 1984; Archer *et al.*, 1988; Skarpe, 1990). In most cases, the change has been attributed to direct and indirect effects of heavy grazing by livestock, perhaps through interaction with changes in climate or atmospheric composition (Emanuel *et al.*, 1985). Walter (1954) and Walker and Noy-Meir (1982) describe bush encroachment as a result of the reduction in resource uptake and competitive ability by grasses under very heavy grazing, resulting in more resources being available for woody vegetation. Shallow-rooted encroacher shrubs use resources in the same rooting zone as grasses (Skarpe, 1990). Reduced water uptake by grasses may also result in more water penetrating to deeper soil layers, where it is exclusively available for woody species. Indirect effects of herbivory facilitating the establishment of woody plants include the decrease in fire impact and, perhaps, changes in soil processes.

Bush encroachment is not linearly related to stocking rate, but seems to occur when grazing pressure exceeds a threshold value (Skarpe, 1990), presumably when grasses are damaged to the extent where their competitive ability is seriously reduced. Herbivory intensity is not homogeneous even in the same paddock, but is spatially variable at different scales (Senft *et al.*, 1987; O'Connor, 1992).

Fig. 9.6. 'Bush encroachment': dense *Acacia mellifera* scrub in heavily grazed range, Botswana.

Weber and Jeltsch (1998) used a spatially explicit model to study dynamics of grasses and woody plants with different grazing pressure and different degrees of spatial variation. They found that the increase in woody plants was positively related both to grazing intensity and to degree of patchiness in the grazing. The model also predicted an increase in the probability for bush encroachment with number of years the grazing continued (Weber and Jeltsch, 1998). The mechanisms behind these findings are not clear, but may be related to grazing-induced spatially and temporally small-scale competition patterns between woody and herbaceous plants.

Large herbivores as ecosystem engineers – some conclusions

Large-herbivore food selection often encompasses different scales, from landscape, in migratory wildlife and nomadic livestock, to ecosystem, patch, plant or plant part (Senft *et al.*, 1987). By selective use, animals may influence the diversity of ecosystems in such different scales. Grazing and browsing large herbivores interact with the abiotic variation in landscapes, and may increase spatial variation by enhancing relatively resource-rich patches and the competitive advantage of fast-growing palatable plants. Under other conditions, large herbivores will reduce the spatial variation, when plant competition following grazing favours slow-growing nutrient-poor plants (Hobbs, 1996). The impact of large herbivores on ecosystem processes is caused both by direct effects on plants, through eating them, and by indirect effects, including impact on soil processes and disturbance regime, e.g. fire. The changes in ecosystem processes have relevance for the competitive hierarchies between plant species and growth forms, and subsequently for future herbivory patterns.

Conclusions

Competition between plants is an important factor, acting at least temporarily, in most vegetation. Large herbivores generally do not kill the plants they feed upon, but they do affect their competitive ability. This may eventually lead to competitive exclusion of some species and to herbivory-induced succession, leading to changes in vegetation composition.

Large herbivores feed selectively in order to maximize the intake rate of digestible nutrients and energy, while avoiding harmful compounds. Many nutrients that are essential for herbivores are also essential for plants, which have developed different traits and strategies to secure resources to maintain competitiveness and fitness in the presence of herbivores. Plants must at the same time adapt to many other environmental factors, such as insect herbivory, drought, fire, heat and frost. Competition will generally favour plants with the most resource-economic adaptive traits for all biotic and non-biotic constraints. Therefore, it is often not possible to point out a single environmental factor as the reason for the development of a certain plant trait. For example, to have a large proportion of biomass and nutrient stores underground may enhance competitive success in relation to both large herbivores and fire and drought. This implies that a plant trait or strategy that gives competitive advantages under intense herbivory in one environment may not do so in another. Rapid compensatory growth may be a good adaptation to herbivory in a resource-rich environment, but not in a poor one. Further, plant adaptations to certain environmental factors may at the same time influence competitiveness under herbivory. Thus plants adapted to dry conditions would compete better under heavy grazing than would tall plants adapted to subhumid conditions.

The evaluation of herbivore impact on plant competition patterns is further complicated by the highly dynamic nature of savannah vegetation in general and of dry savannahs in particular. In such systems, vegetation changes are largely event-driven, triggered, for example, by rainfall, fires or insect outbreaks. The effect of large herbivores on plant competitive patterns interacts with such factors, perhaps as much by preventing change as by inducing it.

Large herbivores influence plant communities by feeding selectively, i.e. by eating certain plants more than others. In that way, they maintain a competitive hierarchy among plants and hence a plant species composition that would change with a change in herbivory regime. After an increase or decrease in herbivory, changes in the competitive patterns between plants will cascade through the community, and the vegetation will undergo a succession until the directed change in species composition ceases and is replaced by non-directional dynamics. Because of indirect effects of herbivory, influencing, for example, the fire regime, nutrient cycling and soil structure, after a return to the original herbivory regime in most cases the succession will not go through the same stages as when herbivory was first changed. Neither will the succession in most cases end with the same vegetation composition as before the first change took place.

Small-scale, short-term effects of grazing and browsing animals on savannah vegetation are fairly well known, even if the processes, including competitive patterns, are often less well understood. Little is known about the significance of large herbivores in a long-term perspective and on an ecosystem or landscape scale. Savannahs have evolved with large herbivores, and it is doubtful whether savannah vegetation, as we know it today, would exist without a history of grazing and browsing large herbivores. Today the grazing intensity in many savannahs is high, perhaps higher than before modern livestock keeping, and the species diversity of herbivores is low. In most cases, one or a few species of ruminant grazers, often cattle and/or sheep, dominate heavily. Many systems have no or few browsers, and most systems lack megaherbivores. Further, animal mobility is becoming more restricted. The implications of these changes in the large herbivore fauna for the competitive patterns in savannah vegetation, particularly related to the woody component and to long-term spatial aspects, are largely unknown.

Acknowledgements

I wish to thank Roger Bergström and the two editors of this volume for comments on the manuscript.

References

Aber, J.D. and Melillo, J.M. (1991) *Terrestrial Ecosystems.* Saunders College, Philadelphia, USA.

Acocks, J.P.H. (1975) *Veld Types of South Africa.* Memoirs of the Botanical Survey of South Africa, No. 40. Botanical Research Institute, Pretoria, South Africa.

Allen, E.B. and Allen, M.F. (1990) The mediation of competition by mycorrhizas in successional and patchy environments. In: Grace, J.B. and Tilman, D. (eds) *Perspectives on Plant Competition.* Academic Press, San Diego, USA, pp. 367–390.

Archer, S., Scifres, C. and Bassham, C.R. (1988) Autogen succession in a subtropical savannah: conversion of grassland to thorn woodland. *Ecological Monographs* 58, 111–127.

Belsky, A.J. (1984) Role of small browsing mammals in preventing woodland regeneration in the Serengeti National Park, Tanzania. *African Journal of Ecology* 22, 271–279.

Berendse, F. (1985) The effect of grazing on the outcome of competition between plant species with different nutrient requirements. *Oikos* 44, 35–39.

Berendse, F. and Elberse, W.Th. (1990) Competition and nutrient availability in heathland and grassland ecosystems. In: Grace, J.B. and Tilman, D. (eds) *Perspectives on Plant Competition.* Academic Press, San Diego, USA, pp. 93–116.

Berendse, F., Elberse, W.Th. and Geerts, R.H.M. (1992) Competition and nitrogen loss from plants in grassland ecosystems. *Ecology* 73, 46–53.

Bergström, R. (1992) Browse characteristics and impact of browsing on trees and shrubs in African savannas. *Journal of Vegetation Science* 3, 315–324.

Bergström, R. and Danell, K. (1987) Effects of simulated browsing by moose on morphology and biomass of two birch species. *Journal of Ecology* 75, 533–544.

Bergström, R., Skarpe, C. and Danell, K. (2000) Plant responses and herbivory following simulated browsing and stem cutting of *Combretum apiculatum. Journal of Vegetation Science* 11, 409–414.

Bond, W.J. and van Wilgen, B.W. (1996) *Fire and Plants.* Chapman and Hall, London.

Botkin, D.B., Mellilo, J.M. and Wu, L.S.-Y. (1981) How ecosystem processes are linked to large mammal population dynamics. In: Fowler, C.W. and Smith, T.D. (eds) *Dynamics of large Mammal Populations.* John Wiley & Sons, New York, USA, pp. 373–387.

Bryant, J.P., Chapin, F.S., III and Klein, D.R. (1983) Carbon/nutrient balance of boreal plants in relation to vertebrate herbivory. *Oikos* 40, 357–368.

Bryant, J.P., Heitkonig, I., Kuropat, P. and Owen-Smith, N. (1991) Effects of severe defoliation on the long-term resistance to insect attack and on leaf chemistry in six woody species in the southern African savannah. *American Naturalist* 137, 50–63.

Cissé, M.I. (1980) Effects of various stripping regimes on foliage production of some browse bushes in the Sudano-Sahelian zone. In: le Houérou, H.N. (ed.) *Browse in Africa. The Current State of Knowledge.* International Livestock Centre for Africa, Addis Ababa, Africa, pp. 211–214.

Coley, P.D. (1983) Herbivory and defence characteristics of tree species in a lowland tropical forest. *Ecological Monographs* 53, 209–233.

Coley, P.D., Bryant, J.P. and Chapin, F.S. (1985) Resource availability and plant antiherbivore defence. *Science* 230, 895–899.

Connell, J.H. (1978) Diversity in tropical rainforests and coral reef. *Science* 199, 1302–1310.

Cooper, S.M. and Owen-Smith, N. (1986) Effects of plant spinescence on large mammalian herbivores. *Oecologia* 8, 446–455.

Coughenour, M.B., McNaughton, S.J. and Wallace, L.L. (1985) Responses of an African graminoid (*Themeda triandra* Forsk.) to frequent defoliation, nitrogen and water: a limit of adaptation to herbivory. *Oecologia* 68, 105–110.

Cox, C.S. and McEvoy, P.B. (1983) Effect of summer moisture stress on the capacity of tansy ragwort (*Senecio jacobaea*) to compensate for defoliation by cinnabar moth (*Tyria jacobaea*). *Journal of Applied Ecology* 20, 225–234.

Crawley, M.J. (1983) *Herbivory: the Dynamics of Animal-Plant Interactions.* University of California Press, Berkeley, USA.

Cumming, D.H.M. (1982) The influence of large herbivores on savanna structure in Africa. In: Huntley, B.J. and Walker, B.H. (eds) *Ecology of Tropical Savannahs.* Ecological Studies 42, Springer Verlag, Berlin, Germany, pp. 217–245.

Danell, K., Bergström, R. and Edenius, L. (1994) Effects of large mammalian browsers on woody plant architecture, biomass and nutrients. *Journal of Mammology* 75, 833–844.

Dangerfield, J.M. and Madukanele, B. (1996) Overcompensation by *Acacia erubescens* in response to simulated browsing. *Journal of Tropical Ecology* 12, 905–908.

Day, T.A. and Detling, J.K. (1990) Grassland patch dynamics and herbivore grazing preference following urine deposition. *Ecology* 71, 180–188.

Dirzo, R. and Harper, J.L. (1982) Experimental studies of slug–plant interactions: 4. The performance of cyanogenic and acyanogenic morphs of *Trifolium repens* in the field. *Journal of Ecology* 70, 119–138.

du Toit, J.T., Bryant, J.P. and Frisby, K. (1990) Regrowth and palatability of acacia shoots following pruning by African savannah browsers. *Ecology* 71, 149–153.

Emanuel, W.R., Shugart, H.H. and Stevenson, M.P. (1985) Climatic change and the broad-scale distribution of terrestrial ecosystem complexes. *Climatic Change* 7, 29–43.

Feeny, P. (1976) Plant apparency and chemical defence. *Recent Advances in Phytochemistry* 10, 1–40.

Fowler, N.L. and Rausher, M.D. (1985) Joint effects of competitors and herbivores on growth and reproduction in *Aristolochia reticulata. Ecology* 66, 1580–1587.

Frost, P., Medina, E., Menaut, J.-C., Solbrig, O., Swift, M. and Walker, B. (1986) *Responses of Savannahs to Stress and Disturbance*. Biology International Special issue 10.

Georgiadis, N.J. and McNaughton, S.J. (1988) Interactions between grazers and a cyanogenic grass, *Cynodon plectostachyus*. *Oikos* 51, 343–350.

Gifford, R.M. and Marshall, C. (1973) Photosynthesis and assimilate distribution in *Lolium multiflorum* following differential tiller defoliation. *Australian Journal of Biological Sciences* 26, 517–526

Gowda, J.H. (1996) Spines of *Acacia tortilis:* what do they defend and how? *Oikos* 77, 279–284.

Gowda, J.H. (1997) Physical and chemical response of juvenile *Acacia tortilis* trees to browsing: experimental evidence. *Functional Ecology* 11, 106–111.

Grace, J.B. (1990) On the relationship between plant traits and competitive ability. In: Grace, J.B. and Tilman, D. (eds) *Perspectives on Plant Competition*. Academic Press, San Diego, USA, pp. 51–66.

Grace, J.B. (1991) A clarification of the debate between Grime and Tilman. *Functional Ecology* 5, 583–587.

Grime, J.P. (1973) Control of species density in herbaceous vegetation. *Journal of Environment Management* 1, 151–167.

Grime, J.P. (1983) *Plant Strategies and Vegetation Processes*, 3rd edn. John Wiley & Sons, Chichester, UK.

Grubb, P.J. (1992) A positive distrust in simplicity – lessons from plant defences and from competition among plants and among animals. *Journal of Ecology* 80, 585–610.

Guy, P.R., Mahlangu, Z. and Charidza, H. (1979) Phenology of some trees and shrubs in the Sengwa Wildlife Research Area, Zimbabwe–Rhodesia. *South African Journal of Wildlife Research* 9, 47–54.

Hacker, R.B. (1984) Vegetation dynamics in grazed mukwa shrubland community. I. The mid-storey shrubs. *Australian Journal of Botany* 32, 239–250.

Harper, J.L. (1977) *Population Biology of Plants*. Academic Press, London, UK.

Hobbie, S.E. (1992) Effects of plant species on nutrient cycling. *Trends in Ecology and Evolution* 7, 336–339.

Hobbs, N.T. (1996) Modification of ecosystems by ungulates. *Journal of Wildlife Management* 60, 695–713.

Hofmann, R.R. (1989) Evolutionary steps of ecophysiological adaptation and diversification of ruminants: a comparative view of their digestive system. *Oecologia* 78, 443–457.

Hofmann, R.R. and Stewart, D.R.M. (1972) Grazer or browser: a classification based on the stomach structure and feeding habits of East African ruminants. *Mammalia* 36, 226–240.

Horner, J.D., Gosz, J.R. and Cates, R.G. (1988) The role of carbon-based plant secondary metabolites in decomposition in terrestrial ecosystems. *American Naturalist* 132, 869–883.

Huntley, B.J. (1982) Southern African savannahs. In: Huntley, B.J. and Walker, B.H. (eds) *Ecology of Tropical Savannahs*. Springer Verlag, Berlin, Germany, pp.101–119.

Huston, M. (1979) A general hypothesis of species diversity. *American Naturalist* 113, 81–101.

Huston, M. (1985) Patterns of species diversity on coral reefs. *Annual Review of Ecology and Systematics* 16, 149–177.

Janzen, D.H. (1979) New horizons in the biology of plant defences. In: Rosenthal, G.A. and Janzen, D.H. (eds) *Herbivores: their Interaction with Secondary Plant Metabolites*. Academic Press, New York, USA, pp. 331–350.

Järemo, J., Nilsson, P. and Tuomi, J. (1996) Plant compensatory growth: herbivory or competition? *Oikos* 77, 238–247.

Jefferies, R.L., Klein, D.R. and Shaver, G.R. (1994) Vertebrate herbivores and northern plant communities: reciprocal influences and responses. *Oikos* 71, 193–206

Keddy, P.A. (1989) *Competition*. Chapman and Hall, London, UK.

Lange, R.T. (1969) The piosphere, sheep track and dung patterns. *Journal of Range Management* 22, 396–400.

Laycock, W.A. (1967) How heavy grazing and protection affect sagebrush–grass ranges. *Journal of Range Management* 29, 206–213.

Le Houérou, H.N. (1989) *The Grazing Land Ecosystem of the African Sahel*. Ecological Studies 75. Springer Verlag, Berlin, Germany.

Lock, L.M. (1972) The effects of hippopotamus grazing on grasslands. *Journal of Ecology* 60, 445–467.

Louda, S.M., Keeler, K.H. and Holt, R.D. (1990) Herbivore influences on plant performance and competitive interactions. In: Grace, J.B. and Tilman, D. (eds) *Perspectives on Plant Competition*. Academic Press, San Diego, USA, pp. 414–444.

Lundberg, P. and Danell, K. (1990) Functional response of browsers to tree exploitation by moose. *Oikos* 58, 378–384.

McNaughton, S.J. (1985) Interactive regulation of grass yields and chemical properties by defoliation, a salivary chemical and inorganic nutrition. *Oecologia* 65, 478–486.

McNaughton, S.J. (1988) Mineral nutrition and spatial concentrations of African ungulates. *Nature* 334, 343–345.

McNaughton, S.J. (1992) The propagation of disturbance in savannahs through food webs. *Journal of Vegetation Science* 3, 301–314.

McNaughton, S.J. and Chapin, F.S., III (1985) Effects of phosphorus nutrition and defoliation on C_4 graminoids from the Serengeti plains. *Ecology* 66, 1617–1629.

McNaughton, S.J., Wallace, L.L. and Coughenour, M.B. (1983) Plant adaptation in an ecosystem context: effects of defoliation, nitrogen and water on growth of an African sedge. *Ecology* 64, 307–318.

Maschinski, J. and Whitham, T.G. (1989) The continuum of plant responses to herbivory: the influence of plant association, nutrient availability, and timing. *American Naturalist* 134, 1–19.

Mazancourt, C. de, Loreau, M. and Abbadie, L. (1998) Grazing optimization and nutrient cycling: when do herbivores enhance plant production? *Ecology* 79, 2242–2252.

Menaut, J.-C., Gignoux, J., Prado, C. and Clobert, J. (1990) Tree community dynamics in a humid savannah of the Côte d'Ivoire: modelling the effects of fire and competition with grass and neighbours. *Journal of Biogeography* 17, 471–481.

Milchunas, D.G. and Lauenroth, W.K. (1993) Quantitative effects of grazing on vegetation and soils over a global range of environments. *Ecological Monographs* 63, 327–366.

Milchunas, D.G., Sala, O.E. and Lauenroth, W.K. (1988) A generalized model of the effects of grazing by large herbivores on grassland community structure. *American Naturalist* 132, 87–106.

Milewski, A.V., Young, T.P. and Madden, D. (1991) Thorns as induced defences: experimental evidence. *Oecologia* 86, 70–75.

Milthorpe, F.L. and Davidson, J.L. (1966) Physiological aspects of regrowth in grasses. In: Milthorpe, F.L. and Ivins J.D. (eds) *The Growth of Cereals and Grasses.* Butterworths, London, UK, pp. 241–255.

Misra, R. (1983) Indian savannas. In: Bourlière, F. (ed.) *Tropical Savannahs.* Ecosystems of the World 13, Elsevier, Amsterdam, pp. 441–462.

Mueggler, W.F. (1972) Influence of competition on the response of bluebunch wheatgrass to clipping. *Journal of Range Management* 25, 88–92.

Murphy, J.S. and Briske, D.D. (1992) Regulation of tillering by apical dominance: Chronology, interpretive value and current perspectives. *Journal of Range Management* 45, 419–429.

Newsome, A.E. (1975) An ecological comparison of the two arid-zone kangaroos of Australia, and their anomalous prosperity since the introduction of ruminant stock to their environment. *Quarterly Review of Biology* 50, 389–424.

Newsome, A.E. (1983) The grazing Australian marsupials. In: Bourlière, F. (ed.) *Tropical Savannahs.* Ecosystems of the World 13. Elsevier, Amsterdam, The Netherlands, pp. 441–462.

Noy-Meir, I. (1979/80) Structure and function of desert ecosystems. *Israel Journal of Botany* 28, 1–19.

Noy-Meir, I. (1982) Stability of plant-herbivore models and possible applications to savannah. In: Huntley, B.J. and Walker, B.H. (eds) *Ecology of Tropical Savannahs.* Springer Verlag, Berlin, Germany, pp. 591–609.

Noy-Meir, I. (1993) Compensating growth of grazed plants and its relevance to the use of rangelands. *Ecological Applications* 3, 32–34.

O'Connor, T.G. (1991) Influence of rainfall and grazing on the compositional change of the herbaceous layer of a sandveld savannah. *Journal of the Grassland Society of Southern Africa* 8, 103–109.

O'Connor, T.G. (1992) Patterns in plant selection by grazing cattle in two savannah grasslands: a plant's eye view. *Journal of the Grassland Society of Southern Africa* 9, 97–104.

Oesterheld, M. and McNaughton, S.J. (1988) Intraspecific variation in response of *Themeda triandra* to defoliation: the effect of time of recovery and growth rates on compensatory growth. *Oecologia* 77, 181–186.

Ojasti, J. (1983) Ungulates and large rodents of South America. In: Bourlière, F. (ed.) *Tropical Savannahs.* Ecosystems of the World 13, Elsevier, Amsterdam, The Netherlands, pp. 427–440.

Olson, B.E. and Richards, J.H. (1988) Annual replacement of the tillers of *Agropyron desertorum* following grazing. *Oecologia* 76, 1–6.

Owen-Smith, N. (1988) *Megaherbivores: the Influence of Very Large Body Size on Ecology.* Cambridge University Press, Cambridge, UK.

Owen-Smith, N. and Cooper, S.M. (1987) Palatability of woody plants to browsing ruminants in a south African savanna. *Ecology* 68, 319–331.

Owen-Smith, N. and Novellie, P. (1982) What should a clever ungulate eat? *American Naturalist* 119, 151–178.

Paine, R.T. (1966) Food web complexity and species diversity. *American Naturalist* 100, 65–75.

Paine, R.T. (1971) A short-term experimental investigation of resource partitioning in a New Zealand rocky intertidal habitat. *Ecology* 52, 1096–1106.

Pastor, J. and Naiman, R.J. (1992) Selective foraging and ecosystem processes in boreal forests. *American Naturalist* 139, 691–705.

Pastor, J., Dewey, B., McInnes, P. and Cohen, Y. (1993) Moose browsing and soil fertility in the boreal forests of Isle Royale National Park. *Ecology* 74, 467–480.

Pellew, R.A. (1983) The impacts of elephant, giraffe and fire upon the *Acacia tortilis* woodlands of the Serengeti. *African Journal of Ecology* 21, 41–74.

Pellew, R.A. (1984) The feeding ecology of a selective browser, the giraffe (*Giraffa camelopardalis tippelskirchi*). *Journal of Zoology* 202, 57–81.

Price, P.W. (1991) The plant vigour hypothesis and herbivore attack. *Oikos* 62, 244-251.

Prins, H.H.T. and Olff, H. (1998) Species-richness of African grazer assemblages: towards a functional explanation. In: Newbery, D.M., Prince, H.H.T. and Brown, N. (eds) *Dynamics of Tropical Communities*. Blackwell Science, Oxford, UK, pp. 449–490.

Rhoades, D.F. and Cates, R.G. (1976) Towards a general theory of plant antiherbivore chemistry. *Recent Advances in Phytochemistry* 19, 168–213.

Ritchie, M.E., Tilman, D. and Knops, J.M.H. (1998) Herbivore effects on plant nitrogen dynamics in oak savannah. *Ecology* 79, 165–177.

Roux, E. (1969) *Grass: a Story of Frankenwald*. Oxford University Press, Cape Town.

Rusch, G.M. and Oesterheld, M. (1997) Relationship between productivity, and species and functional group diversity in grazed and non-grazed pampas grassland. *Oikos* 78, 519–526.

Rutherford, M.C. (1980) Annual plant production–precipitation relations in arid and semi-arid regions. *South African Journal of Science* 76, 53–56.

Sabelis, M.W., Van Balen, M., Bakker, F.M., Bruin, J., Drukker, B., Egas, M., Janssen, A.R.M., Lesna, I.K., Pels, B., Van Rijn, P.C.J. and Scitareanu, P. (1999) The evolution of direct and indirect plant defence against herbivorous arthropods. In: Olff, H., Brown, V.K. and Drent, R.H. (eds) *Herbivores: between Plants and Predators*. Blackwell Science, Oxford, pp. 109–168.

Sampson, A.W. (1919) *Plant Succession in Relation to Range Management*. Bulletin No. 791, United States Department of Agriculture, United States Government Printing Office, Washington, USA.

Sarmiento, G. (1992) Adaptive strategies of perennial grasses in South American savannahs. *Journal of Vegetation Science* 3, 325–336.

Scholes, R.J. and Walker, B.H. (1993) *An African Savannah. Synthesis of the Nylsvley Study*. Cambridge University Press, Cambridge, UK.

Seagle, S.W., McNaughton, S.J. and Reuss, R.W. (1992) Simulated effects of grazing on soil nitrogen and mineralization in contrasting Serengeti grasslands. *Ecology* 73, 1105–1123.

Senft, R.L., Coughenour, M.B., Bailey, D.W., Rittenhouse, L.R., Sala, O.E. and Swift, D.M. (1987) Large herbivore foraging and ecological hierarchies: landscape ecology can enhance traditional foraging theory. *Bioscience* 37, 789–799.

Senock, R.S., Sisson, W.B. and Donart, G.B. (1991) Compensatory photosynthesis of *Sporobolus flexuosus* (Thurb.) Rydb. following simulated herbivory in the northern Chihuahuan desert. *Botanical Gazette* 152, 275–281.

Sinclair, A.R.E. and Arcese, P. (1995) Serengeti in the context of worldwide conservation efforts. In: Sinclair, A.R.E. and Arcese, P. (eds) *Serengeeti II – Dynamics, Management, and Conservation of an Ecosystem*. University of Chicago Press, Chicago, USA, pp. 31–46.

Skarpe, C. (1986) Plant community structure in relation to grazing and environmental changes along a north–south transect in the western Kalahari. *Vegetatio* 68, 3–18.

Skarpe, C. (1990) Shrub layer dynamics under different herbivore densities in an arid savannah, Botswana. *Journal of Applied Ecology* 27, 873–885.

Skarpe, C. (1991) Spatial patterns and dynamics of woody vegetation in an arid savannah. *Journal of Vegetation Science* 2, 565–572.

Skarpe, C. (1992) Dynamics of savannah ecosystems. *Journal of Vegetation Science* 3, 293–300.

Skarpe, C. (1997) Ecology of the vegetation in the draw-down zone of Lake Kariba. In: Moreau, J. (ed.) *Advances in the Ecology of Lake Kariba*. University of Zimbabwe, Harare, Zimbabwe, pp. 120–138.

Skarpe, C. (2000) Desertification, no-change or alternative states: can we trust simple models on livestock impact in dry rangelands? *Applied Vegetation Science* 3, 261–268.

Stoddart, L.A., Smith, A.D. and Box, T.W. (1975) *Range Management*. McGraw-Hill Book Company, New York, USA.

Stuart-Hill, G.C. and Tainton, N.M. (1989) The competitive interaction between *Acacia karroo* and the herbaceous layer and how this is influenced by defoliation. *Journal of Applied Ecology* 26, 285–298.

Teage, W.R. (1985) Leaf growth of *Acacia karroo* trees in response to frequency and intensity of defoliation. In: Tothill, J.C. and Mott, J.J. (eds) *Ecology of the World's Savannahs*. Australian Academy of Science, Canberra, Australia, pp. 220–222.

Tilman, D. (1977) Resource competition between planctonic algae: an experimental and theoretical approach. *Ecology* 58, 338–348.

Tilman, D. (1987) On the meaning of competition and the mechanisms of competitive superiority. *Functional Ecology* 1, 304–315.

van Vegten, J.A. (1983) Thornbush invasion in a savanna ecosystem in eastern Botswana. *Vegetatio* 56, 3–7.

Vickery, M.L. (1984) *Ecology of Tropical Plants*. John Wiley & Sons, Chichester, UK.

Vitousek, P. (1982) Nutrient cycling and nutrient use efficiency. *American Naturalist* 119, 553–572.

Walker, B.H. (ed.) (1987) *Determinants of tropical savannahs*. IUBS Monograph Series 3, Oxford, UK.

Walker, B.H. and Noy-Meir, I. (1982) Aspects of the stability and resilience of savannah ecosystems. In: Huntley, B.J. and Walker, B.H. (eds) *Ecology of Tropical Savannahs*. Springer Verlag, Berlin, Germany, pp. 591–609.

Wallace, L.L., McNaughton, S.J. and Coughenour, M.B. (1985) Effects of clipping and two levels of nitrogen on the gas exchange, growth and productivity of two east African graminoids. *American Journal of Botany* 72, 222–230.

Walter, H. (1939) Grasland, Savanne und Busch der ariden Teilen Afrikas in ihrer ökologischen Bedingtheit. *Jb. Wiss. Bot.* 87, 750–860.

Walter, H. (1954) Die Verbuschung, eine Erscheinung der subtropischen Savannengebiete, und ihre ökologischen Ursachen. *Vegetatio* 5/6, 6–10.

Weber, G.E. and Jeltsch, F. (1998) Spatial aspects of grazing in savannah rangelands – a modelling study of vegetation dynamics. In: Brebbia, J.L. and Power, H. (eds) *Ecosystems and Sustainable Development*. Computer Mechanics Publications, Southampton, UK, pp 427– 437.

Werger, M.J.A. (1977) Effects of game and domestic livestock on vegetation in east and southern Africa. In: Krause, W. (ed.) *Handbook of Vegetation Science*, Part XIII, *Application of Vegetation Science to Grassland Husbandry*. Junk, The Hague, The Netherlands, pp. 149–159.

Wolfson (1999) The response of forage plants to defoliation; grasses. In: Tainton, N. (ed.) *Veld Management in South Africa*. University of Natal Press, Pietermaritzburg, pp. 91–101.

10 Competition and Environmental Stress in Temperate Grasslands

Duane A. Peltzer* and Scott D. Wilson
Department of Biology, University of Regina, Regina, Saskatchewan, Canada

Introduction

There is abundant evidence that competition helps to determine the species composition of temperate grasslands (Harper, 1977; Lauenroth and Aguilera, 1998; Wilson, 1998, 1999). One approach to organizing knowledge about competition is to examine how it varies along environmental gradients. Here we review which traits confer competitive ability to plants and the role of competition along environmental gradients of stress.

Plant competitive ability can be divided into two components: competitive response and competitive effect (Goldberg, 1990). Good response competitors are those species able to resist suppression by competitors. On the other hand, good effect competitors are able to reduce the performance of other species. Traits conferring either aspect of competitive ability, response or effect, may change along stress gradients. For example, plants from unproductive, stressful environments tend to have high below-ground biomass allocation, long-lived tissues, high nutrient retention and high tolerance of water and nutrient stress (Chapin, 1980; Chapin *et al.*, 1993); these traits may allow them to displace plants from more productive environments (see review in Goldberg, 1990).

Both field and garden experiments show that stress-tolerant species can dominate vegetation over the long term, due to higher levels of nutrient retention and tissue longevity (Berendse and Aerts,

1987; Aerts and van der Peijl, 1993; Berendse, 1994). It is not clear, however, whether these traits identify good stress tolerators or good competitors in stressful environments (Grime, 1977; Huston and Smith, 1987; Tilman, 1988; Berendse and Elberse, 1990). Whereas many studies have examined competitive interactions along natural and experimental productivity gradients in grasslands (Wilson and Tilman, 1991, 1993, 1995; Reader and Bonser, 1993; Reader *et al.*, 1994; Peltzer *et al.*, 1998; see also reviews by Goldberg and Barton, 1992; Gurevitch *et al.*, 1992), relatively few studies have explored how suites of traits responsible for stress tolerance affect competitive ability, or which traits should be related to competitive ability along environmental gradients.

How does the role of competition vary along gradients of environmental stress? Here we follow Grime's (1979) definition of stress as any factor that limits biomass production. In grasslands, biomass is typically limited by soil resources: water (Lauenroth *et al.*, 1978; Sala *et al.*, 1988; Silvertown *et al.*, 1994) or nutrients (Tilman, 1987; Berendse, 1994). Peak standing crop (i.e. above-ground plant biomass) is frequently used as a measure of stress in herbaceous grassland communities: low standing crop reflects high stress and low soil resources. Grime (1979) predicted that stress should limit plant growth rates and resource demand. Thus, competition might be relatively unimportant on dry or nutrient-poor soils that support little standing crop. Competition might

*Present address: Landcare Research, Lincoln, New Zealand.

© CAB *International* 2001. *Competition and Succession in Pastures*
(eds P.G. Tow and A. Lazenby)

increase in importance as soil resources increase and standing crop increases. This idea can be traced back to Darwin (Harper, 1977) and is still current (Keddy, 1989). Alternatively, stressful habitats may be characterized by intense competition for the resources that limit growth, such as nutrients and water (Newman, 1973; Tilman, 1988). As resources increase, stress decreases, standing crop increases, shade increases and competition may shift from being mostly below ground to mostly above ground.

We review evidence for the hypothesis that the role of competition in controlling community structure changes along gradients of stress, resource availability and standing crop. Our focus is on field experiments that examine the role of competition and variation in competitive ability along gradients of stress in natural temperate grasslands. We include mid-latitude arid and alpine grasslands but not tropical grasslands or savannahs or Arctic graminoid communities.

Competitive Ability

The relationship between competitive ability and the distribution or abundance of species in communities is still poorly understood, despite many experiments on competitive interactions in the field (see reviews by Goldberg and Barton, 1992; Gurevitch et al., 1992; Goldberg, 1996). Several important questions remain. What is competitive ability? Does competitive ability vary among species or environments? How important is competitive ability relative to abiotic factors, other ecological processes, and historical factors (Givnish, 1986; Welden and Slauson, 1986; Felsenstein, 1988; Keddy, 1989; Harvey and Pagel, 1991; Aarssen, 1992; Underwood and Petraitis, 1993; Westoby et al., 1995; Silvertown and Dodd, 1996)? Much interest has focused on whether the abundance and distribution of species are the result of variation in stress tolerance, competitive ability or niche differentiation among species (Grime, 1977; Grubb, 1977; Chapin, 1980; Keddy, 1989; Smith and Huston, 1989; Austin, 1990; Aarssen, 1992; McLellan et al., 1997). Here we discuss the relationships among competitive ability, stress tolerance and community composition.

Components of competitive ability

Goldberg (1990) distinguishes between the ability of plants to perform well in the presence of neighbours (competitive response), and the ability of plants to reduce the performance of other species (competitive effect). Traits conferring success for both response and effect may change along stress gradients.

Competitive responses

Competitive response ability (sensu Goldberg, 1990) is the ability of a plant to resist suppression by neighbours (i.e. the resident vegetation). This can be measured as the change in performance of a target plant or species in response to the presence of neighbours, either conspecific or interspecific. There are several reasons for using small target plants in existing vegetation (Goldberg, 1990). First, all plants must regenerate and pass through critical early life stages in order to establish; early life stages may be more sensitive to the effects of neighbours and the environment than established adult plants (Grubb, 1977). Thus, individual performance (e.g. seedling survival and growth) can be linked with population processes (e.g. recruitment, population growth, distribution). Secondly, competitive response ability determines which species persist in a habitat to contribute to community-level diversity and productivity (Wilson and Tilman, 1995). Thirdly, good response competitors are able to withstand resource shortages imposed by competing plants and are also likely to persist in stressful habitats.

Plant traits associated with competitive response ability are those which allow a species to persist and perform well in the presence of neighbours, and thus resource shortages or stress. Similarly, the best competitors in Tilman's (1982) resource ratio hypothesis model of plant competition are those species having the lowest resource requirement, or R*. Empirical support for this model was provided by Wedin and Tilman's (1993) experimental mixtures of native prairie grasses. In their study, grasses with the lowest R* values won in competition, regardless of initial planting densities in mixtures. Generally, traits suggested to confer competitive response ability in relatively unproductive systems are identical to those suggested for stress tolerance (Table 10.1). Typically, these species have high root : shoot ratios and low growth rates, are small and have nutrient-conserving mechanisms, such as long-lived tissues, carbon-based defences and storage organs (Grime, 1977, 1979; Chapin, 1980, 1991; Tilman, 1988; Berendse and Elberse, 1990; Chapin et al., 1990).

Table 10.1. Summary of plant traits suggested to confer high competitive response or competitive effect ability at low and high primary productivity. The last column shows traits associated with stress tolerance. Traits for competitive ability are modified from Goldberg (1996).

	Response		Effect		
	Low productivity	High productivity	Low productivity	High productivity	Tolerance
Leaf allocation	Low[abf]	Low[af]	Low	High	Low[de]
Stem allocation	Low[ab]	High[a]	Low	High	Low[de]
Root allocation	High[abf]	Low[af]	High	Low	High[de]
Reproductive allocation	Low[ab]	Low[a]	High	Low[cg]	Low[cdi]
Growth rate	Low[abf]	Low[abf]	High	High[c]	Low[cde]
Litter production	Low	High	High	High[c]	Low[c]
Photosynthetic rate	Low[a]	High	High	High	Low[e]
Height at maturity	Low[a]	High[a]	High	High[c]	Low[c]
Leaf area	Low	High	High	High[g]	Low[ci]
Plant mass	Low	High[a]	High	High[c]	Low
Specific root length	High[a]	High	High	High[h]	High
Leaf area/mass	Low[af]	High[af]	Low	High[cg]	Low
Tissue longevity	High[ag]	Low[adg]	High	Low[cg]	High[cdei]
Plasticity	Low	Low	High	High[c]	Low[cd]
Tissue [N]	Low[abg]	High[bg]	High	High	Low[de]
Rate of nutrient uptake	Low[b]	Low[b]	High	High[c]	Low[e]
Nutrient storage	High	Low	High	Both	High[di]
Nutrient losses or leaching	Low[b]	High[b]	High	High	Low[d]
Nutrient foraging ability	High	Low	High	Both	Low
Nutrient-based defences	High[b]	Low[b]	Low	Low[c]	High[e]
C-based defences	High[e]	Low	Low	Low	High
Mycorrhizal infection	High[g]	Low	Low	Low	High[d]
Shade tolerance	Low[bf]	High[af]	Low	Low[c]	Low
Drought tolerance	High[f]	Low[f]	Low	Low[c]	High
Low-nutrient tolerance	High[bg]	Low[ag]	Low	Low[c]	High

a, Tilman (1988); b, Tilman (1990); c, Grime (1977); d, Chapin (1980); e, Chapin et al. (1993); f, Smith and Huston (1989); g, Berendse and Elberse (1990); h, Caldwell and Richards (1986); i, Grime (1979).

Why might variation in competitive response be important in grasslands? Variation in competitive responses among species may determine species positions within a competitive hierarchy. This in turn may determine their distributions along stress gradients or their relative abundance in a community (Grime, 1979; Keddy, 1989). The relationship between species competitive responses and distributions is relatively well documented in wetlands (e.g. Wilson and Keddy, 1986; Gaudet and Keddy, 1988; Keddy and Shipley, 1989) but less well understood in grasslands (Mitchley and Grubb, 1986; Aarssen, 1988; Wilson and Tilman, 1995). Generally, there is no strong relationship between competitive responses and the distribution or abundance of species along stress gradients (Herben and Krahulec, 1990; Silvertown and Dale, 1991; Shipley and Keddy, 1993). For example, Wilson and Tilman

(1995) measured the competitive responses of eight grassland species (four forbs and four grasses) in an old field in Minnesota. The species showing the weakest response, i.e. the species most suppressed by neighbours, was the numerical dominant of natural vegetation, the perennial grass *Schizachyrium scoparium*. This result was robust across four combinations of nitrogen (N) availability and soil disturbance.

STRESS TOLERANCE AND COMPETITIVE RESPONSES. Because competition is often for resources and many of the traits associated with stress tolerance interact with patterns of resource availability, there should be a close link between competitive response ability and stress tolerance. This is not a new idea; Grime (1977) listed traits associated with stress tolerators and competitive species two decades ago. For Grime (1977), any energy a plant spends coping with stress decreases competitive ability, so that there is a trade-off between stress tolerance and competitive ability. In contrast, other authors define a good competitor as a species able to perform well despite resource shortages (Tilman, 1982, 1988; Berendse and Elberse, 1990; reviewed by Grace 1990, 1991).

The distinction is this: species in productive environments are competitive if they exploit resources as quickly and efficiently as possible (Grime, 1977). Resource competition may also be intense in unproductive habitats (Wilson and Tilman, 1991, 1993, 1995; Wilson, 1993a, b), but the species that lower limiting resources to the lowest level and use them most efficiently may dominate the vegetation (Tilman, 1982). Thus, there may be no trade-off between stress tolerance and competitive ability. Stress tolerance and competitive ability can be conferred by the same traits (Tilman, 1982). Traits predicted to confer both competitive response ability and stress tolerance are summarized in Table 10.1.

WITHIN-SPECIES VARIATION IN COMPETITIVE RESPONSES. Within-species or within-population variability in competitive ability may also contribute to species coexistence and community composition. Within-species variability may diminish among-species differences in competitive ability. Genotypic variability and specificity of interactions with neighbouring plants or among sites have been shown in pastures and old fields (Aarssen, 1988; Mehrhoff and Turkington, 1990; Turkington, 1991). For example, Mehrhoff and Turkington (1990) compared the competitive ability of five populations of *Trifolium repens* from different-aged pastures. They found considerable variation in competitive ability among populations of *T. repens* in pot, garden and pasture competition experiments with grasses.

Aarssen (1992) reviewed several theories regarding variation in competitive abilities at the genotype level; his working hypothesis is that variation in competitive ability among genotypes within a species is as great as variation among species. One implication of this hypothesis is that competitive exclusion will not occur at the level of the species in a community. Cheplick (1997) explored within-species variation in competitive response using 11 genotypes of the rhizomatous perennial grass *Amphibromus scabrivalis* grown in competition with *Lolium perenne*. Genotypes with more widely

spaced ramets ('guerilla' strategy) were more strongly suppressed by *L. perenne* than genotypes producing few, closely spaced ramets ('phalanx' strategy). The role of within-species variation in competitive response in determining species persistence in natural communities deserves further attention.

RESPONSES TO HETEROGENEITY. Species may respond differently to resource heterogeneity. Such heterogeneity may occur in both space (Campbell and Grime, 1989; Campbell *et al.*, 1991; Caldwell and Pearcy, 1994; Casper and Cahill, 1998) and time (Fitter, 1986; Campbell and Grime, 1989; Bilbrough and Caldwell, 1997; Goldberg and Novoplansky, 1997).

Recent studies have shown that plants respond to soil nutrient patchiness independently from resource level (Hutchings, 1988; Casper and Cahill, 1998). There is abundant evidence that the ability to forage for patchy resources differs among species of grasses (Jackson *et al.*, 1990; Humphrey and Pyke, 1997; Reynolds *et al.*, 1997). Further, there may be a trade-off between the precision of foraging and the size of resource patches exploited (Campbell *et al.*, 1991). Large, rhizomatous grasses should be better foragers than tussock grasses, but their foraging precision should be low, due to wider spacing between ramets.

Species may also differ in their responses to temporal heterogeneity (Grime, 1979; Chapin, 1980; Goldberg and Novoplansky, 1997). In natural grasslands, such as the North American Great Plains, soil resources and water may be available only during infrequent pulses (Sala and Lauenroth, 1985; Sala *et al.*, 1992), and differences in ability to exploit temporal patches may contribute to competitive success (Caldwell, 1994). Campbell and Grime (1989) showed that a relatively slow-growing grass, *Festuca ovina*, was able to use short nutrient pulses of from 0.1 to 10 h in duration, while a faster-growing grass, *Arrhenatherum elatius*, could use only longer pulses. Grime (1994) suggests that species occurring in habitats with 'chronic nutrient stress' have large, long-lived root systems, which remain functional throughout the year and are capable of utilizing short resource pulses. In contrast, faster-growing species tend to produce new tissues to capture additional resources, making their responses to pulses much slower. Although resource allocation has received much attention, grasses may also vary in their physiological ability to capture nutrients in response to resource pulses (Mouat, 1983; Caldwell, 1994). Interest in the importance of plant responses to resource heterogeneity is increasing (Shorrocks and Swingland, 1990; Bell and Lechowicz, 1994; Caldwell and Pearcy, 1994; Grime, 1994; Miller *et al.*, 1995; Bilbrough and Caldwell, 1997; Casper and Cahill, 1998), but much remains to be done in grasslands.

Competitive effects

Competitive effect ability (*sensu* Goldberg, 1990) is defined as the ability of a neighbouring plant or species to suppress a focal individual or species. Competitive effects are quantified as the per-plant or per-unit-mass (e.g. per-gram) reduction in performance of the focal plant or species (see review by Goldberg and Scheiner, 1993). Competitive effects are often assumed to be equivalent among neighbour species, because all plants use the same resources (e.g. light, water, nutrients) and these resources are supplied along gradients rather than as discrete packages (Harper, 1965; Goldberg and Werner, 1983; Goldberg, 1996). As noted above, this assumption requires more investigation. Bakker (1996) found differences in competitive effects between introduced and native grasses in mixed-grass prairie. Wedin and Tilman (1993) also found significant differences between early- and late-successional grasses; later grasses lowered soil N availability more than early grasses, which may explain their eventual dominance of old fields.

Competitive effects may occur at several scales. In the short term, plants reduce resource levels and create localized zones of resource depletion (Caldwell, 1994; Huston and DeAngelis, 1994). Over longer time periods, plants may modify nutrient cycles through shoot or root litter quality or plant–soil community feedbacks; other long-term feedbacks may arise through mutualisms, such as mycorrhizas, or through diseases (Allen and Allen, 1990; Fitter, 1991; Read, 1991; Wilson and Agnew, 1992; Wedin and Pastor, 1993; Bever, 1994; Stark, 1994; Bever *et al.*, 1997; Kleb and Wilson, 1997; Wilson, 1998).

EFFECTS ON HETEROGENEITY. Plants can also influence the spatial heterogeneity of soil resources, water and light in grasslands (Hook *et al.*, 1991; Caldwell, 1994). For example, Kleb and Wilson (1997) used a reciprocal soil transplant experiment between mixed-grass prairie and aspen (*Populus*

tremuloides) forest. Over one growing season, prairie grasses significantly lowered the heterogeneity of both available N and moisture. Thus, grasses decreased the patchiness of soil resources.

STRESS AND COMPETITIVE EFFECTS. Stress is the result of both biotic (due to competitive effects) and abiotic (environmental) processes (Goldberg and Novoplansky, 1997). Interactions between biotic and abiotic stress may occur if competitive effect varies with environment. Few studies to date have separated stress caused by neighbouring vegetation from stress imposed by the environment. Such studies are needed to understand the relationship between biotic and abiotic stresses in grasslands. Austin (1990) suggests that environmental influences on species performance are as important as species interactions. It is clear that both influence grasses. For example, in an old field in Minnesota, increasing N availability increased the growth of transplanted grasses by about 90%; removing neighbours increased growth by 40%; and both increasing N and removing neighbours increased growth by 150% (see Fig. 1 in Wilson and Tilman, 1995).

Some questions remain. Are species with larger competitive effects less tolerant of harsh or fluctuating abiotic conditions? The distinction between biotic and abiotic stress is important because they probably act at different scales. Abiotic stress gradients may occur at larger scales, influencing entire communities, and filtering the local species pool by removing those species that are physiologically incapable of tolerating local environmental conditions (Harper, 1977). Variation in competitive effect ability, and thus biotic stress, acts to eliminate species after environmental filtering through competitive exclusion, similar to Diamond's (1975) assembly rules. This model has been applied to wetlands (Keddy, 1992), but has not been tested explicitly in grasslands. Part of the reason may be that there are few contrasting environments with discrete boundaries in grasslands; instead, both biotic and abiotic stresses commonly occur along gradients.

In summary, good response competitors are able to withstand resource shortages imposed by neighbours. On the other hand, good effect competitors are able to strongly suppress the performance of competing species. However, these two components of competitive ability are not necessarily correlated (Goldberg, 1990). One way to compare competitive ability among species or along gradients of stress is by measuring competition intensity, the relative decline in species performance caused by competition.

Measuring Competition in the Field

A simple way to measure the intensity of competition in the field is to compare the performance of transplants in plots without neighbours with that in plots with neighbours. Any difference in performance is presumably attributable to the presence of neighbours. This can be done at sites differing in some other factor of interest, e.g. stress or grazing intensity. It can also be done with different transplant species in order to compare their responses. Comparisons of competition intensity (CI) among sites or species should be standardized for the effect of the sites or species identities on transplant performance (Wilson and Keddy, 1986; Grace, 1995; Miller, 1996) as:

$$CI = (NN - AN) / NN$$

where NN is transplant performance (e.g. survivorship, mass, growth, seed production, tiller number) in plots with no neighbours present and AN is performance with all neighbours present.

Regression analysis can be used to examine relationships between CI and stress or standing crop (e.g. Wilson and Keddy, 1986; Belcher et al., 1995; Grace, 1995; Miller, 1996; Peltzer et al., 1998). Care should be taken, however, not to use transplant performance as a measure of stress: in this case, the same term (e.g. growth in the absence of neighbours) ends up on both sides of the regression equation, producing a spurious correlation.

Alternatively, such experiments can be examined with analysis of variance (ANOVA) of transplant performance: a significant interaction between competition and site treatments would suggests that competition intensity varies among sites (e.g. Platenkamp and Foin, 1990; Goldberg and Scheiner, 1993; Wilson and Tilman, 1995). ANOVA of performance revealing significant interactions between transplant species and competition suggest interspecific differences in competitive ability; three-way interactions, including site, suggest that these differences vary with stress (Wilson and Tilman, 1995).

Does Competition Vary with Standing Crop?

Stress is a common phenomenon in natural grasslands. Plant productivity, usually measured as aboveground biomass, is normally limited by soil resources: water and nutrients (Lauenroth et al., 1978; Tilman, 1987; Sala et al., 1988; Silvertown et al., 1994). Stress can occur along natural gradients – for example, rainfall gradients or north- vs. south-facing slopes. Stress can also vary through time with among-year variation in precipitation and temperature. Here we review how competition varies along natural and experimental stress gradients.

Is standing crop a reliable indicator of stress?

Standing crop should reflect stress from a physiological point of view, but other biotic factors may influence standing crop, making it a poor indicator of stress. Standing crop, of course, reflects rates of biomass removal, as well as production. Thus low standing crop sites may be produced by high disturbance rates. Standing crop in a Minnesota oldfield grassland was enhanced with three levels of N addition; controls made a fourth treatment. All N levels were crossed with four levels of soil disturbance applied with a mechanical tiller, which produced 0, 25, 50 and 100% bare ground at the start of each growing season. Large mammals and abiotic disturbance (e.g. fire) were excluded from the experiment, so that the effects of N availability and disturbance could be studied independently. Competition intensity, measured by transplants of the grass S. scoparium (Wilson and Tilman, 1993) and, in a later experiment, using transplants of eight species (Wilson and Tilman, 1995), decreased significantly with increasing disturbance, but did not vary with N availability. The results suggest that low- standing- crop habitats may have low levels of competition if the low standing crop is partly attributable to disturbance. Removal of biomass by disturbance decreases community demand on resources and decreases competition (Taylor et al., 1990). In a series of studies, Reader (1992, 1993) and Reader and Bonser (1993) performed plant removal experiments (i.e. removing one or more species from intact vegetation (sensu

Keddy, 1989) at varying levels of stress and herbivory. Competition affected transplants only in plots caged to exclude grazers. In open plots, grazers were far more important than competition in controlling plant success.

Natural stress gradients

In natural vegetation, variation in standing crop is often inferred to reflect variation in stress. Reader *et al.* (1994) grew seedlings of *Poa pratensis* in cleared plots and intact vegetation in fields in Australia, Europe and North America. Competition intensity did not increase with standing crop when all sites were examined together. Transplant experiments along a gradient of soil depth and standing crop in Ontario also found no variation in competition intensity (Belcher *et al.*, 1995). Similar results were found in an old field in Michigan (Foster and Gross, 1997); in this case, however, competition intensity increased significantly with neighbourhood litter mass. A meta-analysis of 34 studies found no difference in competition intensity between more stressful habitats (desert and Arctic) and less stressful habitats (prairies, meadows and old fields) (Gurevitch *et al.*, 1992).

   Other experiments, in contrast, have found variation in competition intensity. Del Moral (1983) grew transplants in two alpine grasslands in Washington. In one neighbourhood, with 50 g m^{-2} standing crop, neighbours increased transplant survival. In a second neighbourhood, with 650 g m^{-2} standing crop, no transplants survived in the presence of neighbours. This result suggests that neighbours were facilitative at low standing crop but competitive at high. Gurevitch (1986) followed the fate of naturally establishing seedlings of grasses along a topographic gradient in Arizona. Neighbours decreased seedling performance more in moister, low-lying sites than on dry ridge tops. Reader and Best (1989) removed neighbours in an old field in Ontario and followed the population response of the composite *Hieracium floribundum* in low- and high-standing-crop sites. Initial *Hieracium* densities were made similar by thinning all plots at the start of the experiment. Neighbours reduced *Hieracium* performance at high-standing-crop sites but had no effect at low standing crop. Similar results were found for three species of transplants in the same system (Reader, 1992) and for the grass *Poa pratensis* (Reader and

Bonser, 1993), but not for *P. compressa*, which was equally suppressed in both habitats. Bonser and Reader (1995) used a wider selection of standing-crop values and found competition intensity to increase with standing crop. In Montana, grass neighbours inhibited an annual mustard in a wet year but facilitated it in a dry year (Greenlee and Callaway, 1996).

Experimental stress gradients

Other transplant experiments have been performed along experimental fertility gradients. As in the case of natural gradients, they produce conflicting results about variation in competition intensity. Reader and Best's (1989) *Hieracium* experiment described above found significant variation in competition intensity associated with natural variation in standing crop, but no variation in competition intensity was produced by supplying extra water to the vegetation. Wilson and Shay (1990) removed neighbours from around established grass tussocks in Manitoba mixed-grass prairie and found that neighbour removal increased tussock size to a similar extent in both fertilized and unfertilized plots. Burning also had no effect on competition intensity. Wilson and Tilman (1991) transplanted three grass species into clearings and intact vegetation within a 5-year-old N addition experiment in Minnesota: neighbours suppressed transplant growth to the same extent at all N levels. Similar experiments in a nearby field with one (Wilson and Tilman, 1993), two (Wilson, 1994) and eight species (Wilson and Tilman, 1995) produced the same result. Further, competition did not increase with added N in either undisturbed or tilled plots, indicating that both perennial and annual grass neighbourhoods had similar behaviour along the experimental fertility gradient (Wilson and Tilman, 1993, 1995). DiTommaso and Aarssen (1991) transplanted three grass species into clearings and intact vegetation in fertilized and unfertilized plots in an old field in Ontario and found that fertility had no effect on competition intensity. Grass transplants grown for 3 years in a 5-year-old N addition experiment in an old field in Saskatchewan were equally suppressed at all levels of N availability (Peltzer *et al.*, 1998). Tree and shrub transplants in the same field were equally suppressed, regardless of whether the vegetation was supplied with extra water and N (Li and Wilson, 1998).

One exception to the trend of competition intensity not varying along experimental gradients is given by Reader (1990), who examined the impact of neighbours on *H. floribundum* in unfertilized and fertilized plots: neighbour removal increased *Hieracium* recruitment and survival only in fertilized plots.

Summary of competition intensity and standing crop

With one exception, competition intensity did not vary with stress along experimental gradients. Competition intensity frequently increased with standing crop along natural gradients, but not always. In some cases, the increase in competition intensity was produced by a single point where neighbours had no negative effect at low standing crop (Del Moral, 1983; Bonser and Reader, 1995; Greenlee and Callaway, 1996). Overall, the results support the idea that competition intensity increases with standing crop on natural gradients more than on experimental gradients (Goldberg and Barton, 1992).

The results of any experiment depend on the conditions under which it is carried out, and measurements of competition intensity have been carried out within a wide range of standing crop values (summarized in Belcher *et al.*, 1995). We summarized the results of several studies that gave the growth rate of grass transplants with and without herbaceous neighbours. Some studies reported final transplant mass (Wilson, 1994; Gerry and Wilson, 1995) but were included because they reported initial mass and allowed calculation of growth rates. The studies were performed in fields abandoned from cultivation for at least 10 years. We included studies in which both neighbour roots and shoots were removed and for which standing crop could be determined (Wilson and Tilman, 1991, 1993, 1995; Wilson, 1993a, 1994; Reader *et al.*, 1994; Gerry and Wilson, 1995; Bakker, 1996; Foster and Gross, 1997; Peltzer *et al.*, 1998; S.D. Wilson, unpublished data; D.A. Peltzer and S.D. Wilson, unpublished data). We examined grass transplants with herbaceous neighbours, in order to avoid drastic differences between transplant and neighbour morphology (e.g. grasses vs. trees). We excluded two young sites from Reader *et al.* (1994) because the topsoil had been removed and the plots were undergoing primary succession (H. Olff,

1996, personal communication). Competition intensity (CI) was calculated as above for each species, year and vegetation type.

There was no clear relationship between competition intensity and standing crop (Fig. 10.1), either for the complete data set or for the data set with the large study of Reader *et al.* (1994) excluded. Thus, our summary of other studies corroborates the lack of an obvious pattern in competition intensity found by Reader *et al.* (1994). Although there is always a danger of falsely failing to reject the null hypothesis of no effect, an analysis of the same data set, but including recently disturbed plots, found that competition intensity did increase significantly with field age (Wilson, 1999). This disparity suggests that standing crop is a relatively poor predictor of competition intensity. Habitats which are so stressful that plants cannot maintain live mass are, of course, likely to have little competition (e.g. Grubb, 1992; Belcher *et al.*, 1995; Kadmon, 1995; Goldberg and Novoplansky, 1997), but this description probably does not apply to temperate grasslands.

Lastly, dominant species may differ in their competitive effects regardless of stress, so that in any particular environment, the intensity of competition is determined by the identity of the dominant species and not by stress or resource availability. Bakker (1996) grew transplants with and without neighbours in two sections of a 50-year-old Saskatchewan field. One section had been planted with the introduced pasture grass *Agropyron cristatum*, and the other section had undergone natural succession to native prairie grasses. Competition intensity varied little between dominant vegetation types, but the *F* ratio for the competition term in the ANOVA of transplant growth was twice as high in the *Agropyron*-dominated vegetation than in the native-dominated vegetation, suggesting that *Agropyron* exerted greater competitive effects than did native species (Underwood and Petraitis, 1993). Further, the growth of transplants of the native grass *Bouteloua gracilis* decreased significantly with neighbour mass only when grown in native vegetation, and the growth of transplants of *Agropyron* decreased significantly with neighbour mass only when grown in *Agropyron*-dominated vegetation. These results suggest that competition was most intense in intraspecific pairings and that variation in competitive effects occurred without variation in abiotic stress. Similar results were found for the

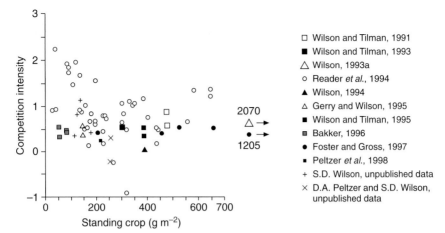

Fig. 10.1. Competition intensity as a function of standing crop. Competition intensity was measured as the relative reduction in grass transplant growth caused by grass neighbours. Standing crop increases with decreasing stress. Data were taken from several studies in natural temperate grasslands.

grass *Anthoxanthum odoratum* in California, where *A. odoratum* is more suppressed by intraspecific neighbours on dry soils than by interspecific neighbours on moister soils (Platenkamp and Foin, 1990). Goldberg *et al.* (1995) outline a method for testing whether differences in competition intensity are attributable to differences in stress or simply to differences among neighbouring species.

Stress and Root Competition

If competition is equally important at all levels of stress, the mechanism of competition may still shift from below ground to above ground as stress decreases and standing crop increases. Field experiments suggest that competition in grasslands occurs mostly below ground and that neighbour shoots have little effect on transplant performance (Cook, 1985; Snaydon and Howe, 1986; Wilson and Tilman, 1991, 1993, 1995; Seager *et al.*, 1992; Wilson, 1993a, b; Belcher *et al.*, 1995; Peltzer *et al.*, 1998; see review by Casper and Jackson, 1997). Cook and Ratcliff (1984) used root exclusion tubes of varying depths to show that the intensity of root competition decreased as neighbour-free soil volume increased.

Shoot competition occurs in some grasslands if standing crop is increased through fertilization. Transplant growth in unfertilized Minnesota old-field plots was controlled entirely by root competition but was influenced by both root and shoot competition

in fertilized plots with less light penetration (Wilson and Tilman 1991, 1993, 1995). This did not occur, however, in a grassland with a smaller range of standing crop (Belcher *et al.*, 1995): in this case, competition was always below ground, regardless of neighbour mass. Competition was also entirely below ground in species-poor experimental plots, in which fertilization did not produce changes in species composition (Peltzer *et al.*, 1998). Taken together, these studies suggest that light competition may occur only if standing crop is relatively high and if tall life-forms are available to colonize fertilized plots and cause shading.

Competitive Ability and Grassland Community Structure

Variation in competitive ability among species occurs at the individual level. A next step is to determine if variation among individual species can be used to predict their performance in a community (Goldberg, 1990; Goldberg *et al.*, 1995). Specifically, does variation in the performance of individuals scale up to the population and community levels? One method to assess individual- vs. community-level competitive ability is the community density series of Goldberg *et al.* (1995). This method manipulates the density of the entire community to levels both below and above those naturally occurring at a site. Very low-density treatments are null communities, which should have no competition occurring among plants. As the community density increases, so do the frequency and importance of species interactions, including competition. The difference in performance of species between low- and high-density treatments is its community-level competitive ability. For example, species whose relative abundance increases with community density would have a higher community-level competitive ability. Other general attempts to scale plant effects at lower scales to phenomena at higher scales have been discussed at length elsewhere (Ehleringer and Field, 1993; Jones and Lawton, 1995; Bazzaz, 1996).

Plant traits may be used to predict which species have good competitive abilities, and to test these predictions empirically (Gaudet and Keddy, 1988; Keddy, 1989, 1992; van der Werf et al., 1993; Grime et al., 1997; Rösch et al., 1997; Reader, 1998). A partial list of these traits is given in Table 10.1. Whole communities can be screened for general patterns of traits associated with abundance and distribution of species, although this is logistically difficult. Experimental evaluations of traits for competitive ability are needed (e.g. Wilson, 1991; Wedin and Tilman, 1993).

One difficulty with trying to scale up variation in competitive ability or with using plant traits to predict competitive ability is that competitive ability may be inconsistent among sites or habitats (Underwood and Petraitis, 1993; Miller, 1994; D.A. Peltzer and S.D. Wilson, unpublished data; see review by Goldberg, 1996). For example, Vinton and Burke (1997) examined the effects of prairie plants on carbon and N cycling at three sites varying in annual precipitation. They found that differences among species were greatest at the two most productive sites. This result was probably due to patchy plant cover at the driest site (the comparison was between bare ground and plant cover) and a more continuous cover at the two wetter sites (the comparison was among species). In a review of studies comparing species' competitive abilities between environments, Goldberg (1996) found that competitive responses tended to be consistent among environments while competitive effects often varied among environments. Further work is needed to examine how species' competitive effects vary with environment and why. Factorial experiments designed to observe species effects × environment interactions are ideal for determining variation in competitive abilities among species (Goldberg and Scheiner, 1993).

Interactions between Grazing and Competition

Grazers have a number of impacts on native grasslands: they can remove 9–57% of net above-ground foliage production (Frank et al., 1998), promote shoot growth by removing older, less productive tissue (Caldwell et al., 1981; McNaughton, 1983, 1984), enhance nutrient availability by increasing rates of nutrient cycling (McNaughton et al., 1989; Day and Detling, 1990; Holland et al., 1992) and create small-scale disturbances by trampling or burrowing (Crawley, 1983; Huntley, 1991; Olff and Ritchie, 1998). In addition, grazers have many indirect effects in grasslands. For example, removal of above-ground biomass may enhance primary production through removal of detritus, increased soil moisture status and plant water-use efficiency, reduced fire frequency and altered species composition (McNaughton, 1984, 1985; Knapp and Seastedt, 1986; Archer, 1995).

At the level of the individual plant, grazers reduce survival, growth and fecundity (Crawley, 1983; Bullock, 1996). Several authors suggest that grazing can promote plant growth through removing senescing tissue and enhancing subsequent growth, although the majority of studies do not support this idea (see review by Belsky, 1986). By damaging plants, herbivores induce defence mechanisms in plants such as altered morphology or increased levels of defence chemicals in leaves (Crawley, 1983; Vicari and Bazely, 1993). Grazing of above-ground tissues also has consequences below ground. For example, classic work by Weaver (1950) showed that intense grazing of above-ground foliage can strongly reduce below-ground biomass in short-grass prairie. However, the effects of above-ground grazing on below-ground productivity vary among grazing systems and the grazing history of the site (Belsky, 1986; Milchunas and Lauenroth, 1993).

Plants have several anti-herbivore defences including high concentrations of silica in leaves, lower-growing meristems, high levels of fibre and low levels of protein and a variety of defence compounds (Vicari and Bazely, 1993). Several plant traits confer resistance to both herbivory and environmental stress. For example, high tissue density, tough, fibrous leaves and high concentrations of secondary compounds confer resistance to herbivores and stress (Chapin, 1980, 1991; Chapin et al., 1990; Grubb, 1992; see Table 10.1). In addition, many traits allowing grasses to withstand grazing may actually be adaptations to a semi-arid environment (Coughenour, 1985).

Grazing may have bigger impacts in stressful habitats than in more productive habitats, for at least two reasons. First, fewer nutrients are usually available for plants to take up and use for regrowth after biomass removal (Grime, 1979). Secondly, shoot herbivory usually results in a decline in critical mycorrhizal mutualists associated with plant roots (Gehring and Whitham, 1994), further reducing the ability of grazed plants to take up nutrients needed for compensatory growth.

Herbivory may alter the outcome of competition between grassland plant species. More specifically, grazers can influence plant competitive ability either by modifying growth and morphology or by influencing plant abundance and distribution (Louda et al., 1990). Further, by removing standing biomass, grazing acts to shift plant competition from shoots (i.e. for light) to roots (i.e. for soil nutrients or water) (Milchunas et al., 1992). A common assertion is that herbivores shift the balance of competition by preferentially consuming dominant competitors, altering the relative competitive ability of species and promoting the abundance of plants that are otherwise rare or excluded (Tansley and Adamson, 1925; McNaughton, 1985; Huntley 1991; Clay et al., 1993; Olff and Ritchie, 1998). This view is supported by a number of studies. For example, selective grazing of the late-successional grass S. scoparium results in the increase of two grasses that are normally poorer competitors, Bothriochloa saccharoides and Stipa leucotricha (Anderson and Briske, 1995). Grazing in a southern English pasture increased the abundance of the N-fixing forb T. repens and decreased L. perenne, because the taller Lolium lost more above-ground mass to grazers (Parsons et al., 1991). For two grasses growing in low-nutrient soils, clipping both caused greater reductions in the growth of A. elatius than in that of Festuca rubra and accelerated the replacement of Arrhenatherum by Festuca (Berendse et al., 1992). Below-ground insect herbivory may also accelerate succession to a grass-dominated sward by reducing the size and growth of perennial forbs and perhaps their competitive ability (Brown and Gange, 1990, 1992). Thus, herbivore-induced changes of competitive interactions can drive species replacement in grasslands.

The argument that selective grazing causes the competitive replacement of palatable species by unpalatable species assumes that a trade-off should exist between competitive ability and grazing tolerance (Moretto and Distel, 1997). For example, plant height should increase a plant's competitive ability for light, but may also increase its proneness to grazing (Gaudet and Keddy, 1988; Oksanen, 1993). A trade-off between competitive ability and grazing tolerance should occur for competitive effect ability, but not necessarily for competitive response ability; this is because many of the plant traits conferring competition response or stress tolerance ability also confer resistance to herbivores (see Table 10.1).

Grazers may be essential for the restoration of native grasslands. Fire is often used in prairie restoration across North America, but repeated burning can lower plant species diversity (Collins et al., 1995). Recent work by Collins et al. (1998) found that both mowing and grazing by bison enhanced species diversity after fire by selectively reducing the abundance of dominant C_4 (warm-season) grasses, thus reducing their competitive effects and allowing subordinate C_3 (cool-season) grasses and forbs to persist and contribute to grassland diversity. Restoration of diversity by grazers may be successful at small spatial scales (i.e. < 10–100 m^2), but recent studies suggest that grazing reduces plant diversity over larger spatial scales (i.e. > 1000 m^2) (Olff and Ritchie, 1998; Stohlgren et al., 1999a, b).

As discussed above, we know that grazing or herbivory strongly affects the survival, growth and reproduction of plants, but how does herbivory modify the outcome of competition along gradients of stress (Connell, 1975; Menge and Sutherland, 1987; Louda, 1989; Louda et al., 1990)? Most studies have examined the effects of grazers on range condition or species composition, or have used manual clipping in pot experiments to examine the effects of grazers on plant competition; relatively few studies have quantified the effects of herbivory on plant competition under natural conditions. Further, the effects of different kinds of herbivores on plant interactions is essentially unknown, i.e. what are the effects of large-bodied ungulates vs. small burrowing mammals vs. insects, or native grazers vs. cattle in the same grassland system? For example, even though below-ground herbivory can greatly reduce plant productivity, few studies have considered the role of invertebrate grazers on roots, even though most grassland production is below ground (Stanton, 1988; Brown and Gange, 1990). In short-grass prairie, herbivory by white grubs, the larvae of June beetles (including *Phyllophaga fimbripes*), strongly depresses the abundance of the dominant grass, *B. gracilis,* for up to 14 years after an insect outbreak (Coffin *et al.,* 1998). Clearly, further work is required to understand how different kinds of above- and belowground herbivory interact with competition to determine the productivity and species composition of grasslands.

Non-competitive Interactions

Many plant species can coexist in grasslands despite having large differences in competitive abilities (e.g. Mitchley and Grubb, 1986; Silvertown *et al.,* 1994), suggesting that other factors either prevent the competitive exclusion of species or promote their coexistence (Bullock, 1996).

Grasses might avoid competition by exploiting different niches. If this happens, then competition might contribute little to grassland patterns (Leibold, 1995). It is difficult to imagine how plants can divide up resources into distinct niches, because all plants use the same resources (Harper, 1977; Tilman, 1982; Goldberg and Werner, 1983). Several studies have explicitly examined resource partitioning in plant communities (e.g. Werner and Platt, 1976; Russell *et al.,* 1985), but are often inconclusive. For example, Mahdi *et al.* (1988) did not find significant niche separation among eight British grassland species using either six niche dimensions based on soil nutrient levels and phenology or ratios of N : P as a test of Tilman's (1982) resource ratio hypothesis.

In North American prairies, many species coexist in relatively few habitat types. In these grasslands, two niche axes are commonly studied: water use and phenology. Werner and Platt (1976) showed niche partitioning for populations of six goldenrod species (*Solidago* spp.) along a soil moisture gradient in an Iowa prairie, although the same species did not show niche partitioning in an old field in Michigan. Many grassland species display phenological divergence in growth, which is related to photosynthetic pathway (Ehleringer and Monson, 1993). For example, Kemp and Williams (1980) examined photosynthesis, respiration and growth in two common prairie grasses, *Agropyron smithii* (C_3) and *B. gracilis* (C_4). *Bouteloua gracilis* grew best in warmer conditions, but both species

responded similarly to water stress. The authors suggest that niche separation may occur along seasonal temperature gradients rather than seasonal moisture gradients for these two grasses. Sala *et al.* (1982) found that *A. smithii* took up soil moisture earlier in the growing season than did *B. gracilis.* Similar results were reported for the same two grasses by Christie and Detling (1982), using a replacement series experiment. Similarly, many authors suggest that the success of the invasive grass *Bromus tectorum* in large areas of the western USA is at least partly due to its growth earlier in the season than native grasses and shrubs (Mack, 1986). Niche space can also be divided up either through different regeneration niches (Grubb, 1977) or through ratios of limiting nutrients (Tilman, 1982).

Because there is no theory predicting which niche axes species should segregate along, it is impossible to determine if no niche segregation occurs among species or whether the wrong niche axes were measured. Thus, there are difficulties with interpreting what niche differences mean when they are found and whether they are the cause or the result of species coexistence (Arthur, 1982; Silvertown, 1983). Leibold (1995) linked traditional theory of the niche with mechanistic models of community interactions (e.g. Tilman, 1982; Holt *et al.,* 1994), which may be a more promising approach to testing niche theory.

More recent alternative explanations of species coexistence have been developed which do not invoke niche differentiation. Silvertown and Law (1987) review some of these, including spatial aggregation in populations (Yodzis, 1986), non-transitive competition (Keddy, 1989; Shipley, 1993), a variety of non-equilibrium theories (Shmida and Ellner, 1984; Chesson and Huntly, 1997) and density-independent mortality (Huston, 1979). For example, Ågren and Fagerström (1984) demonstrated that two species with similar juvenile competitive abilities and lifespan can coexist if seed production (i.e. reproductive events) varies stochastically. Other models have demonstrated that spatial patchiness or environmental heterogeneity promotes species coexistence, at least for annual communities (Pacala and Tilman, 1994). In a recent review, Chesson (1991) suggests that we look for trade-offs that can determine coexistence and stability in communities – for example, dispersal vs. competitive ability in grassland plants (Tilman, 1997).

Conclusions

Field experiments suggest that competition controls plant performance at all levels of stress in natural grasslands (see Fig. 10.1). Further, competition is primarily among roots for below-ground resources, such as water and nutrients. The relative importance of shoot competition can increase as stress decreases, shoot mass increases and light penetration decreases. In addition, results from previous studies indicate that competition intensity changes more along natural stress gradients than along experimental gradients: we need studies of the differences between natural and experimental stress gradients in the same system. Particularly for natural gradients, we need to separate the effects of abiotic and biotic stress, in order to determine the relative role of environment and species interactions in controlling the species composition in grassland vegetation.

Using plant traits to predict individual- and community-level competitive ability is a powerful approach to testing current competition theory. Screening experiments using several species in environments differing in stress or productivity are needed; interactions among species, competition and environment should be examined to determine how competitive ability changes with abiotic and biotic stresses. Results to date indicate that many of the traits that characterize stress-tolerant species should also enable such species to be effective competitors in stressful habitats.

Field experiments have failed to show that an ability to perform well in the presence of neighbours (i.e. good competitive response ability) is linked to the abundance or distribution of species. In contrast, grasses that are best able to reduce resources (i.e. good effect competitors) tend to dominate nutrient-poor soils, suggesting that competitive effect ability may determine the relative abundance of species in grasslands.

In the process of resource consumption, grasses may also affect the patchiness of resources. The scale of this patchiness is in turn likely to influence other plants. Our understanding of the relative contributions of competition and stress to pasture community composition will be enhanced by closer study of the dynamics and heterogeneity of below-ground processes.

Understanding how competitive ability and environment interact is increasingly important because grassland environments are often altered by humans through fertilization, nutrient pollution, mowing, grazing and the introduction of new species. Moderate levels of grazing can interact with plant competition to maintain the diversity of grasslands and offset declines in plant diversity caused by frequent fires. Thus, grazing may be an important tool for the restoration of grassland diversity. However, more work is needed on comparing the effects of different grazers on the outcome of competition in the same system.

Acknowledgements

We thank the Natural Sciences and Engineering Research Council of Canada for support and P. Tow and A. Lazenby for helpful comments on an earlier draft of the manuscript.

References

Aarssen, L.W. (1988) 'Pecking order' of four plant species from pastures of different ages. *Oikos* 51, 3–12.

Aarssen, L.W. (1992) Causes and consequences of variation in competitive ability in plant communities. *Journal of Vegetation Science* 3, 165–174.

Aerts, R. and van der Peijl, M.J. (1993) A simple model to explain the dominance of low-productive perennials in nutrient-poor environments. *Oikos* 66, 144–147.

Ågren, G.I. and Fagerström, T. (1984) Limiting dissimilarity in plants: randomness prevents exclusion of species with similar competitive abilities. *Oikos* 43, 369–375.

Allen, E.B. and Allen, M.F. (1990) The mediation of competition by mycorrhizae in successional and patchy environments. In: Grace, J.B. and Tilman, D. (eds) *Perspectives on Plant Competition.* Academic Press, San Diego, USA, pp. 367–389.

Anderson, V.J. and Briske, D.D. (1995) Herbivore-induced species replacement in grasslands: is it driven by herbivory tolerance or avoidance? *Ecological Applications* 5, 1014–1024.

Archer, S. (1995) Tree-grass dynamics in a *Prosopis*-thornscrub savanna parkland: reconstructing the past and predicting the future. *Ecoscience* 2, 83–99.

Arthur, W. (1982) The evolutionary consequences of interspecific competition. In: MacFayden, A. and Ford, E.D. (eds) *Advances in Ecological Research*. Academic Press, London, UK, pp. 127–187.

Austin, M.P. (1990) Community theory and competition in vegetation. In: Grace, J.B. and Tilman, D. (eds) *Perspectives on Plant Competition*. Academic Press, New York, USA, pp. 215–238.

Bakker, J.D. (1996) Competition and the establishment of native grasses in crested wheatgrass fields. MSc thesis, University of Regina, Saskatchewan, Canada.

Bazzaz, F.A. (1996) *Plants in Changing Environments: Linking Physiological, Population, and community Ecology*. Cambridge University Press, New York, USA.

Belcher, J.W., Keddy, P.A. and Twolan-Strutt, L. (1995) Root and shoot competition intensity along a soil depth gradient. *Journal of Ecology* 83, 673–682.

Bell, G. and Lechowicz, M.J. (1994) Spatial heterogeneity at small scales and how plants respond to it. In: Caldwell, M.M. and Pearcy, R.W. (eds) *Exploitation of Environmental Heterogeneity by Plants*. Academic Press, San Diego, USA, pp. 391–414.

Belsky, A.J. (1986) Does herbivory benefit plants? A review of the evidence. *American Naturalist* 127, 870–892.

Berendse, F. (1994) Competition between plant populations at low and high nutrient supplies. *Oikos* 71, 253–260.

Berendse, F. and Aerts, R. (1987) Nitrogen-use efficiency: a biologically meaningful definition. *Functional Ecology* 1, 293–296.

Berendse, F. and Elberse, W. (1990) Competition and nutrient availability in the heathland and grassland ecosystems. In: Grace, J.B. and Tilman, D. (eds) *Perspectives in Plant Competition*. Academic Press, San Diego, USA, pp. 93–116.

Berendse, F., Elberse, W.Th. and Geerts, R.H.M.E. (1992) Competition and nitrogen loss from plants in grassland ecosystems. *Ecology* 73, 46–53.

Bever, J.D. (1994) Feedback between plants and their soil communities in an old field community. *Ecology* 75, 1965–1977.

Bever, J.D., Westover, K.M. and Antonovics, J. (1997) Incorporating the soil community into plant population dynamics: the utility of the feedback approach. *Journal of Ecology* 85, 561–573.

Bilbrough, C.J. and Caldwell, M.M. (1997) Exploitation of springtime ephemeral N pulses by six Great Basin plant species. *Ecology* 78, 231–243.

Bonser, S.P. and Reader, R.J. (1995) Plant competition and herbivory in relation to vegetation biomass. *Ecology* 76, 2176–2183.

Brown, V.K. and Gange, A.C. (1990) Insect herbivory belowground. *Advances in Ecological Research* 20, 1–58.

Brown, V.K. and Gange, A.C. (1992) Secondary plant succession: how is it modified by insect herbivory? *Vegetatio* 101, 3–13.

Bullock, J.M. (1996) Plant competition and population dynamics. In: Hodgson, J. and Illius, A.W. (eds) *The Ecology and Management of Grazing Systems*. CAB International, Wallingford, UK, pp. 69–100.

Caldwell, M.M. (1994) Exploiting nutrients in fertile soil microsites. In: Caldwell, M.M. and Pearcy, R.W. (eds) *Exploitation of Environmental Heterogeneity by Plants*. Academic Press, San Diego, USA, pp. 325–347.

Caldwell, M.M. and Pearcy, R.W. (1994) *Exploitation of Environmental Heterogeneity by Plants: Ecophysiological Processes Above and Below Ground*. Academic Press, New York, USA.

Caldwell, M.M. and Richards, J.H. (1986) Competing root systems: morphology and models of absorption. In: Givnish, T.J. (ed.) *On the Economy of Plant Form and Function*. Cambridge University Press, Cambridge, UK, pp. 251–273.

Caldwell, M.M., Richards, J.H., Johnson, D.A., Nowak, R.S. and Dzuree, R.S. (1981) Coping with herbivory: photosynthetic capacity and resource allocation in two semiarid *Agropyron* bunchgrasses. *Oecologia* 50, 14–24.

Campbell, B.D. and Grime, J.P. (1989) A comparative study of plant responsiveness to the duration of episodes of mineral nutrient enrichment. *New Phytologist* 112, 261–267.

Campbell, B.D., Grime, J.P. and Mackey, J.M.L. (1991) A tradeoff between scale and precision in resource foraging. *Oecologia* 87, 532–538.

Casper, B.B. and Cahill, J.F. (1998) Population level responses to nutrient heterogeneity and density by *Abutilon theophrasti* (Malvaceae): an experimental neighborhood approach. *American Journal of Botany* 85, 1680–1687.

Casper, B.B. and Jackson, R.B. (1997) Plant competition underground. *Annual Review of Ecology and Systematics* 28, 545–570.

Chapin, F.S., III (1980) The mineral nutrition of wild plants. *Annual Review of Ecology and Systematics* 11, 233–260.

Chapin, F.S., III (1991) Effects of multiple environmental stresses on nutrient availablity and use. In: Mooney, H.A., Winner, W.E. and Pell, E.J. (eds) *Response of Plants to Multiple Stresses*. Academic Press, San Diego, USA, pp. 67–88.

Chapin, F.S., III, Schultze, E.-D. and Mooney, H.A. (1990) The ecology and economics of storage in plants. *Annual Review of Ecology and Systematics* 21, 423–447.

Chapin, F.S., III, Autumn, K. and Pugnaire, F. (1993) Evolution of suites of traits in response to environmental stress. *American Naturalist* 142, 78–92.

Cheplick, G.P. (1997) Responses to severe competitive stress in a clonal plant: differences between genotypes. *Oikos* 79, 581–591.

Chesson, P. (1991) A need for niches? *Trends in Ecology and Evolution* 6, 26–28.

Chesson, P. and Huntly, N. (1997) The roles of harsh and fluctuating conditions in the dynamics of ecological communities. *American Naturalist* 150, 519–553.

Christie, E.K. and Detling, J.K. (1982) Analysis of interference between C_3 and C_4 grasses in relation to temperature and soil nitrogen supply. *Ecology* 63, 1277–1284.

Clay, K., Marks, S. and Cheplick, G.P. (1993) Effects of insect herbivory and fungal endophyte infection on competitive interactions among grasses. *Ecology* 74, 1767–1777.

Coffin, D.P., Laycock, W.A. and Lauenroth, W.K. (1998) Disturbance intensity and above- and belowground herbivory effects on long-term (14 y) recovery of a semiarid grassland. *Plant Ecology* 139, 221–233.

Collins, S.L., Glenn, S.M. and Gibson, D.J. (1995) Experimental analysis of intermediate disturbance and initial floristic composition: decoupling cause and effect. *Ecology* 76, 486–492.

Collins, S.L., Knapp, A.K., Briggs, J.M., Blair, J.M. and Steinauer, E.M. (1998) Modulation of diversity by grazing and mowing in native tallgrass prairie. *Science* 280, 745–747.

Connell, J.H. (1975) Some mechanisms producing structure in natural communities: a model and evidence from field experiments. In: Cody, M.L. and Diamond, J.M. (eds) *Ecology and Evolution of Communities.* Belknap Press, Cambridge, UK, pp. 460–490.

Cook, S.J. (1985) Effect of nutrient application and herbicides on root competition between green panic seedlings and a *Heteropogon* grassland sward. *Grass and Forage Science* 40, 171–175.

Cook, S.J. and Ratcliff, D. (1984) A study of the effects of root and shoot competition on the growth of green panic (*Panicum maximum* var. *trichoglume*) seedlings in an existing grassland using root exclusion tubes. *Journal of Applied Ecology* 21, 971–982.

Coughenour, M.B. (1985) Graminoid responses to grazing by large herbivores: adaptations, exaptation, and interacting processes. *Annals of the Missouri Botanical Garden* 72, 852–863.

Crawley, M.J. (1983) *Herbivory: the dynamics of animal-plant interations.* Studies in Ecology Vol. 10, University of California Press, Berkeley, California, USA.

Day, T.A. and Detling, J.K. (1990) Grassland patch dynamics and herbivore grazing preference following urine deposition. *Ecology* 71, 180–188.

Del Moral, R. (1983) Competition as a control mechanism in subalpine meadows. *American Journal of Botany* 70, 232–245.

Diamond, J.M. (1975) Assembly of species communities. In: Cody, M.L. and Diamond, J.M. (eds) *Ecology and Evolution of Communities.* Belknap Press, Cambridge, UK, pp. 342–444.

DiTommaso, A. and Aarssen, L.W. (1991) Effect of nutrient level on competition intensity in the field for three coexisting grass species. *Journal of Vegetation Science* 2, 513–522.

Ehleringer, J.R. and Field, C.B. (1993) *Scaling Physiological Processes: Leaf to Globe.* Academic Press, Orlando, USA.

Ehleringer, J.R. and Monson, R.K. (1993) Evolutionary and ecological aspects of photosynthetic pathway variation. *Annual Review of Ecology and Systematics* 24, 411–439.

Felsenstein, J. (1988) Phylogenies and quatitative methods. *Annual Review of Ecology and Systematics* 19, 445–471.

Fitter, A.H. (1986) Aquisition and utilization of resources. In: Crawley, M.J. (ed.) *Plant Ecology.* Blackwell, Oxford, UK, pp. 375–405.

Fitter, A.H. (1991) Costs and benefits of mycorrhizas: implications for functioning under natural conditions. *Experientia* 47, 350–355.

Foster, B.L. and Gross, K.L. (1997) Partitioning the effects of plant biomass and litter on *Andropogon gerardii* in old-field vegetation. *Ecology* 2091–2104,

Frank, D.A., McNaughton, S.J. and Tracy, B.F. (1998) The ecology of the earth's grazing ecosystems. *BioScience* 48, 513–521.

Gaudet, C.L. and Keddy, P.A. (1988) A comparative approach to predicting competitive ability from plant traits. *Nature* 334, 242–243.

Gehring, C.A. and Whitham, T.G. (1994) Interactions between aboveground herbivores and the mycorrhizal mutualists of plants. *Trends in Ecology and Evolution* 9, 251–255.

Gerry, A.K. and Wilson, S.D. (1995) The influence of initial size on the competitive responses of six plant species. *Ecology* 76, 272–279.

Givnish, T.J. (1986) Economics of biotic interactions. In: Givnish, T.J. (ed.) *On the Economy of Plant Form and Function.* Cambridge University Press, Cambridge, UK, pp. 667–680.

Goldberg, D.E. (1990) Components of resource competition in plant communities. In: Grace, J.B. and Tilman, D. (eds) *Perspectives on Plant Competition*. Academic Press, San Diego, USA, pp. 27–49.

Goldberg, D.E. (1996) Competitive ability: definitions, contingency and correlated traits. *Philosophical Proceedings of the Royal Society of London B* 351, 1377–1385.

Goldberg, D.E. and Barton, A.M. (1992) Patterns and consequences of interspecific competition in natural communities: a review of field experiments with plants. *American Naturalist* 139, 771–801.

Goldberg, D. and Novoplansky, A. (1997) On the relative importance of competition in unproductive environments. *Journal of Ecology* 85, 409–418.

Goldberg, D.E. and Scheiner, S.M. (1993) ANOVA and ANCOVA: field competition experiments. In: Scheiner, S.M. and Gurevitch, J. (eds) *Design and Analysis of Ecological Experiments*. Chapman and Hall, New York, USA, pp. 69–93.

Goldberg, D.E. and Werner, P.A. (1983) Equivalence of competitors in plant communities: a null hypothesis and a field experimental approach. *American Journal of Botany* 70, 1098–1104.

Goldberg, D.E., Turkington, R. and Olsvig-Whittaker, L. (1995) Quantifying the community-level consequences of competition. *Folia Geobotanica et Phytotaxonomica* 30, 231–242.

Grace, J.B. (1990) On the relationship between plant traits and competitive ability. In: Grace, J.B. and Tilman, D. (eds) *Perspectives on Plant Competition*. Academic Press, San Diego, USA, pp. 51–65.

Grace, J.B. (1991) A clarification of the debate between Grime and Tilman. *Functional Ecology* 5, 583–587.

Grace, J.B. (1995) On the measurement of plant competition intensity. *Ecology* 76, 305–307.

Greenlee, J.T. and Callaway, R.M. (1996) Abiotic stress and the importance of interference and facilitation in montane bunchgrass communities in western Montana. *American Naturalist* 148, 386–396.

Grime, J.P. (1977) Evidence for the existence of three primary strategies in plants and its relevance to ecological and evolutionary theory. *American Naturalist* 111, 1169–1194.

Grime, J.P. (1979) *Plant Strategies and Vegetation Processes*. John Wiley & Sons, Chichester, UK.

Grime, J.P. (1994) The role of plasticity in exploiting environmental heterogeneity. In: Caldwell, M.M. and Pearcy, R.W. (eds) *Exploitation of Environmental Heterogeneity by Plants*. Academic Press, San Diego, USA, pp. 1–19.

Grime, J.P., Thompson, K., Hunt, R., Hodgson, J.G., Cornelissen, J.H.C., Rorison, I.H., Hendry, G.A.F., Ashenden, T.W., Askew, A.P., Band, S.R., Booth, R.E., Bossard, C.C., Campbell, B.D., Cooper, J.E.L., Davison, A.W., Gupta, P.L., Hall, W., Hand, D.W., Hannah, M.A., Hillier, S.H., Hodkinson, D.J., Jalili, A., Liu, Z., Mackey, J.M.L., Matthews, N., Mowforth, M.A., Neal, A.M., Reader, R.J., Reiling, K., Ross-Fraser, W., Spencer, R.E., Sutton, F., Tasker, D.E., Thorpe, P.C. and Whitehouse, J. (1997) Integrated screening validates primary axes of specialisation in plants. *Oikos* 79, 259–281.

Grubb, P.J. (1977) The maintenance of species-richness in plant communities: the importance of the regeneration niche. *Biological Reviews* 52, 107–145.

Grubb, P.J. (1992) A positive distrust in simplicity – lessons from plant defences and from competition among plants and among animals. *Journal of Ecology* 80, 585–610.

Gurevitch, J. (1986) Competition and the local distribution of the grass *Stipa neomexicana*. *Ecology* 67, 46–57.

Gurevitch, J., Morrow, L.A., Wallace, A. and Walsh, J.S. (1992) A meta-analysis of competition in field experiments. *American Naturalist* 140, 539–572.

Harper, J.L. (1965) The nature and consequences of the interference amongst plants. *Proceedings of the International Conference on Genetics* 11, 465–481.

Harper, J.L. (1977) *Population Biology of Plants*. Academic Press, London, UK.

Harvey, P.H. and Pagel, M.D. (1991) *The comparative Method in Evolutionary Biology*. Oxford University Press, Oxford, UK.

Herben, T. and Krahulec, F. (1990) Competitive hierarchies, reversals of rank order and the de Wit approach: are they compatible? *Oikos* 58, 254–256.

Holland, E.A., Parton, W.J., Detling, J.K. and Coppock, D.L. (1992) Physiological responses of plant populations to herbivory and their consequences for ecosystem nutrient flow. *American Naturalist* 140, 685–706.

Holt, R.D., Grover, H. and Tilman, D. (1994) Simple rules for interspecific dominance in systems with exploitative and apparent competition. *American Naturalist* 144, 741–771.

Hook, P.B., Burke, I.C. and Lauenroth, W.K. (1991) Heterogeneity of soil and plant N and C associated with individual plants and openings in North American shortgrass steppe. *Plant and Soil* 138, 247–256.

Humphrey, L.D. and Pyke, D.A. (1997) Clonal foraging in perennial wheatgrasses: a strategy for exploiting patchy soil nutrients. *Journal of Ecology* 85, 601–610.

Huntley, N.J. (1991) Herbivores and the dynamics of communities and ecosystems. *Annual Review of Ecology and Systematics* 22, 477–503.

Huston, M. (1979) A general hypothesis of species diversity. *American Naturalist* 113, 81–101.

Huston, M. and Smith, T. (1987) Plant succession: life history and competition. *American Naturalist* 130, 168–198.

Huston, M.A. and DeAngelis, D.L. (1994) Competition and coexistence: the effects of resource transport and supply rates. *American Naturalist* 144, 954–977.

Hutchings, M.J. (1988) Differential foraging for resources and structural plasticity in plants. *Trends in Ecology and Evolution* 3, 200–204.

Jackson, L.E., Strauss, R.B., Firestone, M.K. and Bartolome, J.W. (1990) Influence of tree canopies on grassland productivity and nitrogen dynamics in deciduous oak savanna. *Agricultural Ecosystems and Environments* 32, 89–105.

Jones, C. and Lawton, J.H. (1995) *Linking Species and Ecosystems.* Chapman and Hall, New York, USA.

Kadmon, R. (1995) Plant competition along soil moisture gradients: a field experiment with the desert annual *Stipa capensis. Journal of Ecology* 83, 253–262.

Keddy, P.A. (1989) *Competition.* Chapman and Hall, London, UK.

Keddy, P.A. (1992) Assembly and response rules: two goals for predictive community ecology. *Journal of Vegetation Science* 3, 157–164.

Keddy, P.A. and Shipley, B. (1989) Competitive hierarchies in herbaceous plant communities. *Oikos* 54, 234–241.

Kemp, P.R. and Williams, G.J. (1980) A physiological basis for niche separation between *Agropyron smithii* (C_3) and *Bouteloua gracilis* (C_4). *Ecology* 61, 846–858.

Kleb, H.R. and Wilson, S.D. (1997) Vegetation effects on soil resource heterogeneity in prairie and forest. *American Naturalist* 150, 283–298.

Knapp, A.K. and Seastedt, T.R. (1986) Detritus accumulation limits productivity of tallgrass prairie. *BioScience* 36, 662–668.

Lauenroth, W.K. and Aguilera, M.O. (1998) Plant-plant interactions in grasses and grasslands. In: Cheplick, G.P. (ed.) *Population Biology of Grasses.* Cambridge University Press, Cambridge, UK, pp. 209–230.

Lauenroth, W.K., Dodd, J.L. and Sims, P.L. (1978) The effects of water- and nitrogen-induced stresses on plant community structure in a semiarid grassland. *Oecologia* 36, 211–222.

Leibold, M.A. (1995) The niche concept revisited: mechanistic models and community coexistence. *Ecology* 76, 1371–1382.

Li, X. and Wilson, S.D. (1998) Facilitation among woody plants establishing in an old field. *Ecology* 79, 2694–2705.

Louda, S. (1989) Differential predation pressure: a general mechanism for structuring plant communities along complex gradients? *Trends in Ecology and Evolution* 4, 158–159.

Louda, S.M., Keeler, K.H. and Holt, R.D. (1990) Herbivore influences on plant performance and competitive interactions. In: Grace, J. and Tilman, D. (eds) *Perspectives in Plant Competition.* Academic Press, San Diego, USA, pp. 414–444.

Mack, R.N. (1986) Alien plant invasion into the intermountain west: a case history. In: Mooney, H.A. and Drake, J. (eds) *Ecology of Biological Invasions of North America and Hawaii.* Springer-Verlag, New York, USA, pp. 191–213.

McLellan, A.J., Law, R. and Fitter, A.H. (1997) Response of calcareous grassland plant species to diffuse competition: results from a removal experiment. *Journal of Ecology* 85, 479–490.

McNaughton, S.J. (1983) Compensatory plant growth as a response to herbivory. *Oikos* 40, 329–336.

McNaughton, S.J. (1984) Grazing lawns: animals in herds, plant form, and coevolution. *American Naturalist* 124, 863–886.

McNaughton, S.J. (1985) Ecology of a grazing ecosystem: the Serengeti. *Ecological Monographs* 55, 259–294.

McNaughton, S.J., Oesterheld, M., Frank, D.A. and Williams, K.J. (1989) Ecosystem-level patterns of primary productivity and herbivory in terrestrial habitats. *Nature* 341, 142–144.

Mahdi, A., Law, R. and Willis, A.J. (1988) Large niche overlaps among coexisting plant species in a limestone grassland community. *Journal of Ecology* 76, 386–400.

Mehrhoff, L.A. and Turkington, R. (1990) Microevolution and site-specific outcomes of competition among pasture plants. *Journal of Ecology* 78, 745–756.

Menge, B.A. and Sutherland, J.P. (1987) Community regulation: variation in disturbance, competition, and predation in relation to environmental stress and recruitment. *American Naturalist* 130, 730–757.

Milchunas, D.G. and Lauenroth, W.K. (1993) Quantitative effects of grazing on vegetation and soils over a global range of environments. *Ecological Monographs* 63, 327–366.

Milchunas, D.G., Lauenroth, W.K. and Chapman, P.L. (1992) Plant competition, abiotic, and long- and short-term effects of large herbivores on demography of opportunistic species in a semiarid grassland. *Oecologia* 92, 520–531.

Miller, R.E., Ver Hoeff, J.M. and Fowler, N.L. (1995) Spatial heterogeneity in eight central Texas grasslands. *Journal of Ecology* 83, 919–928.

Miller, T.E. (1994) Direct and indirect species interactions in an early old-field plant community. *American Naturalist* 143, 1007–1025.

Miller, T.E. (1996) On quantifying the intensity of competition across gradients. *Ecology* 77, 978–981.

Mitchley, J. and Grubb, P.J. (1986) The control of relative abundance of perennials in chalk grassland in southern England. I. Constancy of rank order and results of pot and field experiments on the role of interference. *Journal of Ecology* 74, 1139–1166.

Moretto, A.S. and Distel, R.A. (1997) Competitive interactions between palatable and unpalatable grasses native to a temperate semi-arid grassland of Argentina. *Plant Ecology* 130, 155–161.

Mouat, M.C.H. (1983) Competitive adaptation by plants to nutrient shortage through modification of root growth and surface change. *New Zealand Journal of Agricultural Science* 26, 327–332.

Newman, E.I. (1973) Competition and diversity in herbaceous vegetation. *Nature* 244, 310.

Oksanen, L. (1993) Plant strategies and environmental stress: a dialectic approach. In: Fowden, L., Mansfield, T. and Stoddart, J. (eds) *Plant Adaptation to Environmental Stress*. Chapman and Hall, London, UK, pp. 313–333.

Olff, H. and Ritchie, M.E. (1998) Effects of herbivores on grassland plant diversity. *Trends in Ecology and Evolution* 13, 261–265.

Pacala, S.W. and Tilman, D. (1994) Limiting similarity in mechanistic and spatial models of plant competition in heterogeneous environments. *American Naturalist* 143, 222–257.

Parsons, A.J., Harvey, A. and Johnson, I.R. (1991) Plant–animal interactions in a continuously grazed mixture. 2. The role of differences in the physiology of plant growth and of selective grazing on the performance and stability of species in a mixture. *Journal of Applied Ecology* 28, 635–658.

Peltzer, D.A., Wilson, S.D. and Gerry, A.K. (1998) Competition intensity along a productivity gradient in a low-diversity grassland. *American Naturalist* 151, 465–476.

Platenkamp, G.A.V. and Foin, T.C. (1990) Ecological and evolutionary importance of neighbors in the grass *Anthoxanthum odoratum*. *Oecologia* 83, 201–208.

Read, D.J. (1991) Mycorrhizas in ecosystems. *Experientia* 47, 376–391.

Reader, R.J. (1990) Competition constrained by low nutrient supply: an example involving *Hieracium floribundum* Wimm and Grab. (Compositae). *Functional Ecology* 4, 573–577.

Reader, R.J. (1992) Herbivory as a confounding factor in an experiment measuring competition among plants. *Ecology* 73, 373–376.

Reader, R.J. (1993) Control of seedling emergence by ground cover and seed predation in relation to seed size for some old-field species. *Journal of Ecology* 81, 169–176.

Reader, R.J. (1998) Relationship between species relative abundance and plant traits for an infertile habitat. *Plant Ecology* 134, 43–51.

Reader, R.J. and Best, B.J. (1989) Variation in competition along an environmental gradient: *Hieracium floribundum* in an abandoned pasture. *Journal of Ecology* 77, 673–684.

Reader, R.J. and Bonser, S.P. (1993) Control of plant frequency on an environmental gradient – effects of abiotic variables, neighbours, and predators on *Poa pratensis* and *Poa compressa* (Gramineae). *Canadian Journal of Botany* 71, 592–597.

Reader, R.J., Wilson, S.D., Belcher, J.W., Wisheu, I., Keddy, P.A., Tilman, D., Morris, E.C., Grace, J.B., McGraw, J.B., Olff, H., Turkington, R., Klein, E., Leung, Y., Shipley, B., van Hulst, R., Johansson, M.E., Nilsson, C., Gurevitch, J., Grigulis, K. and Beisner, B.E. (1994) Intensity of plant competition in relation to neighbor biomass: an intercontinental study with *Poa pratensis*. *Ecology* 75, 1753–1760.

Reynolds, H.L., Hungate, B.A., Chapin, F.S., III and D'Antonio, C.M. (1997) Soil heterogeneity and plant competition in an annual grassland. *Ecology* 78, 2076–2090.

Rösch, H., Van Rooyen, M.W. and Theron, G.K. (1997) Predicting competitive interactions between pioneer plant species by using plant traits. *Journal of Vegetation Science* 8, 489–499.

Russell, P.J., Flowers, T.J. and Hutchings, M.J. (1985) Comparison of niche breadths and overlaps of halophytes on salt marshes of differing diversity. *Vegetatio* 61, 171–178.

Sala, O.E. and Lauenroth, W.K. (1985) Root profiles and the ecological effect of light rainshowers in arid and semiarid regions. *American Midland Naturalist* 114, 406–408.

Sala, O.E., Lauenroth, W.K. and Reid, C.P.P. (1982) Water relations: a new dimension for niche separation between *Bouteloua gracilis* and *Agropyron smithii* in North American semiarid grasslands. *Journal of Applied Ecology* 19, 647–657.

Sala, O.E., Parton, W.J., Joyce, L.A. and Lauenroth, W.K. (1988) Primary production of the central grassland region of the United States. *Ecology* 69, 40–45.

Sala, O.E., Lauenroth, W.K. and Parton, W.J. (1992) Long-term soil water dynamics in the shortgrass steppe. *Ecology* 73, 1175–1181.

Seager, N.G., Kemp, P.D. and Chu, A.C.P. (1992) Effect of root and shoot competition from established hill country pasture on perennial ryegrass. *New Zealand Journal of Agricultural Research* 35, 359–363.

Shipley, B. (1993) A null model for competitive hierarchies in competition matrices. *Ecology* 74, 1693–1699.

Shipley, B. and Keddy, P.A. (1993) Evaluating the evidence for competitive hierarchies in plant communities. *Oikos* 69, 340–345.

Shmida, A. and Ellner, S. (1984) Coexistence of plant species with similar niches. *Vegetatio* 58, 29–55.

Shorrocks, B. and Swingland, I.R. (1990) *Living in a Patchy Environment*. Oxford University Press, Oxford, UK.

Silvertown, J. (1983) The distribution of plant species in limestone pavement: tests of species interaction and niche separation against null hypotheses. *Journal of Ecology* 71, 819–828.

Silvertown, J. and Dale, P. (1991) Competitive hierarchies and the structure of herbaceous plant communities. *Oikos* 61, 441–444.

Silvertown, J. and Dodd, M. (1996) Comparing plants and connecting traits. *Philisophical Transactions of the Royal Society of London B* 351, 1233–1239.

Silvertown, J. and Law, R. (1987) Do plants need niches? Some recent developments in plant community ecology. *Trends in Ecology and Evolution* 2, 24–26.

Silvertown, J., Dodd, M.E., McConway, K., Potts, J. and Crawley, M. (1994) Rainfall, biomass variation, and community composition in the Park Grass Experiment. *Ecology* 75, 2430–2437.

Smith, T.M. and Huston, M.A. (1989) A theory of the spatial and temporal dynamics of plant communities. *Vegetatio* 83, 49–69.

Snaydon, R.W. and Howe, D.C. (1986) Root and shoot competition between established ryegrass and invading grass seedlings. *Journal of Applied Ecology* 23, 667–674.

Stanton, N.L. (1988) The underground in grasslands. *Annual Review of Ecology and Systematics* 19, 573–589.

Stark, J.M. (1994) Causes of soil nutrient heterogeneity at different scales. In: Caldwell, M.M. and Pearcy, R.W. (eds) *Exploitation of Environmental Heterogeneity by Plants*. Academic Press, San Diego, USA, pp. 255–284.

Stohlgren, T.J., Binkley, D., Chong, G.W., Kalkhan, M.A., Schell, L.D., Bull, K.A., Otsuki, Y., Newman, G., Bashkin, M. and Son, Y. (1999a) Exotic plant species invade hot spots of native plant diversity. *Ecological Monographs* 69, 25–46.

Stohlgren, T.J., Schell, L.D. and Vanden Heuvel, B. (1999b) How grazing and soil quality affect native and exotic plant diversity in Rocky Mountain grasslands. *Ecological Applications* 9, 45–64.

Tansley, A.G. and Adamson, R.S. (1925) Studies of the vegetation of the English chalk. III. The chalk grasslands of the Hampshire–Sussex border. *Journal of Ecology* 13, 177–223.

Taylor, D.R., Aarssen, L.W. and Loehle, C. (1990) On the relationship between r/K selection and environmental carrying capacity: a new habitat templet for plant life history strategies. *Oikos* 58, 239–250.

Tilman, D. (1982) *Resource Competition and Community Structure*. Princeton University Press, Princeton, New Jersey, USA.

Tilman, D. (1987) Secondary succession and the pattern of plant dominance along experimental nitrogen gradients. *Ecological Monographs* 57, 189–214.

Tilman, D. (1988) *Plant Strategies and the Dynamics and Structure of Plant Communities*. Princeton University Press, Princeton, New Jersey, USA.

Tilman, D. (1990) Constraints and tradeoffs: toward a predictive theory of competition and succession. *Oikos* 58, 3–15.

Tilman, D. (1997) Community invasibility, recruitment limitation, and grassland biodiversity. *Ecology* 78, 81–92.

Turkington, R. (1991) Rapid change in a patchy environment – the 'world' from a plant's-eye view. In: Dudley, E.C. (ed.) *The Unity of Biology*. Dioscorides, Portland, Oregon, pp. 194–200.

Underwood, A.J. and Petraitis, P.S. (1993) Structure of intertidal assemblages in different locations: how can local processes be compared. In: Ricklefs, R.E. and Schluter, D. (eds) *Species Diversity in Ecological Communities*. University of Chicago Press, Chicago, USA, pp. 39–51.

van der Werf, A., Vannuenen, M., Visser, A.J. and Lambers, H. (1993) Contribution of physiological and morphological plant traits to a species' competitive ability at high and low nitrogen supply: a hypothesis for inherently fast-growing and slow-growing monocotyledonous species. *Oecologia* 94, 434–440.

Vicari, M. and Bazely, D.R. (1993) Do grasses fight back? The case for antiherbivore defences. *Trends in Ecology and Evolution* 8, 137–141.

Vinton, M.A. and Burke, I.C. (1997) Contingent effects of plant species on soils along a regional moisture gradient in the Great Plains. *Oecologia* 110, 393–402.

Weaver, J.E. (1950) Effects of different intensities of grazing on depth and quantity of roots of grasses. *Journal of Range Management* 3, 100–113.

Wedin, D.A. and Pastor, J. (1993) Nitrogen mineralization dynamics in grass monocultures. *Oecologia* 96, 186–192.

Wedin, D.A. and Tilman, D. (1993) Competition among grasses along a nitrogen gradient: initial conditions and mechanisms of competition. *Ecological Monographs* 63, 199–229.

Welden, C.W. and Slauson, W.L. (1986) The intensity of competition versus its importance: an overlooked distinction and some implications. *Quarterly Review of Biology* 61, 23–44.

Werner, P.A. and Platt, W.J. (1976) Ecological relationships of co-occurring goldenrods (*Solidago*: Compositae). *American Naturalist* 110, 959–971.

Westoby, M., Leishman, M. and Lord, J. (1995) Issues of interpretation after relating comparative datasets to phylogeny. *Journal of Ecology* 83, 892–893.

Wilson, J.B. and Agnew, A.D.Q. (1992) Positive feedback switches in plant communities. *Advances in Ecological Research* 23, 263–336.

Wilson, S.D. (1991) Plasticity, morphology and distribution in twelve lakeshore plants. *Oikos* 62, 292–298.

Wilson, S.D. (1993a) Competition and resource availability in heath and grassland in the Snowy Mountains of Australia. *Journal of Ecology* 81, 445–451.

Wilson, S.D. (1993b) Belowground competition in forest and prairie. *Oikos* 68, 146–150.

Wilson, S.D. (1994) Initial size and the competitive responses of two grasses at two levels of soil nitrogen: a field experiment. *Canadian Journal of Botany* 1349–1354.

Wilson, S.D. (1998) Competition between grasses and woody plants. In: Cheplick, G.P. (ed.) *Population Biology of Grasses*. Cambridge University Press, Cambridge, UK, pp. 231–254.

Wilson, S.D. (1999) Plant interactions during secondary succession. In: Walker, L.R. (ed.) *Ecosystems of Disturbed Ground*. Elsevier, Amsterdam, The Netherlands, pp. 629–650.

Wilson, S.D. and Keddy, P.A. (1986) Measuring diffuse competition along an environmental gradient: results from a shoreline plant community. *American Naturalist* 127, 862–869.

Wilson, S.D. and Shay, J.M. (1990) Competition, fire and nutrients in a mixed-grass prairie. *Ecology* 71, 1959–1967.

Wilson, S.D. and Tilman, D. (1991) Components of plant competition along an experimental gradient of nitrogen availability. *Ecology* 72, 1050–1065.

Wilson, S.D. and Tilman, D. (1993) Plant competition in relation to disturbance, fertility and resource availability. *Ecology* 74, 599–611.

Wilson, S.D. and Tilman, D. (1995) Competitive responses of eight old-field plant species in four environments. *Ecology* 76, 1169–1180.

Yodzis, P. (1986) Competition, mortality and community structure. In: Diamond, J. and Case, T.J. (eds) *Community Ecology*. Harper and Row, New York, USA, pp. 480–491.

11 Interaction of Competition and Management in Regulating Composition and Sustainability of Native Pasture

D.L. Garden[1] and T.P. Bolger[2]

[1]NSW Agriculture, Canberra, Australia; [2]CSIRO Plant Industry, Canberra, Australia

Ecological factors associated with agricultural production have significantly modified the composition of grasslands and woodlands throughout the world (Mack, 1989; Hobbs and Hopkins, 1990; Taylor, 1990; Moore, 1993; McDougall, 1994; Prober and Thiele, 1995). In Australia, Moore (1993) has noted that these changes have been more marked in the southern temperate regions. In these areas, the combined effects of clearing (removal of trees), grazing, application of fertilizers and introduction (accidental or deliberate) of exotic species have produced a wide range of grassland types. The resultant composition of these grasslands has been influenced by the interaction of these factors and their effects on competitive relationships within the various grassland communities. This review seeks to describe these competitive relationships, using the temperate grasslands and woodlands of south-east Australia as an example.

Vegetation Changes in South-eastern Australian Temperate Grasslands

Past changes

Botanical change in Australian grasslands is closely linked to agricultural development since settlement in 1788. Because of isolation and lack of ungulate fauna, a unique vegetation originally developed in Australia, which was very sensitive to changes in grazing pressure, fertility and the introduction of exotic species (Moore, 1967; Lodge and Whalley, 1989; Mack, 1989; Hobbs and Hopkins, 1990). This vegetation was characterized in grassland areas by a dominance of taller (mainly warm-season) perennial grasses, together with a wide range of forbs (Norton, 1971; Doing, 1972; Benson and Wyse Jackson, 1994; McDougall, 1994; Prober, 1996; Eddy et al., 1998). There were few legumes present (Moore, 1970) and, consequently, soil nitrogen (N) levels were likely to have been low (Moore, 1967; Whalley et al., 1978). This vegetation was adapted to frequent but irregular burning and light grazing by marsupials (Norton, 1971; Fox, 1999; Groves, 1999).

After settlement, livestock and agricultural practices were imported from Europe (Hobbs and Hopkins, 1990; Fox, 1999). To the European eye, many grassland areas were seen as ideal areas to graze cattle and sheep (Norton, 1971; Benson and Wyse Jackson, 1994). Open grasslands and woodlands were favoured because these areas did not require removal of trees to provide adequate grass for grazing (Donald, 1970; Moore, 1970; Groves, 1999). However, as these areas were settled, the areas of grassland were expanded by the practice of felling or ring-barking trees, which allowed the grassy understorey to increase and provide sufficient herbage to make grazing by sheep and cattle

© CAB International 2001. Competition and Succession in Pastures
(eds P.G. Tow and A. Lazenby)

feasible. Even allowing for areas that were already open grasslands or woodlands, the area cleared of trees for agriculture is substantial (Wells *et al.*, 1984; Hobbs and Hopkins, 1990; Fox, 1999; Groves, 1999).

The effects of grazing by sheep and cattle (and by introduced rabbits) on the grasslands and woodlands of Australia were profound (Hobbs and Hopkins, 1990; Wilson, 1990; Groves, 1999). In temperate south-east Australia, Moore (1970, 1993) has recorded the change from predominantly tall, warm-season to predominantly short, cool-season perennial grasses by this means alone. However, these changes were accelerated by the accidental introduction of cool-season annual grasses, legumes and forbs (Donald, 1970; Moore, 1970, 1993; Groves, 1986; Whalley and Lodge, 1987; Lodge and Whalley, 1989) and of a changed pattern of burning (Norton, 1971; Gill, 1975; Fox, 1999). The annual grasses (particularly those from the genera *Vulpia*, *Bromus*, *Critesion* and *Avena*), with their high seed production and rapid growth (Moore, 1967; Wallace, 1998), were very successful invaders of both disturbed and undisturbed grasslands and woodlands (Donald, 1970; Cocks, 1994). Annual legumes (e.g. *Trifolium glomeratum*, *Trifolium campestre*, *Trifolium dubium*) and weeds (e.g. species of the genera *Taraxacum*, *Hypochoeris*, *Rumex*, *Arctotheca*, *Cirsium* and *Silybum*) were also successful invaders (Moore, 1993; McDougall, 1994). Many of these species have become naturalized and are common components of native[1] and sown pastures (Doing, 1972; McIntyre *et al.*, 1995).

Since early settlement, species of European origin have been deliberately introduced into agricultural areas (Donald, 1970; Groves, 1986, 1999; Fox, 1999). In highly fertile areas, these legumes and grasses flourished and became important components of the pastures. However, in less fertile areas, it was not until the use of superphosphate became widespread that highly productive pastures were developed (Donald, 1970). Two types of pastures became common. First, legumes (principally the annual *Trifolium subterraneum* and perennial *Trifolium repens*) were oversown into existing grassland with superphosphate, with the composition of the resultant pastures depending on the interaction between fertility, grazing pressure and competition between existing and invading species. Secondly, mixtures of introduced legumes and grasses (e.g. *Lolium rigidum*, *Lolium perenne*, *Phalaris aquatica*, *Dactylis glomerata*, *Festuca arundinacea*) were sown with superphosphate following disturbance by cultivation. In this case, most native perennial grasses were removed and a different set of competitive relationships ensued. In each of these pasture types, the original native perennial grasses were regarded as having limited value, because of poor quality for grazing (Donald, 1970; Donnelly, 1972) or poor adaptation to the changed conditions (Cocks, 1994). However, recent evidence suggests that, while this is generally so for most warm-season grasses, certain year-long green native perennial grasses show remarkable resilience in the face of introduced European agriculture and have been able to become dominants in both types of pastures (Magcale-Macandog and Whalley, 1991; Garden *et al.*, 1996, 2000a).

Patterns of change

Botanical change in the temperate grasslands of south-east Australia has been described by Moore (1970, 1993), Whalley *et al.* (1978); Whalley and Lodge (1987) and Lodge and Whalley (1989). Over time, native perennial grass composition has changed from a dominance by taller caespitose (mainly warm-season) genera to one with considerably higher amounts of cool-season or year-long green genera. However, the extent of change in botanical composition is more complex than this, and depends on several factors, including original composition, climate, soil type, fertilizer and degree of disturbance (cultivation, sowing of exotic species, grazing pressure). Consequently, the present composition of native grass-based pastures in temperate south-east Australia varies widely.

The direction and speed of change in botanical composition of temperate grasslands depend on whether or not cultivation and sowing of exotic species have occurred. Many descriptions of change (e.g. Moore, 1970; Whalley and Lodge, 1987) consider only the situation where the original grasslands were altered by grazing, fertilizer and invasion of cool-season annuals. Figures 11.1 and 11.2 show the separate (but interrelated) pathways of change that occur for pastures on the southern tablelands of New South Wales (NSW) which have been either originally unsown (Fig. 11.1) or cultivated and sown (Fig. 11.2). Common dominants in the original pastures were likely to have been *Themeda triandra*, *Poa* spp. and *Austrostipa* spp[2] (Moore,

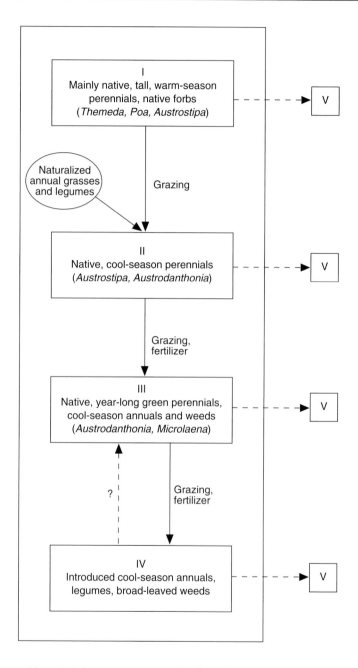

Fig. 11.1. Sequence of botanical changes in native grasslands on the southern tablelands of New South Wales without cultivation. Species listed are dominants only. Links to Stage V refer to Fig. 11.2.

1993; Benson and Wyse Jackson, 1994). However, there is likely to have been local variation, according to soil type and climate, with *Aristida ramosa*, *Bothriochloa macra* and *Joycea pallida* (syn. *Chionochloa pallida*) being dominants in some cases

(Donald, 1970; Doing, 1972; Whalley *et al.*, 1978; Moore, 1993). The first major change in composition (to Stage II (see Fig. 11.1)) was from the taller warm-season perennials to shorter cool-season perennials, such as *Austrodanthonia* spp.[3] and

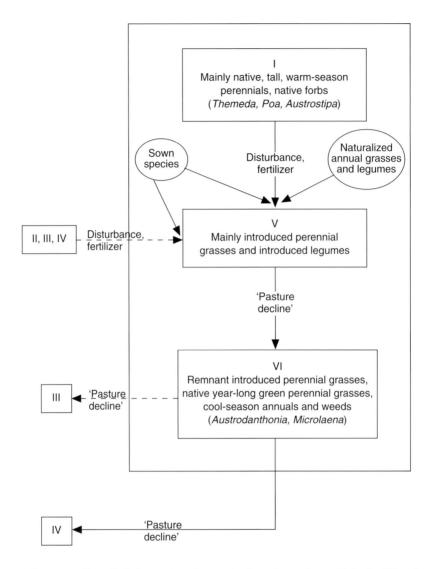

Fig. 11.2. Sequence of botanical changes in native grasslands on the southern tablelands of New South Wales following cultivation and sowing of exotic species. Species listed are dominants only. Links to Stages II, III and IV refer to Fig. 11.1.

Austrostipa spp. (Donald, 1970; Moore, 1970, 1993; Whalley and Lodge, 1987; Lodge and Whalley, 1989). At the same time, accidentally introduced annual grasses and legumes became part of the communities.

The next change in composition (to Stage III) was caused by further grazing and the application of fertilizer (Fig. 11.1). Legumes and annual grasses, increased as a result of the increased soil fertility. While *Austrostipa* spp. remained, the balance of native perennial grasses shifted more towards

Austrodanthonia spp. At the same time, *Microlaena stipoides*, which was probably a minor component of the original communities, appeared to increase (Lodge and Whalley, 1989), provided rainfall was adequate (Magcale-Macandog and Whalley, 1991). On the tablelands of NSW, there are many pastures with this composition today (Fig. 11.3; Garden *et al.*, 1993, 2000a). Further degeneration of pastures to the final stage shown by Moore (1970) is indicated by the path from III to IV, and results in dominance by annual grasses, legumes and broad-leaved weeds. Although Fig. 11.1 allows for the possibility of reversion from Stage IV to III, the mechanisms are unclear and it is unlikely to be easily achieved.

Moore (1970, 1993) suggests that during the changes described above, there was an ingress of shorter warm-season perennial species from drier regions to the west. There is no question that these species (e.g. *Chloris truncata*, *Panicum effusum*, *Enneapogon nigricans*) are part of communities today, although they may be more important in northern areas (e.g. Whalley and Lodge, 1987). However, whether these species did migrate during the last 200 years or were subdominants in the orig-inal vegetation is not clear. Also, the status of *M. stipoides* is unclear, as it is not mentioned in earlier descriptions of botanical change (e.g. Moore, 1970, 1993). There is little suggestion in the literature that *M. stipoides* was a dominant in the original grass-lands, except in the cooler, moister areas to the east of the tablelands in NSW (Whalley *et al.*, 1978). However, it appears to have become more common following a severe drought between 1979 and 1982 in eastern Australia (Lodge and Whalley, 1989).

Where cultivation occurs and introduced species are sown with fertilizer (commonly termed 'pasture improvement'), grasslands from Stages I, II, III and IV (see Fig. 11.1) can be converted to Stage

(a) *Microlaena stipoides* dominant

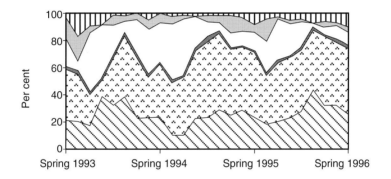

(b) *Austrodanthonia* spp. dominant

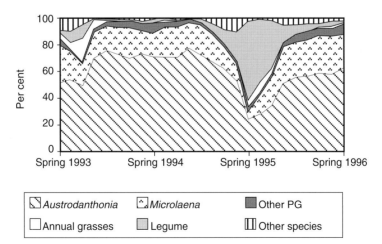

Fig. 11.3. Changes is species composition from 1993 to 1996 in native grass-based pastures on the tablelands of New South Wales. 'Other PG' includes other native and introduced perennial grasses. 'Other species' is mainly broad-leaved weeds. (From Garden *et al.*, 2000a.)

V (see Fig. 11.2). Depending on soil conditions (e.g. soil acidity) and the success of establishment, resultant pastures can range from highly successful sown pastures to those with only low proportions of exotic species present (Doing, 1972; Kemp and Dowling, 1991; Garden et al., 1993; Dowling et al., 1996). There is some evidence that pasture composition can move from Stage V to VI (see Fig. 11.2) under the influence of increased soil acidity, reduced fertilizer, drought and overgrazing (Hutchinson, 1992; Garden et al., 1993; Hutchinson and King, 1999). These effects, loosely described as 'pasture decline', have concerned researchers and others for some time (Cook et al., 1978; Archer et al., 1993; Wilson and Simpson, 1993; Cocks, 1994; Lodge, 1994; Martyn, 1995; Dowling et al., 1996; Jones, 1996; Bolger and Garden, 1998; Hutchinson and King, 1999), since sowing pastures of exotic species is an expensive exercise (Vere et al., 1997). However, Hutchinson (1992) has argued that if sown perennial grasses can be maintained in pastures, they can provide strong competitive exclusion of annuals, allowing the pasture to be maintained in Stage V.

Despite this last point, Stage VI pastures represent a situation that is common on the tablelands of NSW (Kemp et al., 1996), and it is often difficult to tell the difference from Stage III, unless paddock history is known. While there are some remaining introduced perennial grasses, the proportion is low (Doing, 1972; Dowling et al., 1996). The bulk of the perennial species may be made up of native year-long green grasses (e.g. *Austrodanthonia* spp., *M. stipoides*) (Magcale-Macandog and Whalley, 1991, 1993; Munnich et al., 1991; Garden et al., 2000a). Depending on season and fertilizer, intertussock spaces are filled with introduced cool-season annual grasses and legumes and broad-leaved weeds (Dowling et al., 1996; Kemp et al., 1996). Such pastures are stable (Bolger and Garden, 1998) and productive under a wide range of conditions, providing the perennial : annual ratio is high, as for the sown perennial situation described by Hutchinson (1992). Wilson and Simpson (1993) suggest that there is a need for a different view of sown pastures, because, in reality, few pastures are entirely composed of the species sown and levels of 25% sown grass and 25% legume might be considered satisfactory.

Progress to Stage IV (see Fig. 11.1) from VI (see Fig. 11.2) results when a pasture is destabilized, perennial grasses are reduced and bare areas are created, which allow recruitment sites for annual grasses (Cocks, 1994; Prober and Thiele, 1995; Dowling et al., 1996; Jones, 1996; Wallace, 1998). This situation can be created by overgrazing caused by drought and high stocking rates (Cook et al., 1978; Kemp and Dowling, 1991; Hutchinson, 1992; Hutchinson and King, 1999). Stage IV is an inherently unstable system (Wilson and Simpson, 1993; Kemp et al., 1996; Bolger and Garden, 1998), as, during normal dry seasons and extended droughts, there is limited ground cover, making changes in botanical composition unpredictable and reliant on timing of rainfall events.

Factors in Botanical Change

Vulnerability to invasion

Mack (1989) has noted that the temperate grasslands of Australia, South America and western North America show a common pattern of vulnerability to invasion by exotic plants, in contrast to grasslands in Eurasia, Southern Africa and central North America. The characteristics of the former grasslands which make them vulnerable to invasion appear to be related to the morphology and phenology of the dominant grasses and the grazing pattern under which these grasses evolved (Mack, 1989). The vulnerable grasslands are dominated by caespitose (tussock) grasses, which develop by intervaginal tillering, in which emerging tillers remain erect inside the leaf sheath and are therefore more exposed to grazing by ungulates than tillers of stoloniferous or rhizomatous (non-caespitose) grasses. Since caespitose grasses also rely more on sexual reproduction than non-caespitose grasses, a reduction in flowering tillers by grazing and the sensitivity of seedlings to grazing mean that these grasses are more at risk (Noble and Slatyer, 1980; Mack, 1989).

Another factor that is common in the vulnerable grasslands is the lack of large mammalian grazers in their evolution. These animals affect perennial grasses by both selective grazing and trampling, situations where caespitose grasses are more affected than non-caespitose grasses. A further factor is the effect of trampling on intertussock plants and the crust of mosses and lichens, which occurs in the intertussock spaces. While this disturbance is damaging to the original plants and their seedlings, it provides an ideal environment for exotic plants that have evolved under these conditions. Thus, the combined effects of introduction of ungulates into vulnerable grasslands together with exotic plants have been the destruction of native caespitose grasses, the dispersal of exotic plants and the preparation of a suitable seed-bed for plants that are more suited to the changed conditions (Mack, 1989).

Climate

Climatic effects are generally long-term. However, subtle shifts in, say, rainfall pattern can have effects over a shorter period (e.g. increased summer rain may increase the proportion of summer-growing species). Climate frequently interacts with other factors affecting composition. For example, grazing pressure on pastures increases dramatically during drought, as stock attempt to survive on less and less feed. Perhaps one of the greatest effects on the composition of grasslands in temperate south-east Australia has been the extreme grazing pressure applied during droughts (e.g. Hutchinson, 1992). Although normal grazing may have been moderate on a particular area, the need to maintain animals on pastures that have little or no growth for long periods may result in extreme pressure being applied to plants. In addition, following drought, plants that are best able to adapt to the changed conditions are often annual species and weeds, which, with their higher seed production, are better

able to colonize bare areas (Pettit *et al.*, 1995; Hutchinson and King, 1999) that have been created during the drought. Figure 11.4 shows the number of seedlings of a range of species germinating from undisturbed soil cores taken from a grazed native grass-based pasture (*Austrodanthonia* spp. dominant) on the southern tablelands of NSW. Exotic species are normally minor components of this pasture (Garden *et al.*, 2000a), but, following a drought in 1994/95 and above-average rainfall in 1995/96, the numbers of annual legume and grass and broad-leaved weed seeds in the cores increased dramatically. The importance of episodic events (e.g. droughts, unseasonal rainfall) in controlling botanical change in grasslands should not be underestimated.

Grazing

The reason grazing pressure affects composition so dramatically is that certain plants are better adapted to grazing than others, mainly through their structure (location of buds and growing points) or the way they utilize stored reserves to recover from grazing (Whalley *et al.*, 1978). In general, the taller tussock-forming grasses do not tolerate grazing as well as shorter, stoloniferous species (Mack, 1989) or introduced species that have evolved under close grazing. A summary of the response to grazing of a number of perennial native grasses is given by Lodge and Whalley (1989). The caespitose C_4

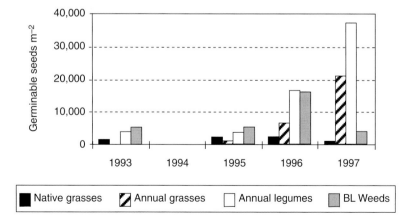

Fig. 11.4. Germination of seedlings of native grasses, annual grasses and legumes and broad-leaved weeds recorded from soil cores taken from a native grass-based pasture (*Austrodanthonia* spp. dominant) on the southern tablelands of New South Wales in autumn and maintained in a moist condition for 12 months. Soil cores were not taken in 1994. Note: these numbers indicate potential (not actual) recruitment.

grasses (e.g. *Dichanthium, Sorghum, Eulalia* and *Themeda*) decrease with grazing, compared with shorter species that have more protected growing points (e.g. *Bothriochloa, Chloris, Austrodanthonia, Panicum* and *Microlaena*) (Whalley *et al.*, 1978).

Jones (1996) has highlighted the possible evolutionary and reproductive strategies of taller caespitose grasses (e.g. *T. triandra*) and the shorter *M. stipoides* and *Austrodanthonia* spp. as mechanisms that may explain the outcome of competition under grazing. This is based on the observations of Pettit *et al.* (1995) that perennial species that are adversely affected by grazing and rely on seed for regeneration are at a disadvantage compared with 'facultative seeder/resprouters', which are able to resprout from protected buds or recruit from seed. Noble and Slatyer (1980) have also discussed these mechanisms and noted that species may become locally extinct if the adult population is lost and no propagules (seeds) remain to allow regeneration. Jones (1996) suggests that *M. stipoides* and some species of *Austrodanthonia* may be 'facultative seeder/resprouters', and this may explain their dominance in current pastures that have been heavily grazed. This is supported to some extent by the observation that *Austrodanthonia racemosa* appears to have extravaginal tillers (P. Linder, personal communication), allowing tillers to grow horizontally along the ground under heavy grazing, where they

are more protected. *Microlaena stipoides* is also reputed to have a rhizomatous or stoloniferous habit under grazing (Whalley *et al.*, 1978; McIntyre *et al.*, 1995). Certainly, both these species appear to form 'grazing lawns' (Jones, 1996), as described by McNaughton (1984).

Data from a survey on the southern tablelands of NSW (Garden *et al.*, 2000b) suggest that introduced perennial grasses and the native species *M. stipoides* and *Austrodanthonia* spp. generally tolerate grazing better than other native grasses (Fig. 11.5). The main species in this latter group were *T. triandra, Poa sieberiana, Austrostipa* spp., *B. macra* and *A. ramosa*. While these data indicate possible different responses to grazing by different species, there are very few experimental data to confirm this. Lodge and Whalley (1989) have summarized the responses to grazing in a number of experiments in south-east Australia and concluded that there were at least two reasons why significant changes were not apparent. First, many experiments were only relatively short-term, whereas changes in species composition may be of a more long-term nature (Wilson and Simpson, 1993). Allied with this is the possibility that changes may be dependent on episodic events, such as droughts or favourable rainfall. Secondly, all the reviewed experiments were carried out on pastures that had already been extensively modified (Moore, 1993) and therefore

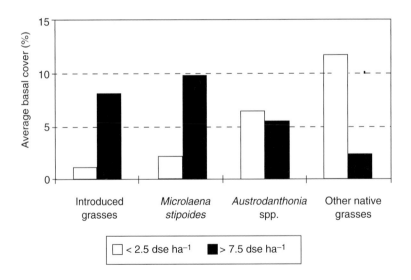

Fig. 11.5. Effect of paddock stocking rates on the basal cover of introduced species, *Microlaena stipoides, Austrodanthonia* spp. and other native grasses present in grazed pastures on the central and southern tablelands of New South Wales. dse, dry sheep equivalents.

treatments may have had to have been more severe than (or of a different nature from) those that had previously given rise to the existing plant communities. Some experimental results (e.g. Garden *et al.*, 2000a) indicate that pastures in Stages III and VI (see Figs 11.1 and 11.2), and dominated by year-long green C_3 native grasses (see Fig. 11.3) are stable under a wide range of grazing regimes.

Although the data in Fig. 11.5 suggest that introduced perennial grasses are tolerant of grazing, they do not normally recruit well from seed in grazed pastures under Australian conditions (Jones, 1996; Virgona and Bowcher, 1998; Hutchinson and King, 1999). This may partly explain the phenomenon of 'pasture decline', as any mortality of existing plants cannot be replaced by recruitment of new plants from seed (Noble and Slatyer, 1980). This is shown by the data of Hutchinson (1992), who found in a long-term experiment that basal cover of *P. aquatica* was reduced during major droughts, especially at high stocking rates. Without the ability to recruit from seed, the population of *P. aquatica* plants stabilized at successively lower levels following each drought. Spaces formerly occupied by phalaris were taken by annual grasses, which suggests that the pasture was moving from Stage V or VI towards IV (see Figs 11.1 and 11.2). Hutchinson and King (1999) use the term 'annualization' to describe these changes.

While allowing for the possibility of poor climatic adaptation of introduced perennial grasses, Hutchinson and King (1999) attribute the loss of introduced perennial grasses mainly to a combination of reduced N economy, as legumes decline with time, and strong preferential grazing of these grasses. Clearly, this makes them more vulnerable under high stocking rates, even if they are generally tolerant of grazing. Wilson and Simpson (1993) also concluded that continuous grazing by sheep, especially at high stocking rates, shifts the balance in favour of annual species and weeds, mainly by increased selection of the perennial grasses by animals.

Considerable effort has been expended in investigating the use of grazing management to reverse the decline in perennial grass content in both natural and sown pastures (Lodge, 1994; Fitzgerald and Lodge, 1997; Mason *et al.*, 1997; Lodge *et al.*, 1998). While some success has been achieved (e.g. Dowling *et al.*, 1996; Kemp *et al.*, 1996), there has been no evidence that pastures can be successfully moved from Stage VI to V, let alone from Stage IV to any other state. However, Lodge *et al.* (1998)

and Hutchinson and King (1999) remind us that, whatever grazing system is used, adequate fertilizer is required for persistence of sown perennial pastures.

Fertilizer

Indigenous Australian plants have evolved in soils which are low in N and P and, in general, many of these plants are unable to tolerate higher levels of fertility (Groves, 1999). However, introduced perennial and annual grasses can utilize increased nutrient levels successfully. Their greater growth allows them to crowd or shade competing plants, thus leading to changed composition. There are also differences in this regard between species of native perennial grasses. For example, data from the southern tablelands of NSW show that *M. stipoides* and *Austrodanthonia* spp. respond better to increased fertility (superphosphate plus legume N) than many of the other native grasses (Fig. 11.6). However, while Robinson *et al.* (1993) found an association between abundance of *Austrodanthonia* spp. and applied fertilizer in a similar environment, there was no such association for *M. stipoides*, and neither of these groups was affected by soil phosphorus level. On the northern tablelands of NSW, Whalley *et al.* (1978) found that frequency of *Austrodanthonia* spp. (but not other native species) increased with superphosphate application, although *Austrodanthonia* spp. could be adversely affected by competition from clover. At a finer level, Bolger and Garden (1999) found that responses to applied P and N by eight species of *Austrodanthonia* differed, in both absolute terms and critical values (nutrient required for 90% of maximum growth). *Austrodanthonia racemosa* had among the highest critical values for P and N, suggesting that this species has higher requirements for both these nutrients.

Robinson (1976) carried out experiments to determine the response of native grasses to N fertilizer and found that, under these conditions, *A. racemosa* achieved similar dry-matter production to *P. aquatica* and other introduced perennial grasses. When compared with C_4 native grasses (*T. triandra*, *B. macra*), a proper comparison was not possible due to the fact that plants of these species died at the higher N levels. However, using available data, there were substantial differences in favour of *A. racemosa*, which suggests that it is better able to respond to N than the C_4 grasses.

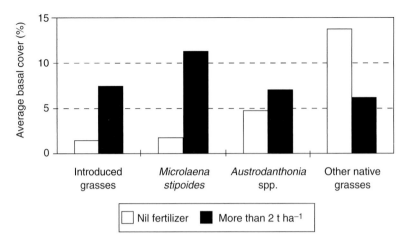

Fig. 11.6. Effect of total superphosphate applied on the basal cover of introduced species, *Microlaena stipoides*, *Austrodanthonia* spp. and other native grasses present in grazed pastures on the central and southern tablelands of New South Wales.

Groves *et al.* (1973) and Fisher (1974) investigated competition between *T. triandra* and *Poa labillardieri* when N and P fertilizers were added, and noted the competitive advantage of *Poa* over *Themeda* under higher fertility. Although *T. triandra* did respond to increased fertility, it appears to be disadvantaged when species are present that can better respond to increased nutrition. Therefore, *T. triandra* is likely to decrease in fertilized pastures, when competing species, such as introduced perennial grasses, cool-season annual grasses, some C_3 native perennial grasses (e.g. *Austrodanthonia* spp., *M. stipoides*, *Poa* spp.) and broad-leaved weeds, are present.

The main purpose of applying fertilizer to pastures is usually to encourage sown exotic species. If application is continued, the proportion of sown species increases and, hopefully, a stable pasture phase is reached. However, many sown pastures remain unstable and can be reinvaded by native species, annual grasses and weeds at any time (Lodge and Whalley, 1989), especially if fertility declines. This is shown in Figs 11.1 and 11.2 by the paths between Stages V and VI and between Stages VI and III and IV, and corresponds with the circumstances termed 'pasture decline'.

The data in Fig. 11.6 indicate that application of phosphorus fertilizer (together with a legume) is likely to increase the proportions of *M. stipoides*, *Austrodanthonia* spp. and any sown species in a pasture and to decrease other native species. Lodge and

Whalley (1989) also cite many instances where application of fertilizer and legumes has altered composition. However, there are no reports of changes caused by reduced fertilizer input. Despite this, there is no doubt that much so-called 'pasture decline' is a result of this practice. For example, Garden *et al.* (1993) reported many instances where landholders described paddocks as 'improved', but admitted that little or no fertilizer had been applied in the last 10 years. In most cases, these paddocks were a mixture of native perennial grasses (principally *Austrodanthonia* spp. and *M. stipoides*) and naturalized annuals, while sown perennial grasses had largely disappeared. While other soil factors may be important (e.g. soil acidity), declining fertility is likely to be as important as increasing fertility in contributing to botanical change.

Fire

Australian native vegetation, including grasses, has evolved under a higher fire frequency than at present and is generally tolerant of fire (Gill, 1975; Fox, 1999). Initially fires were caused naturally (e.g. lightning strikes), but the arrival of Aboriginal people increased the frequency of fires, as they used fire as a tool to manage vegetation for their own purposes. In fact, there is evidence that, by frequent burning, Aborigines helped to develop and maintain grasslands (Christensen and Burrows, 1986).

However, the much later arrival of European settlers and their different attitude to the use of fire meant that there was a change in the frequency and intensity of fires, and this change again influenced composition of vegetation (Fox, 1999). The assumption has been that fire frequency has been less after European settlement than before. Also, intensity of fires altered as areas were alternately protected (for safety), but then burnt by wildfires (Christensen and Burrows, 1986).

Effects of fire vary depending on species composition, amount of herbage present, season of the year, frequency and intensity of burning and how plants are grazed following burning (Gill, 1975; Lodge and Whalley, 1989). Literature on effects of fire on vegetation frequently mentions the concept of 'sprouters' and 'non-sprouters' (seeders) (Gill, 1975, 1981; Christensen and Burrows, 1986; Fox, 1999). Plants that are adapted to fire are often good sprouters from protected buds. In the case of perennial grasses, buds can be protected within large tussocks, allowing recovery from fire.

Wedin (1995) has likened fire to grazing and has used this analogy, together with C/N and lignin/N ratios, to show that fire favours the dominance of C_4 grasses, in this case, in North American tall-grass prairie. The core of his argument is that N controls the dynamics of grassland, and fire, by volatilizing N, maintains low N status, thereby favouring the C_4 dominants (see section on ecological processes underlying botanical change below).

On the other hand, fire can be a potent agent in allowing invasion of exotic species into grasslands and other native communities, depending on community structure, the nature of the invaders and the fire regime (Christensen and Burrows, 1986). Groves (1999) has highlighted the interaction between grazing and fire in affecting composition of grasslands. Essentially, if continuous grazing occurs soon after a fire, plant reserves may be depleted in some grasses, to the point where mortality occurs and botanical composition is changed. In southern Australia, this usually favours C_3 grasses compared with the original C_4 dominants (Groves, 1999).

Disturbance

The general concept of disturbance as a factor in botanical change is well accepted. Fox (1986), in a well-argued review, concluded that there can be no invasion of natural communities without distur-

bance. However, the degree of invasion of a community depends on its susceptibility (Fox, 1986; Mack, 1989) and the intensity of disturbance. One of the most dramatic ways of changing pasture composition is disturbance by cultivation (or herbicide) and the sowing of introduced species. Effects can be very long-lasting and, often, irreversible over the shorter term. Most native species do not tolerate disturbance of this type, although the effects may vary. For example, Munnich et al. (1991) found that cultivation reduced the abundance of *Austrodanthonia* spp. but not of *M. stipoides*, although this may just reflect the ability of *M. stipoides* to recover from disturbance, as measurements were made several years after cultivation. There may also be differences between different species of *Austrodanthonia*, as Scott and Whalley (1982) found that *A. bipartita* (syn. *Danthonia linkii* var. *linkii*) was little affected by cultivation, in contrast to other species of *Austrodanthonia*.

Survey data (Garden et al., 1993) reveal that, despite cultivation, many previously sown pastures can recover with time and be recolonized by native perennial grasses. However, there are two important caveats. First, many of these areas are on shallow, infertile soils and there is no guarantee that an adequate establishment of introduced species was obtained at sowing. Secondly, almost without exception, these areas are not recolonized by the original taller caespitose C_4 dominants (e.g. *T. triandra*), but by the same species that become dominants in Stage III (i.e. year-long green C_3 species). In our view, the most destructive effects of cultivation occur if it is repeated. While recovery can occur from one or more cultivations in a single season for pasture establishment, repeated cultivation over a number of years (e.g. as part of a cropping rotation) can cause rapid 'annualization' of pastures (Robinson et al., 1993).

In south-east Australia, herbicides are used as a substitute for cultivation for pasture establishment by 'direct drilling' (Simpson and Langford, 1996) or as a means of reducing seed set of annual grasses by the practice of 'spray-topping' (Wallace, 1998). While direct drilling is not a common practice on the tablelands of NSW (Garden et al., 2000b), it is likely to create the same effect as cultivation, depending on the herbicide used and whether single or repeated applications are made. *Austrodanthonia* spp. are susceptible to glyphosate at normal rates for pasture establishment, although *M. stipoides* is quite resistant to this herbicide and some others used against grasses (Lodge and McMillan, 1994; Campbell and Van de Ven, 1996; Simpson and Langford, 1996). However, both species are highly susceptible to flupropanate, the herbicide commonly used for control of serrated tussock (*Nassella trichotoma*) (Campbell and Vere, 1995; Campbell and Van de Ven, 1996). On the other hand, *T. triandra* and *B. macra* can tolerate relatively high levels of flupropanate and glyphosate, especially during winter, when vegetation is frosted and growth is low. The tolerance of native grasses to many of the herbicides recommended for annual grass control (Wallace, 1998) is unknown.

Invasion of annual species

Apart from deliberate introduction of legumes and grasses, the main source of exotic species in Australia has been the accidental introduction in ships' ballast, stock fodder and imported seed of a range of annual grasses, legumes and weeds (Groves, 1986; Wallace, 1998). Most are from the Mediterranean region and have become successfully naturalized in Australia. Annual legumes (mainly *Trifolium* spp.) have changed the N economy of grassland soils, allowing the better-adapted C_3 annual grasses and weeds (and native grasses) to increase and compete strongly with the existing C_4 grasses (Moore, 1967). While some genera of annual grasses (e.g. *Aira*, *Briza*) are adapted to low-fertility soils, others have a wider tolerance (e.g. *Vulpia*) and still others (e.g. *Bromus*, *Critesion* syn. *Hordeum*) have high nutrient requirements (Whalley *et al.*, 1978). Therefore, there are few areas in temperate Australia that do not have some of these species present. Also, many agricultural practices (e.g. cultivation, application of fertilizer, grazing) encourage their spread, particularly those with moderate to high fertility requirements.

Whalley *et al.* (1978) have noted the importance of grazing in the spread of naturalized species. In the original grasslands, there was probably a closed canopy because of the low grazing pressure. However, when livestock were introduced, the normal accumulation of plant material during the summer period was reduced, allowing the creation of bare areas in autumn for the germination of cool-season species with the commencement of autumn/winter rains (Whalley *et al.*, 1978). While grazing management is regarded as a control mea-

sure for annual grasses, timing is critical (Wallace, 1998). Also, the assumption that maintaining a high stocking rate at all times will control annual grasses is invalid, as Hutchinson and King (1999) found that this practice invariably results in 'annualization' of pastures.

Interactions

It is axiomatic that the preceding factors do not operate in isolation and, at any time, there are complex interactions between the various factors. For example, invasion of annual species is affected by the amount of disturbance that has occurred, the current and preceding climate, soil nutrient levels (as determined by edaphic factors, applied fertilizer and recycling of nutrients by animals) and the pattern and amount of grazing. For this reason, the outcome of competition between perennial and annual plants is often difficult to predict. However, in the types of grasslands described here, the net direction of change appears to be from the original perennial C_4 dominants towards year-long green C_3 grasses and annual species.

Ecological Processes Underlying Botanical Change

Very little research has been undertaken to reveal the ecological processes that underlie the patterns of change in Australian temperate grasslands as outlined in Fig. 11.1, and this situation is largely true for plant community ecology in general. However, recently, the roles of two complementary ecological processes, competition and recruitment limitation, have been recognized as important determinants of plant community structure (Tilman, 1997; Zobel, 1997; Hubbell *et al.*, 1999). Competition is recognized as having strong effects on plant community structure within a site.

Recruitment limitation (the inability of species to disperse, germinate or establish successfully) is considered to be a major factor determining species composition, abundance and diversity between sites. Several recent papers (Tilman, 1997; Zobel, 1997; Hubbell *et al.*, 1999) emphasize the interplay of within-site competitive interactions and site-to-site recruitment limitation as joint determinants of local and regional species composition and diversity. Recruitment limitation is hypothesized to reduce the effects of within-site competitive inter-actions by generating sites in which inferior competitors escape from competition with their superiors. Interspecific trade-offs between recruitment and competitive ability (Tilman, 1990, 1994) can explain the coexistence of many competing species.

Competition and recruitment limitation may interact with grazing and other management and environmental factors to determine patterns of change in Australian temperate grasslands (see Fig. 11.1). Nitrogen is often the most limiting resource in many grasslands (Wedin, 1995). The original dominants of Australian temperate grasslands were warm-season perennial grasses, which are generally good competitors for N and tend to maintain soil mineral N at low levels (Moore, 1967; Wedin and Tilman, 1993). Wedin and Tilman (1993) found that the competitive dominants in North American tall-grass prairie are the species that can reduce the concentration of the limiting resource to the lowest level. Reductions in soil mineral N levels under monoculture swards of component species were used to predict competitive outcomes and dominance. Field experiments of two-species mixtures showed that predictions of competitive outcome based on this resource reduction model of plant competition were correct in all cases. Native C_4 grasses reduce soil mineral N levels compared with C_3 species, in part by tying up N in their slowly decomposing litter (Wedin and Tilman, 1990). These species thus create the low-N conditions in which they are superior competitors. This positive feedback between plant competition and N cycling may be an important process in maintaining the stability of many grassland plant communities (Moore, 1993; Wedin, 1995), including the Stage I grasslands referred to in Figs 11.1 and 11.2. The similar situation with fire and N levels has been mentioned previously.

Ungulate grazing was the factor that precipitated the change to Stage II pastures dominated by native, cool-season perennial grasses (see Fig. 11.1). The native, warm-season perennials were poorly adapted to grazing, as discussed earlier. Grazing reduced the cover and production of these dominants, caused soil disturbance which created open sites for invasion of other species, and increased the rate of nutrient cycling via the animal pathway (e.g. Wedin, 1995). Under this new regime, the previously subordinate cool-season perennial grasses, such as *Austrodanthonia* spp.,

which were more tolerant of grazing and better recruiters, became dominant. Low levels of soil mineral N under Stage I pastures, dominated by *T. triandra*, prevented invasion by naturalized annual grass species, while Stage II pastures, dominated by *Austrodanthonia* spp., had higher levels of soil mineral N, allowing the naturalized annual species to invade (Moore, 1993). Because of trade-offs between recruitment and competitive ability (Tilman, 1990, 1994), the dominant species are actually more vulnerable to (local) extinction with changing conditions (e.g. disturbance, habitat fragmentation, extreme climatic events) than less abundant subordinate species (Tilman *et al.*, 1994). Depending on the circumstances, these extinctions can be catastrophic or may occur long after conditions change – so in this sense they are an 'extinction debt' that comes due in the future. This line of reasoning suggests that the change from Stage I to Stage II pastures is largely irreversible, due to recruitment limitation.

Further grazing and the application of fertilizer caused the change from Stage II to Stage III pastures, with year-long green native grasses increasing and naturalized annual grasses and legumes becoming significant components of the pasture. For competitive exclusion via the resource reduction mechanism to occur, plants must be able to deplete resources to low levels and maintain them at those levels (Wedin and Tilman, 1993). Thus, nutrient enrichment, via fertilizer application and legume N fixation, further restricts the ability of the native, year-long green perennials to outcompete the naturalized annual species. Along nutrient gradients, competition remains an important force, but the mechanism shifts from being mainly below-ground competition for nutrients in low-nutrient conditions, towards above-ground competition for light in high-nutrient conditions (Wilson and Tilman, 1991; Bolger, 1998). There appears to be a trade-off in the ability of species to compete for soil nutrients or light, due to associated differences in allocation patterns, with good N competitors having high allocation to roots, low tissue N and low allocation to seed (Tilman, 1990). These dynamics make it easy to understand why changes in species composition occur along nutrient gradients.

Greater fluctuations in soil mineral N under Stage III pastures, in combination with climatic variation, can result in large fluctuations in the proportions of perennials and annuals within and across years (see Fig. 11.3). These transient and reversible fluctuations in species composition do not generally cross thresholds and thus these Stage III pastures are considered to be stable under a wide range of conditions (Garden *et al.*, 2000a). However, continued grazing and inputs of fertilizer, perhaps in combination with an extreme climatic event, such as drought, can destabilize these Stage III pastures and shift them to a state dominated by less stable and less productive annual species (Stage IV). The ecological processes that may cause the loss of the year-long green perennial grasses from Stage III pastures are less clear. The direct effects of grazing on survival of the perennial grasses is one possibility. Another possibility is nutrient eutrophication causing the perennials to be outcompeted by the annuals for light. But there is also the possibility of a 'filter' effect (*sensu* Torssell and McKeon, 1976; Grime, 1998) of the annual species, with competition acting to prevent ('filter') recruitment of new perennial plants (e.g. Lodge, 2000). This recruitment limitation could result in slow or catastrophic loss of the perennial grasses, depending on conditions. It is unclear if this change from Stage III to Stage IV pastures is reversible. However, as the year-long green perennials are good recruiters, it seems plausible that removing the filter effect of the annuals by reducing fertility, or appropriate climatic conditions, or both, could result in the successful recruitment of year-long green perennials, allowing reversion to Stage III.

These ecological processes are hypothesized to operate in a similar manner in determining species composition in cultivated situations, as outlined in Fig. 11.2. The main difference is that the introduced perennial grasses are poor recruiters, so changes in pasture states (Stage V to Stage VI, and Stage VI to Stage III or Stage IV) due to loss of these species are considered essentially irreversible, due to recruitment limitation.

Although we have provided some support for the role of these ecological processes in determining patterns of change in Australian (native) temperate grasslands, much of this section remains hypothetical, awaiting experimental testing. Nevertheless, the recognized importance of the complementary roles of competition and recruitment limitation in structuring such diverse plant communities as wetlands (Weiher and Keddy, 1995), calcareous grasslands (Burke and Grime, 1996), tall-grass prairie (Tilman, 1997) and tropical forests (Hubbell *et al.*, 1999) suggests that these processes are generic and likely to be important in Australian temperate grasslands as well.

Vegetation Effects on Ecosystem Function and Sustainability

Ecosystem function can be defined as species–environment interactions in the transformation and flux of energy and matter (carbon, water, mineral nutrients) as determined by ecological processes, such as primary productivity, decomposition, evapotranspiration, competition and recruitment. The functioning of ecosystems is widely recognized to depend on the identities of the species that the ecosystems contain (Huston, 1997; Tilman *et al.*, 1997; Grime, 1998). Further, it has been argued that a relatively small set of plant and animal species and abiotic processes is critical for determining ecosystem function (Holling, 1992; Walker, 1992), and these species and processes necessarily vary in different ecosystems. When ecosystems become degraded by overexploitation to the point where formerly dominant species are eliminated or largely replaced, it is often possible to demonstrate a causal connection between changes in species composition (or losses in biodiversity) and declines in ecosystem function and its benefits, such as clean water and air, which are required by society (Chapin *et al.*, 1998; Grime, 1998). The effects of species identity and species richness on ecosystem function has been a current and controversial topic in the recent ecological literature (Huston, 1997; Tilman *et al.*, 1997; Chapin *et al.*, 1998; Grime, 1998). A reason for the controversy is that the underlying mechanisms for these effects are not clear. Do more species simply increase the probability of having a single productive species that controls ecosystem function? Or do more species allow the community to tap more resources because these species differ in the timing or rooting depth at which they acquire resources?

A 'mass ratio' hypothesis recognizes that, even in species-rich vegetation, most of the plant biomass may be accounted for by a small number of dominant species, and proposes that the immediate effects on ecosystem function are in proportion to contributions to plant biomass, are determined largely by the traits and functional diversity of the dominant plant species and are relatively insensitive to the richness of subordinate species (Grime, 1998). However, species whose effects on ecosystem processes are similar often differ in their responses to environmental variations (e.g. a C_3 and a C_4 perennial grass), or they would be unlikely to coexist in the community. Consequently, a higher species diversity would increase the probability that the processes will be sustained, even if a particular species is lost in response to some extreme event or to a directional change in the environment (Tilman *et al.*, 1997; Chapin *et al.*, 1998). In this case, the co-occurrence of species in a community with similar ecological effects does not necessarily imply that these species are 'redundant' (*sensu* Walker, 1992). Further, in habitats that are spatially heterogeneous, no species would be competitively superior throughout the entire habitat. Different species dominating in different patches would lead to better resource capture and increase total community biomass (Tilman *et al.*, 1997). The effect of species diversity on the proportion of environmental conditions 'covered' by at least one species depends on the size of the niche of each species relative to the range of environmental conditions in the habitat (Tilman *et al.*, 1997).

Attribution of immediate control of ecosystem function to dominant plant species does not exclude subordinate species from involvement in the determination of ecosystem function and sustainability. Consistent associations between certain dominant and subordinate species may reflect a complementary exploitation of habitat, resulting in a more complete capture of resources and minor benefits to productivity. Evidence of complementarity between dominant and subordinate members of a plant community was presented by Campbell *et al.* (1991). Dominance was achieved by the development of a coarse-grained architecture, in which main roots and shoots spread rapidly through a large volume of habitat, with rather imprecise concentration in resource-rich sectors. A complementary foraging mechanism was recognized in subordinates, where resource capture was achieved by a precise but local concentration of roots and shoots in resource-rich patches.

Perhaps more importantly, subordinate species are suspected to play a crucial, if intermittent, role by influencing the recruitment of dominant species (Grime, 1998). Subordinate species may act as a filter, influencing regeneration by different potential dominant species following changes in management, disturbance or extreme climatic events. According to this hypothesis, the significance of plant species diversity in relationship to deterioration of ecosystem function may arise primarily from its effects on the recruitment of dominant species, rather than any immediate effects of species richness *per se*.

The key ecosystem functions relating to sustainable land use which are under threat in southern Australia due to land use and vegetation changes are deep drainage, N leaching and loss of ground cover. Ecosystem degradation results where deep drainage leads to dryland salinity, N leaching leads to soil acidification or loss of ground cover leads to soil erosion and declining water quality. The presence and persistence of a substantial perennial grass component in these grasslands are a key factor in stabilizing these ecosystem functions and maintaining sustainable grazing enterprises (Bolger and Garden, 1998).

Dryland salinity results from the removal of deep-rooted perennial vegetation (often trees and shrubs) and replacement with shallow-rooted (usually annual) crops and pastures. This leads to increased drainage of water below the root zone, rising water tables, mobilization of salt stored in the regolith and intersection of saline groundwater tables with the soil surface in lower parts of the landscape. Soil acidification is caused by the mineralization of fertilizer ammonium and N_2-fixation by legumes, followed by leaching of nitrate below the plant root zone. Loss of vegetative ground cover, especially of perennial species, results in increased soil erosion and loss of nutrients from the land surface and in declines in stream water quality from soil particles and attached nutrients.

The patterns of change in Australian temperate grasslands (see Figs 11.1 and 11.2) can be related to changes in these ecosystem functions. Stage I grasslands dominated by warm-season grasses, such as *Themeda*, are considered to be deep-rooted, and their summer-active growth habit results in a greater depletion of soil water in autumn, before the start of the winter–spring growing season (Johnston *et al.*, 1999). This can result in reduced water loss to deep drainage, which should reduce the risk of dryland salinity. In addition, grasslands dominated by *Themeda* maintain soil mineral N at low levels (Moore, 1967), which results in reduced N leaching and soil acidification. The maintenance of low soil mineral N levels by Stage I grasslands prevents invasion of annual species (Moore, 1993), and thus these grasslands have a high proportion of perennial ground cover, which leads to low rates of soil erosion. Thus the attributes of the perennial, summer-active vegetation, which dominates Stage I grasslands, tend to stabilize these key ecosystem functions.

At the other extreme are Stage IV grasslands, dominated by annual species, where perennial species are absent or a minor component. With the loss of perennial species, there is little or no growth in response to summer and early-autumn rains and there are low levels of ground cover for much of the year. These attributes can lead to greater deep drainage and N leaching and increased risks of dryland salinity, soil acidification and soil erosion. These Stage IV grasslands can be characterized as being unstable in regard to these ecosystem functions, and thus unsustainable.

The effects of intermediate stages of vegetation change, such as Stage II, III and V grasslands, on these ecosystem functions are less clear. To the degree that these grasslands have a dominant component of perennial species, it is reasonable to hypothesize that their effect on these ecosystem functions is closer to that of Stage I rather than Stage IV grasslands. However, the identity and particular attributes of the dominant perennial species may still be important. For example, *M. stipoides*, which is a dominant species in Stage III grasslands, is more summer-active than *Austrodanthonia* spp., and introduced perennial grasses dominant in Stage V grasslands are less summer-active and tend to have higher soil mineral N levels in autumn than *Themeda* or *Austrodanthonia* spp. (Moore, 1967). A better recognition and understanding of the role of vegetation in key ecosystem processes will assist in developing grassland systems and management practices that are both environmentally and economically sustainable.

Conclusions

The temperate grasslands of south-east Australia provide an interesting example of the effects of competition on changes in botanical composition. The initial factors triggering change in grassland structure were the susceptibility of the grasslands to invasion and the introduction of exotic species. Competition and recruitment limitation, influenced by climate, fire, grazing, fertilizer, disturbance and their interactions, then produced changes that were in many cases irreversible, at least in the short term. The role of N, as described by Wedin (1999), appears to have been pivotal in these changes.

Grasslands in south-east Australia currently exist in many states, but rarely without a proportion of annual species present. These introduce instability into such grasslands, as their influence on the dynamics of perennial grass populations can be large. The proportion of year-long green native perennial grasses has been maintained or enhanced in many grasslands compared with the undisturbed state, and it is the balance between these species and annual species that governs grassland stability. The proportion of perennial species also contributes to sustainability by reducing deep drainage, nitrate leaching and soil erosion and by providing persistent forage for grazing animals.

In this chapter, the value of understanding the underlying processes controlling changes on botanical composition has been emphasized, but these mechanisms have not been tested experimentally in the grasslands of temperate south-east Australia. It is important that work is done at this level to understand the principles involved so that management systems for these grasslands, both for agricultural production and for conservation, can be developed on a sound ecological basis.

Notes

1. Grasslands which have not been deliberately sown to introduced species. However, they may have naturalized species present.
2. The nomenclature of Jacobs and Everett (1996) is used throughout for Australian species formerly in the genus *Stipa*.
3. The nomenclature of Linder and Verboom (1996), as corrected by Linder (1997), is used throughout for Australian species formerly in the genus *Danthonia*.

References

Archer, K.A., Read, J. and Murray, G. (1993) Pasture decline – real or imagined? In: Michalk, D. (ed.) *Proceedings 8th Annual Conference, Grassland Society of NSW*. Grassland Society of NSW, Orange, Australia, pp. 425–436.

Benson, J. and Wyse Jackson, M. (1994) The Monaro region. In: McDougall, K. and Kirkpatrick, J.B. (eds) *Conservation of Lowland Native Grasslands in South-eastern Australia*. World Wide Fund for Nature, Sydney, Australia, pp. 13–43.

Bolger, T.P. (1998) Aboveground and belowground competition among pasture species. In: Michalk, D.L. and Pratley, J.E. (eds) *Proceedings 9th Australian Agronomy Conference*. Australian Society of Agronomy, Melbourne, Australia, pp. 282–285.

Bolger, T.P. and Garden, D.L. (1998) Ecological sustainability – species persistence in grazing systems. *Animal Production in Australia* 22, 65–67.

Bolger, T.P. and Garden, D.L. (1999) Nutrient responses of wallaby grass (*Danthonia* spp.) from the New South Wales tablelands. In: Eldridge, D. and Freudenberger, D. (eds) *Proceedings VI International Rangeland Congress*, Vol. 1. VI International Rangeland Congress Inc., Townsville, Queensland, Australia, pp. 269–271.

Burke, M.J.W. and Grime, J.P. (1996) An experimental study of plant community invasibility. *Ecology* 77, 776–790.

Campbell, B.D., Grime, J.P. and Mackey, J.M.L. (1991) A trade-off between scale and precision in resource foraging. *Oecologia* 87, 532–538.

Campbell, M.H. and Van de Ven, R. (1996) Tolerance of native grasses to Frenock® and Roundup®. In: Virgona, J. and Michalk, D. (eds) *Proceedings 11th Annual Conference, Grassland Society of NSW*. Grassland Society of NSW, Orange, Australia, pp. 120–121.

Campbell, M.H. and Vere, D.T. (1995) *Nassella trichotoma* (Nees) Arech. In: Groves, R.H., Shepherd, C.H. and Richardson, R.G. (eds) *The Biology of Australian Weeds*, Vol. 1. R.G. and F.J. Richardson, Melbourne, Australia, pp. 189–202.

Chapin, F.S., III, Sala, O.E., Burke, I.C., Grime, J.P., Hooper, D.U., Lauenroth, W.K., Lombard, A., Mooney, H.A., Mosier, A.R., Naeem, S., Pacala, S.W., Roy, J., Steffen, W.L. and Tilman, D. (1998) Ecosystem consequences of changing biodiversity. *Bioscience* 48, 45–52.

Christensen, P.E. and Burrows, N.D. (1986) Fire: an old tool with a new use. In: Groves, R.H. and Burdon, J.J. (eds) *Ecology of Biological Invasions: an Australian Perspective*. Australian Academy of Science, Canberra, Australia, pp. 97–105.

Cocks, P.S. (1994) Colonisation of a South Australian grassland by invading Mediterranean annual and perennial pasture species. *Australian Journal of Agricultural Research* 45, 1063–1076.

Cook, S.J., Blair, G.J. and Lazenby, A. (1978) Pasture degeneration. II. The importance of superphosphate, nitrogen and grazing management. *Australian Journal of Agricultural Research* 29, 19–29.

Doing, H. (1972) *Botanical Composition of Pasture and Weed Communities in the Southern Tablelands Region, South-eastern Australia*. CSIRO Division of Plant Industry Technical Paper No. 30, CSIRO, Melbourne, Australia.

Donald, C.M. (1970) The pastures of southern Australia. In: Leeper, G.W. (ed.) *The Australian Environment*. CSIRO and Melbourne University Press, Melbourne, Australia, pp. 68–82.

Donnelly, J.R. (1972) The grazing of native pastures in Tasmania. In: Leigh, J.H. and Noble, J.C. (eds) *Plants for Sheep in Australia*. Angus and Robertson, Sydney, Australia, pp. 39–40.

Dowling, P.M., Kemp, D.R., Michalk, D.L., Klein, T.A. and Millar, G.D. (1996). Perennial grass response to seasonal rests in naturalised pastures of central New South Wales. *Rangeland Journal* 18, 309–326.

Eddy, D., Mallinson, D., Rehwinkel, R. and Sharp, S. (1998) *Grassland Flora, a Field Guide for the Southern Tablelands (NSW & ACT)*. World Wide Fund for Nature, Australia, Australian National Botanic Gardens, NSW National Parks and Wildlife Service, Environment ACT, Canberra, Australia.

Fisher, H.J. (1974) Effect of nitrogen fertiliser on a kangaroo grass (*Themeda australis*) grassland. *Australian Journal of Experimental Agriculture and Animal Husbandry* 14, 526–532.

FitzGerald, R.D. and Lodge, G.M. (1997) *Grazing Management of Temperate Pastures: Literature Reviews and Grazing Guidelines for Major Species*. NSW Agriculture Technical Bulletin 47, NSW Agriculture, Orange, Australia.

Fox, M.D. (1986) The susceptibility of natural communities to invasion. In: Groves, R.H. and Burdon, J.J. (eds) *Ecology of Biological Invasions: an Australian Perspective*. Australian Academy of Science, Canberra, Australia, pp. 58–66.

Fox, M.D. (1999) Present environmental influences on the Australian flora. In: Orchard, A.E. and Thompson, H.S. (eds) *Flora of Australia* Vol. 1, *Introduction*, 2nd edn. ABRS/CSIRO, Melbourne, Australia, pp. 205–249.

Garden, D.L., Dowling, P.M. and Eddy, D.A. (1993) The potential for native grasses on the central and southern tablelands. In: Michalk, D. (ed.) *Proceedings 8th Annual Conference, Grassland Society of NSW*. Grassland Society of NSW, Orange, Australia, pp. 115–116.

Garden, D.L., Jones, C., Friend, D., Mitchell, M. and Fairbrother, P. (1996) Regional research on native grasses and native grass-based pastures. *New Zealand Journal of Agricultural Research* 39, 471–485.

Garden, D.L., Lodge, G.M., Friend, D.A., Dowling, P.M. and Orchard, B.A. (2000a). Effects of grazing management on botanical composition of native grass-based pastures in temperate south-east Australia. *Australian Journal of Experimental Agriculture* 40, 225–245.

Garden, D.L., Dowling, P.M., Eddy, D.A. and Nicol, H.I. (2000b) A Survey of Farms on the Central, Southern and Monaro Tablelands of NSW: management practices, farmer knowledge of native grasses, and extent of native grass areas. *Australian Journal of Experimental Agriculture* 40, 1081–1088.

Gill, A.M. (1975) Fire and the Australian flora: a review. *Australian Forestry* 38, 4–25.

Gill, A.M. (1981) Adaptive responses of Australian vascular plant species to fire. In: Gill, A.M., Groves, R.H. and Noble, I.R. (eds) *Fire and the Australian Biota*. Australian Academy of Science, Canberra, Australia, pp. 243–272.

Grime, J.P. (1998) Benefits of plant diversity to ecosystems: immediate, filter and founder effects. *Journal of Ecology* 86, 902–910.

Groves, R.H. (1986) Plant invasions of Australia: an overview. In: Groves, R.H. and Burdon, J.J. (eds) *Ecology of Biological Invasions: an Australian Perspective*. Australian Academy of Science, Canberra, Australia, pp. 137–149.

Groves, R.H. (1999) Present vegetation types. In: Orchard, A.E. and Thompson, H.S. (eds) *Flora of Australia*, Vol. 1, *Introduction*, 2nd edn. ABRS/CSIRO, Melbourne, Australia, pp. 369–401.

Groves, R.H., Keraitis, K. and Moore, C.W.E. (1973) Relative growth of *Themeda australis* and *Poa labillardieri* in pots in response to phosphorus and nitrogen. *Australian Journal of Botany* 21, 1–11.

Hobbs, R.J. and Hopkins, A.J.M. (1990) From frontier to fragments: European impact on Australia's vegetation. In: Saunders, D.A., Hopkins, A.J.M. and How, R.A. (eds) Australian ecosystems: 200 years of utilisation, degradation and reconstruction. *Proceedings Ecological Society of Australia* 16, 93–114.

Holling, C.S. (1992) Cross-scale morphology, geometry and dynamics of ecosystems. *Ecological Monographs* 62, 447–502.

Hubbell, S.P., Foster, R.B., O'Brien, S.T., Harms, K.E., Condit, R., Wechsler, B., Wright, S.J. and Loo de Lao, S. (1999) Light-gap disturbances, recruitment limitation, and tree diversity in a neotropical forest. *Science* 283, 554–557.

Huston, M.A. (1997) Hidden treatments in ecological experiments: re-evaluating the ecosystem function of biodiversity. *Oecologia* 110, 449–460.

Hutchinson, K.J. (1992) The grazing resource. In: Hutchinson, K.J. and Vickery, P.J. (eds) *Proceedings, 6th Australian Agronomy Conference*. Australian Society of Agronomy, Melbourne, Australia, pp. 54–60.

Hutchinson, K.J. and King, K. (1999) Sown temperate pasture decline – fact or fiction? In: Garden, D., Lloyd Davies, H., Michalk, D. and Dove, H. (eds) *Proceedings 14th Annual Conference, Grassland Society of NSW*. Grassland Society of NSW, Orange, Australia, pp. 78–86.

Jacobs, S.W.L. and Everett, J. (1996) *Austrostipa*, a new genus, and new names for Australasian species formerly included in *Stipa* (Gramineae). *Telopea* 6, 579–595.

Johnston, W.H., Clifton, C.A., Cole, I.A., Koen, T.B., Mitchell, M.L. and Waterhouse, D.B. (1999) Low input grasses useful in limiting environments (LIGULE). *Australian Journal of Agricultural Research* 50, 29–53.

Jones, C.E. (1996) Pastoral value and production from native pastures. *New Zealand Journal of Agricultural Research* 39, 449–456.

Kemp, D.R. and Dowling, P.M. (1991) Species distribution within improved pastures over central NSW in relation to rainfall and altitude. *Australian Journal of Agricultural Research* 42, 647–659.

Kemp, D.R., Dowling, P.M. and Michalk, D.L. (1996) Managing the composition of native and naturalised pastures with grazing. *New Zealand Journal of Agricultural Research* 39, 569–578.

Linder, H.P. (1997) Nomenclatural corrections in the *Rytidosperma* complex (Danthonieae, Poaceae). *Telopea* 7, 269–274.

Linder, H.P. and Verboom, G.A. (1996) Generic limits in the *Rytidosperma* (Danthonieae, Poaceae) complex. *Telopea* 6, 597–627.

Lodge, G.M. (1994) The role and future use of perennial native grasses for temperate pastures in Australia. *New Zealand Journal of Agricultural Research* 37, 419–426.

Lodge, G.M. (2000) Competition among seedlings of perennial grasses, subterranean clover, white clover, and annual ryegrass in replacement series mixtures. *Australian Journal of Agricultural Research* 51, 377–383.

Lodge, G.M. and McMillan, M.G. (1994) Effects of herbicides on wallaby grass (*Danthonia* spp.). 2. Established plants. *Australian Journal of Experimental Agriculture* 34, 759–764.

Lodge, G.M. and Whalley, R.D.B. (1989) *Native and Natural Pastures on the Northern Slopes and Tablelands of New South Wales: a Review and Annotated Bibliography*. Technical Bulletin 35, NSW Agriculture and Fisheries, Sydney, Australia.

Lodge, G.M., Scott, J.M., King, K.L. and Hutchinson, K.J. (1998) A review of sustainable pasture production issues in temperate native and improved pastures. *Animal Production in Australia* 22, 79–89.

McDougall, K. (1994) Grassland flora and significant species. In: McDougall, K. and Kirkpatrick, J.B. (eds) *Conservation of Lowland Native Grasslands in South-eastern Australia*. World Wide Fund for Nature, Australia, pp. 4–7.

McIntyre, S., Lavorel, S. and Tremont, R.M. (1995) Plant life-history attributes: their relationship to disturbance response in herbaceous vegetation. *Journal of Ecology* 83, 31–44.

Mack, R.N. (1989). Temperate grasslands vulnerable to plant invasions: characteristics and consequences. In: Drake, J.A., Mooney, H.A., di Castri, F., Groves, R.H., Kruger, F.J., Rejmánek, M. and Williamson, M. (eds) *Biological Invasions: a Global Perspective*. Scope 37, John Wiley & Sons, Chichester, UK, pp. 155–179.

McNaughton, S.J. (1984) Grazing lawns: animals in herds, plant form, and coevolution. *The American Naturalist* 124, 863–886.

Magcale-Macandog, D.B. and Whalley, R.D.B. (1991) Distribution of *Microlaena stipoides* and its association with introduced perennial grasses in a permanent pasture on the Northern Tablelands of New South Wales. *Australian Journal of Botany* 39, 295–303.

Magcale-Macandog, D.B. and Whalley, R.D.B. (1993) Factors affecting the distribution of *Microlaena stipoides* on the Northern Tablelands of New South Wales. In: *Proceedings XVII International Grassland Congress*. New Zealand Grassland Association, Palmerston North, New Zealand, pp. 313–315.

Martyn, S.R. (1995) Permanent pasture, fact or fiction? An economic perspective. In: *Proceedings 5th Annual Conference, Tasmanian Branch of the Victorian Grassland Society*. Grassland Society of Victoria, Melbourne, Australia, pp. 14–20.

Mason, W., Kay, G. and Lodge, G. (1997) Sustainable grazing systems – a program to deliver improved temperate pastures in Australia. In: *Proceedings XVIII International Grassland Congress*, Session 24. Association Management Centre, Calgary, Canada, pp. 13–14.

Moore, R.M. (1967) The naturalisation of alien plants in Australia. In: *Towards a New Relationship of Man and Nature in Temperate Lands*. Part III. *Changes Due to Introduced Species*. IUCN Publications New Series No. 9, International Union for Conservation of Nature and Natural Resources/UNESCO, Morges, Switzerland, pp. 82–97.

Moore, R.M. (1970) South-eastern temperate woodlands and grasslands. In: Moore, R.M. (ed.) *Australian Grasslands*. ANU Press, Canberra, Australia, pp. 169–190.

Moore, R.M. (1993) Grasslands of Australia. In: Coupland, R.T. (ed.) *Ecosystems of the World. 8B. Natural Grasslands. Eastern Hemisphere and Résumé*. Elsevier, Amsterdam, The Netherlands, pp. 315–360.

Munnich, D.J., Simpson, P.C. and Nicol, H.I. (1991) A survey of native grasses in the Goulburn district and factors influencing their abundance. *Rangelands Journal* 13, 118–129.

Noble, I.R. and Slatyer, R.O. (1980) The use of vital attributes to predict successional changes in plant communities subject to recurrent disturbances. *Vegetatio* 43, 5–21.

Norton, B.E. (1971) The grasslands of the New England tableland in the nineteenth century. *Armidale and District Historical Society Journal and Proceedings* 15, 1–13.

Pettit, N.E., Froend, R.H. and Ladd, P.G. (1995) Grazing in remnant woodland vegetation: changes in species composition and life form groups. *Journal of Vegetation Science* 6, 121–130.

Prober, S.M. (1996) Conservation of the grassy white box woodlands: rangewide floristic variation and implications for reserve design. *Australian Journal of Botany* 44, 57–77.

Prober, S.M. and Thiele, K.R. (1995) Conservation of the grassy white box woodlands: relative contributions of size and disturbance to floristic composition and diversity of remnants. *Australian Journal of Botany* 43, 349–366.

Robinson, G.G. (1976) Productivity and response to nitrogen fertiliser of the native grass *Danthonia racemosa* (wallaby grass). *Australian Rangeland Journal* 1, 49–52.

Robinson, J.B., Munnich, D.J., Simpson, P.C. and Orchard, P.W. (1993) Pasture associations and their relation to environment and agronomy in the Goulburn district. *Australian Journal of Botany* 41, 627–636.

Scott, A.W. and Whalley, R.D.B. (1982) The distribution and abundance of species of *Danthonia* D.C. on the New England Tablelands (Australia). *Australian Journal of Ecology* 7, 239–248.

Simpson, P. and Langford, C. (1996) Whole-farm management of grazing systems based on native and introduced species. *New Zealand Journal of Agricultural Research* 39, 601–609.

Taylor, S.G. (1990) Naturalness: the concept and its application to Australian ecosystems. In: Saunders, D.A., Hopkins, A.J.M. and How, R.A. (eds) Australian ecosystems: 200 years of utilisation, degradation and reconstruction. *Proceedings Ecological Society of Australia* 16, 411–418.

Tilman, D. (1990) Constraints and tradeoffs: toward a predictive theory of competition and succession. *Oikos* 58, 3–15.

Tilman, D. (1994) Competition and biodiversity in spatially structured habitats. *Ecology* 75, 2–16.

Tilman, D. (1997) Community invasibility, recruitment limitation, and grassland biodiversity. *Ecology* 78, 81–92.

Tilman, D., May, R.M., Lehman, C.L. and Nowak, M.A. (1994) Habitat destruction and the extinction debt. *Nature* 371, 65–66.

Tilman, D., Lehman, C.L. and Thomson, K.T. (1997) Plant diversity and ecosystem productivity: theoretical considerations. *Proceedings of the National Academy of Sciences of the USA* 94, 1857–1861.

Torssell, B.W.R. and McKeon, G.M. (1976) Germination effects on pasture composition in a dry monsoonal climate. *Journal of Applied Ecology* 13, 593–603.

Vere, D.T., Campbell, M.H. and Kemp, D.R. (1997) *Pasture Improvement Budgets for Conventional Cultivation, Direct Drilling and Aerial Seeding in the Central and Southern Tablelands of NSW*. NSW Agriculture Industry Economics Unit Bulletin, NSW Agriculture, Orange, Australia.

Virgona, J. and Bowcher, A. (1998) Effects of pasture management on germinable seed bank in a degraded phalaris pasture. In: Michalk, D.L. and Pratley, J.E. (eds) *Proceedings 9th Australian Agronomy Conference*. Australian Society of Agronomy, Melbourne, Australia, pp. 178–180.

Walker, B.H. (1992) Biodiversity and ecological redundancy. *Conservation Biology* 6, 18–23.

Wallace, A. (1998) *Vulpia bromoides* (L.) S.F. Gray and *V. myuros* (L.) C.C. Gmelin. In: Panetta, F.D., Groves, R.H. and Shepherd, R.C.H. (eds) *The Biology of Australian Weeds*, Vol. 2. R.G and F.J. Richardson, Melbourne, Australia, pp. 291–308.

Wedin, D.A. (1995) Species, nitrogen and grassland dynamics: the constraints of stuff. In: Jones, C.G. and Lawton, J.H. (eds) *Linking Species and Ecosystems*. Chapman and Hall, New York, USA, pp. 253–262.

Wedin, D.A. (1999) Nitrogen availability, plant–soil feedbacks and grassland stability. In: Eldridge, D. and Freudenberger, D. (eds) *Proceedings VI International Rangeland Congress*, Vol 1. VI International Rangeland Congress Inc., Townsville, Australia, pp. 193–197.

Wedin, D.A. and Tilman, D. (1990) Species effects on nitrogen cycling: a test with perennial grasses. *Oecologia* 84, 433–441.

Wedin, D.A. and Tilman, D. (1993) Competition among grasses along a nitrogen gradient: initial conditions and mechanisms of competition. *Ecological Monographs* 63, 199–229.

Weiher, E. and Keddy, P. (1995) The assembly of experimental wetland plant communities. *Oikos* 73, 323–335.

Wells, K.F., Wood, N.H. and Laut, P. (1984) *Loss of Forests and Woodlands in Australia: a Summary by State, Based on Rural Local Government Areas*. CSIRO Division of Water and Land Research Technical Memorandum 84/4, CSIRO, Melbourne, Australia.

Whalley, R.D.B. and Lodge, G.M. (1987) Use of native and natural pastures. In: Wheeler, J.L., Pearson, C.J. and Robards, G.E. (eds) *Temperate Pastures: their Production, Use and Management*. Australian Wool Corporation/CSIRO, Melbourne, Australia, pp. 533–550.

Whalley, R.D.B., Robinson, G.G. and Taylor, J.A. (1978) General effects of management and grazing by domestic livestock on the rangelands of the northern tablelands of New South Wales. *Australian Rangeland Journal* 1, 174–190.

Wilson, A.D. (1990) The effect of grazing on Australian ecosystems. In: Saunders, D.A., Hopkins, A.J.M. and How, R.A. (eds) Australian ecosystems: 200 years of utilisation, degradation and reconstruction. *Proceedings Ecological Society of Australia* 16, 235–244.

Wilson, A.D. and Simpson, R.J. (1993) The pasture resource base: status and issues. In: Kemp, D.R. and Michalk, D.L. (eds) *Pasture Management: Technology for the 21st Century*. CSIRO, Melbourne, Australia, pp. 1–25.

Wilson, S.D. and Tilman, D. (1991) Components of plant competition along an experimental gradient of nitrogen availability. *Ecology* 72, 1050–1065.

Zobel, M. (1997) The relative role of species pools in determining plant species richness: an alternative explanation of species coexistence? *Trends in Ecology and Evolution* 12, 266–269.

12 Global Climate Change Effects on Competition and Succession in Pastures

B.D. Campbell and D.Y. Hunt

AgResearch, Grasslands Research Centre, Palmerston North, New Zealand

Introduction

The balance of scientific evidence suggests that human activities are changing the atmosphere and climate of the earth (Santer *et al.*, 1995). These observed atmospheric changes include increases in atmospheric carbon dioxide (CO_2) concentration, decreases in ozone (O_3) concentration, increases in ultraviolet (UV)-B radiation, increases in nitrous oxide (N_2O) concentration and increases in methane (CH_4) (Schimel *et al.*, 1995). These increasing gas concentrations are expected to alter climatic conditions (the 'enhanced greenhouse effect'), resulting in increased global atmosphere and sea-surface temperatures, changes in rainfall distribution and intensity and changes in the frequency of extreme climatic events such as droughts and storms (Kattenberg *et al.*, 1995). Significant questions are therefore raised about how ecosystems and food production activities will be affected.

Several excellent recent reviews have examined the implications of global change for terrestrial vegetation (Reilly *et al.*, 1995; Körner and Bazzaz, 1996; Walker and Steffen, 1996; Lumsden, 1997; Rozema *et al.*, 1997a; Walker *et al.*, 1998). In many vegetation types, including pastures, competition remains a key unresolved issue in predicting vegetation responses to global climate change. Our chapter specifically considers how four global climatic change factors – atmospheric CO_2, stratospheric O_3 temperature and rainfall – might affect competition and succession in pastures. The predominant emphasis to date in global climate change research has been on the long-term effects of climate change, with relatively little emphasis on the importance of human interference in modifying climate change impacts. For the specific case of intensive pasture systems, such decisions as choice of seed mixture, fertilizer regime or stocking pressure can be of overriding importance in determining competitive outcomes and vegetation succession. Therefore, where possible, we examine the implications of these changes for management decisions that influence competition in pastures.

A Framework for Analysing Effects

The long-term and complex nature of these changes call for a comprehensive predictive framework for identifying the effects of global climate changes on competition within pasture ecosystems. We propose here a simple model (Fig. 12.1), which recognizes that plant development is primarily driven by resources (e.g. energy, CO_2, mineral nutrients) and disturbance agents (e.g. herbivores, invertebrates, pathogens) in the immediate habitat. Through the respective resource inputs and disturbance events, these factors determine the resource capture, resource allocation, growth, senescence, reproduction and recruitment of different plants within a habitat. The expanding individuals or populations of these plants result in competition for resources. The processes of resource inputs, disturbance events and competition act as drivers and feedback loops linking the different boxes in Fig. 12.1.

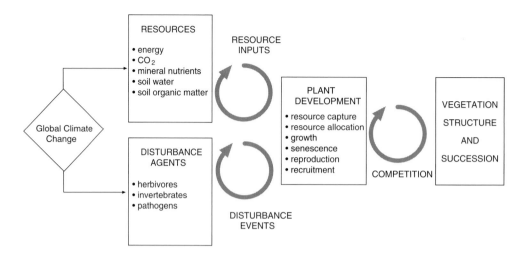

Fig. 12.1. Conceptual model of effects of global climate change on factors influencing competition in pastures.

The model is consistent with current competition theory. Many conceptual models of competition now recognize the fundamental importance of habitat resource levels and herbivores or other disturbance agents in driving competitive interactions (e.g. Grime, 1977, 1979; Tilman, 1982, 1988, 1997; Campbell *et al.*, 1991; Wilson and Tilman, 1991; Campbell and Grime, 1992; Grace, 1993; Goldberg, 1994; Kadmon, 1995; Twolan-Strutt and Keddy, 1996; Goldberg and Novoplansky 1997; Tilman *et al.*, 1997).

Resource levels strongly determine both the intensity of competition (the amount of decrease in growth rate, fecundity, size or fitness due to competition) and the importance of competition (the relative degree to which competition contributes to the overall decrease in growth rate, fecundity, size or fitness below its optimal condition) by having a direct impact on plant growth and development (Campbell and Grime, 1992). For example, decreased energy supply lowers the temperature of leaf and root tissues and reduces growth rates. Nutrient or water limitations reduce the rates of carbon uptake and growth, reduce rates of capture of light, water and mineral nutrients, and alter the balance of carbon allocation more in favour of roots (Hunt and Lloyd, 1987; Tilman, 1988). Chronically low nutrient availability interspersed with short nutrient pulses favours genotypes that conserve nutrients in long-lived leaves and root systems (Campbell and Grime, 1989). Spatial patchiness in mineral nutrients or light favours genotypes with high morphological plasticity and active

foraging by roots or shoots (Campbell *et al.*, 1991). Disturbance reduces competition by removing biomass of certain plants (Wardle and Barker, 1997), creating gaps with higher levels of light and nutrients. Subsequently, this reduced competition can result in increased plant recruitment and coexistence (Lavorel and Chesson, 1995; Dear *et al.*, 1998), vegetative expansion (McLellan *et al.*, 1997) and increased rates of secondary succession (Fraser and Grime, 1998).

Some species respond better than others to changes in resource levels and disturbance, due to their particular combinations of traits (Campbell and Grime, 1992). These differential responses by the component plants within the vegetation can change the intensity and importance of competitive interactions and result in significant shifts in vegetation structure and succession. This altered vegetation structure can in turn result in longer-term shifts in resource availability and/or disturbance agents within the habitat, by feedbacks through competition, resource inputs and disturbance events (see Fig. 12.1).

We can begin to test understanding of the sensitivity of competition in pastures to global climate change by using this model (see Fig. 12.1). The model recognizes that global climate change does not act directly on competition, but rather acts indirectly through several different processes, involving both resource inputs and disturbance events. This approach permits mechanistic analysis of the relative importance of these different processes in determining competition and succession.

Effects of Increasing Atmospheric CO_2

Background

The concentration of CO_2 in the atmosphere is currently rising as a result of the burning of fossil fuels and deforestation. This increase is observed to be about 2 p.p.m. $year^{-1}$ (Schimel et al., 1995), representing the imbalance between carbon release into the atmosphere and the capacity of the terrestrial biosphere and oceans to absorb this extra CO_2. Recent international agreements, such as the Kyoto Conference in 1997, have identified targets for reducing CO_2 emissions from all countries of the world, but these targets will still result in a continuing significant increase in CO_2 in the atmosphere.

The primary concern about the rising CO_2 concentrations is that this gas can act as a heat trap in the atmosphere (Schimel et al., 1995), resulting in an increase in heat energy retention and a consequent elevation in the temperature at the earth surface (the so-called 'enhanced greenhouse effect'). However, elevated CO_2 also has another direct implication for competition processes, because it has the potential to directly alter plant growth and change water and nutrient cycling and disturbance.

Potential effects on pastures

Up until the early 1990s there was considerable uncertainty about the way terrestrial ecosystems would respond to elevated CO_2, as most of the predictions were based on the results of laboratory experiments. Recently, the implications for ecosystems, including pastures, have become clearer (Mooney et al., 1999), largely as a result of a worldwide concentration of effort on research into the effects of elevated CO_2 on intact ecosystems.

Overall, the most recent evidence suggests that grasslands are likely to be less affected by elevated CO_2 than was originally believed. On average, grassland primary production is increased by about 15% for a doubling of CO_2 (Mooney et al., 1999), although individual systems vary more widely. This is a significant downward revision of previous estimates of about 30% (Kimball, 1983; Newton, 1991).

This recent research has identified several potential effects of elevated CO_2 on resources, disturbance agents and plant development. The elevation of atmospheric CO_2 represents an upward shift in carbon resources available to plants (see Fig. 12.1) and can produce direct effects on several plant development processes (Table 12.1).

A number of studies have recorded differences in sensitivity to CO_2 between different genotypes and species (Campbell et al., 1993; Poorter, 1993; Poorter et al., 1996). Although varied responses to CO_2 have been measured for C_3 genotypes and species, these differences are not currently well predicted by any existing conventional plant functional type (PFT) scheme (Campbell and Grime, 1993; Vasseur and Potvin, 1998). Several plant traits are likely to be responsible, including photosynthesis type (Greer et al., 1995; Bowes, 1996), sink strength (Diaz, 1995), relative growth rate (Poorter et al., 1996), specific leaf area (Roumet and Roy, 1996), competitive attributes (Hunt et al., 1991, 1993) and carbon demand associated with nitrogen- (N-) fixing (Lüscher et al., 1998) or mycorrhizal symbioses (Norby et al., 1987; Diaz et al., 1993; Poorter, 1993). Both structural and physiological responses to CO_2 are likely to influence the competitive success of species. Teughels et al. (1995) identified leaf area and leaf photosynthesis as two important shoot factors determining competitive success of Lolium perenne and Festuca arundinacea in simple, clipped mixtures. More complex sets of characters might be expected for other plant types and ecosystems.

The potential plant developmental responses in Table 12.1 are highly dependent on the prevailing habitat and/or seasonal conditions. The relative response to CO_2 can be enhanced by higher temperatures (Campbell et al., 1993; Newton et al., 1994), N availability (Owensby et al., 1996a; Schenk et al., 1996; Soussana et al., 1996), light level (Zangerl and Bazzaz, 1984) and water limitation (Knapp et al., 1993; Jackson et al., 1995).

There are follow-on effects on N availability, nutrient cycling and water use as a result of these changes in plant development (see Table 12.1). Increases in leaf carbon : nitrogen (C : N) ratios may potentially slow litter decomposition rates, resulting in less available N. However, recent evidence suggests that this is not necessarily universally observed as hypothesized, because litter C : N ratios can differ from tissue ratios, due to reabsorption of N before senescence (Mooney et al., 1999).

Table 12.1. Some potential effects of elevated CO_2 on plant development, resources and disturbance.

Plant development	Resources	Disturbance
Increased CO_2 leading to: • Increased photosynthesis rate and decreased photorespiration (Bowes, 1996) • Decreased stomatal conductance (Owensby et al., 1997) and increased water use efficiency (Amthor, 1995; Schapendonk et al., 1997) • Increased tillering and leaf area (Ferris et al., 1996) • Increased biomass production (Mooney et al., 1999) • Increased root production (Jongen et al., 1995; Fitter et al., 1997) and root : shoot ratio (Schenk et al., 1996) • Increased flowering and reproduction (Ackerly and Bazzaz, 1995) • Altered phenology (Navas et al., 1997) • Decreased Rubisco and leaf N content (Cotrufo et al., 1998) and increased non-structural carbohydrate content (Casella and Soussana, 1997; Read et al., 1997)	• Increased symbioses leading to increased N_2-fixation by legumes (Zanetti et al., 1996) and nutrient acquisition through mycorrhizal colonization (Klironomos et al., 1996; Rillig et al., 1998), with implications for longer-term nutrient availability • Increased carbon in the rhizosphere (Canadell et al., 1996) leading to immobilization of N in microbial biomass in some cases (Diaz et al., 1993) but not in others (Hungate et al., 1997) • Increased water use efficiency leading to altered water use, extended seasonal availability of water to plants (Fredeen et al., 1997; Owensby et al., 1997; Jackson et al., 1998), as well as conse-quent increases in microbial activity and nutrient availability in soils • Changes in plant tissue chemistry, growth and decomposition rates (van Ginkel et al., 1996), and increased rhizosphere carbon and soil microbial biomass activity (Schenk et al., 1995), leading to increased carbon turnover and/or storage (Cotrufo and Ineson, 1996; Luo et al., 1996; Niklaus and Körner 1996; Thornley and Cannell, 1997), N limitation (Soussana et al., 1996) and changes in nutrient cycling • Possibly lesser effect on longer-term or equilibrium nutrient availability, due to greater organic matter pool and fewer losses (Thornley and Cannell, 1997) and/or gradual mobilization of unavailable forms of N and P over decades to centuries (Gifford et al., 1996a)	• Changes in plant tissue chemistry leading to increased tissue consumption rates by insects (Roth and Lindroth, 1995), increased pathogen damage (O'Neill, 1994) and altered insect and pathogen population sizes • Changes in plant tissue chemistry and growth rates (Schenk et al., 1997) leading to reduced livestock intake, growth and reproduction (Owensby et al., 1996b) and changes in ruminant selective grazing and stocking rates

For water-limited systems, elevated CO_2 can result in greater water availability for longer in the growing season, especially if there is not an increase in leaf transpiration surface per unit of ground area (Schapendonk et al., 1997). The hydrological consequences of elevated CO_2 in water-limited systems are often greater than the direct CO_2 fertilization effect on photosynthesis (Mooney et al., 1999). However, in systems with low leaf area index, large physiological responses to CO_2 will probably not alter total ecosystem water budgets dramatically (Jackson et al., 1998).

The changes in plant tissue composition are also expected to alter herbivore activities (see Table 12.1). Here there is evidence that the responses of invertebrate herbivores and ruminants may differ.

Implications for competition

Some significant changes in the intensity and importance of competition for light, nutrients and water in competitive relations in pasture ecosystems would be predicted based on the effects of CO_2 outlined in Table 12.1, including the following:

- An increased competitive advantage to CO_2-responsive species, with a generally greater advantage to C_3 than to C_4 species.
- An increased competitive advantage to late-season species, as a result of CO_2-induced increases in seasonal water availability.
- An increased competitive advantage to legumes and mycorrhizal species in situations where elevated CO_2 reduces long-term N availability, with differing responses depending on resource inputs.
- An increased intensity of competition for light, water and nutrients as a result of increased biomass, root : shoot allocation and resource capture resulting in greater rates of competitive exclusion, but a reduced intensity in cases where there is an increased proportion of leaf material removed by herbivores.

Experimental evidence for testing these predictions is still rather limited, as only a handful of CO_2 experiments on pastures include both monoculture and mixture treatments to allow analysis of the effects of competition.

The major prediction from analysis of photosynthesis physiology is that C_3 plants should show greater responses to elevated CO_2 than C_4 species

(Bowes, 1996). Here it might be expected that the increased growth of C_3 species relative to C_4 species (resulting from a greater stimulation of photosynthesis) would result in a shift in the relative competitive advantage. However, recent evidence from water-limited grasslands shows that the predicted greater comparative advantage of C_3 plants in mixed C_3–C_4 ecosystems under elevated CO_2 is not universally observed in ecosystems. In the Kansas tall-grass prairie, C_4 species have been shown to be more responsive to CO_2 in this mixed C_3–C_4 community. It is suggested that this results from reduced stomatal conductance, leading to slower rates of soil water depletion and consequently a greater amount of soil water remaining available later in the growing season (Owensby et al., 1997). Here, the alleviation of water limitations at higher CO_2 was more important than differences in CO_2 effects on leaf photosynthesis rates in determining competitive outcomes. Similarly, in California annual grasslands, there is evidence that more soil water remains at the end of the growing season at elevated CO_2 (Fredeen et al., 1997). As a consequence, the greater late-season growth of annuals (especially in dry years) can result in altered community composition (Field et al., 1996).

This evidence confirms that shifts in competition and succession in ecosystems due to elevated CO_2 are not simply predictable from the responses of plants grown in isolation (Bazzaz, 1990; Körner, 1995). The strongest effects of elevated CO_2 on plant–plant interactions are likely to arise indirectly through effects on plant growth and competition for other resources (Bazzaz and McConnaughay, 1992), as well as through effects on ecosystem resource availability and disturbance rates. Significant shifts in competitive interactions are possible (e.g. Wray and Strain, 1987; Bazzaz and Garbutt, 1988).

To date, the best-documented effects of CO_2 on competitive interactions in improved pasture ecosystems are those occurring between ryegrass (L. perenne) and white clover (Trifolium repens). Recent results from a free air carbon dioxide experiment (FACE) on Swiss pasture (Hebeisen et al., 1997) have specifically examined monocultures and mixtures to test the competitive interactions between these two species under elevated CO_2 (350 p.p.m. vs. 600 p.p.m.) with different N fertilizer (10–14 vs. 42–56 g N m^{-2} year^{-1}) and cutting regimes (four vs. seven to eight cuts) over a 3-year period. Elevated CO_2 resulted in an increase in the competitive ability of *T. repens* relative to *L. perenne*, such that the yield response of *T. repens* to elevated CO_2 was +17% in monoculture and +64% (averaged over 3 years) in mixture. In contrast, the +7% response to CO_2 of *L. perenne* in monoculture was reduced to a −1% response (averaged over 3 years) in mixture.

In these grass–clover mixtures, the positive response of symbiotic N_2-fixation to elevated CO_2 can increase subsequent soil N availability, and reduce an N limitation to the grass that would otherwise occur as a result of the imbalance between C and N cycles (Soussana and Hartwig, 1996). The significant CO_2 response of *T. repens* in monocultures was independent of N, whereas elevated CO_2 induced a yield stimulation in *L. perenne* monocultures only at high N (Hebeisen *et al.*, 1997). In the mixtures, elevated CO_2 resulted in a higher proportion of *T. repens* at all cutting and N treatments, but the magnitude of the effect differed depending on N and cutting. It was suggested in this case that elevated CO_2 may have increased the N limitation to growth through increased denitrification and sequestration of N into litter, soil organic matter and soil microorganisms (Diaz *et al.*, 1993; Schenk *et al.*, 1995). Under the high-phosphorus (P) conditions in this experiment (1–6 mg P kg^{-1} soil plus P fertilizer), the N-fixing legume *T. repens* was not so dependent on soil N supply and responded to the reduced availability of N with an increased N_2-fixation (Zanetti *et al.*, 1996). The N yield derived from N_2-fixation in the *T. repens* monocultures was increased by 25%, due to elevated CO_2. This increase in the proportion of clover could lead to a new equilibrium between clover and grasses. However, the higher proportion of legumes could also be expected to increase the N availability in soil over time and thus lead to an increase in the competitive ability of the grasses (Thornley *et al.*, 1995).

Similarly, Schenk *et al.* (1997) examined CO_2 effects on competition in mini-swards of *T. repens* and *L. perenne* grown as monocultures and mixtures in open-topped chambers. They determined that white clover was enhanced by CO_2 enrichment with either no added N or with 200 kg N ha^{-1}, but a suppression of ryegrass in mixed swards was only observed under low-N conditions (Schenk *et al.*, 1997). This suppression was attributed to intensified competition for light. It was again suggested that the effect of elevated CO_2 on the balance of species and the outcome of competition in grass–clover swards is mainly dependent on the N

status. Stewart and Potvin (1996) also found that elevated CO_2 increased the strength and number of plant–plant interactions in pasture, with *T. repens* benefiting more than *Poa pratensis* from elevated CO_2. The result was strongly dependent on plant density. Elevated CO_2 increased competition but also increased opportunities for invasion of this plant community by *T. repens*. It was suggested that changes in invasiveness could be better predicted by traits related to the acquisition of space (such as a stoloniferous growth habit), rather than traits related to growth or height.

This apparently greater competitive advantage of *T. repens* at elevated CO_2 in pastures is likely to occur only when soil fertility and climatic conditions allow the N_2-fixation advantage of legumes to be expressed. For example, in low-P soils of Swiss meadows there is no obvious advantage to *Trifolium* species due to elevated CO_2; instead, the greatest stimulation of yield in this mixed community was observed for *Poa alpina* (Schäppi and Körner, 1996) or *Carex flacca* (Rötzel *et al.*, 1997). In relation to temperature, Campbell and Hart (1996) found that elevated CO_2 favoured competitive suppression of *T. repens* by grass at day/night temperatures of 18/13°C, resulting in a decreased yield of *T. repens* with elevated CO_2. However, the competitive suppression was reduced at temperatures of 28/23°C, so that CO_2 resulted in a stimulation of *T. repens* content in swards. This shift is consistent with the higher temperature optimum for growth of *T. repens* compared with grasses, and suggests that elevated CO_2 may favour differing components of the pasture as temperature changes during the growing season.

Similarly, where grazing management restricts certain species, the potential CO_2-induced changes in competition may not be expressed. In the Swiss FACE, infrequent defoliation significantly increased the yield of *L. perenne* in monocultures under elevated CO_2, whereas with frequent defoliation no CO_2-induced yield stimulation occurred (Hebeisen *et al.*, 1997). No such effect of defoliation on the CO_2 response was observed for *T. repens* in monocultures. A strong decline in clover proportion was observed in the mixture under infrequent grazing and/or high N fertilization, and it was concluded that the competitive ability of *T. repens* was highest in frequently defoliated mixtures with low N fertilization. However, Taylor and Potvin (1997) found no significant interaction between effects of disturbance and CO_2 on the overall diversity of a Canadian pasture.

In summary, these examples confirm an increased advantage to certain pasture species with elevated CO_2, with consequences for competition and succession. However, the examples of shifts in seasonal water availability and the effects of N fertilizer and grazing emphasize the importance of resource availability and disturbance regimes in determining the final outcome of the effect of elevated CO_2 on competition. For example, the predicted advantage to C_3 species or legumes can be modified by the water availability or P availability of the ecosystem. So far, these short–medium-term (1–5-year) ecosystem-level experiments on pastures have not allowed conclusive testing of the predictions that long-term (decades to centuries) alterations in soil resources and disturbance will result in long-term shifts in the intensity of competition and rates of succession in pastures.

Effects of Decreasing Stratospheric O_3

Background

A layer of O_3 in the stratosphere above the earth absorbs damaging UV-C radiation and also part of the UV-B radiation coming from the sun. Recently, it has been recognized that stratospheric O_3 is being depleted by reactions involving chlorine and bromine compounds in the stratosphere, and solar UV-B reaching the surface of the earth is currently increasing (Madronich, 1993). This depletion of O_3 has been attributed to a number of human activities, including emissions of chlorinated compounds (Pyle, 1997). The intensity of UV-B radiation is higher at mid- to low-latitude regions than at high latitudes, and decreases in O_3 are expected to result in greatest relative increases in UV-B dose rates at the high-latitude and mid-latitude regions (Pyle, 1997). International agreements have been put in place to limit the emissions of these compounds, but the depletion of O_3 and increase in UV-B radiation are expected to continue into next century because of the persistent nature of these chemicals in the atmosphere.

Potential effects on pastures

A primary effect of O_3 depletion is a change in incident solar energy resources (see Fig. 12.1), manifested as an increase in the intensity of UV-B radi-

ation striking the plant. A variety of plant developmental responses have been observed to result from elevated UV-B radiation (Table 12.2). The biological impact of UV-B radiation on plant development is a combined function of damage, repair and acclimatization processes within the plant. Plants are continually adjusting adaptive mechanisms to minimize UV-B radiation damage (Jansen et al., 1998). As with elevated CO_2, the effects of UV-B radiation on plant development are highly dependent on the prevailing levels of other environmental parameters, and must be interpreted in the context of interactions with other stresses (Jordan, 1993).

Species and genotypes do not all show similar responses when exposed to elevated UV-B radiation. Graminoids tend to be more resistant to UV-B damage than herbaceous dicotyledonous plants (Caldwell and Flint, 1994), and legumes appear to be quite sensitive to UV-B. Differences between genotypes and species in the tolerance of UV-B radiation appear to be associated with various developmental features, including epidermal hairs (Bornman et al., 1997), UV-B-absorbing compounds (Caldwell and Flint, 1993; Rozema et al., 1997a), antioxidants (Bornman et al., 1997), leaf thickness (Day et al., 1993), leaf longevity (Björn et al., 1997), photosynthetic pathway (Van and Garrard, 1976; van de Staaij et al., 1990; Ernst et al., 1997) and architectural parameters (Corlett et al., 1997).

Changes in resource availability are expected to result from these changes in development (see Table 12.2). An increased synthesis of tannins and lignin with enhanced UV-B radiation has important consequences for decomposition processes, as well as herbivory, and could alter nutrient cycles. There is evidence that elevated UV-B radiation can reduce decomposition rates by increasing leaf lignin content; however, there can also be a direct effect of elevated UV-B radiation in increasing the photodegradation of lignin (Rozema et al., 1997d). These are two significant but opposite effects of enhanced UV-B radiation on decomposition of plant material, illustrating the complexity of predicting long-term effects on resource availability.

The implications of elevated UV-B levels for disturbance (see Table 12.2) are also difficult to predict, as our understanding of the effects of UV-B radiation on higher plant–consumer interactions allows only very tentative conclusions to be drawn (Paul et al., 1997). A considerable range of plant responses is anticipated, with both positive and negative

Table 12.2. Some potential effects of ozone depletion and consequent UV-B radiation increase on plant development, resources and disturbance.

Plant development	Resources	Disturbance
Increased UV-B radiation leading to: • Damage to DNA, membranes, photosystem II (Caldwell et al., 1989) • Increased flavonoids (Tevini et al., 1991; Beggs and Wellman, 1994; Caldwell et al., 1995), antioxidants, phenolic compounds, UV-B-absorbing compounds (Tosserams et al., 1997a, b), tannins and lignins (Rozema et al., 1997a) • Decreased height (Tosserams, 1997b), leaf length, leaf area (Tevini and Teramura, 1989; Antonelli et al., 1997; Björn et al., 1997), increased axillary branching and altered leaf angle and canopy architecture (Barnes et al., 1995) • Altered cell division (Bornman et al., 1997; Corlett et al., 1997), increased cuticle thickness (Manetas et al., 1997), epicuticular wax (Steinmuller and Tevini, 1985) and leaf thickness (Rozema et al., 1997a, b) • Increased or decreased stomatal conductance (Teramura, 1983; Teramura et al., 1984; Dai et al., 1995), reduced water use efficiency (Runeckles and Krupa, 1994) and increased water stress (Teramura et al., 1984) • Decreased yield (Tevini and Teramura, 1989; Tevini et al., 1990; Runeckles and Krupa, 1994) • Altered symbioses with microorganisms (Newsham et al., 1998) • Altered phenology (Mark et al., 1996), senescence and seed production (Newsham et al., 1998)	• Changes in plant tissue chemistry and growth rates, increases in tannin and lignin contents and lower decomposition rates (Newsham et al., 1997), leading to altered nutrient uptake and nutrient return through litter (Paul et al. 1997), with consequences for longer-term nutrient availability • Increased litter degradation due to direct effects of elevated UV-B radiation (Rozema et al., 1997d) • Changes in plant symbioses leading to changes in nutrient uptake and availability • Changes in stomatal conductance and growth rates leading to altered water use (Manetas et al., 1997) and cycling, with consequences for seasonal pattern of water availability	• Changes in plant tissue chemistry and altered insect and pathogen population levels (Hatcher and Paul, 1994; Ayres et al., 1996; Paul et al., 1997), leading to altered insect and pathogen damage • Changes in plant tissue chemistry and growth rates leading to changes in ruminant selective grazing and stocking rates

consequences. The direct effects of UV-B on consumers can also be both positive and negative, and the nature and magnitude of these effects will differ.

Overall, the current evidence from short- to medium-term ecosystem-level experiments provides little indication that enhanced UV-B radiation will markedly depress plant growth and primary production of terrestrial ecosystems (Caldwell *et al.*, 1995; Rozema *et al.*, 1997a). However, the consequences of changes in morphology and plant chemistry for resource levels, carbon, nutrient and water cycling, symbiotic relationships with soil-based microorganisms and plant–herbivore relationships suggest that there are likely to be significant effects for competitive relationships and succession.

Implications for competition

Some significant shifts in competitive relations in pasture ecosystems would be predicted on the basis of the effects of ozone depletion outlined in Table 12.2, including the following:

- A similar overall intensity of competition but with shifts in competitive interactions in favour of UV-B-tolerant species, with a greater depressive effect on legumes than on graminoids.
- Altered competitive interactions for light, due to changes in the relative distributions of leaves and capture of resources and altered competition for below-ground resources as a result of shifts in root : shoot ratios.
- Shifts in competitive interactions seasonally, due to interactions of UV-B with other stress factors, such as water stress, temperature and nutrient stress.
- Altered long-term grass : legume competitive interactions as a result of long-term reductions in N availability due to decline in legumes, but with oscillations in grass : clover dominance.
- Altered competition as a result of UV-B-induced changes in herbivore and pathogen levels, but knowledge is too limited to predict the direction change.

It is not currently possible to test these predicted consequences for competition and succession in pastures for two reasons: (i) there are almost no data available on the effects of UV-B radiation on pasture species or intact pasture ecosystems; and (ii) there are few studies of competition and ecosystem properties for any ecosystems at all. This omission

was recognized over 17 years ago by Gold and Caldwell (1983) in reviewing the effects of UV-B radiation on plant competition in terrestrial ecosystems, and it still largely exists today.

In the absence of this information for pastures, one of the few examples of an analysis of potential changes in competition is provided by a simple model of a wheat–wild oat community reported by Barnes *et al.* (1995) and Caldwell (1997). Experimental evidence from a 7-year experiment for this community suggests that elevated UV-B radiation causes no changes in yield in monocultures, but in mixtures there is a shift in the competitive balance of 50 : 50 mixtures in favour of wheat. Here it has been hypothesized that elevated UV-B radiation acts on a specific flavin or flavin-like compound as a photoreceptor controlling morphological changes, including allocation. The specific prediction was that a more pronounced reduction in internode length in wild oat than in wheat would shift the relative display of foliage with height in the canopy of the mixture, and that the shift in foliage display by the two species in the mixture given extra UV-B radiation was sufficient to shift the balance of competition for light to drive photosynthesis. A validated canopy model (developed to evaluate leaf area position changes in terms of leaf area index, photosynthetic photon flux density and net canopy photosynthesis) predicted that small change in foliage position can be amplified disproportionately to a shift in light interception, carbon gain and competition.

Pastures often contain species of similar graminoid origin but have a much increased complexity, due to perennial life histories, the presence of the grazing animal, a greater diversity of species and often an important role for N_2-fixation by legumes. The effects of these additional factors are not tested by this wheat–wild oat example. In summary, the effects of UV-B on competition remain to be tested for pastures, including assessment of long-term changes in pasture development and community structure associated with shifts in soil resource availability and disturbance.

Effects of Changing Temperature

Background

The atmospheric concentrations of CO_2, CH_4, nitrous oxide and other so-called 'greenhouse gases' regulate global average temperature (Schimel et al., 1995). Over the past century, the global temperature has risen by 0.3–0.6°C (Nicholls et al., 1995). It is not yet possible to attribute this unequivocally to human influences. Nevertheless, the Intergovernmental Panel on Climate Change (IPCC) has concluded that 'the balance of evidence suggests that there is a discernible human influence on global climate' (IPCC, 1995).

Climate scientists cannot currently give precise predictions of how climate will change in the future. For the greenhouse gas emission scenarios adopted by IPCC in 1995, incorporating possible effects of anthropogenic aerosols, the projected increase in mean global temperature lies between 1°C and 3.5°C by the year 2100 (Kattenberg et al., 1995). Confidence in predicting the climatic effects of increasing greenhouse gases is higher at the hemispheric to continental scale than at the regional or local scales.

From the perspective of pastoral agriculture, there are two important dimensions to the projected changes in global temperature: (i) changes in mean temperatures, with implications for growth rates and productivity; and (ii) changes in the frequency of extreme temperature events (e.g. frosts or hot days) and temperature variability patterns, with significant consequences for the survival of different species.

Potential effects on pastures

The effects of temperature on pasture plant growth and development have been well studied and reported elsewhere (e.g. McWilliam, 1978), so we shall not cover them in detail. Temperature determines the rates of many important enzymatic reactions and biological processes in pasture plant development (Table 12.3). In many respects, temperature acts as a switch on plant development and the role of temperature assumes increasing importance as it deviates from 20°C (McWilliam, 1978). Chilling or freezing temperatures can have significant implications for plant growth and development, and special adaptations can be recognized which protect against these extremes, such as cryoprotectant soluble sugar adjustment (Leegood and Edwards, 1996) and membrane lipid protection. Similarly, adaptations can also be recognized to cope with extreme heating temperatures, such as heat-shock proteins and dormancy mechanisms.

Plant species with the C_4 photosynthetic pathway are at a selective advantage in environments experiencing high temperatures, high irradiance and limited supplies of water during active growth (McWilliam, 1978).

Increased temperature can have an impact on soil resources through altered nutrient mineralization rates, symbiotic associations and higher rates of evapotranspiration leading to increased soil water removal (see Table 12.3), but these are generalized relationships only. Most carbon cycle models (e.g. McGuire et al., 1992; Parton et al., 1995) assume that temperature will have a direct effect on plant growth, but recent evidence (Fitter et al., 1998) showing that root production and respiration can be a function of change in the length of the growing season, rather than temperature, suggests this may need to be reassessed. Furthermore, the prediction of effects of increasing temperature on N mineralization is more complex than a simple linear relation between temperature and N release. For example, highest rates of N mineralization are observed with fluctuating warm/cold temperatures in some soils (Campbell et al., 1973).

Changes in disturbance are also anticipated through altered incidences of pests and diseases and altered herbivore feeding (see Table 12.3). In temperate zones, there could be an increased prevalence of invasive subtropical pests of both pastures and livestock (Sutherst et al., 1996). The response of grassland to higher temperatures is also predicted to be highly dependent on grazing management (Thornley and Cannell, 1997).

The general results observed from ecosystem-level warming experiments in a range of vegetation types (Mooney et al., 1999) are that: (i) ecosystem responses to warming are greatest in cooler environments; (ii) ecosystem responses are less pronounced and have longer time-lags than physiological and growth responses; and (iii) winter warming has greater influence than summer warming.

Implications for competition

Significant shifts in competitive relations in pasture ecosystems would be predicted on the basis of effects of elevated temperatures outlined in Table 12.3, including the following:

- An increased competitive advantage to C_4 species and warm-season species, due to higher temperatures and fewer frosts.

Table 12.3. Some potential effects of increasing temperature on plant development, resources and disturbance.

Plant development	Resources	Disturbance
Increased heat energy under suboptimal temperature conditions leading to: • Increased seed germination, increased photosynthesis rate (Falk et al., 1996) and increased respiration (Larigauderie and Körner, 1995) • Increased tillering and leaf area and biomass production (McWilliam, 1978), and increased senescence • Increased root : shoot ratio • Increased N_2-fixation by legumes (McWilliam, 1978) • Increased flowering and longer growing period (McWilliam, 1978) • Increased leaf protein content, increased lignin • Increased endophytic alkaloid contents • Decreased frequency of chilling and freezing temperature events leading to less cold-temperature injury to plants, especially C_4 plants of subtropical and tropical origin (Etherington, 1982) Increased heat energy under supra-optimal temperature conditions leading to reversal of some effects above: • Increased heat stress on plants in warmer environments, favouring C_4 plants	• Increased soil temperatures leading to increased decomposition rates and N mineralization rates (Thornley and Cannell, 1997) with increased short-term N availability but decreased long-term N availability in wet soils due to losses • Increased soil temperatures leading to increased soil respiration and loss of carbon (Thornley and Cannell, 1997)	• Increased day temperatures and decreased frequency of freezing temperature events leading to build-up of insect and pathogen populations (Sutherst et al., 1996) • Increased temperatures leading to more heat stress events in grazing livestock and lower consumption rates • Increased populations of pests and diseases of livestock (Sutherst et al., 1996), leading to reduced livestock growth and consumption rates

- An increased competitive advantage to legumes due to higher spring temperatures, but decreased advantage where there is higher N availability and winter temperatures.
- An increased intensity of competition for light, water and nutrients due to higher growth rates and generally higher nutrient availability, higher evapotranspiration rates and earlier plant water deficits, resulting in greater rates of competitive exclusion.
- An increased competitive advantage to disease- and insect-tolerant genotypes, where there is an increased prevalence of diseases and insects.

Temperature is recognized as a principal control on the geographical distribution of C_3 and C_4 species (Teeri and Stowe, 1976; Epstein *et al.*, 1998). Therefore, an increase in the competitive ability of C_4 species relative to C_3 species is one of the most significant changes in competitive relations in pastures predicted to result from an upward shift in mean temperature and accompanying reductions in the frequency of frost and/or incidence of hot days (Campbell *et al.*, 1996). This prediction is supported by some specific tests of changes in competition in response to increased temperature, and it is also apparent that the rate at which seasonal switches in competitive dominance of C_3 and C_4 species occur can be strongly dependent on the prevailing fertility or disturbance regime. For example, Harris *et al.* (1981a, b) found marked switches in the competitive ability of *Paspalum dilatatum* and *L. perenne* when transferred between high (24°C day/18°C night) and low (14°C day/8°C night) temperatures, with the greater advantage to *P. dilatatum* in the warm-temperature regime. However, because of marked suppression at low temperature under infrequent cutting, *P. dilatatum* showed only a slow increase in competitive dominance over ryegrass after transfer from low to high temperatures, whereas under frequent cutting *P. dilatatum* became rapidly dominant. Similarly, Tow *et al.* (1997a, b, c) recorded consistently superior growth of *Digitaria eriantha* at high temperatures and of *Medicago sativa* at low temperatures in both mixtures and monocultures, giving rise to complementary growth patterns in the field. Both the transfer of N from the legume to the grass and seasonal switches in dominance associated with changing temperature were identified as factors allowing coexistence of these species (Tow *et al.*, 1997c). However, *M. sativa* was least competitive

under very wet or dry conditions at summer temperatures, a situation aggravated by summer dominance of *D. eriantha*. Here it was emphasized that grazing management should be used to reduce grass dominance and ensure a satisfactory grass–legume balance. Cook *et al.* (1976) also found that, even though *L. perenne* (C_3) and *Bothriochloa macra* (C_4) had markedly different responses to temperature (with *L. perenne* superior at 16/10°C and 23/17°C and *B. macra* superior at 31/25°C), the *L. perenne* competed more successfully with *B. macra* only under conditions of high fertility and high moisture availability up to 23/17°C.

In existing temperate–subtropical climate transition zones, there is therefore likely to be an increasing prevalence of tropical grasses and more restricted opportunities for introducing improved temperate legumes and grasses into pastures. In existing temperate zones, there is likely to be an increasing prevalence of subtropical grass and consequently a greater suppression of clover and desirable temperate grass species (Campbell *et al.*, 1996; Coffin and Lauenroth, 1996). There is already evidence of an increased subtropical grass distribution in New Zealand associated with warmer temperatures (Field and Forde, 1990), a trend possibly contributed to by lower fertilizer inputs over the same period. Without specific intervention by managers, this will mean a greater dominance of C_4 species in spring and late summer periods, where at present the temperature conditions favour a switch in competitive balance back to temperate species. The competitive ability of C_4 species would also be favoured over C_3 species where there were more rapid rates of soil drying at high temperatures, due to the higher water use efficiency of C_4 species.

We anticipate that there will be a change in management of intensive temperate pastures to reduce these subtropical grass invasions, with managers increasing the frequency of pasture renewal with temperate species, altering grazing management to reduce stock damage in winter and maintaining a greater herbage mass in summer, using more N fertilizer to maintain grass tiller densities and using summer-active pasture species, such as *F. arundinacea*, *Dactylis glomerata* and *Cichorium intybus*, as well as pasture plant breeders developing more competitive ryegrass species (Campbell, 1996).

In legume-based pastures, the increase in C_4 grasses would be likely to result in a competitive suppression of clover, with a consequent decline in N_2-fixation rates and a reduction in the long-term availability of symbiotically fixed N (Campbell *et al.*, 1996). Additional changes in competitive interactions are expected to result from altered plant phenology, due to a longer growing period, an altered period of flowering and damaging effects of supra-optimal temperatures on temperate species in warm seasons. For example, additional heating of 2.5°C above fluctuating ambient levels reduced the productivity of *L. perenne* in Swiss grassland by 52%, and resulted in enhanced N concentration in tissue but no reduction in CO_2 uptake capacity (Nijs *et al.*, 1996). In temperate pastures, these effects will enhance existing seasonal switches in competitive balance in favour of warm-season species, such as clover.

Increased soil N availability associated with higher mineralization rates and higher N_2-fixation rates at warmer soil temperatures would increase the dominance of grass over legumes (Schwinning and Parsons, 1996a, b). A higher herbage mass with more intense competitive interactions would result unless there were an accompanying increase in defoliation pressure (Campbell and Grime, 1992). The longer-term trends in soil resource availability resulting from soil warming are complex and difficult to predict. The system could also experience higher N losses and smaller N pools (Thornley and Cannell, 1997). Here effects of elevated CO_2 and changes in water availability on N cycling will determine the resulting competition in ecosystems.

Increased herbage consumption rates due to a higher incidence of subtropical pests would be predicted to lower competitive interactions due to removal of standing biomass. Here, though, differential sensitivity of some species is expected to alter competitive balances in favour of certain more tolerant species. New pests are likely to spread into some regions (Sutherst *et al.*, 1996), increasing the costs of pest management in farming systems. Counterbalancing this effect, an accompanying increase in the incidence of pests and diseases of livestock may reduce grazing pressure. The magnitude of these effects is potentially large, but there is insufficient evidence to predict the ultimate consequences for pasture ecosystems.

In summary, there is considerable evidence to support the prediction of increased C_4 competitive dominance in response to increases in temperature. There may be several management tools that can be used to offset or optimize these changes in pastures by manipulating fertility or disturbance. The longer-term implications of increased prevalence of

diseases and insects and changes in the role of legumes at higher temperatures have not been studied in similar detail. Furthermore, the overall change in competition intensity with higher temperatures and accompanying longer-term changes in resource and disturbance levels in pastures have not been determined.

Effects of Changing Rainfall

Background

Warmer temperatures are predicted to lead to a more vigorous global hydrological cycle (Kattenberg *et al.*, 1995). This is expected to mean more severe floods and/or droughts in some places and less severe floods and/or droughts in other places. Several climate models indicate an increase in precipitation intensity, suggesting a possibility for more extreme rainfall events. Knowledge is currently not sufficient to provide precise predictions for regions; however, there is an expectation of increased precipitation and soil moisture at high latitudes in winter, and in most cases the increases extend well into mid-latitudes (Kattenberg *et al.*, 1995). The expected changes in precipitation are smaller if it is assumed that there will be an increased production of aerosols (minute particles in the atmosphere) resulting from fossil-fuel burning.

Soil moisture is a more relevant parameter for assessing the impacts of changes in the hydrological cycle on vegetation than precipitation, since it incorporates the integrated effects of changes in precipitation, evaporation and runoff. However, current models do not predict global soil moisture well, because of the oversimplified land-surface parameterization schemes in these models. Most models predict increased soil moisture in high northern latitudes in winter and a drier surface in summer in northern mid-latitudes (Kattenberg *et al.*, 1995).

Potential effects on pastures

As with temperature, the importance of water availability for pastures is well documented and the reader is referred to reviews on general aspects of water relations for further background (e.g. Turner and Begg, 1978; Kramer and Boyer, 1995). This availability of water is a critical factor determining the natural distribution and production of grasslands (Le Houerou and Hoste, 1977; Webb *et al.*, 1983; Sala *et al.*, 1988; Stephenson, 1990) and has a major impact on the annual productivity and composition of sown pastures. Therefore uncertainties about future rainfall patterns significantly constrain our capacity to predict future change in competition and succession in pastures. In the absence of precise prediction of future seasonal and regional patterns of rainfall and soil water availability, the implications of both decreases (Table 12.4a) and increases (Table 12.4b) in rainfall must be assessed.

In general, under water stress, pasture plants show more rapid adjustments in morphology, such as reduced leaf expansion rates or tillering, than in physiological processes, such as photosynthesis rates (Turner and Begg, 1978; Cornic and Massacci, 1996). Following severe drought, there is an opportunity for increased recruitment of seeds from the seed bank in gaps (Specht and Clifford, 1991). The population could also be shifted more towards drought-tolerant or annual species. In contrast, an increase in rainfall and soil water content can increase plant growth and development rates (see Table 12.4b), particularly in existing seasonally dry environments. Here, there is the likelihood of a shift to species characteristic of more mesic conditions and a shift from annuals to short-lived perennials following high summer rainfall events (Robertson, 1987).

Generally, C_4 species display greater water use efficiency (Kramer and Boyer, 1995), giving a selective advantage relative to C_3 species when exposed to limited water-supplies. Other adaptations can be recognized in pasture plants conferring tolerance of water deficits, including rapid curtailment of leaf area expansion in response to water deficit in order to limit water loss, leaf shedding, sclerophylly, leaf reorientation or rolling, hairiness, waxiness, root : shoot ratio and osmotic potential adjustment and prostrate growth (Turner and Begg, 1978; Turner, 1994). In contrast, flooding resistance usually includes arenchyma in roots, reduced root : shoot ratio, elongation to avoid submergence and increased plant height (Jackson and Drew, 1984; Blom *et al.*, 1990). Although in general these adaptations to tolerating drought and flooding are distinct, there is some evidence of a positive relationship between tolerance of both drought and flooding for populations of *P. dilatatum* from the pampas of Argentina (Loreti and Oesterheld, 1996).

Reduced soil water availability also results in a reduction in the uptake of N, P and other nutrients by plants (see Table 12.4a), and reduced growth observed as a result of moderate water deficits can, in part, arise from reduced availability of mineral nutrients in dry upper layers of soil (Turner and Begg, 1978). With extra rainfall and the probability of more intense rainfall events, there is also the potential for flooding and hypoxia reducing plant growth rates, as well as increased treading damage and soil erosion on steeper land (see Table 12.4b).

Pasture growth rates are reduced as a result of water stress, resulting in an imbalance of livestock demand to feed supply. During drought, there can be excessive utilization rates in grazed pasture systems, with long-term damage to pasture plants under prolonged drought (see Table 12.4a). Also, a reduction in feed supply can accentuate the damage caused by insect populations.

Implications for competition

Some shifts in competitive relations in pasture ecosystems predicted on the basis of effects of altered rainfall outlined in Table 12.4 include the following:

- A decreased intensity of competition for light, water and nutrients in seasons where altered rainfall contributes to greater moisture stress or waterlogging, but an increased intensity where extra rainfall reduces summer moisture deficit.
- An increased competitive advantage to C_4 species and drought-tolerant species and a decreased competitive advantage to clover with an increased frequency of mild water stress.
- A decreased competitive suppression of seedling recruitment in autumn with lower summer water deficits, and a similar decrease in spring with increased trampling damage in wet winter conditions.

Where warm-season rainfall decreases in a region that already experiences a summer moisture deficit, we predict a lower overall live herbage mass and N availability and therefore reduced competitive interactions in the sward. There is evidence of lower herbage mass from numerous experiments manipulating drought, and also some evidence of reduced competition (e.g. Tow *et al.*, 1997b). Small-statured individual pasture plant genotypes or species experiencing competitive suppression can

Table 12.4. Some potential effects of changes in rainfall on plant development, resources and disturbance.

Plant development	Resources	Disturbance
(a) Decreased rainfall Decreased plant-available water leading to: • Reduced cell expansion and division (Turner and Begg, 1978) • Reduced stomatal conductance, decreased transpiration (Kramer and Boyer, 1995) • Osmotic adjustment and increased cell solute concentration (Kramer and Boyer, 1995) • Reduced leaf area development, reduced tillering and reduced productivity (Turner and Begg, 1978) • Reduced root growth, increased root : shoot ratio and deeper rooting (Turner and Begg, 1978) • Reduced protein synthesis (Kramer and Boyer, 1995) • Reduced photosynthesis and respiration with severe stress (Cornic and Massacci, 1996) • Reduced N_2-fixation (Huang et al., 1975a, b; Kramer and Boyer, 1995) • Increased senescence • Accelerated phenology and flowering under mild water stress but reductions in severe stress (White et al., 1997) • Increased recruitment from seed bank **(b) Increased rainfall** Increased plant-available water, especially in summer-dry environments, leading to: • Increased leaf area development, increased tillering • Increased root growth • Reduced root : shoot ratio • Increased productivity • Shallower rooting • Greater rates of recruitment from seed banks	• Reduction in soil water leading to reduced availability and uptake of N and other mineral nutrients (Kramer and Boyer, 1995) • Reductions in water in upper soil layers leading to decreased microbial activity and decomposition rates • Increased nutrient supply following rainfall in drier seasons • Flooding of soils during wet periods leading to decreased diffusion of O_2 to roots, hypoxia and shoot dehydration (Kramer and Boyer, 1995)	• Excessive grazing pressure during droughts unless there is destocking • Increased insect herbivore damage with low herbage mass during drought • Soil erosion • Increased damage to plants by livestock treading and pugging in wet conditions

show the most variable response to fluctuating moisture stress and show less increase in yield as growth conditions improve (Harris and Sedcole, 1974). Consequently, competitive interactions will probably change to involve root competition at greater depths in the soil profile and altered shoot architecture, light distribution patterns and shoot competition. The specific effects on nutrient availability and long-term changes require further study.

The predicted shift in favour of drought-tolerant or avoiding species following drought is expected to occur both through natural succession and through human intervention. An increase in warm-season rainfall usually favours C_4 species over C_3 species, with a resulting increasing competitive suppression of white clover (Campbell et al., 1996). In areas experiencing an increased frequency of drought, change is expected to be implemented in the form of greater use of drought-tolerant species, altered grazing management, greater use of conserved feed, modified farming systems to better fit seasonal feed availability and, in some cases, more extensive use of irrigation. Competition for water between annual clover and grass can reduce clover establishment, especially in semi-arid pastures. A possible strategy for reducing this competition under low water-supply is to graze the perennial to minimize transpiration and thus extend the availability of water to the clover seedlings. For example, there is evidence that growth of clover seedlings in phalaris swards is reduced by a combination of competition for water, nitrate and light, but that clover can be released from this competition by defoliation of the grass (Dear et al., 1998). Here, grazing reduced scavenging for water by perennial grass roots and was an important factor in slowing the drying of the soil surface in both March and May. In annual Mediterranean-type pastures, the conditions favouring clover appear to be high autumn rainfall and low biomass in winter, reducing competition from non-legume species, followed by heavy rain in early spring (Arnold and Anderson, 1987).

A larger proportion of gaps in pasture cover following drought would also be expected to increase seedling recruitment and reduce competitive suppression of seedlings in autumn following rain. Perennial grasses can withstand droughts by assuming dormancy, whereas annual species avoid moisture stress by completing life cycles during brief mesic conditions resulting from occasional rain, and then surviving the remaining period as seeds

(Aguado-Santacruz and Garcia-Moya, 1998). In contrast, if rainfall increases in the cool season in winter wet soil, the increase in damage to vegetation and disruption of soil structure would be anticipated to reduce pasture production and competition intensity overall in the spring. This could result in a greater proportion of clover in the sward and of spring-germinating species.

In summary, there is some evidence to support the predicted changes in competition with altered rainfall but the final outcome on pasture systems is highly dependent on the precise pattern of rainfall change and changes in management in response to drought. As noted above, the effects of changes in rainfall on competition in pastures will also be modified by prevailing changes in elevated CO_2 and temperature. The evidence suggests that an increased intensity of rainfall events and greater rainfall variability in temperate and subtropical regions would increase the proportions of C_4 species in pastures, reinforcing effects of higher temperatures. Given the long history of managing drought in pastoral systems, farming industries are likely to be highly responsive to changes in rainfall patterns in some cases, and this altered management will significantly modify the consequences of rainfall change for competition and succession.

Sensitivities of Different Pasture Systems

From the foregoing, it is apparent that there are several current weaknesses in predicting the consequences of global climate change for competition in pastures. Of overriding importance is the knowledge that, over time, pasture managers will modify their management inputs to attempt to minimize negative impacts of climate change and maximize any potential benefits (Gifford et al., 1996b; Dale, 1997), and these adaptations will have implications for competition and succession processes. As a consequence, the final impacts will differ from those where there was no adjustment in management.

In addition, there is a significant lack of data to test long-time-scale changes in resource availability and disturbance resulting from changes in global climate. The response of these complex systems cannot easily be predicted on the basis of oversimplified experimental set-ups; see, for example, discussion of effects of elevated CO_2 on microbial biomass (Kampichler et al., 1998). Inertia is expected to result in a considerable delay in the full effects being expressed (Milchunas and Lauenroth, 1995). In this regard, there have been virtually no tests of the combined effects of changes in CO_2, UV-B, temperature and rainfall on competition or succession in pastures or other terrestrial ecosystems (Krupa and Kickert, 1993; Sullivan, 1997). For example, plant competition might change markedly under current climatic change with simultaneous enhancement of atmospheric CO_2 and solar UV-B radiation (Rozema et al., 1997c). These interactions are not easily predicted (Caldwell et al., 1995) and the parameterization of competitive outcomes remains a key uncertainty in all current pasture models predicting global climate change effects. This issue is especially important, given the overriding importance of environmental conditions in determining the magnitudes of responses. Additionally, there is still significant uncertainty surrounding regional climate scenarios with global climate change, and this limits our current capacity to forecast precise effects on competition and succession for specific pastoral systems.

Based on the foregoing analyses, and with these limitations firmly in mind, we have made some tentative predictions of the sensitivity of certain sown pasture systems and the implications for management and succession (Table 12.5). Overall, in the absence of management intervention, we would predict an increasing intensity of competition for light, water and nutrients in communities where there is increased resource supply and/or decreased disturbance levels due to the combined effects of global climate change. This would result in higher rates of competitive exclusion, but may or may not cause an increased dominance and a reduction in species richness, as the outcome is dependent on whether dominant or subordinate species are favoured most by the particular climatic change (see, for example, Potvin and Vasseur, 1997). Shifts in competitive advantage to certain groups of species are also predicted.

It is evident from Table 12.5 that some of the concurrent global climate changes would have opposite effects on competition and succession; for example, elevated CO_2 would encourage legume performance in temperate systems, whereas elevated UV-B and decreased rainfall would have the opposite effect. Our prediction of an increased intensity of competition is consistent with CENTURY model predictions of a general increase in worldwide grassland productivity and reduced grassland

C losses as a result of the combined effects of CO_2 and climate change (Parton et al., 1995).

As stated, pastoral managers are expected to intervene, using resources and disturbance to manipulate some of the direct effects of global climate change, and in some cases to dampen the potential effects on competition and succession (Campbell et al., 1997). We predict (see Table 12.5) that pasture managers will act to modify many of the resource inputs and grazing disturbance inputs under their control. In order to manage competition to optimize returns from a farming enterprise in future, these managers will require a better understanding of how resource inputs and grazing disturbance would need to be modified with climatic change. This analysis of potential implications of global climate change for resources, disturbance and competition and succession must be extended in the future to identify the specific changes required in these management inputs. For example, additional N fertilizer or manipulation of grazing management can be used to alter resource availability and the carbon and water balances of pastures (Thornley and Cannell, 1997; Ham and Knapp, 1998). For economic and environmental reasons, the input of N fertilizer in intensive European pastures is expected to decrease in the future, favouring clover and accentuating some CO_2-induced stimulation of clover growth in mixed swards (Schenk et al., 1997).

Managers also have the potential to select pasture cultivars that are better adapted to climate change, to optimize production and to prevent less desirable plants from invading productive pastures (e.g. Gifford et al., 1996b). This is expected to include improvement of the temperature, UV-B and drought tolerance of species of Lolium, Festuca, Dactylis, Bromus, Trifolium and other temperate species, as well as improved forage quality and agronomic suitability of C_4 species. This introduction of improved plant types might better maximize resource use efficiency of mixtures (Vandermeer et al., 1998). The effects of climate change identified in Tables 12.1–12.4 point to particular components of plant development that could be modified through forage plant genetic improvement to result in cultivars better suited to future climates.

In analysing which of the systems in Table 12.5 will be particularly sensitive in the future to global climate change, it is helpful to note that a basic trade-off exists between the productivity and resilience of plant-based systems. The most produc-

Table 12.5. Predicted sensitivity of some sown pasture systems, changes in competition and consequences for management, vegetation structure and succession.

System	Some potential changes in competition	Possible management intervention	Some potential changes in vegetation structure and succession
Legume-based intensive lowland pastures	Increased intensity of competition and increased competitive advantage to legumes due to elevated CO_2 and temperature, but offset to unknown extent by elevated UV-B radiation and increased water stress at sites with declining rainfall. Increase in competitive ability of invading C_4 species due to altered temperature, rainfall and UV-B radiation	Increased stocking rate Sowing of UV-tolerant germ-plasm Irrigation	Shift to greater legume proportion and production, but offset by oscillations in grass : legume due to fluctuating N availability
Multispecies pasture systems	Increased intensity of competition due to elevated CO_2, and increased long-term N availability due to higher temperature; greater competitive advantage to C_3 in mesic sites due to elevated CO_2	Increased stocking rate	Shift to dominance by fewer species with consequent reduction in resilience to extreme events
Semi-arid annual pastures	Increased intensity of competition and increased competitive advantage to late-season species due to CO_2-induced increases in seasonal water-supply	Increased stocking rate	Shift from annuals to short-lived perennials
Low-fertility hill-land pastures	Possible reduction in intensity of competition due to increased N limitation and increased soil loss in higher-rainfall areas	Reduced stocking rate	Decline in productivity, increase in pasture diversity, increase in woody-weed invasion
Subtropical sown pastures	Increased intensity of competition and increased competitive advantage to C_4 grasses due to higher temperature, especially in drier areas or areas with an increase in summer rainfall	Increased frequency of cultivation to establish temperate grasses and legumes Higher fertilizer inputs, especially N Increased stocking rate	Increased prevalence of C_4 species without management intervention

tive agroecosystems are often the simplest (Vandermeer *et al.*, 1998), but these can be the least resilient to disturbance and perturbation, because of increased economic and ecological instability (e.g. Clawson, 1985). Given the need for highly resilient agroecosystems to feed an expanding human population, there is likely to be a trend to more complex pastoral ecosystems in the future (Vandermeer *et al.*, 1998), and this will include an increase in the area of pastures where multispecies mixtures are used in complex pastures. We suggest that competition will become an increasingly important issue as more species and genotypes are included in pasture ecosystems, and there must be greater emphasis on management to prevent competitive exclusion of the component species. Based on our analysis above, these multispecies pasture systems will be among the most sensitive systems to effects of global climate change in terms of competition and succession, as they have the most precise specifications for maintaining high species richness. We also predict a high sensitivity to climate change in legume-based lowland and semi-arid annual pastures, because the fastest rates of successional change due to climate change are predicted for productive but intensively disturbed communities (Campbell and Grime, 1993). This sensitivity arises from the high rates of turnover of individuals and frequent perturbations to dominance in these systems.

Conclusions

Global climate change is likely to significantly alter competition and succession in pastures, both through initial effects on plant development and through longer-term effects on soil resources and disturbance agents. The effects on resource supply and disturbance are predicted to be of greater overall consequence for ecosystem structure and function. The short-term responses of plant development to climate change variables are better understood than the long-term effects on resources and disturbance.

Based on available evidence, we predict an increased intensity of competition for light, water and nutrients in pastures as a result of increasing resource availability due to increased CO_2, rainfall and temperature, but this will be offset to some extent by effects of increased UV-B radiation and water stress. This increase in competition intensity is predicted to generally reduce species richness in

pastures and to result in the loss of some desirable species from pastures, reducing the resilience of the communities to subsequent perturbations.

Shifts in competitive interactions between pasture components are also predicted, with a general increase in competitive success of legumes and C_4 grasses. Rates of change are expected to be fastest in productive, intensively managed pastures. The effects of some global climate change variables appear to operate in different directions from others in this respect. For example, clover competitive ability is predicted to be increased by elevated CO_2 and temperature, but this might be limited by an accompanying increase in UV-B radiation and declining rainfall.

Driving climate variables are not yet sufficiently predictable to make reliable forecasts of future ecosystem responses (e.g. Scholes and van Breemen, 1997). Changes in rainfall patterns remain a key unknown variable. These rainfall effects might be of overriding importance in determining pasture productivity, competitive interactions and succession in future, but they are the least well defined at present.

Additional progress in this area can be achieved by: (i) making use of the expanding theoretical understanding of the fundamental effects of resources, disturbance and plant development on competition processes; and (ii) undertaking further research to reduce significant uncertainties surrounding interactive effects of global climate change on resources, disturbance and plant development. Mathematical models will be critical in assessing the long-term effects of changes in resource availability and disturbance on competitive interactions and succession, as these cannot be assessed by short-term tests (Rastetter, 1996). Data are needed on the effects of interacting global climate change variables (e.g. $CO_2 \times$ water; UV-B $\times$ water) on resource and disturbance levels in intact ecosystems. These integrated experiments will assist the modelling of long-term vegetation change processes, including competition and succession.

Human intervention will be a key determinant of future competitive interactions in pastures. A primary objective must be to identify how management inputs of resources (e.g. fertilizer), plant types (e.g. improved germ-plasm, novel species mixtures) and disturbance (e.g. grazing and land management) can be manipulated to optimize the outcomes of competitive interactions in pastures with ongoing global climate change. We cannot yet provide precise specifications of what is needed to optimize these competitive outcomes in pastures, because significant uncertainty remains about the magnitude of the effects of global climate change on competition and succession. This uncertainty must be reduced if pastoral industries are to adapt effectively to the future challenges that global climate change presents to the economic efficiency and security of food and fibre production.

Acknowledgements

This work contributes at the core research level to the Global Change and Terrestrial Ecosystems (GCTE) project of the International Geosphere–Biosphere Programme (IGBP). Funding was provided by the New Zealand Foundation for Research, Science and Technology (Contract Number C10632).

References

Ackerly, D.D. and Bazzaz, F.A. (1995) Plant growth and reproduction along CO_2 gradients: non-linear responses and implications for community change. *Global Change Biology* 1, 199–207.

Aguado-Santacruz, G.A. and Garcia-Moya, E. (1998) Environmental factors and community dynamics at the southernmost part of the North American Graminetum. *Plant Ecology* 135, 13–29.

Amthor, J.S. (1995) Terrestrial higher-plant response to increasing atmospheric $[CO_2]$ in relation to the global carbon cycle. *Global Change Biology* 1, 243–274.

Antonelli, F., Grifoni, D., Sabatini, F. and Zipoli, G. (1997) Morphological and physiological responses of bean plants to supplemental UV radiation in a Mediterranean climate. *Plant Ecology* 128, 127–136.

Arnold, G.W. and Anderson, G.W. (1987) The influence of nitrogen level, rainfall, seed pools, and pasture biomass on the botanical composition of annual pastures. *Australian Journal of Agricultural Research* 38, 339–354.

Ayres, P.G., Gunasekera, T.S., Rasanayagam, M.S. and Paul, N.D. (1996) Effects of UV-B radiation (280–320nm) on foliar saprotrophs and pathogens. In: Frankland, J.C., Magan, N. and Gadd, G.M. (eds) *Fungi and Environmental Change.* Cambridge University Press, Cambridge, UK, pp. 32–50.

Barnes, P.W., Flint, S.D. and Caldwell, M.M. (1995) Early-season effects of supplemented solar UV-B radiation on seedling emergence, canopy structure, simulated stand photosynthesis and competition for light. *Global Change Biology* 1, 43–53.

Bazzaz, F.A. (1990) The response of natural ecosystems to the rising global CO_2 levels. *Annual Review of Ecology and Systematics* 21, 167–196.

Bazzaz, F.A. and Garbutt, K. (1988) The response of annuals in competitive neighbourhoods: effects of elevated CO_2. *Ecology* 69, 937–946.

Bazzaz, F.A. and McConnaughay, K.D.M. (1992) Plant–plant interactions in elevated CO_2 environments. *Australian Journal of Botany* 40, 547–563.

Beggs, C.J. and Wellman, E. (1994) Photocontrol of flavonoid biosynthesis. In: Kendrick, R.E. and Kronenberg, G.H.M. (eds) *Photomorphogenesis in Plants.* Kluwer Academic Publishers, Dordrecht, The Netherlands, pp. 733–750.

Björn, L.O., Callaghan, T., Gehrke, C., Gunnarsson, T., Holmgren, B., Johanson, U., Snogerup, S., Sonesson, M., Sterner, O. and Yu, S.-G. (1997) Effects on subarctic vegetation of enhanced UV-B radiation. In: Lumsden, P.J. (ed.) *Plants and UV-B: Responses to Environmental Change.* Cambridge University Press, Cambridge, UK, pp. 233–246.

Blom, C.W.P.M., Boegemann, G.M., Laan, P., Van der Sman, A.J.M., Van der Steeg, H.M. and Voesenek, L.A.C.J. (1990) Adaptations to flooding in plants from river areas. *Aquatic Botany* 38, 29–47.

Bornman, J.F., Reuber, S., Cen, Y.-P. and Weissenbock, G. (1997) Ultraviolet radiation as a stress factor and the role of protective pigments. In: Lumsden, P.J. (ed.) *Plants and UV-B: Responses to Environmental Change.* Cambridge University Press, Cambridge, UK, pp. 157–168.

Bowes, G. (1996) Photosynthetic responses to changing atmospheric carbon dioxide concentration. In: Baker, N.R. (ed.) *Photosynthesis and the Environment.* Kluwer Academic Publishers, Dordrecht, The Netherlands, pp. 387–407.

Caldwell, M.M. (1997) Alterations in competitive balance. In: Lumsden, P.J. (ed.) *Plants and UV-B: Responses to Environmental Change.* Cambridge University Press, Cambridge, UK, pp. 305–315.

Caldwell, M.M. and Flint, S.D. (1993) Implications of increased solar UV-B for terrestrial vegetation. In: Chanin, M.L. (ed.) *The Role of the Stratosphere in Global Change.* Springer-Verlag, Heidelberg, Germany, pp. 495–516.

Caldwell, M.M. and Flint, S.D. (1994) Stratospheric ozone reduction, solar UV-B radiation and terrestrial ecosystems. *Climatic Change* 28, 375–394.

Caldwell, M.M., Teramura, A.H. and Tevini, M. (1989) The changing solar ultraviolet climate and the ecological consequences for higher plants. *Trends in Ecology and Evolution* 4, 363–367.

Caldwell, M.M., Teramura, A.H., Tevini, M., Bornman, J.F., Björn, L.O. and Kulandaivelu, G. (1995) Effects of increased solar ultraviolet radiation on terrestrial plants. *Ambio* 24, 166–173.

Campbell, B.D. (1996) Climate change and pastures. In: *Proceedings of the Ruakura Dairy Farmers Conference.* Hamilton, New Zealand, pp. 67–74.

Campbell, B.D. and Grime, J.P. (1989) A comparative study of plant responsiveness to the duration of episodes of mineral nutrient enrichment. *New Phytologist* 112, 261–267.

Campbell, B.D. and Grime, J.P. (1992) An experimental test of plant strategy theory. *Ecology* 73, 15–29.

Campbell, B.D. and Grime, J.P. (1993) Prediction of grassland plant responses to global change. In: *Proceedings of the XVII International Grassland Congress.* Palmeston North, New Zealand, pp. 1109–1118.

Campbell, B.D. and Hart, A.L. (1996) Competition between grasses and *Trifolium repens* with elevated atmospheric CO_2. In: Körner, C. and Bazzaz, F. (eds) *Carbon Dioxide, Populations, and Communities.* Academic Press, San Diego, USA, pp. 301–331.

Campbell, B.D., Grime, J.P., Mackey, J.M.L. and Jalili, A. (1991) The quest for a mechanistic understanding of resource competition in plant communities: the role of experiments. *Functional Ecology* 5, 241–253.

Campbell, B.D., Laing, W.A. and Newton, P.C.D. (1993) Variation in the response of pasture plants to carbon dioxide. In: *Proceedings of the XVII International Grassland Congress.* Palmerston North, New Zealand, pp. 1125–1126.

Campbell, B.D., McKeon, G.M., Gifford, R.M., Clark, H., Stafford Smith, D.M., Newton, P.C.D. and Lutze, J.L. (1996) Impacts of atmospheric composition and climate change on temperate and tropical pastoral agriculture. In: Bouma, W.J., Pearman, G.I. and Manning, M.R. (eds) *Greenhouse: Coping with Climate Change.* CSIRO Publishing, Melbourne, Australia, pp. 171–189.

Campbell, B.D., Stafford Smith, D.M. and McKeon, G.M. (1997) Elevated CO_2 and water supply interactions in grasslands: a pastures and rangelands management perspective. *Global Change Biology* 3, 177–187.

Campbell, C.A., Biederbeck, V.O. and Warder, F.G. (1973) Influence of simulated fall and spring conditions on the soil system: III. Effect of method of simulating spring temperatures on ammonification, nitrification and microbial populations. *Soil Science Society of America Proceedings* 37, 382–386.

Canadell, J.G., Pitelka, L.F. and Ingram, J.S.I. (1996) The effects of elevated $[CO_2]$ on plant–soil carbon below-ground: a summary and synthesis. *Plant and Soil* 187, 391–400.

Casella, E. and Soussana, J.F. (1997) Long-term effects of CO_2 enrichment and temperature increase on the carbon balance of a temperate grass sward. *Journal of Experimental Botany* 48, 1309–1321.

Clawson, D.L. (1985) Harvest security and intraspecific diversity in traditional tropical agriculture. *Economic Botany* 39, 56–67.

Coffin, D.P. and Lauenroth, W.K. (1996) Transient responses of North-American grasslands to changes in climate. *Climatic Change* 34, 269–278.

Cook, S.J., Lazenby, A. and Blair, G.J. (1976) Comparative responses of *Lolium perenne* and *Bothriochloa macra* to temperature, moisture, fertility and defoliation. *Australian Journal of Agricultural Research* 27, 769–778.

Corlett, J.E., Stephen, J., Jones, H.G., Woodfin, R., Mepsted, R. and Paul, N.D. (1997) Assessing the impact of UV-B radiation on the growth and yield of field crops. In: Lumsden, P.J. (ed.) *Plants and UV-B: Responses to Environmental Change.* Cambridge University Press, Cambridge, UK, pp. 195–211.

Cornic, G. and Massacci, A. (1996) Leaf photosynthesis under drought stress. In: Baker, N.R. (ed.) *Photosynthesis and the Environment.* Kluwer Academic Publishers, Dordrecht, The Netherlands, pp. 347–366.

Cotrufo, M.F. and Ineson, P. (1996) Elevated CO_2 reduces field decomposition rates of *Betula pendula* Roth. leaf litter. *Oecologia* 106, 525–530.

Cotrufo, M.F., Ineson, P. and Scott, A. (1998) Elevated CO_2 reduces the nitrogen concentration of plant tissues. *Global Change Biology* 4, 43–54.

Dai, Q., Peng, S., Chavez, A.Q. and Vergara, B.S. (1995) Effects of UV-B radiation on stomatal density and opening in Rice (*Oryza sativa* L.). *Annals of Botany* 76, 65–70.

Dale, V.H. (1997) The relationship between land-use change and climate change. *Ecological Applications* 7, 753–769.

Day, T.A., Martin, G. and Vogelmann, T.C. (1993) Penetration of UV-B radiation in foliage: evidence that the epidermis behaves as a non-uniform filter. *Plant, Cell and Environment* 16, 735–741.

Dear, B.S., Cocks, P.S., Wolfe, E.C. and Collins, D.P. (1998) Established perennial grasses reduce the growth of emerging subterranean clover seedlings through competition for water, light, and nutrients. *Australian Journal of Agricultural Research* 49, 41–51.

Diaz, S. (1995) Elevated CO_2 responsiveness, interactions at the community level and plant functional types. *Journal of Biogeography* 22, 289–295.

Diaz, S., Grime, J.P., Harris, J. and McPherson, E. (1993) Evidence of a feedback mechanism limiting plant response to elevated carbon dioxide. *Nature* 364, 616–617.

Epstein, H.E., Lauenroth, W.K., Burke, I.C. and Coffin, D.P. (1998) Regional productivities of plant species in the Great Plains of the United States. *Plant Ecology* 134, 173–195.

Ernst, W.H.O., van de Staaij, J.W.M. and Nelissen, H.J.M. (1997) Reaction of savanna plants from Botswana on UV-B radiation. *Plant Ecology* 128, 162–170.

Etherington, J.R. (1982) *Environment and Plant Ecology*, 2nd edn. John Wiley & Sons, Chichester, UK.

Falk, S., Maxwell, D.P., Laudenbach, D.E. and Huner, N.P.A. (1996) Photosynthetic adjustment to temperature. In: Baker, N.R. (ed.) *Photosynthesis and the Environment.* Kluwer Academic Publishers, Dordrecht, The Netherlands, pp. 367–385.

Ferris, R., Nijs, I., Behaeghe, T. and Impens, I. (1996) Contrasting CO_2 and temperature effects on leaf growth of perennial ryegrass in spring and summer. *Journal of Experimental Botany* 47, 1033–1043.

Field, C.B., Chapin, F.S., III, Chiariello, N.R., Holland, E.A. and Mooney, H.A. (1996) The Jasper Ridge CO_2 experiment: design and motivation. In: Koch, G.W. and Mooney, H.A. (eds) *Carbon Dioxide and Terrestrial Ecosystems.* Academic Press, London, UK, pp. 121–145.

Field, T.R.O. and Forde, M.B. (1990) Effect of climate warming on the distribution of C4 grasses in New Zealand. *Proceedings of the New Zealand Grasslands Association* 51, 47–50.

Fitter, A.H., Graves, J.D., Wolfenden, J., Self, G.K., Brown, T.K., Bogie, D. and Mansfield, T.A. (1997) Root production and turnover and carbon budgets of two contrasting grasslands under ambient and elevated atmospheric carbon dioxide concentrations. *New Phytologist* 137, 247–255.

Fitter, A.H., Graves, J.D., Self, G.K., Brown, T.K., Bogie, D.S. and Taylor, K. (1998) Root production, turnover and respiration under two grassland types along an altitudinal gradient: influence of temperature and solar radiation. *Oecologia* 114, 20–30.

Fraser, L.H. and Grime, J.P. (1998) Top-down control and its effect on the biomass and composition of three grasses at high and low soil fertility in outdoor microcosms. *Oecologia* 113, 239–246.

Fredeen, A.L., Randerson, J.T., Holbrook, N.M. and Field, C.B. (1997) Elevated atmospheric CO_2 increases water availability in a water-limited grassland ecosystem. *Journal of the American Water Resources Association* 33, 1033–1039.

Gifford, R.M., Barrett, D.J., Lutze, J.L. and Samarakoon, A.B. (1996a) Agriculture and global change: scaling direct carbon dioxide impacts and feedbacks through time. In: Walker, B.H. and Steffen, W.L. (eds) *Global Change and Terrestrial Ecosystems.* Cambridge University Press, Cambridge, UK, pp. 229–259.

Gifford, R.M., Campbell, B.D. and Howden, S.M. (1996b) Options for adapting agriculture to climate change: Australian and New Zealand examples. In: Bouma, W.J., Pearman, G.I. and Manning, M.R. (eds) *Greenhouse: Coping with Climate Change.* CSIRO Publishing, Melbourne, Australia, pp. 399–416.

Gold, W.G. and Caldwell, M.M. (1983) The effects of ultraviolet-B radiation on plant competition in terrestrial ecosystems. *Physiologia Plantarum* 58, 435–444.

Goldberg, D. (1994) Influence of competition at the community level: an experimental version of the null models approach. *Ecology* 75, 1503–1506.

Goldberg, D. and Novoplansky, A. (1997) On the relative importance of competition in unproductive environments. *Journal of Ecology* 85, 409–418.

Grace, J.B. (1993) The effects of habitat productivity on competition intensity. *Trends in Ecology and Evolution* 8, 229–230.

Greer, D.H., Laing, W.A. and Campbell, B.D. (1995) Photosynthetic responses of thirteen pasture species to elevated CO_2 and temperature. *Australian Journal of Plant Physiology* 22, 713–722.

Grime, J.P. (1977) Evidence for the existence of three primary strategies in plants and its relevance to ecological and evolutionary theory. *American Naturalist* 111, 1169–1194.

Grime, J.P. (1979) *Plant Strategies and Vegetation Processes.* John Wiley & Sons, Chichester, UK.

Ham, J.M. and Knapp, A.K. (1998) Fluxes of CO_2, water vapor, and energy from a prairie ecosystem during the seasonal transition from carbon sink to carbon source. *Agricultural and Forest Meteorology* 89, 1–14.

Harris, W. and Sedcole, J.R. (1974) Competition and moisture stress effects on genotypes in three grass populations. *New Zealand Journal of Agricultural Research* 17, 443–454.

Harris, W., Forde, B.J. and Hardacre, A.K. (1981a) Temperature and cutting effects on the growth and competitive interaction of ryegrass and paspalum I. Dry matter production, tiller numbers, and light interception. *New Zealand Journal of Agricultural Research* 24, 299–307.

Harris, W., Forde, B.J. and Hardacre, A.K. (1981b) Temperature and cutting effects on the growth and competitive interaction of ryegrass and paspalum II. Interspecific competition. *New Zealand Journal of Agricultural Research* 24, 309–320.

Hatcher, R.D. and Paul, N.D. (1994) The effect of UV-B radiation on herbivory of pea by *Autographa gamma.* *Entomologia Experimentalis et Applicata* 71, 227–233.

Hebeisen, T., Lüscher, A., Zanetti, S., Fischer, B.U., Hartwig, U.A., Frehner, M., Hendrey, G.R., Blum, H. and Nösberger, J. (1997) Growth response of *Trifolium repens* L. and *Lolium perenne* L. as monocultures and bi-species mixture to free air CO_2 enrichment and management. *Global Change Biology* 3, 149–160.

Huang, C.Y., Boyer, J.S. and Vanderhoef, L.N. (1975a) Acetylene reduction (nitrogen fixation) and metabolic activities of soybean having various leaf and nodule water potentials. *Plant Physiology* 56, 222–227.

Huang, C.Y., Boyer, J.S. and Vanderhoef, L.N. (1975b) Limitation of acetylene reduction (nitrogen fixation) by photosynthesis in soybean having low water potentials. *Plant Physiology* 56, 228–232.

Hungate, B.A., Chapin, F.S., III, Zhong, H., Holland, E.A. and Field, C.B. (1997) Stimulation of grassland nitrogen cycling under carbon dioxide enrichment. *Oecologia* 109, 149–153.

Hunt, R. and Lloyd, P.S. (1987) Growth and partitioning. *New Phytologist* 103, 235–250.

Hunt, R., Hand, D.W., Hannah, M.A. and Neal, A.M. (1991) Responses to CO_2 enrichment in 27 herbaceous species. *Functional Ecology* 5, 410–420.

Hunt, R., Hand, D.W., Hannah, M.A. and Neal, A.M. (1993) Further responses to CO_2 enrichment in British herbaceous species. *Functional Ecology* 7, 661–668.

IPCC (1995) Summary for policymakers. In: Houghton, J.T., Meira Filho, L.G., Callander, B.A., Harris, N., Kattenberg, A. and Maskell, K. (eds) *Climate Change 1995: the Science of Climate Change*. Cambridge University Press, Cambridge, UK, pp. 3–7.

Jackson, M.B. and Drew, M.C. (1984) Effects of flooding on growth and metabolism of herbaceous plants. In: Kozlowski, T.T. (ed.) *Flooding and Plant Growth*. Academic Press, London, UK, pp. 47–128.

Jackson, R.B., Luo, Y., Cardon, Z.G., Sala, O.E., Field, C.B. and Mooney, H.A. (1995) Photosynthesis, growth and density for the dominant species in a CO_2-enriched grassland. *Journal of Biogeography* 22, 221–225.

Jackson, R.B., Sala, O.E., Paruelo, J.M. and Mooney, H.A. (1998) Ecosystem water fluxes for two grasslands in elevated CO_2: a modeling analysis. *Oecologia* 113, 537–546.

Jansen, M.A.K., Gaba, V. and Greenberg, B.M. (1998) Higher plants and UV-B radiation: balancing damage, repair and acclimation. *Trends in Plant Science: Reviews* 3, 131–135.

Jongen, M., Jones, M.B., Hebeisen, T., Blum, H. and Hendry, G. (1995) The effects of elevated CO_2 concentrations on the root growth of *Lolium perenne* and *Trifolium repens* grown in a FACE system. *Global Change Biology* 1, 361–371.

Jordan, B.R. (1993) The molecular biology of plants exposed to ultraviolet-B radiation and the interaction with other stresses. In: Jackson, M.B. and Black, C.R. (eds) *Interacting Stresses on Plants in a Changing Climate*. Springer-Verlag, Heidelberg, Germany, pp. 153–170.

Kadmon, R. (1995) Plant competition along soil moisture gradients: a field experiment with the desert annual *Stipa capensis*. *Journal of Ecology* 83, 253–262.

Kampichler, C., Kandeler, E., Bardgett, R.D., Jones, T.H. and Thompson, L.J. (1998) Impact of elevated atmospheric CO_2 concentration on soil microbial biomass and activity in a complex, weedy field model ecosystem. *Global Change Biology* 4, 335–346.

Kattenberg, A., Giorgi, F., Grassl, H., Meehl, G.A., Mitchell, J.F.B., Stouffer, R.J., Tokioka, T., Weaver, A.J. and Wigley, T.M.L. (1995) Climate models – projections of future climate. In: Houghton, J.T., Filho Meira, L.G., Callander, B.A., Harris, N., Kattenberg, A. and Maskell, K. (eds) *Climate Change 1995: the Science of Climate Change*. Cambridge University Press, Cambridge, UK, pp. 285–357.

Kimball, B.A. (1983) Carbon dioxide and agricultural yield: an assemblage and analysis of 430 prior observations. *Agronomy Journal* 75, 779–788.

Klironomos, J.N., Rillig, M.C. and Allen, M.F. (1996) Below-ground microbial and microfaunal responses to *Artemisia tridentata* grown under elevated atmospheric CO_2. *Functional Ecology* 10, 527–534.

Knapp, A.K., Hamerlynck, E.P. and Owensby, C.E. (1993) Photosynthetic and water relations responses to elevated CO_2 in the C_4 grass *Andropogon gerardii*. *International Journal of Plant Sciences* 154, 459–466.

Körner, C. (1995) Towards a better experimental basis for upscaling plant responses to elevated CO_2 and climate warming. *Plant, Cell and Environment* 18, 1101–1110.

Körner, C. and Bazzaz, F.A. (1996) *Carbon Dioxide, Populations, and Communities*. Academic Press, San Diego, USA.

Kramer, P.J. and Boyer, J.S. (1995) *Water Relations of Plants and Soils*. Academic Press, San Diego, USA.

Krupa, S.V. and Kickert, R.N. (1993) The greenhouse effect: the impacts of carbon dioxide (CO_2), ultraviolet-B (UV-B) radiation and ozone (O_3) on vegetation (crops). *Vegetatio* 104/105, 223–238.

Larigauderie, A. and Körner, C. (1995) Acclimation of leaf dark respiration to temperature in alpine and lowland plant species. *Annals of Botany* 76, 245–252.

Lavorel, S. and Chesson, P. (1995) How species with different regeneration niches coexist in patchy habitats with local disturbances. *Oikos* 74, 103–114.

Leegood, R.C. and Edwards, G.E. (1996) Carbon metabolism and photorespiration: temperature dependence in relation to other environmental factors. In: Baker, N.R. (ed.) *Photosynthesis and the Environment*. Kluwer Academic Publishers, Dordrecht, The Netherlands, pp. 191–221.

Le Houerou, H.N. and Hoste, C.H. (1977) Rangeland production and annual rainfall relations in the Mediterranean basin in the African Sahelo-Sudanian zone. *Journal of Range Management* 30, 181–189.

Loreti, J. and Oesterheld, M. (1996) Intraspecific variation in the resistance to flooding and drought in populations of *Paspalum dilatatum* from different topographic positions. *Oecologia* 108, 279–284.

Lumsden, P. (1997) *Plants and UV-B: Responses to Environmental Change*, Cambridge University Press, Cambridge, UK.

Luo, Y., Jackson, R.B., Field, C.B. and Mooney, H.A. (1996) Elevated CO_2 increases below ground respiration in California grasslands. *Oecologia* 108, 130–137.

Lüscher, A., Hendrey, G.R. and Nösberger, J. (1998) Long-term responsiveness to free air CO_2 enrichment of functional types, species and genotypes of plants from fertile permanent grassland. *Oecologia* 113, 37–45.

McGuire, A.D., Melillo, J.M., Joyce, L.A., Kicklighter, D.W., Grace, A.L., Moore, B. and Vorosmarty, C.J. (1992) Interactions between carbon and nitrogen dynamics in estimating net primary productivity for potential vegetation in North America. *Global Biogeochemical Cycles* 6, 101–124.

McLellan, A.J., Law, R. and Fitter, A.H. (1997) Response of calcareous grassland plant species to diffuse competition: results from a removal experiment. *Journal of Ecology* 85, 479–490.

McWilliam, J.R. (1978) Response of pasture plants to temperature. In: Wilson, J.R. (ed.) *Plant Relations in Pastures*. CSIRO Publishing, Melbourne, Australia, pp. 17–34.

Madronich, S. (1993) The atmosphere and UV-B radiation at ground level. In: Young, A.R., Björn, L.O., Moan, J. and Nultsch, W. (eds) *Environmental UV Photobiology*. Plenum Press, New York, pp. 1–39.

Manetas, Y., Petropoulou, Y., Stamatakis, K., Nikolopoulos, D., Levizou, E., Psarus, G. and Karabourniotis, G. (1997) Beneficial effects of enhanced UV-B radiation under field conditions: improvement of needle water relations and survival capacity of *Pinus pinea* L. seedlings during the dry Mediterranean summer. *Plant Ecology* 128, 100–108.

Mark, U., Saile-Mark, M. and Tevini, M. (1996) Effects of solar UVB radiation on growth, flowering and yield of central and southern European maize cultivars (*Zea mays* L.). *Photochemistry and Photobiology* 64, 457–463.

Milchunas, D.G. and Lauenroth, W.K. (1995) Inertia in plant community structure: state changes after cessation of nutrient-enrichment stress. *Ecological Applications* 5, 452–458.

Mooney, H.A., Canadell, J., Chapin, F.S., Ehleringer, J., Körner, Ch., McMurtrie, R., Parton, W.J., Pitelka, L. and Schultze, E.-D. (1999) Ecosystem physiology responses to global change. In: Walker, B.H., Steffen, W.L., Canadell, J. and Ingram, J.S.I. (eds) *The Terrestrial Biosphere and Global Change: Implications for Natural and Managed Ecosystems*. IGBP Book Series No. 4, Cambridge University Press, Cambridge, UK, pp. 141–189.

Navas, M.-L., Sonie, L., Richarte, J. and Roy, J. (1997) The influence of elevated CO_2 on species phenology, growth and reproduction in a Mediterranean old-field community. *Global Change Biology* 3, 523–530.

Newsham, K.K., McLeod, A.R., Roberts, J.D., Greenslade, P.D. and Emmett, B.A. (1997) Direct effects of elevated UV-B radiation on the decomposition of *Quercus robur* leaf litter. *Oikos* 79, 592–602.

Newsham, K.K., Lewis, G.C., Greenslade, P.D. and McLeod, A.R. (1998) *Neotyphodium lolii*, a fungal leaf endophyte, reduces fertility of *Lolium perenne* exposed to elevated UV-B radiation. *Annals of Botany* 81, 397–403.

Newton, P.C.D. (1991) Direct effects of increasing carbon dioxide on pasture plants and communities. *New Zealand Journal of Agricultural Research* 34, 1–24.

Newton, P.C.D., Clark, H., Bell, C.C., Glasgow, E.M. and Campbell, B.D. (1994) Effects of elevated CO_2 and simulated seasonal changes in temperature on the species composition and growth rates of pasture turves. *Annals of Botany* 73, 53–59.

Nicholls, N., Gruza, G.V., Jouzel, J., Karl, T.R., Ogallo, L.A. and Parker, D.E. (1995) Observed climate variability and change. In: Houghton, J.T., Meira Filho, L.G., Callander, B.A., Harris, N., Kattenberg, A. and Maskell, K. (eds) *Climate Change 1995: the Science of Climate Change*. Cambridge University Press, Cambridge, UK, pp. 132–192.

Nijs, I., Teughels, H., Blum, H., Hendrey, G. and Impens, I. (1996) Simulation of climate change with infrared heaters reduces the productivity of *Lolium perenne* L. in summer. *Environmental and Experimental Botany* 36, 271–280.

Niklaus, P.A. and Körner, C. (1996) Responses of soil microbiota of a late successional alpine grassland to long term CO_2 enrichment. *Plant and Soil* 184, 219–229.

Norby, R.J., O'Neill, E.G., Hood, W.G. and Luxmore, R.J. (1987) Carbon allocation, root exudation and mycorrhizal colonization of *Pinus echinata* seedlings grown under CO_2 enrichment. *Tree Physiology* 3, 203–210.

O'Neill, E.G. (1994) Responses of soil biota to elevated atmospheric carbon dioxide. *Plant and Soil* 165, 55–65.

Owensby, C.E., Ham, J.M., Knapp, A., Rice, C.W., Coyne, P.I. and Auen, L.M. (1996a) Ecosystem-level responses of tallgrass prairie to elevated CO_2. In: Koch, G.W. and Mooney, H.A. (eds) *Carbon Dioxide and Terrestrial Ecosystems*. Academic Press, London, UK, pp. 147–162.

Owensby, C.E., Cochran, R.C. and Auen, L.M. (1996b) Effects of elevated carbon dioxide on forage quality for ruminants. In: Körner, C. and Bazzaz, F.A. (eds) *Carbon Dioxide, Populations, and Communities*. Academic Press, San Diego, USA, pp. 363–371.

Owensby, C.E., Ham, J.M., Knapp, A.K., Bremer, D. and Auen, L.M. (1997) Water vapour fluxes and their impact under elevated CO_2 in a C_4-tallgrass prairie. *Global Change Biology* 3, 189–195.

Parton, W.J., Scurlock, J.M.O., Ojima, D.S., Schimel, D.S., Hall, D.O. and Scopegram Group Members (1995) Impact of climate change on grassland production and soil carbon worldwide. *Global Change Biology* 1, 13–22.

Paul, N.D., Rasanayagam, S., Moody, S.A., Hatcher, P.E. and Ayres, P.G. (1997) The role of interactions between trophic levels in determining the effects of UV-B on terrestrial ecosystems. In: Rozema, J., Gieskes, W.W.C., van de Geijn, S.C., Nolan, C. and de Boois, H. (eds) *UV-B and Biosphere*. Kluwer Academic Publishers, Dordrecht, The Netherlands, pp. 296–308.

Poorter, H. (1993) Interspecific variation in the growth response of plants to an elevated ambient CO_2 concentration. *Vegetatio* 104/105, 77–97.

Poorter, H., Roumet, C. and Campbell, B.D. (1996) Interspecific variation in the growth response of plants to elevated CO_2: a search for functional types. In: Körner, Ch. and Bazzaz, F.A. (eds) *Carbon Dioxide, Populations, and Communities*. Academic Press, San Diego, USA, pp. 375–412.

Potvin, C. and Vasseur, L. (1997) Long-term CO_2 enrichment of a pasture community: species richness, dominance, and succession. *Ecology* 78, 666–677.

Pyle, J.A. (1997) Global ozone depletion: observations and theory. In: Lumsden, P.J. (ed.) *Plants and UV-B: Responses to Environmental Change*. Cambridge University Press, Cambridge, UK, pp. 3–11.

Rastetter, E.B. (1996) Validating models of ecosystem response to global change. How can we best assess models of long-term global change? *BioScience* 46, 190–198.

Read, J.J., Morgan, J.A., Chatterton, N.J. and Harrison, P.A. (1997) Gas exchange and carbohydrate and nitrogen concentrations in leaves of *Pascopyrum smithii* (C_3) and *Bouteloua gracilis* (C_4) at different carbon dioxide concentrations and temperatures. *Annals of Botany* 79, 197–206.

Reilly, J., Baethgen, W., Chege, F.E., van de Geijn, S.C., Erda, L., Iglesias, A., Kenny, G., Patterson, D., Rogasik, J., Rötter, R., Rosenzweig, C., Sombroek, W. and Westbrook, J. (1995) Agriculture in changing climate: impacts and adaptation. In: Watson, R.T., Zinyowera, M.C. and Moss, R.H. (eds) *Climate Change 1995: Impacts, Adaptations and Mitigation of Climate Change: Scientific–Technical Analyses*. Cambridge University Press, Cambridge, UK, pp. 427–467.

Rillig, M.C., Allen, M.F., Klironomos, J.N., Chiariello, N.R. and Field, C.B. (1998) Plant species-specific changes in root-inhabiting fungi in a California annual grassland: responses to elevated CO_2 and nutrients. *Oecologia* 113, 252–259.

Robertson, G. (1987) Effect of drought and high summer rainfall on biomass and composition of grazed pastures in western New South Wales. *Australian Rangeland Journal* 9, 79–85.

Roth, S.K. and Lindroth, R.L. (1995) Elevated atmospheric CO_2: effects on phytochemistry, insect performance and insect-parasitoid interactions. *Global Change Biology* 1, 173–182.

Rötzel, C., Leadley, P.W. and Körner, C. (1997) Non-destructive assessments of the effects of elevated CO_2 on plant community structure in a calcareous grassland. *Acta Oecologia* 18, 231–239.

Roumet, C. and Roy, J. (1996) Prediction of the growth response to elevated CO_2: a search for physiological criteria in closely related grass species. *New Phytologist* 134, 615–621.

Rozema, J., van de Staaij, J., Björn, L.O. and Caldwell, M. (1997a) UV-B as an environmental factor in plant life: stress and regulation. *Trends in Ecology and Evolution* 12, 22–28.

Rozema, J., Chardonnens, A., Tosserams, M., Hafkenscheid, R. and Bruijnzeel, S. (1997b) Leaf thickness and UV-B absorbing pigments of plants in relation to an elevational gradient along the Blue Mountains, Jamaica. *Plant Ecology* 128, 150–59.

Rozema, J., Lenssen, G.M., van de Staaij, J.W.M., Tosserams, M., Visser, A.J. and Broekman, R.A. (1997c) Effects of UV-B radiation on terrestrial plants and ecosystems: interaction with CO_2 enrichment. *Plant Ecology* 128, 182–191.

Rozema, J., Tosserams, M., Nelissen, H.J.M., van Heerwaarden, L., Broekman, R.A. and Flierman, N. (1997d) Stratospheric ozone reduction and ecosystem processes: enhanced UV-B radiation affects chemical quality and decomposition of leaves of the dune grassland species *Calamagrostis epigeios*. *Plant Ecology* 128, 284–294.

Runeckles, V.C. and Krupa, S.V. (1994) The impact of UV-B radiation and ozone on terrestrial vegetation. *Environmental Pollution* 83, 191–213.

Sala, O.E., Parton, W.J., Joyce, L.A. and Lauenroth, W.K. (1988) Primary production of the central grasslands region of the United States. *Ecology* 69, 40–45.

Santer, B.D., Wigley, T.M.L., Barnett, T.P. and Anyamba, E. (1995) Detection of climate change and attribution of causes. In: Houghton, J.T., Meira Filho, L.G., Callander, B.A., Harris, N., Kattenberg, A. and Maskell, K. (eds) *Climate Change 1995: the Science of Climate Change*. Cambridge University Press, Cambridge, UK, pp. 407–443.

Schapendonk, A.H.C.M., Dijkstra, P., Groenwold, J., Pot, C.S. and Van De Geijn, S.C. (1997) Carbon balance and water use efficiency of frequently cut *Lolium perenne* L. swards at elevated carbon dioxide. *Global Change Biology* 3, 207–216.

Schäppi, B. and Körner, C. (1996) Growth responses of an alpine grassland to elevated CO_2. *Oecologia* 105, 43–52.

Schenk, U., Manderscheid, R., Hugen, J. and Weigel, H.-J. (1995) Effects of CO_2 enrichment and intraspecific competition on biomass partitioning, nitrogen content and microbial biomass carbon in soil of perennial ryegrass and white clover. *Journal of Experimental Botany* 46, 987–993.

Schenk, U., Jäger, H.J. and Weigel, H.-J. (1996) Nitrogen supply determines responses of yield and biomass partitioning of perennial ryegrass to elevated atmospheric carbon dioxide concentrations. *Journal of Plant Nutrition* 19, 1423–1440.

Schenk, U., Jäger, H.J. and Weigel, H.-J. (1997) The response of perennial ryegrass/white clover swards to elevated atmospheric CO_2 concentrations. 1. Effects on competition and species composition and interaction with N supply. *New Phytologist* 135, 67–79.

Schimel, D., Alves, D., Enting, I., Heimann, M., Joos, F., Raynaud, D., Wigley, T., Prather, M., Derwent, R., Ehhalt, D., Fraser, P., Sanhueza, E., Zhou, X., Jonas, P., Charlson, R., Rodhe, H., Sadasivan, S., Shine, K.P., Fouquart, Y., Ramaswamy, V., Solomon, S., Srinivasan, J., Albritton, D., Derwent, R., Isaksen, I., Lal, M. and Wuebbles, D. (1995) Radiative forcing of climate change. In: Houghton, J.T., Meira Filho, L.G., Callander, B.A., Harris, N., Kattenberg, A. and Maskell, K. (eds) *Climate Change 1995: the Science of Climate Change*. Cambridge University Press, Cambridge, UK, pp. 65–131.

Scholes, R.J. and van Breemen, N. (1997) The effects of global change on tropical ecosystems. *Geoderma* 79, 9–24.

Schwinning, S. and Parsons, A.J. (1996a) Analysis of the coexistence mechanisms for grasses and legumes in grazing systems. *Journal of Ecology* 84, 799–813.

Schwinning, S. and Parsons, A.J. (1996b) A spatially explicit population model of stoloniferous N-fixing legumes in mixed pasture with grass. *Journal of Ecology* 84, 815–826.

Soussana, J.F. and Hartwig, U.A. (1996) The effects of elevated CO_2 on symbiotic N_2 fixation: a link between the carbon and nitrogen cycles in grassland ecosystems. *Plant and Soil* 187, 321–332.

Soussana, J.F., Casella, E. and Loiseau, P. (1996) Long-term effects of CO_2 enrichment and temperature increase on a temperate grass sward. II. Plant nitrogen budgets and root fraction. *Plant and Soil* 182, 101–114.

Specht, R.L. and Clifford, H.T. (1991) Plant invasion and soil seed banks: control by water and nutrients. In: Groves, R.H. and di Castri, F. (eds) *Biogeography of Mediterranean Invasions*. Cambridge University Press, Cambridge, UK, pp. 191–205.

Steinmuller, D. and Tevini, M. (1985) Action of ultraviolet radiation (UV-B) upon cuticular waxes in some crop plants. *Planta* 164, 557–564.

Stephenson, N.L. (1990) Climatic control of vegetation distribution: the role of the water balance. *American Naturalist* 135, 649–670.

Stewart, J. and Potvin, C. (1996) Effects of elevated CO_2 on an artificial grassland community: competition, invasion and neighbourhood growth. *Functional Ecology* 10, 157–166.

Sullivan, J.H. (1997) Effects of increasing UV-B radiation and atmospheric CO_2 on photosynthesis and growth: implications for terrestrial ecosystems. *Plant Ecology* 128, 194–206.

Sutherst, R.W., Yonow, T., Chakraborty, S., O'Donnell, C. and White, N. (1996) A generic approach to defining impacts of climate change on pests, weeds and diseases in Australasia. In: Bouma, W.J., Pearman, G.I. and Manning, M.R. (eds) *Greenhouse: Coping with Climate Change*. CSIRO Publishing, Melbourne, Australia, pp. 281–307.

Taylor, K. and Potvin, C. (1997) Understanding the long-term effect of CO_2 enrichment on a pasture: the importance of disturbance. *Canadian Journal of Botany* 75, 1621–1627.

Teeri, J.A. and Stowe, L.G. (1976) Climatic patterns and the distribution of C_4 grasses in North America. *Oecologia* 23, 1–12.

Teramura, A.H. (1983) Effects of ultraviolet-B radiation on the growth and yield of crop plants. *Physiologia Plantarum* 58, 415–427.

Teramura, A.H., Forseth, I.N. and Lydon, J. (1984) Effects of ultraviolet-B radiation on plants during mild water stress. The insensitivity of soybean internal water relations to ultraviolet-B radiation. *Physiologia Plantarum* 62, 373–380.

Teughels, H., Nijs, I., Van Hecke, P. and Impens, I. (1995) Competition in a global change environment: the importance of different plant traits for competitive success. *Journal of Biogeography* 22, 297–305.

Tevini, M. and Teramura, A.H. (1989) UV-B effects on terrestrial plants. *Photochemistry and Photobiology* 50, 479–487.

Tevini, M., Mark, U. and Saile, M. (1990) Plant experiments in growth chambers illuminated with natural sunlight. In: Payer, H.D., Pfirrman, T. and Mathy, P. (eds) *Environmental Research with Plants in Closed Chambers*. Air Pollution Research Report. Commission of the European Communities, Dordrecht, The Netherlands, pp. 240–251.

Tevini, M., Braun, J. and Fieser, F. (1991) The protective function of the epidermal layer of rye seedlings against ultraviolet-B radiation. *Photochemistry and Photobiology* 53, 329–333.

Thornley, J.H.M., Bergelson, J. and Parsons, A.J. (1995) Complex dynamics in a carbon–nitrogen model of a grass–legume pasture. *Annals of Botany* 75, 79–94.

Thornley, J.H.M. and Cannell, M.G.R. (1997) Temperate grassland responses to climate change: an analysis using the Hurley Pasture Model. *Annals of Botany* 80, 205–221.

Tilman, D. (1982) *Resource Competition and Community Structure.* Princeton University Press, Princeton, USA.

Tilman, D. (1988) *Plant Strategies and the Dynamics and Structure of Plant Communities.* Princeton University Press, Princeton, USA.

Tilman, D. (1997) Community invasibility, recruitment limitation, and grassland biodiversity. *Ecology* 78, 81–92.

Tilman, D., Lehman, C.L. and Thomson, K.T. (1997) Plant diversity and ecosystem productivity: theoretical considerations. *Proceedings of the National Academy of Sciences of the United States of America* 94, 1857–1861.

Tosserams, M., Bolink, E. and Rozema, J. (1997a) The effect of enhanced ultraviolet-B radiation on germination and seedling development of plant species occurring in a dune grassland ecosystem. *Plant Ecology* 128, 138–147.

Tosserams, M., Magendans, E. and Rozema, J. (1997b) Differential effects of elevated ultraviolet-B radiation on plant species of a dune grassland ecosystem. *Plant Ecology* 128, 266–281.

Tow, P.G., Lovett, J.V. and Lazenby, A. (1997a) Adaptation and complementarity of *Digitaria eriantha* and *Medicago sativa* on a solodic soil in a subhumid environment with summer and winter rainfall. *Australian Journal of Experimental Agriculture* 37, 311–322.

Tow, P.G., Lazenby, A. and Lovett, J.V. (1997b) Effects of environmental factors on the performance of *Digitaria eriantha* and *Medicago sativa* in monoculture and mixture. *Australian Journal of Experimental Agriculture* 37, 323–333.

Tow, P.G., Lazenby, A. and Lovett, J.V. (1997c) Relationships between a tropical grass and lucerne on a solodic soil in a subhumid, summer–winter rainfall environment. *Australian Journal of Experimental Agriculture* 37, 335–342.

Turner, I.M. (1994) Sclerophylly: primarily protective? *Functional Ecology* 8, 669–675.

Turner, N.C. and Begg, J.E. (1978) Responses of pasture plants to water deficits. In: Wilson, J.R. (ed.) *Plant Relations in Pastures.* CSIRO Publishing, Melbourne, Australia, pp. 50–66.

Twolan-Strutt, L. and Keddy, P.A. (1996) Above- and below-ground competition intensity in two contrasting wetland plant communities. *Ecology* 77, 259–270.

Van, T.K. and Garrard, L.A. (1976) Effects of UV-B radiation on net photosynthesis of some C_3 and C_4 plants. *Proceedings of the Soil and Crop Science Society of Florida* 35, 1–3.

Vandermeer, J., van Noordwijk, M., Anderson, J., Ong, C. and Perfecto, I. (1998) Global change and multi-species agroecosystems: concepts and issues. *Agriculture, Ecosystems and Environment* 67, 1–22.

van de Staaij, J.W.M., Rozema, J. and Stroetenga, M. (1990) Expected changes in Dutch coastal vegetation resulting from enhanced levels of solar UV-B. In: Beukema, J.J. (ed.) *Expected Effects of Climate Change on Marine Coastal Ecosystems.* Kluwer Academic Press, Dordrecht, The Netherlands, pp. 211–217.

van Ginkel, J.H., Gorissen, A. and van Veen, J.A. (1996) Long-term decomposition of grass roots as affected by elevated atmospheric carbon dioxide. *Journal of Environmental Quality* 25, 1122–1128.

Vasseur, L. and Potvin, C. (1998) Natural pasture community response to enriched carbon dioxide atmosphere. *Plant Ecology* 135, 31–41.

Walker, B. and Steffen, W.L. (1996) *Global Change and Terrestrial Ecosystems.* Cambridge University Press, Cambridge, UK.

Walker, B.H., Steffen, W.L., Canadell, J. and Ingram, J.S.I. (1998) *Implications of Global Change for Natural and Managed Ecosystems: A Synthesis of GCTE and Related Research.* IGBP Book Series No. 4, Cambridge University Press, Cambridge, UK.

Wardle, D.A. and Barker, G.M. (1997) Competition and herbivory in establishing grassland communities: implications for plant biomass, species diversity and soil microbial activity. *Oikos* 80, 470–480.

Webb, W.L., Lauenroth, W.K., Szarek, S.T. and Kinerson, R.S. (1983) Primary production and abiotic controls in forests, grasslands and desert. *Ecology* 64, 134–151.

White, T.A., Campbell, B.D. and Kemp, P.D. (1997) Invasion of temperate grassland by a subtropical annual grass across an experimental matrix of water stress and disturbance. *Journal of Vegetation Science* 8, 847–854.

Wilson, S.D. and Tilman, D. (1991) Components of plant competition along an experimental gradient of nitrogen availability. *Ecology* 72, 1050–1065.

Wray, S.M. and Strain, B.R. (1987) Competition in old-field perennials under CO_2 enrichment. *Ecology* 68, 1116–1120.

Zanetti, S., Hartwig, U.A., Lüscher, A., Hebeisen, T., Frehner, M., Fischer, B.U., Hendrey, G.R., Blum, H. and Nösberger, J. (1996) Stimulation of symbiotic N_2 fixation in *Trifolium repens* L. under elevated atmospheric pCO_2 in a grassland ecosystem. *Plant Physiology* 112, 575–583.

Zangerl, A.R. and Bazzaz, F.A. (1984) The response of plants to elevated CO_2 II. Competitive interactions among annual plants under varying light and nutrients. *Oecologia* 62, 412–417.

13 Competition and Succession in Re-created Botanically Diverse Grassland Communities

Ross Chapman*

CSIRO Plant Industry, Centre for Mediterranean Agricultural Research, Wembley, Australia

Introduction

Since clearing of native forests began in prehistoric times, species-rich grasslands have typically constituted a major habitat within the landscape of temperate regions of Europe (Green, 1990; Rychnovská *et al.*, 1994). The maintenance of this distinct successional stage or sere depends upon continuous human intervention to arrest or deflect otherwise natural successional processes. This intervention usually occurs in the form of either grazing with domestic livestock or haymaking. Because their evolution and preservation are directly linked to human activities, these grasslands are described as plagioclimax or semi-natural communities.

Such grasslands have traditionally supported the production of a number of livestock-based agricultural commodities, including meat, milk and wool. In addition, these communities provide a range of important environmental functions for the wider ecosystem (Fig. 13.1). They provide a habitat not only for a wide range of constituent flora (Rodwell, 1991), but also for a diverse selection of vertebrate and invertebrate fauna (Curry, 1987; Green, 1990). Furthermore, these grasslands enhance the aesthetic qualities of the rural landscape and so increase the recreational and amenity values of the countryside (Green, 1990; Hopkins and Hopkins, 1994; Fig. 13.2).

Despite their acknowledged environmental values, agricultural intensification over the 20th century has led to the widespread loss of these habitats from many areas. This loss of habitat has been particularly acute in lowland regions and has endangered many of the once widespread constituent flora and fauna (Perring and Farrel, 1983; Shirt, 1987; Batten *et al.*, 1990). In recent years, there has been growing concern about the implications of the loss of such diversity. In 1992, both the European Community (Council Directive 92/43/EEC) and the United Nations (UNCED, 1993) independently took steps to protect and restore biodiversity in threatened habitats. Within Great Britain, recognition of both the environmental value and the endangered status of traditional species-rich grassland led to the establishment of specific schemes to protect and re-create these communities within the agricultural landscape (MAFF, 1992). However, the process of agricultural intensification has changed many characteristics of both the soil and the landscape. These modified conditions alter the competitive and successional processes that occur within grassland communities, and this creates substantial problems for the successful restoration of such botanically diverse grassland communities.

Engineering works, such as those associated with mineral extraction and the construction of road and rail networks, are also responsible for a

*Present address: c/o Orchard House, Carlton Scroop, Grantham, UK

© CAB *International* 2001. *Competition and Succession in Pastures* (eds P.G. Tow and A. Lazenby)

substantial environmental disturbance and habitat loss. Within Great Britain, the Department of the Environment has recognized the potential for nature conservation within strategies for the reclamation of land disturbed by such operations (DOE, 1989a,b, 1991). As a response, many restoration plans now include a component for nature conservation, and this frequently constitutes the re-creation of species-rich grasslands. However, soils and landscapes of reclaimed sites are typically heavily disturbed. This creates a number of problems for both the successful establishment and the subsequent maintenance of these grasslands that are very different from those experienced within intensive agricultural systems.

This chapter reviews the problems imposed by competition and succession in botanically diverse grasslands re-created in temperate north-western Europe on both extensified agricultural land and sites restored after disturbance during engineering operations. These problems arise from not only the internal conditions of the modified ecosystems, but also the nature of the linkages between that ecosystem and the surrounding landscape. The environmental and ecophysiological origins of these constraints are discussed and some potential solutions are offered.

The Impact of Habitat Productivity and Grazing Livestock on the Composition and Diversity of a Community

Habitat productivity

Habitat productivity may have a profound effect upon both the botanical composition and the diversity of a grassland community.

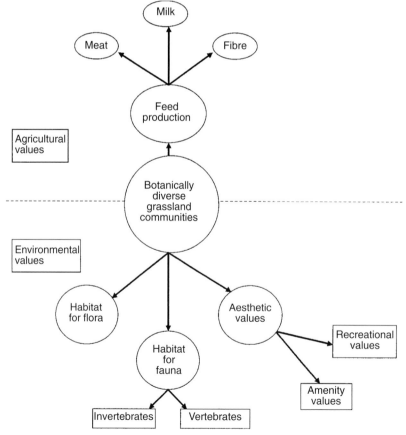

Fig. 13.1. Schematic representation of the agricultural and environmental values of botanically diverse grassland communities.

Fig. 13.2. A visually attractive species-rich grassland community in the Yorkshire Dales, northern England.

Botanical composition

The type of species present within any grassland community is greatly influenced by the fertility and productivity of that particular habitat (Grime, 1979). Under conditions of high productivity, as typically occur on fertile soils with high levels of nutrient supply, communities are typically dominated by fast-growing species, which rapidly establish a tall canopy, such as *Dactylis glomerata* and *Lolium perenne*. These species have been described as possessing a 'competitive' vegetative strategy (Grime, 1979). In contrast, habitats of very low productivity, such as occur on very infertile soils, tend to be dominated by very slow-growing, 'stress-tolerating' species, such as *Molinia caerulea* and *Festuca ovina* (Grime, 1979). Between these two extremes lies a continuum of productivity niches.

Although the mechanisms involved in the dispersal of species along this productivity gradient are still debated (Wilson and Tilman, 1993; Goldberg and Novoplansky, 1997; Vázquez de Aldana and Berendse, 1997), it is clear that some degree of competition is involved. Willems (1983) asserted that fast- and tall-growing species quickly come to dominate the canopy under productive conditions associated with high fertility. This leads to the rapid extinction of photosynthetically active radiation as it passes through the upper layers of the canopy. As a consequence, less radiation is available for species of shorter stature, which exist lower in the canopy, and the community becomes dominated by just a few tall and fast-growing 'competitor' species. Furthermore, it has been alleged that these competitor species may be quick to establish an extensive root system, which, even under conditions of high fertility, will efficiently deplete all available nutrients, thus causing less aggressive species to be subjugated through nutrient competition (Grime, 1979).

The physiological basis by which 'stress tolerators' come to dominate under conditions of low productivity is still to be established. Some have argued that these species have a superior capacity to compete for nutrients when under very low levels of supply (Tilman, 1982; Goldberg and Novoplansky, 1997). Others, in contrast, have maintained that their dominance is simply due to the adoption of a growth strategy that creates a reduced demand for these resources while minimizing losses (Grime, 1979; Vázquez de Aldana and Berendse, 1997).

Botanical diversity

Following observation from numerous habitats, Al-Mufti *et al.* (1977) proposed that a humpback relationship existed between the productivity and the botanical diversity of a vegetative community (Fig. 13.3). According to this model, at very high productivity (greater than 1000 g dry matter (standing crop plus litter) m^{-2}), aggressive competition excludes all but the most competitive species and this imparts a very low level of diversity to the community. At reduced levels of productivity, stresses create gaps within the matrix of competitor species which can be exploited by less competitive species. At extremely low levels of productivity (less than 200 g dry matter m^{-2}), only the most stress-tolerant species can survive, which again induces very low levels of diversity in the community. The greatest diversity occurs at low to intermediate levels of productivity (350–750 g dry matter m^{-2}), where the greatest overlap of 'competitor' and 'stress tolerator' niches occurs.

Various workers have subsequently attempted to verify this relationship in grassland communities. Smith (1994) combined data from three contrasting sites and found no link between botanical diversity and productivity or fertility. Oomes (1992), in contrast, compared botanical diversity and productivity in 27 grassland sites, and found a humpback response very similar to that proposed by Al-Mufti *et al.* (1977). This relationship is further substantiated by the findings of Vermeer and Berendse (1983) and Willems (1983).

When considering the relationship between botanical diversity and productivity, Marrs (1993) and Pegtel *et al.*, (1996) both noted that, while the humpback model of Al-Mufti *et al.* (1977) may generally prevail, the exact relationship may be site- or vegetation type-specific and that the form of the response may well be modified by such influences

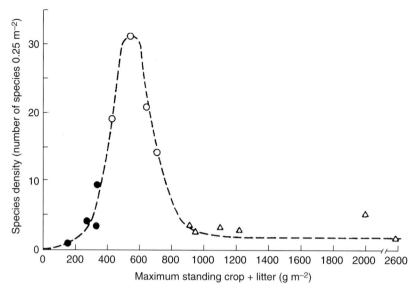

Fig. 13.3. The relationship between maximum standing crop plus litter and species density of herbs at 14 sites in northern England. ○ grasslands; ● woodlands; △ tall herbs. (Reproduced from Al-Mufti *et al.*, 1977.)

as soil pH levels and the intensity and frequency of disturbances.

The impact of the grazing animal

The composition and diversity of the community may also be profoundly influenced by grazing livestock. In particular, animals may significantly influence the physical structure of the community through the partial or complete destruction of the canopy of competitive and dominant species. This may arise either directly, through the effects of defoliation, or indirectly, through the effects of treading, urine scorch and burrowing or scraping by livestock (Grime, 1979). The effect of this disturbance is to create a niche which opportunistic 'ruderal' species can exploit free from competition (Grime, 1979; Smith and Rushton, 1994). These ruderal species are typically fast-growing and yet relatively uncompetitive annuals or short-lived perennials that are capable of producing high seed yields, such as *Bromus mollis* or *Medicago lupulina*. Maintenance of maximum diversity in a grassland community therefore requires regular disturbance by grazing animals in order for ruderal strategists to be retained alongside the competitor and stress tolerator species (Smith and Rushton, 1994).

Grazing animals can also influence the spatial diversity and botanical composition of a pasture

community. That animals are selective in their grazing habit is well established (e.g. Gibb *et al.*, 1989). This behaviour creates spatial heterogeneity in the canopy architecture of the community, with the mean canopy height being shorter in the more frequently grazed patches than in the less grazed patches. This difference in canopy architecture creates niches for plants of contrasting growth habit (Putman *et al.*, 1991) and leads to the establishment of a mosaic of subhabitats within the pasture communities. Furthermore, excretion of dung and urine by grazing animals leads to a localized accumulation of nutrients. The elevated fertility of the affected areas will favour more competitive species, such as *L. perenne* and *Trifolium repens*. The patchy distribution of excreta will create further subhabitats, which add to the spatial diversity of the pasture community.

Conclusions

It has been argued that the peak in botanical diversity occurs when the physical structure of the sward allows the niches of competitor, stress tolerator and ruderal species to overlap. This occurs under conditions of low to moderate productivity coupled with regular livestock-induced disturbances. Under these conditions, competitor species are able to form an extensive but spatially incomplete matrix across the community. Interstices are maintained within this matrix by a combination of stresses, such as nutrient deficiencies, and disturbances, such as poaching by grazing animals. It is these gaps that provide the niches which stress-tolerating and ruderal species exploit. Maintenance of the highest levels of diversity therefore requires that fertility and disturbance levels are finely balanced to ensure that species displaying all three vegetative strategies can coexist. The restoration and subsequent maintenance of sustainable botanically diverse grasslands will thus require careful management of both fertility and grazing to achieve the correct level of productivity and vegetation disturbance.

The Re-creation of Botanically Diverse Grasslands on Formerly Intensive Agricultural Land

The impact of agricultural intensification on botanically diverse grassland communities

The 20th century has witnessed a catastrophic loss of botanically diverse semi-natural grasslands from temperate regions of north-western Europe. This has largely been as a consequence of agricultural intensification. Fuller (1987) has accurately detailed this loss of habitat from lowland England and Wales. In 1932 there existed approximately 7.2 million ha of unimproved semi-natural grassland habitat within this region. By 1984 there were only 0.6 million ha of unimproved lowland grasslands (Fig. 13.4), which represents a habitat loss of some 92%. The process of intensification has affected the grassland habitats in several ways.

Fertilizer applications

Traditional management of grasslands excluded the use of mineral fertilizers (Green, 1982; Archer, 1985; Fuller, 1987). Instead, the avoidance of nutrient depletion depended essentially on biological nitrogen- (N_2-) fixation, the recycling of dung and urine from stock grazing on pastures and, in some instances, on the return to meadows of farmyard manure collected from stock housed over winter (Smith and Jones, 1991; Younger and Smith, 1994). This practice tended to maintain habitat fertility at the levels appropriate for the maintenance of high levels of botanical diversity (Wilkins and Harvey, 1994).

Over recent decades, the adoption of mineral fertilizers as a tool to manage grasslands has become more widespread and it has been estimated that by

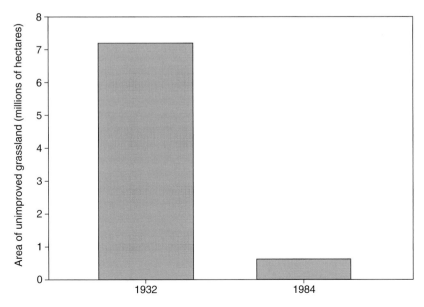

Fig. 13.4. The decline in unimproved semi-natural grassland in England and Wales between 1932 and 1984 (data extracted from Fuller, 1987).

the mid-1980s over 85% of all grasslands from this lowland England and Wales had been affected by mineral fertilizer use (Elsmere, 1986). The increased use of mineral fertilizers clearly benefited the agricultural productivity of these grasslands. However, applications of even small quantities of mineral fertilizers to these botanically diverse grassland communities encourages the dominance of fast-growing, aggressive and competitive species. In particular, the abundance of tall-growing grasses, such as *Brachypodium pinnatum*, *Arrhenatherum elatius*, *D. glomerata* and *L. perenne*, is increased (Kirkham *et al.*, 1996; Smith *et al.*, 1996; Willems and van Nieuwstadt, 1996). Aggressive growth by these competitor species quickly excludes slower-growing, less aggressive species, thus reducing the botanical diversity of the community (Van Hecke *et al.*, 1981; Berendse *et al.*, 1992; Mountford *et al.*, 1993; Smith, 1994; Kirkham *et al.*, 1996; Smith *et al.*, 1996).

Drainage

A fundamental step in the intensification and 'improvement' of the agricultural landscape has frequently been the installation of a field drainage system (Fuller, 1987). An effective drainage system will lower the water-table beneath the grassland community. A direct consequence of this will be the loss of any species associated with wet or waterlogged conditions, such as *Caltha palustris* and *Glyceria fluitans*. Furthermore, the lower water-table will result in increased levels of soil aeration. This will enhance the rate of soil microbial activity, organic matter breakdown and the cycling of nutrients in general and N in particular (Berryman 1975; Oomes *et al.*, 1997). The consequent increase in nutrient supply raises the productivity of the community. This will lead to the displacement of slow-growing, stress-tolerating species by more competitive species, thus suppressing the diversity of the habitat.

Defoliation practice

Traditional grassland management practices in temperate Europe typically included the production of a summer hay crop, which was conserved and used as winter feed. This practice provided a defoliation-free period, lasting from late spring through to summer. This allowed many of the constituent species to flower and set seed, a process vital for the maintenance of many species that depend upon

sexual reproduction to sustain their populations (Smith and Jones, 1991). Modernization of grassland management has seen silage replacing hay as the more common method for conserving winter feed (Hopkins and Hopkins, 1994). Silage production characteristically involves defoliating the sward very much earlier. This practice of early and more frequent defoliation prevents many species from successfully setting seed and leads to their gradual extinction, thereby reducing the community's botanical diversity (Smith *et al.*, 1996).

Extensification of grassland management

The adoption of more intensive systems of grassland production, such as the application of mineral fertilizers, the installation of field drainage systems and the adoption of silage making have seriously degraded the botanical diversity of grassland communities. The adoption of more extensive management practices will therefore be fundamental to any attempt to re-create these communities. However, experiments investigating the impact of extensification, such as the cessation of all applications of fertilizers or manures, brought little immediate improvement in the diversity of the communities studied (Olff and Bakker, 1991; Mountford *et al.*, 1994). These failures were caused by: (i) residual fertility maintaining productivity and competition at undesirable levels; and (ii) slow rates of successional change, due to limitations in the rate of recruitment of additional species into the extensified community. These studies suggest that a successful restoration plan will require the inclusion of practices that actively reduce productivity, through manipulation of nutrient cycles and soil fertility, and the enhancement of the recruitment and successional processes. The problems and potential solutions associated with productivity adjustment will be discussed individually in the next section. The impact this has on successional development and community composition will be discussed immediately afterwards.

Manipulation of nutrient cycles and soil fertility

Excess fertility may be reduced by instigating a net flow of nutrients out of the restored ecosystem. This may be achieved effectively through the removal of either vegetation or nutrient-rich topsoil. Alternatively, a similar effect may be achieved by slowing the rate at which nutrients are cycled within the ecosystem. Various management options are available to meet these objectives.

Vegetation management

Supply of nutrients to any plant community may be considered as a dynamic process, influenced by the rates of nutrient input from external sources, internal nutrient cycling through constituent plant, animal and microbial components of the system, and nutrient exports. Perhaps the simplest way to reduce the chemical fertility and excess productivity of a grassland habitat would be to instigate a net export of nutrients out of the system by both stopping all further inputs, while simultaneously cutting and removing all herbage produced at that site. Several experiments have investigated the efficiency with which such practices reduce the productivity of grassland communities. Bakker (1989) successfully demonstrated that the cutting and removal of a hay crop from a formerly intensively managed grassland led gradually, over a number of years, to a decline in productivity. Olff and Bakker (1991) confirmed these observations and reported that practices that involved taking two cuts of hay per year gave a greater rate of productivity decline than those based around the production of a single hay crop. Similar findings were again reported by Berendse *et al.* (1992). The latter authors, however, found that the rate of decline in productivity induced by a cut-and-clear management practice varied according to soil type. In their experiment, annual herbage productivity was compared on two contrasting sites, one on a sandy soil and one on a clay-on-peat soil. Both sites

yielded approximately 12 t dry matter ha^{-1} $year^{-1}$ at the start of the experiment. After 10 years of unfertilized cut-and-clear management, productivity on the sandy soil had declined to approximately 5 t dry matter ha^{-1} $year^{-1}$, while on a clay-on-peat soil productivity remained between 6 and 8 t dry matter ha^{-1} $year^{-1}$. The successional responses that followed these changes in production will be discussed later.

Other investigations have examined the impact of cut-and-clear practices on the particular nutrient cycling processes operating within contrasting communities in more detail (Table 13.1). These results collectively show that the cutting and clearing of vegetation from unfertilized but formerly intensively managed grasslands led to a fertility decline as a result of nutrient export. The general trend was for yield-limiting deficiencies of potassium (K) and N to be induced within 2–9 years after the introduction of these practices, but the development of yield-limiting phosphorus (P) deficiencies took about 20 years. These contrasting responses may be attributed to the differing methods by which nutrients are cycled within grassland systems and are discussed below.

POTASSIUM CYCLING.   The cation exchange sites of a soil may readily adsorb K ions from the soil solution and retain them in a form that is immobile and yet still available for plant uptake (Russel, 1973). However, soils with a low number of cation exchange sites, such as those with a low clay content, have a limited ability to retain K in this form; any excess K ions remaining in the soil solution are therefore prone to leaching from the soil profile (Olff and Pegtel, 1994).

Vegetation management also has a profound

Table 13.1. Soil impoverishment and the induction of production-limiting nutrient deficiencies by cut-and-clear practices.

Study	Authors' soil description	Apparent induction of nutrient deficiencies	Nutrients remaining unlimited
Oomes (1991)	Sand	K after 9 years	P and N
Olff and Pegtel (1994)	Sand	K and N after 2–6 years; P after 19 years	
Oomes *et al.* (1996)	Peat	K after 10 years	P and N[a]
Pegtel *et al.* (1996)	Gley podzol	N after 3 years; K (moderate) after 3 years; P after 20 years	

[a]N remained unlimited in this study; this is possibly due to the installation of a field drainage system artificially enhancing the N mineralization rate in peaty soils (Oomes, 1991).

impact on the cycling of K. The uptake of K by plants is regulated principally by supply rather than demand; thus, when this nutrient is available in sufficient quantities, the vegetation will readily absorb more than is required for growth (Robson *et al.*, 1989), a phenomenon known as 'luxury uptake'. Under many management practices, such as grazing (Holmes, 1989), mulching (Oomes, 1991; Oomes *et al.*, 1996) and manure recycling (Lecomte, 1980; MAFF, 1982), this plant-absorbed K is returned to the soil and retained within the grassland ecosystem. However, if the nutrients contained within the vegetation are exported from the system entirely, the phenomenon of luxury uptake will lead to the rapid depletion of plant-available K, giving the observed rapid establishment of production-limiting deficiencies.

NITROGEN CYCLING. The cycling of N within grassland systems is a complex process (Whitehead, 1995). A large proportion of soil N may be retained in the immobile and unavailable organic fraction of the soil. Organic N is gradually degraded by microbial activity to release ammonium (NH_4^+). Like K, NH_4^+ is freely adsorbed by soil cation exchange sites, where it may be protected from leaching and remain available for plant uptake. This NH_4^+ may, however, be oxidized to form nitrate (NO_3^-). This form of N is poorly adsorbed by soil particles and is very prone to leaching losses.

Plants are able to absorb N only from the soil's pool of mineral N (NH_4^+ and NO_3^-). The fast growing and highly productive species that are common in intensified grassland communities are able to accumulate N in large quantities within leaf and stem tissues. Because of this, cut-and-clear practices are therefore able to rapidly remove large quantities of mineral N from recently extensified grassland communities that are dominated by more competitive species (Olff and Pegtel, 1994).

A combination of high plant uptake of mineral N and leaching of NO_3^- may therefore lead to a rapid depletion of the soil mineral N pool. However, the main determinant of soil N fertility is the size and mineralizability of the soil organic N pool, which may return as much as 200 kg N ha^{-1} year^{-1} to the mineral N pool (Berendse *et al.*, 1992). Reducing the soil N supply is therefore a longer-term process, which depends upon the depletion of the soil organic N pool. Furthermore, pollutants contaminating rainwater may return an

additional 20 kg mineral N ha^{-1} year^{-1} in some industrialized regions of Europe (Oomes *et al.*, 1997). These processes will affect the time it takes for the cut-and-clear process to deplete the soil N supply.

The processes by which K and N are cycled within grassland ecosystems are therefore quite different. The relative speed with which either nutrient reaches a state of production-limiting deficiency under a cut-and-clear management will depend upon the initial size of the nutrient pool at the start of the process and the ability of the site to buffer N losses. These phenomena are likely to be site-specific and may explain why some studies found cut-and-clear processes to induce K deficiencies before N (e.g. Oomes, 1990), while others found the opposite to be the case (e.g. Pegtel *et al.*, 1996).

PHOSPHORUS CYCLING. Mineral P applied in fertilizers to grassland ecosystems is rapidly but reversibly converted into several forms that are unavailable for plant uptake (Fig. 13.5). Some P may be incorporated into inorganic complexes, some may be adsorbed on to colloidal surfaces and some may become incorporated into the soil organic matter through plant and microbial processes (Gough and Marrs, 1990). These processes are largely reversible and the equilibrium point depends critically upon the concentration of P in the soil solution. Intensive agricultural management, therefore, tends to establish a substantial pool of unavailable P. Following the cessation of fertilizer applications and the commencement of a cut-and-clear management, the vegetation will begin to deplete the soil's pool of plant-available P. As this occurs, additional P will tend to be returned to solution from the various unavailable pools within the soil which effectively buffers the decline in the pool of plant-available P (see Fig. 13.5; Marrs, 1993). Because prolonged intensive agricultural management establishes considerable reserves of unavailable P within the soil, a cut-and-clear process may take many years before it is able to induce P productivity limitations within extensified grasslands (e.g. Marrs, 1993; Pegtel *et al.*, 1996).

Topsoil removal

Plant nutrients are typically concentrated in the surface soil; concomitantly, the microbial popula-

(a)

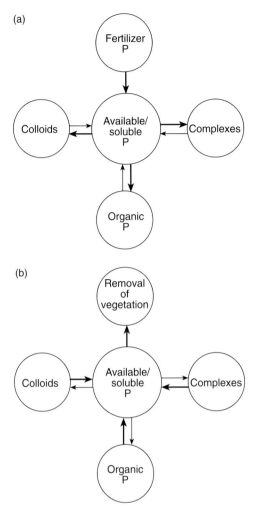

(b)

Fig. 13.5. Schematic representation of the fluxes of P between the soluble, complexed, colloidal and organic pools within grassland soils following (a) fertilizer application and (b) removal of vegetation.

Table 13.2. Dry-matter production and soil P uptake from a hay crop taken in June from a soil with the upper 5 cm removed (top soil removed) and a soil with an intact soil profile (intact soil) (data extracted from Oomes *et al.*, 1996).

	Topsoil removed	Intact soil
Dry-matter yield (t ha^{-1})	2.7	4.6
Uptake of P (kg ha^{-1})	7.29	13.5
Uptake of N (kg ha^{-1})	47.25	75.9

tions, which are critical to nutrient cycling, are also concentrated in the same upper layers of soil (Leeper, 1973; Russel, 1973). Several workers have explored the inclusion of topsoil removal in restoration schemes as a tool for lowering soil fertility and community productivity. Berendse *et al.*, (1992) found that stripping the topsoil from a previously intensive agricultural grassland did indeed lower the productivity of the habitat.

A similar study reported by Oomes *et al.* (1996) examined the impact of topsoil removal on nutrient cycling. They reported that supplies of both N and P were substantially reduced (Table 13.2) and that, unlike cut-and-clear processes, this reduction in chemical fertility was almost immediate. This was probably achieved through both the removal of a significant component of the soil nutrient reserve and a reduction in the rates of cycling through modified levels of microbial activity. Furthermore, this reduction in nutrient supply was accompanied by a lower level of biomass production (see Table 13.2).

While topsoil removal may be an effective way to rapidly reduce the productivity of a site, the operation will require substantial use of plant and equipment. Because of the costs incurred, this process is unlikely to be a practical method of manipulating fertility, unless either the restoration value of a particular habitat is extremely high or the topsoil may be sold off-site to reduce costs.

Manipulation of the water-table

As discussed previously, the installation of a field drainage system lowers the water-table, increases microbial respiration and enhances nutrient cycling. The potential therefore exists to slow the rate of nutrient cycling, reduce the productivity and lessen the level of competition in extensified grasslands by raising the water-table. This may also increase N losses from the system by increasing rates of denitrification. Various experiments have investigated this possibility. An experiment by Oomes (1991) artificially raised the water-table under a grassland on a peaty soil. This was achieved by extracting groundwater from wells and pumping it into ditches surrounding experimental plots. The water then penetrated the soil within the plots by flowing along irrigation pipes spaced at 5 m intervals and buried at 50 cm depth. This allowed the water-table to be raised by approximately 20 cm in the summer and by 25–30 cm in winter. The elevated water-table successfully reduced the supply of K and, to a lesser extent, P to the vegetation over a 3 year period. This indicated that changing the site hydrological condition effectively modified the nutrient cycles.

Similar reports by Berendse *et al.* (1994) and Oomes *et al.* (1996) discussed the effect of raising the water-table by 10–40 cm on a humic clay-over-peat soil. In that instance, the elevated water-table reduced both annual N mineralization and above-ground N accumulation by approximately 20%, indicating that the treatment had successfully manipulated both the soil's hydrological condition and the N cycle. Extractable P and K in the top 0–10 cm soil layer also declined markedly as a result of raising the water-table. Oomes *et al.* (1996) speculated that the decline in P may have been due to an enhanced rate of fixation into insoluble forms by the calcium-rich groundwater. The change in K availability in the upper soil layers is likely to have been caused by root distribution patterns. The elevated water-table concentrated a greater proportion of the roots in the upper, drier layers of soil. Uptake of K would therefore have largely been limited to a shallow depth of soil, causing an appreciable decline in availability over the course of this investigation. Furthermore, Oomes *et al.* (1996) observed that a reduction in productivity was associated with a significant change in community structure. Gradually, taller 'competitor' species were replaced by less aggressive species of shorter stature that were better adapted to conditions of low nutrient availability.

Conclusions

The re-creation of botanically diverse communities in extensified agricultural ecosystems is likely to be severely constrained by high levels of competition from relatively few species induced by excess soil fertility. However, either instigating a net flow of nutrients out from the extensified community or slowing the rate of nutrient cycling within the community presents potential opportunities to reduce this fertility. This will lead to the establishment of levels of productivity and competitive intensity appropriate for the re-establishment of botanically diverse grassland communities. The subsequent re-creation of such a community will depend upon a successful successional response to the habitat's modified fertility status.

Successional responses to reduced competition in extensified agricultural grasslands

Lowering the productivity of extensified agricultural grassland communities will create gaps within the matrix of competitor species within the grassland community. This will generate niches suitable for the re-establishment of less competitive species, which, if colonized successfully, should lead to the natural regeneration of a botanically diverse grassland community. Indeed, Bakker (1989), Olff and Bakker (1991), Olff and Pegtel (1994) and Willems and van Nieuwstadt (1996) all found that practices which successfully suppressed community productivity were also associated with simultaneous increases in botanical diversity. Thus, as new niches were re-established in the extensified community, they were rapidly colonized by previously absent species.

This, however, has not always been the result. A restoration attempt by Oomes *et al.* (1996) successfully lowered the productivity within a grassland community, but this was not associated with any change in botanical diversity. In this study, therefore, the community was clearly limited in its ability to recruit new species into the re-created niches. Berendse *et al.* (1992) and Pegtel *et al.* (1996) similarly reported reductions in productivity that failed to yield the anticipated increase in botanical diversity.

These apparently contradictory results may be explained by considering the seed-bank dynamics and propagule dispersal patterns of grassland species, along with the proximity of the experimental site to external sources for colonizing populations. Many grassland species may be described as generally possessing either a transient or a short-lived seed bank (Thompson and Grime, 1979; Thompson, 1987; Grime *et al.*, 1988). As discussed above, intensification of grassland management excludes all but the most competitive species from the vegetation. Under such management, any less competitive species will be unable to regularly supplement their seed bank and will be quickly eliminated from the system as their seed banks are exhausted (Hutchings and Booth, 1996; Bekker *et al.*, 1997; Kirkham and Kent, 1997).

The recruitment of species into niches re-created in extensified grassland communities will thus depend upon the dispersal of propagules from alternative sources in nearby vegetation. Seeds of many grassland species are generally not well adapted for dispersal, e.g. *Agrostis capillaris*, *Poa pratensis* and *Phleum pratense* (Thompson, 1987), and will typically be dispersed within a radius close to the maternal plant (Carey and Watkinson, 1993; Hutchings and Booth, 1996). Migration of grassland species along vegetative corridors therefore tends to be extremely slow; van Dorp (1993) estimated that dispersal of populations of grassland species along one-dimensional corridors linking different grassland habitats in The Netherlands was limited to a maximum of 4 m year^{-1}.

It may therefore be anticipated that the establishment of natural successional processes in extensified grasslands will depend critically upon the spatial connections between the restored site and the surrounding landscape. If the restored site is close to suitable undisturbed vegetation (as in the experiments of Bakker, 1989; Olff and Bakker, 1991; Olff and Pegtel, 1994; Willems and van Nieuwstadt, 1996), there will be an effective rain of propagules on to the site and natural successional processes may proceed rapidly. In contrast, the experiments of Berendse *et al.* (1992), Oomes *et al.* (1996) and Pegtel *et al.* (1996) were all conducted on sites isolated from other botanically diverse grassland communities. The geographical isolation of these communities would prevent an influx of propagules into the restored site, despite the creation of appropriate niches for colonization.

The role of the grazing animal

Once the level of productivity has been adjusted and recruitment commenced, correct successional development depends upon the adoption of an appropriate defoliation regime. This is well illustrated in an experiment conducted by Gibson and Brown (1992), which investigated the effect of grazing intensity on the re-creation of a calcicolous species-rich pasture on a former arable field. In particular, these workers were investigating whether the role of the grazing animal in creating a plagioclimax pasture community is: (i) to arrest a linear successional process at a specific seral stage; or (ii) to divert the successional process along an entirely separate seral trajectory. In this experiment, only vegetation under the heaviest grazing treatment began to develop a species composition similar to that of the undisturbed pasture 'target' community. Vegetation that was grazed less intensively not only became dominated by competitor and ruderal

species atypical of undisturbed grasslands, but also demonstrated a high degree of shrub encroachment. These results clearly indicate that alternative seral pathways exist once a grassland is re-created. The seral succession necessary for the re-establishment of the distinctive plagioclimax grassland community will only occur under appropriate grazing pressure and livestock management. The actions of the grazing animal are therefore essential for the successional re-development of species-rich pasture communities.

Conclusion

The successional development of botanically diverse grassland communities therefore requires effective species recruitment and appropriate grazing management. Natural successional processes require the establishment of an open ecosystem with strong linkages with appropriate external plant communities. If these conditions are not established, the restored ecosystem will remain isolated and effectively closed. Succession will then be limited by the extremely poor rate of migration from external populations. Under such circumstances, the restoration practitioner must either accept the re-establishment of incomplete communities by purely natural means (van Dorp, 1993) or must aid the process of recruitment by artificially assisting the process of succession (Hutchings and Booth, 1996; Watt *et al.*, 1996; Stevenson *et al.*, 1997). Once recruitment limitations have been overcome, it is vital that an appropriate livestock management regime is adopted to ensure that successional development proceeds along the correct trajectory.

The Re-creation of Botanically Diverse Grasslands on Land Disturbed by Engineering Operations

The re-creation of botanically diverse grasslands on land restored following engineering operations poses a number of problems. As with sites affected by intensive agricultural practices, these will include problems with plant competition, recruitment and succession. However, the environmental context of disturbed sites will typically be very different from agricultural sites. As a consequence, the exact nature of the problems faced may differ substantially.

The environmental conditions of disturbed sites

Soil nutrient availabilities

Many engineering processes, such as opencast mining and road verge construction, conclude with the reconstruction of the soil profile using original top- and subsoils. This soil will, however, have been severely disturbed. This disturbance will certainly include damage induced by the mechanical stripping and replacement of the medium, and may also entail prolonged storage in stockpiles. While these processes may not adversely reduce the availability of soil P and K (Scullion, 1994b; Chapman and Younger, 1995), they may change many of the soil physical qualities, including bulk density, structure and drainage, (Abdul-Kareem and McRae, 1984; Baker *et al.*, 1988; King, 1988). However, perhaps the greatest damage will be inflicted on the soil N cycle. A study by Davies *et al.* (1995) indicated that as much as 0.25 g N m^{-2} may be lost from a soil reinstated from a stockpile. A detailed analysis of the N dynamics during the soil handling process revealed that these losses were due to transformations that occurred during the storage and soon after replacement. N supply rates in restored ecosystems have been shown to be closely correlated with total soil N content (Skeffington and Bradshaw, 1981); thus such a major loss of total soil N will severely compromise the soil's overall fertility.

Many industrial processes generate substantial quantities of inert skeletal wastes, such as colliery spoil, Leblanc waste, fly ash and chalk marl (Bradshaw and Chadwick, 1980; Ash *et al.*, 1994; Mitchley *et al.*, 1996). Natural weathering processes release some P and K from these substances and availabilities of these nutrients may vary from deficient to adequate (Bradshaw, 1997). In contrast, N is virtually absent from these substrates; the major limitation to community productivity therefore typically arises from extreme N deficiencies (Bradshaw, 1997).

Thus deficiencies of macronutrients, and especially of N, severely limit plant growth and vegetation establishment on both restored soils and skeletal substrates remaining following industrial disturbance.

Nitrogen cycling in restored soils

The N availability of any system depends not only on the total amount of N present, but also on the efficiency with which it is cycled through that system. A simplified representation of this N cycle is presented in Fig. 13.6. Engineering operations are likely to disturb this N cycle by disrupting the microbial processes and invertebrate populations.

SOIL MICROBIAL ACTIVITY. Mechanical handling of soils during the stripping and replacement processes leads to a severe loss of soil macropores (King, 1988). Such poor soil structure will compromise the mineralizability of any organic N that remains in the substrate after restoration or is subsequently incorporated (Hassink, 1992; Hassink *et al.*, 1993; Killham *et al.*, 1993). Because of the low soil organic matter content, skeletal substrates are also likely to display poor structural stability and pore size distribution and this will further inhibit N mineralization (Bradshaw, 1997).

Fungi are essential for not only the efficient breakdown of the soil organic matter, but also the recovery and maintenance of the soil structural stability. Stockpiling soils significantly suppresses the soil fungal populations (Harris *et al.*, 1993), and this will further impede the cycling of N in restored soils.

SOIL INVERTEBRATE POPULATIONS. Detailed studies have indicated that populations of earthworms and other decomposer invertebrates (e.g. *Isopoda* and *Diplopoda*) are unusually low on restored sites (Scullion *et al.*, 1988a,b; Scullion, 1994a; Wheater and Cullen, 1997). This slows the rate of decomposition of plant litter and other organic matter (Majer, 1997) and so directly reduces the efficiency with which N is cycled within the restored ecosystem. Furthermore, low populations of earthworms inhibit soil structural development and this in turn suppresses the ability of soil microorganisms to mineralize organic N.

CONCLUSION. Low levels of activity from both microbial and invertebrate decomposers are likely to inhibit the breakdown of organic matter in restored ecosystems. This will impede the efficient cycling of N through the restored ecosystem (see Fig. 13.6) and cause a greater proportion of the N pool to accumulate in the macro-organic and humic N pools. This will significantly reduce the proportion of the ecosystem's already depleted N pool becoming mineralized and plant uptake will

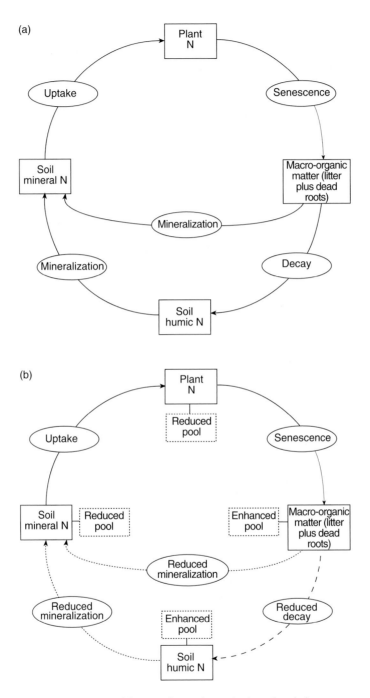

Fig. 13.6. A simplified representation of the N cycling pathways in the soil and plant components of an ecosystem in (a) an undisturbed and (b) a restored environment, showing reduced rates of decay and mineralization leading to a reduced level of plant uptake and an accumulation of N within the macro-organic fraction.

be severely reduced. The implications that this has for ecosystem productivity will be discussed later.

Landscape conditions

Many interactive processes that link ecosystems operate over relatively short distances; these include the dispersal of seeds from grassland species and the migration of populations of wingless invertebrates. Industrial disturbances may occupy great areas of landscape; for example, opencast coal pits in England frequently involve drastic disturbance over hundreds of hectares (Fig. 13.7; British Coal Opencast Executive, 1991). Disturbances of this scale will effectively prevent the establishment of many linkages between the restored and undisturbed ecosystems. An ecosystem developing within a disturbed site will therefore often be geographically isolated, with few opportunities for the recruitment of absent species. This restricts the rate of colonization of appropriate plant species and so compromises natural successional processes. Furthermore, the recruitment of invertebrates vital for efficient N cycling is likely to be inhibited, thus restricting ecosystem fertility (Scullion, 1994a).

Natural successional processes in disturbed ecosystems

Natural successional processes can be extremely successful in the re-creation of botanically diverse grasslands on disturbed sites. An example of this is Millers Dale in Derbyshire, UK. This dale was abandoned after extensive disturbance during mineral extraction in ancient times. Since then, natural processes have led to the establishment of an

Fig. 13.7. An aerial view of an opencast coal pit, showing massive and widespread disturbance to the landscape.

extremely diverse community. This process of natural succession, however, occurred over a period of several centuries. This time-scale will be prohibitively long for most practitioners. Bradshaw (1997) recently reviewed the natural successional processes that operate on disturbed land. Examination of the successional dynamics has demonstrated that ecosystem establishment is constrained principally by the processes of both plant recruitment and fertility development.

Successional development of a restored ecosystem commences with the establishment of a vegetative cover. Skeletal soil media are naturally void of any form of seed bank. In other instances, such as following mineral extraction or road building, the substrate may be composed of heavily disturbed soil. The combined effects of earlier land-use practices and soil handling processes during the engineering operations will typically eliminate species characteristic of botanically diverse grasslands from the seed bank. Natural colonization will therefore depend entirely upon migration of propagules from other, undisturbed sites. However, disturbed sites are often in locations far removed from appropriate undisturbed vegetation, and this typically restricts the rate of immigration. An examination of the flora of several disturbed habitats by Ash et al. (1994) demonstrated that, even after 100 years of natural colonization, ecosystem development was still principally restricted by low rates of species recruitment.

The first plants to naturally colonize disturbed habitats are typically, although not exclusively, slow-growing stress-tolerating species, which are well adapted to nutrient deficiencies but are not necessarily typical of botanically diverse grassland communities (Ash et al., 1994). Further natural development of botanically diverse grassland communities requires the development of ecosystem fertility, followed by the recruitment of a more appropriate complement of species. Because of the low growth rate of the primary colonizing species, soil organic matter accumulation will occur slowly. In addition, the litter and organic residues produced by these species are highly resistant to microbial degradation (Palmer and Chadwick, 1985; Berendse et al., 1989; Robles and Burke, 1997). The development of ecosystem fertility and productivity will therefore only occur slowly. Furthermore, as habitat productivity approaches the critical low to moderate level, the recruitment of further species more typical of botanically diverse grasslands will be constrained by poor linkages with external and undisturbed grassland communities. Thus natural recruitment of species to disturbed sites of enhanced fertility will be a slow process (Chapman and Younger, 1995).

Natural succession will therefore be too slow for most restoration schemes and some level of artificial assistance will be necessary for the re-creation of botanically diverse grassland communities on disturbed sites.

Assisted succession and competitive interactions in disturbed ecosystems

Artificial introduction of propagules

As the establishment of an initial vegetative cover on a disturbed site is recruitment-limited, a fundamental step in an assisted successional development will be the sowing and establishment of an artificially prepared seed mixture. An approach for devising a suitable seed mixture for re-creating botanically diverse grassland communities in disturbed ecosystems has been reported by Hodgson (1989). This method allows the preparation of mixtures of species that are both typical of species-rich grasslands and suitably adapted for conditions of low nutrient availability.

Competitive relationships on nutrient-deficient soils

The successful re-creation of a botanically diverse grassland community in disturbed ecosystems will demand medium- and long-term sustainability, especially in the face of severe N deficiencies. This N deficiency severely reduces the productivity of re-created botanically diverse grasslands on these sites (Table 13.3). In contrast to extensified agricultural land, reduced productivity prevents the more competitive grasses from dominating the community

and reducing the overall diversity. However, the extreme N deficiency creates a niche where legumes, with their independent N supply, can gain a significant competitive advantage. Under such conditions, legumes may quickly come to dominate species less well adapted to N deficiencies and so may severely threaten the diversity of the re-created communities (Figs 13.8 and 13.9; Hodgson, 1989; Chapman and Younger, 1995). Clearly, therefore, the successful re-creation of botanically diverse grasslands on disturbed soil will require measures to curb the competitive advantage of legumes.

Manipulating the N supply of the restored ecosystem

The strong competitive ability of legumes in restored ecosystems may be moderated by manipulating the N cycle. Legume N_2-fixation will cause the level of soil N to gradually build up over time. As ecosystem fertility increases, non-legumes will compete more effectively with legumes, and community diversity and sustainability should increase. However, given that initial N levels may be as much as 2.5 t ha^{-1} less than equilibrium values (Davies *et al.*, 1995), this may take an unacceptable length of time. Alternatively, soil N may be supplemented by the addition of organic amendments, e.g. sludge (Bradshaw, 1997). Sludge, however, may be rich in heavy metals and the impact of these on competition within botanically diverse communities is currently unknown. Alternative sources of N may come from applications of farmyard manure or slurry. Application of slurry direct to the sward runs the risk of suppressing broad-leaved species through scorch damage to the leaves (Anon., 1985; Chapman, 1988); this may be overcome by injecting the slurry directly into the soil. Farmyard manure has long been applied to many traditionally managed botanically diverse communities (Younger

Table 13.3. Hay yields from a traditionally managed botanically diverse grassland on an undisturbed site in Teesdale and two similar communities re-created on land reclaimed following opencast coal mining, all in northern England.

Teesdale (undisturbed)[a]	Butterwell (restored following opencast coal mining)[b]	Acklington (restored following opencast coal mining)[c]
5.4 t ha^{-1}	2.25 t ha^{-1}	2.2 t ha^{-1}

[a]From Younger and Smith, 1994.
[b]R. Chapman, unpublished data.
[c]From Chapman and Younger, 1995.

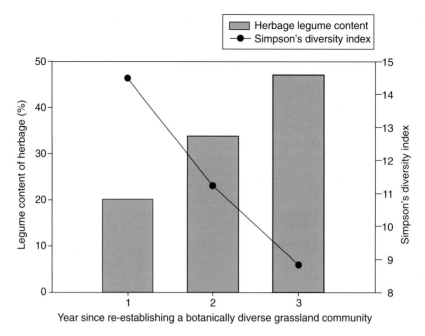

Fig. 13.8. The relationship between herbage legume content and Simpsons's diversity index over the first 3 years following the establishment of a botanically diverse grassland on land restored after opencast coal mining (data extracted from Chapman and Younger, 1995).

Fig. 13.9. A species-rich grassland community re-created on a former opencast coal site, demonstrating aggressive growth by legumes, induced by low soil N content, which is suppressing overall community diversity.

and Smith, 1994) and so may allow an increase in the N capital of the ecosystem with little risk to the community's broad-leaved components.

An experiment by Chapman *et al.* (1996) investigated the possibility of controlling the aggressive legume component by increasing the total N supply with mineral fertilizers. Applications of N increased the abundance of some of the more com-

petitive species, such as *L. perenne* and *Rumex crispus*, while reducing the abundance of *T. repens*. Overall community diversity, however, was unaffected. The small response to the applied N may be due to the timing and/or the concentration of the application.

TIMING OF APPLICATION.   Grassland species display different phenological abilities to respond to pulses of N (McKane *et al.*, 1990; Bilbrough and Caldwell, 1997). It is possible that at the time of application, only a minority of species were in a condition to utilize the enhanced N supply. The remaining species will, at best, have been unaffected by the N applications or even suppressed by the increased vigour from the positively responding species.

CONCENTRATION   OF   APPLICATION.   Grassland species have been shown to differ in their ability to utilize contrasting pulses of N (Campbell and Grime, 1989; Bilbrough and Caldwell, 1997). In particular, competitor species are best able to respond to the high-concentration N pulses that are associated with the style of mineral fertilizer application used in this experiment. This will have given the observed positive response from a minority of competitive species with no actual enhancement of overall community diversity.

Both phenological and N pulse problems may be overcome by spreading the N application over time. This will ensure that increased N supply benefits species from a wider range of phenological groups. Furthermore, individual N applications will be of lower concentration than those used in this experiment. This may enable the less competitive species to utilize the applied N with increased efficiency, thereby obtaining a positive response from a wider range of constituent species and improving the community's overall diversity.

Defoliation management

A study by Chapman *et al.* (1996) investigated the potential for strategic grazing by sheep to suppress the aggressive legume growth in restored ecosystems. Measurements made included herbage production and Simpson's equitability index, which measures the evenness with which constituent species are represented in the community. The grazing treatment effectively reduced legume biomass and this was associated with an increase in the equitability index. Defoliation by sheep, therefore, reduced the productivity of the legumes and allowed other species to compete more successfully in the re-created community. That initial study, however, failed to identify the mechanism underlying this change in competitive relationships. A subsequent investigation, which focused on the most abundant legumes in the community, demonstrated that the impact of strategic grazing arose from the selective defoliation of *T. repens* (Chapman and Younger, 1997). This significantly reduced the subsequent growth of that species and simultaneously released companion grasses from competitive suppression and so enhanced the diversity of the overall community.

It is possible that herbivory by invertebrates may provide similar benefits. Wardle and Barker (1997) reported that invertebrate herbivores selectively impeded the growth of *T. repens* in botanically diverse grasslands in New Zealand. If applied to disturbed ecosystems, this could be expected to reduce the competitiveness of the legume component and stabilize the diversity of re-created grassland communities. The opportunity to stabilize these communities through the managed reintroductions of herbivorous invertebrate populations is clearly an area that requires further research.

Conclusions

The fundamental objective that underlies all ecological restoration projects is the re-creation of stable and sustainable habitats. However, the successful re-establishment of botanically diverse grassland communities on both agricultural and disturbed sites is severely constrained by issues of competition and succession. Both the origin of these problems and their potential solutions relate to a combination of the conditions within the restored ecosystem and the nature of the linkages between that ecosystem and the external environment.

Competition problems arise from inappropriate soil fertility, and so relate directly to the internal conditions of the restored ecosystem. Typically, agricultural habitats will be characterized by excess nutrient availability, while disturbed habitats will tend to be associated with nutrient deficiencies. Nutrient availabilities may be manipulated by managing the cycling of nutrients within the ecosystem. Enhancing the rate of nutrient turnover may increase the proportion of the total nutrient pool available for plant uptake, and so effectively enhance the productivity of an otherwise infertile environment. This may be achieved on disturbed sites by increasing the populations of decomposer invertebrates and enhancing the activity of the soil microorganisms. In contrast, slowing the rate of nutrient cycling might retain more nutrients within fractions unavailable for plant uptake, thus reducing the fertility of the environment. Raising the water-table presents one opportunity for achieving this in intensified agricultural environments. Ecosystem fertility may also be effectively manipulated by importing or exporting nutrients from or to external sources. Nutrient export may be effectively achieved through the removal of vegetation or soil from the restored ecosystem. The depletion of nutrients through the cutting and clearing of vegetation is easy to implement, but may take many years to achieve the desired outcome. Topsoil removal, in contrast, may require greater initial cost and effort from the restoration practitioner, but will yield an immediate benefit in reduced site fertility. Nutrients may be readily imported through the application of either mineral fertilizers or organic manures. While the pool of N in disturbed sites may be increased following biological N_2-fixation by legumes, this may take an unacceptable length of time. Furthermore, aggressive legume growth may severely endanger the stability of botanically diverse grasslands re-created in nutrient-poor disturbed environments. However, research has shown that strategic grazing may successfully check the growth of the legumes and maintain the stability of the re-created community.

The successional development of botanically diverse grassland communities on restored ecosystems depends upon both the recruitment of appropriate propagules and the establishment of an appropriate successional trajectory. Seeds of appropriate species will typically be absent from the seed bank. Because of this, recruitment will depend upon the immigration of propagules from sources outside the restored ecosystem. This process will be greatly influenced by the landscape context of the restored ecosystem. Location of the restored ecosystem adjacent to an appropriate and undisturbed botanically diverse grassland community will allow the effective exchange of propagules. However, many grassland species are able to disperse seeds over only a very short range. Increasing the distance between the restored ecosystem and an undisturbed botanically diverse grassland community will quickly weaken and sever these linkages. Successional development then becomes constrained by the slow rate of natural recruitment. Under these circumstances, the successful re-creation of a botanically diverse grassland community may require the enhancement of succession through the artificial introduction of appropriate propagules using specially prepared seed mixtures.

If all the restrictions on propagule recruitment are overcome, there exist a number of alternative successional trajectories which the restored ecosystem might follow. The adoption of the appropriate trajectory is a vital prerequisite for the successful re-creation of a species-rich grassland. The application of the correct grazing and defoliation management therefore plays an essential role in achieving this objective.

The successful re-creation of a stable and sustainable botanically diverse grassland community therefore requires consideration of competitive and successional processes within the restored ecosystem. These are related to not only those ecological processes that are internal to the restored ecosystem, but also those processes that link that system to the external environment. Managing these processes may require manipulation of nutrient pools and nutrient cycles, careful attention to grazing and defoliation practices and controlling the colonization of the restored habitat by appropriate plant species.

Acknowledgements

I would like to thank Dr Amanda Ellery and Dr David Strong for providing useful comments on the original manuscript.

References

Abdul-Kareem, A.W. and McRae, S.G. (1984) The effects on topsoil of long term storage in stockpiles. *Plant and Soil* 76, 357–363.

Al-Mufti, M.M., Sydes, C.L., Furness, S.B., Grime, J.P. and Band, S.R. (1977) A quantitative analysis of shoot phenology and dominance in herbaceous vegetation. *Journal of Ecology* 65, 759–791.

Anon. (1985) *Conservation and Management of Old Grassland*. Farming and Wildlife Advisory Group, Sandy, UK.

Archer, J.R. (1985) Grassland manuring, past and present. In: Cooper, J.P. and Raymond, W.F. (eds) *Grassland Manuring*. Occasional Symposium No. 20, British Grassland Society, Hurley, UK, pp. 5–14.

Ash, H.J., Gemmel, R.P. and Bradshaw, A.D. (1994) The introduction of native plant species on industrial waste heaps: a test of immigration and other factors affecting primary succession. *Journal of Applied Ecology* 31, 74–84.

Baker, A-M., Younger, A. and King, J.A. (1988) The effect of drainage on herbage growth and soil development. *Grass and Forage Science* 43, 317–334.

Bakker, J.P. (1989) *Nature Management by Grazing and Cutting*. Kluwer Academic Publishers, Dordrecht, The Netherlands.

Batten, L.A., Bibby, C.J., Clement, P., Elliot, D.G. and Porter, R.F. (1990) *Red Data Birds in Britain*. Poyser, London, UK.

Bekker, R.M., Verweij, R.E.N., Smith, R.E.N., Reine, R., Bakker, R.P. and Schneider, S. (1997) Soil seed banks in European grasslands: does land use effect regeneration perspectives? *Journal of Applied Ecology* 34, 1293–1310.

Berendse, F., Bobbink, R. and Rouwenhorst, G. (1989) A comparative study of nutrient cycling in wet heathland systems. II. Litter decomposition and nutrient mineralisation. *Oecologia* 78, 338–348.

Berendse, F., Oomes, M.J.M., Altena H.J. and Elberse, W.Th. (1992) Experiments on the restoration of species-rich meadows in The Netherlands. *Biological Conservation* 62, 59–65.

Berendse, F., Oomes, M.J.M., Altena, H.J. and de Visser, W. (1994) A comparative study of nitrogen flows in two similar meadows affected by different ground water tables. *Journal of Applied Ecology* 31, 40–48.

Berryman, C. (1975) *Improved Production from Drained Grassland.* Technical Bulletin, No. 75/7, Field Drainage Experimental Unit.

Bilbrough, C.J. and Caldwell, M.M. (1997) Exploitation of spring time ephemeral N pulses by six Great Basin plant species. *Ecology* 78, 231–243.

Bradshaw, A.D. (1997) The importance of soil ecology in restoration science. In: Urbanska, K.M., Webb, N.R. and Edwards, P.J. (eds) *Restoration Ecology and Sustainable Development.* Cambridge University Press, Cambridge, UK, pp. 33–64.

Bradshaw, A.D. and Chadwick, M.J. (1980) *The Restoration of Land.* University of California Press, Berkeley, USA.

British Coal Opencast Executive (1991) *Opencast Coal in Northern, Central North-west, South Wales Regions and Scotland.* British Coal Opencast Executive, Mansfield, UK.

Campbell, B.D. and Grime, J.P. (1989) A comparative study of plant responsiveness to the duration of episodes of mineral nutrient enrichment. *New Phytology* 112, 261–267.

Carey, P.D. and Watkinson, A.R. (1993) The dispersal and fates of seeds of winter annual grass *Vulpia ciliata. Journal of Ecology* 81, 759–767.

Chapman, R. (1988) The effect of slurry in the maintenance of the clover component in mixed grass/clover swards. PhD thesis, University of Aberdeen.

Chapman, R. and Younger, A. (1995) The establishment and maintenance of a species-rich grassland on a reclaimed opencast coal site. *Restoration Ecology* 3, 39–50.

Chapman, R. and Younger, A. (1997) Spring grazing to manipulate the composition of a re-created species rich grassland habitat. In: *Proceedings of the XVIII International Grassland Congress, Winnipeg and Saskatoon, Canada,* Vol. 2. pp. 16.5–16.6.

Chapman, R. Collins, J. and Younger, A. (1996) Control of legumes in a species-rich meadow re-created on land restored after opencast coal mining. *Restoration Ecology,* 4, 407–411.

Curry, J.P. (1987) The invertebrate fauna of grassland and its influence on productivity. II. Factors affecting the abundance and composition of the fauna. *Grass and Forage Science* 42, 197–212.

Davies, R., Hodgkinson, R., Younger, A. and Chapman, R. (1995) Nitrogen loss from a soil restored after surface mining. *Journal of Environmental Quality* 24, 1215–1222.

DOE (1989a) *A Review of Derelict Land Policy.* Department of the Environment, London, UK.

DOE (1989b) *Cost Effective Managemnt of Reclaimed Derelict Sites.* HMSO, London, UK.

DOE (1991) *Derelict Land Grant Advice: Derelict Land Grant Policy* (DLGA1), HMSO, London, UK.

Elsmere, J.I. (1986) Use of fertilisers in England and Wales, 1985. In: *Rothamsted Experimental Station Report for 1985.* Lawes Agricultural Trust, Harpenden, UK, 245–251.

Fuller, R.M. (1987) The changing extent and conservation interest of lowland grassland in England and Wales: a review of grassland surveys 1930–84. *Biological Conservation* 40, 281–300.

Gibb, M.J., Baker, R.D. and Sayer, A.M.E. (1989) The impact of grazing severity on perennial ryegrass/white clover swards stocked continuously with beef cattle. *Grass and Forage Science* 44, 315–328.

Gibson, C.W.D. and Brown, V.K. (1992) Grazing and vegetation change: deflected or modified succession. *Journal of Applied Ecology* 29, 120–131.

Goldberg, D. and Novoplansky, A. (1997) On the relative importance of competition in unproductive environments. *Journal of Ecology* 85, 409–418.

Gough, M.W. and Marrs, R.H. (1990) A comparison of soil fertility between semi-natural and agricultural plant communities: implications for the creation of species rich grassland on abandoned agicultural land. *Biological Conservation* 51, 83–96.

Green, B.H. (1990) Agricultural intensification and the loss of habitat, species and amenity in British grasslands: a review of historical change and assessment of future prospects. *Grass and Forage Science* 45, 365–372.

Green J.O. (1982) *A Sample Survey of Grassland in England and Wales 1970–72.* Grassland Research Institute, Hurley, UK.

Grime, J.P. (1979) *Plant Strategies and Vegetation Processes.* John Wiley & Sons, London, UK.

Grime, J.P., Hodgson, J.G. and Hunt, R. (1988) *Comparative Plant Ecology: a Functional Approach to Common Plant Species.* Unwin Hyman, London, UK.

Harris, J.A., Birch, P. and Short, K.C. (1993) The impact of storage of soils during opencast mining on the microbial community: a strategists theory interpretation. *Restoration Ecology* 1, 88–100.

Hassink, J. (1992) Effects of soil texture and structure on C and N mineralization in grassland soils. *Biology and Fertility of Soils* 14, 126–134.

Hassink, J., Bouwman, L.A., Zwart, K.B. and Brussaard, L. (1993) Relationship between habitable pore space, soil biota and mineralization rates in grassland soils. *Soil Biology and Biochemistry* 25, 47–55.

Hodgson, J.G. (1989) Selecting and managing plant materials used in habitat reconstruction. In: Buckley, G.P. (ed.) *Biological Habitat Reconstruction*. Belhaven, London, UK.

Holmes, W. (1989) Grazing management. In: Holmes, W. (ed.) *Grass: Its Production and Utilization*. Blackwell, Oxford, UK, pp. 130–172.

Hopkins, A. and Hopkins, J.J. (1994) UK grasslands now: agricultural production and nature conservation. In: Haggar, R.J. and Peel, S. (eds) *Grassland Management and Conservation*. Occasional Symposium No. 28, British Grassland Society, Reading, UK, pp. 10–19.

Hutchings, M.J. and Booth, K.D. (1996) Studies on the feasability of re-creating chalk grassland vegetation on ex-arable land. I. The potential roles of the seed bank and the seed rain. *Journal of Applied Ecology* 33, 1171–1181.

Killham, K., Amato, M. and Ladd, J.N. (1993) Effect of substrate location in soil and soil pore-water regime on carbon turnover. *Soil Biology and Biochemistry* 25, 57–62.

King, J.A. (1988) Some physical features of soil after opencast mining. *Soil Use and Management* 4, 23–30.

Kirkham, F.W. and Kent, M. (1997) Soil seed bank composition in relation to the above ground vegetation in fertilised and unfertilised hay meadows on a Somerset peat moor. *Journal of Applied Ecology* 34, 889–902.

Kirkham, F.W., Mountford, J.O. and Wilkins, R.J. (1996) The effects of nitrogen, potassium and phosphorus addition on the vegetation of a Somerset peat moor under cutting management. *Journal of Applied Ecology* 33, 1013–1029.

Lecomte, R. (1980) The influence of agronomic applications of slurry on the yield and and composition of arable crops and grassland and on changes in soil properties. In: Gasser, J.K.R. (ed.) *Effluents from Livestock*. Applied Science Publishers, London, UK, pp. 139–180.

Leeper, G.W. (1973) *Introduction to Soil Science*. Melbourne University Press, Melbourne, Australia.

McKane, R.B., Grigal, D.F. and Russelle, M.P. (1990) Spatiotemporal differences in ^{15}N uptake and the organization of an old field plant community. *Ecology* 71, 1126–1132.

MAFF (1982) *Profitable Utilisation of Livestock Manures*. Ministry of Agriculture Fisheries and Food Booklet 2081. Ministry of Agriculture, Fisheries and Food, London, UK.

MAFF (1992) *Environmentally Sensitive Areas: the Pennine Dales*. Ministry of Agriculture, Fisheries and Food, London, UK.

Majer, J.D. (1997) Invertebrates assist the restoration process: an Australian perspective. In: Urbanska, K.M., Webb, N.R. and Edwards, P.J. (eds) *Restoration Ecology and Sustainable Development*. Cambridge University Press, Cambridge, UK, pp. 212–237.

Marrs, R.H. (1993) Soil fertility and nature conservation in Europe: theoretical considerations and practical solutions. *Advances in Ecological Research* 24, 241–300.

Mitchley, J., Buckley, G. and Helliwell, D. (1996) Vegetation establishment on chalk marl spoil: the role of nurse grass species and fertiliser applications. *Journal of Vegetation Science* 7, 543–548.

Mountford, J.O., Lakhani, K.H. and Kirkham, F.W. (1993) Experimental assessment of the effects of nitrogen addition under hay cutting and aftermath grazing on the vegetation of meadows on a Somerset peat moor. *Journal of Applied Ecology* 30, 321–332.

Mountford, J.O., Tallowin, J.R.B., Kirkham, F.W. and Lakhani, K.H. (1994) Effects of inorganic fertilisers in flower rich meadows on the Somerset levels. In Haggar, R.J. and Peel, S. (eds) *Grassland Management and Conservation*. Occasional Symposium No. 28, British Grassland Society, Reading, UK, pp. 74–85.

Olff, H. and Bakker, J.P. (1991) Long term dynamics of standing crop and species composition after the cessation of fertilizer application to mown grassland. *Journal of Applied Ecology* 28, 1040–1052.

Olff, H. and Pegtel, D.M. (1994) Characterisation of the type and extent of nutrient limitation in grassland vegetation using bioassay with intact sods. *Plant and Soil* 163, 217–224.

Oomes, M.J.M. (1990) Changes in dry matter and nutrient yields during the restoration of species-rich grasslands. *Journal of Vegetation Science* 1, 333–338.

Oomes, M.J.M. (1991) Effects of ground water level and the removal of nutrients on the yield of non-fertilized grassland. *Acta Oecologia* 12, 461–469.

Oomes, M.J.M. (1992) Yield and species density of grasslands during restoration management. *Journal of Vegetation Science* 3, 271–274.

Oomes, M.J.M., Olff, H. and Altena, H.J. (1996) Effects of vegetation management and raising the water table on nutrient dynamics and vegetation change in a wet grassland. *Journal of Applied Ecology* 33, 576–588.

Oomes, M.J.M., Kuikman, P.J. and Jacobs, F.H.H. (1997) Nitrogen availability and uptake by grassland in mesocosms at two water levels and two water qualities. *Plant and Soil* 192, 192–259.

Palmer, J.P. and Chadwick, M.J. (1985) Factors affecting the accumulation of nitrogen in colliery spoil. *Journal of Applied Ecology* 22, 249–257.

Pegtel, D.M., Bakker, J.P., Verweij, G.L. and Fresco, L.F.M. (1996) N, K and P deficiency in chronosequential cut summer-dry grasslands on gley podzol after the cessation of fertilizer application. *Plant and Soil* 178, 121–131.

Perring, F.H. and Farrel, L. (1983) *British Red Data Books. 1. Vascular Plants*. RSNS, Lincoln, UK.

Putman, R.J., Fowler, A.D. and Tout, S. (1991) Patterns of use of ancient grassland by cattle and horses and effects on vegetational composition and structure. *Biological Conservation* 56, 329–347.

Robles, M.D. and Burke, I.C. (1997) Legume, grass and conservation reserve program effects on soil organic matter recovery. *Ecological Applications* 7, 345–357.

Robson, M.J., Parsons, A.J. and Williams, T.E. (1989) Herbage production: grasses and legumes. In: Holmes, W. (ed.) *Grass: its Production and Utilization*. Blackwell, Oxford, UK, pp. 7–88.

Rodwell, J.S. (1991) *British Plant Communities*, Vol. 3, *Grasslands and Montane Communities*. Cambridge University Press, Cambridge, UK.

Russel, E.W. (1973) *Soil Conditions and Plant Growth*. Longman, Harlow, UK.

Rychnovská, M., Blazková, D. and Hrabe, F. (1994) Conservation and development of floristically diverse grasslands in Central Europe. In: t'Mannetje, L. and Frame, J. (eds) *Grassland and Society. Proceedings of the 15th General Meeting of the European Grassland Federation*. Wageningen Pers, Wageningen, The Netherlands, pp 266–277.

Scullion, J. (1994a) Earthworms and opencast mining. In: Scullion, J. (ed.) *Restoring Farmland after Coal*. British Coal Opencast, Mansfield, UK, pp. 22–30.

Scullion, J. (1994b) Manurial inputs. In: Scullion, J. (ed.) *Restoring Farmland after Coal*. British Coal Opencast. Mansfield, UK, pp. 31–35.

Scullion, J., Mohammed, A.R.A. and Ranshaw, G.A. (1988a) Changes in earthworm populations following cultivation of undisturbed and former opencast coal mining land. *Agriculture, Ecosystems and Environment* 20, 289–302.

Scullion, J., Mohammed, A.R.A. and Richardson, H. (1988b) Effect of storage and reinstatement procedures on earthworm populations in soils affected by opencast coal mining. *Journal of Applied Ecology* 25, 233–240.

Shirt, D.B. (1987) *British Red Data Books: 2. Insects*. NCC, Peterborough, UK.

Skeffington, R.A. and Bradshaw, A.D. (1981) Nitrogen fixation by plants grown on relaimed china clay waste. *Journal of Applied Ecology* 17, 469–477.

Smith, R.S. (1994) Effects of fertilisers on plant species composition and conservation interest of UK grasslands. In Haggar, R.J. and Peel, S. (eds) *Grassland Management and Conservation*. Occasional Symposium No. 28, British Grassland Society, Reading, UK, pp. 64–73.

Smith, R.S. and Jones, L. (1991) The phenology of mesotrophic grasslands in the Pennine Dales, Northern England: historic cutting dates, vegetation variation and plant species phenologies. *Journal of Applied Ecology* 28, 42–59.

Smith, R.S. and Rushton, S.P. (1994) The effects of grazing management on the vegetation of mesotrophic (meadow) grassland in Northern England. *Journal of Applied Ecology* 31, 13–24.

Smith, R.S., Buckingham, H., Bullard, M.J., Sheil, R.S. and Younger, A. (1996) The conservation management of mesotrophic (meadow) grassland in northern England. I. Effects of cutting date and fertilizer application on the vegetation of a traditionally managed sward. *Grass and Forage Science* 51, 278–291.

Stevenson, M.J., Ward, L.K. and Rywell, R.F. (1997) Re-creating semi-natural communities: vacuum harvesting and hand collection of seed on calcareous grassland. *Restoration Ecology* 5, 66–76.

Thompson, K. (1987) Seeds and seed banks. *New Phytologist* 106, 23–34.

Thompson, K. and Grime, J.P. (1979) Seasonal variation in the seed banks of herbaceous species in ten contrasting habitats. *Journal of Ecology* 67, 893–921.

Tilman, D. (1982) *Plant Strategies and the Dynamics and Structure of Plant Communities*. Monographs in Population Biology 26, Princeton University Press, Princeton, USA.

UNCED (1993) *The Earth Summit: the United Nations Conference on Environment and Development*. Graham Trotman/Martinus Nijhoff, London, Boston.

van Dorp, D. (1993) Zaaddispersie: een onderbelicht proces in het herstelbeheer. *De Levende Natuur* 6, 205–209.

Van Hecke, P., Impens, I. and Beheaghe, T.J. (1981) Temporal variation of species composition and species diversity in permanent grassland plots with different fertiliser treatments. *Vegetatio* 47, 221–232.

Vázquez de Aldana, B.R. and Berendse, F. (1997) Nitrogen use efficiency in six perennial grasses from contrasting habitats. *Functional Ecology* 11, 619–626.

Vermeer, J.G. and Berendse, F. (1983) The relationship between nutrient availability, shoot biomass and species richness in grassland and wetland communities. *Vegetatio* 53, 121–126.

Wardle, D.A. and Barker, G.M. (1997) Competiton and herbivory in establishing grassland communities: implications for plant biomass, species diversity and soil microbial activity. *Oikos* 80, 470–480.

Watt, T.A., Treweek, J.R. and Woolmer, F.S. (1996) An experimental study of the impact of seasonal sheep grazing on formerly fertilized grassland. *Journal of Vegetation Science* 7, 535–542.

Wheater, C.P. and Cullen, W.R. (1997) The flora and invertebrate fauna of abandoned limestone quarries in Derbyshire, United Kingdom. *Restoration Ecology* 5, 77–84.

Whitehead, D.C. (1995) *Grassland Nitrogen*. CAB International, Wallingford, UK.

Wilkins, R.J. and Harvey, H.J. (1994) Management options to acheive agricultural and nature conservation objectives. In: Haggar, R.J. and Peel, S. (eds) *Grassland Management and Conservation*. Occasional Symposium No. 28, British Grassland Society, Reading, UK, pp. 86–94.

Willems, J.H. (1983) Species composition and above ground phytomass in chalk grassland with different management. *Vegetatio* 52, 171–180.

Willems, J.H. and van Nieuwstadt, M.G.L. (1996) Long term effects of fertilization on above-ground phytomass and species diversity in calcareous grassland. *Journal of Vegetation Science* 7, 177–184.

Wilson, S.D. and Tilman, D. (1993) Plant competition and resource availability in response to disturbance and fertilization. *Ecology* 74, 599–611.

Younger, A. and Smith, R.S. (1994) Hay meadow management in the Pennine Dales, Northern England. In: Hagger, R.J. and Peel, S. (eds) *Grassland Management and Nature Conservation. Proceedings of a Joint Meeting between the British Grassland Society and the British Ecological Society Held at Leeds University*. British Grassland Society, Reading, UK, pp. 137–143.

14 Implications of Competition between Plant Species for the Sustainability and Profitability of a Virtual Farm Using a Pasture–Wheat Rotation

B.R. Trenbath

Centre for Legumes in Mediterranean Agriculture, University of Western Australia, Nedlands, Australia

Introduction

Farmer experience and the results of many experiments show the impact of competition between plant species on the productivity of pastures and crops (Donald, 1963; Harper, 1977). For instance, the competitive balance between legume and grass components of annually regenerating pastures in a Mediterranean environment affects not only present production but also that of succeeding crops and pasture through its effect on soil N level and weeds (Willoughby, 1954; Watson and Lapins, 1964; McCown *et al.*, 1987; Latta and Carter, 1998). Many farmers try to carefully regulate this balance. Again, it seems likely that most farmers also know that, by managing a crop to maximize its competitive ability, yield loss due to weeds will be minimized. Although both these types of competitive effect have implications for the financial viability of mixed farms, I have found no bioeconomic study that has estimated the monetary value of plant competition in a whole-farm context.

In this chapter, a simulation model is used to explore the likely long-term biological and financial outcomes of different ways of managing competition between subterranean clover and ryegrass in the annual pasture phase and between wheat and these pasture species in the cropping phases of a simplified, theoretical mixed farm. This farm is set in the central wheat-belt of Western Australia, with an all-arable area of 1800 ha of a single soil type, Wongan loamy sand. On this arable area, the model assumes that a single rotation is regularly practised, having 1 year of annual pasture alternating with 1 year of wheat. The pasture and wheat stubble are grazed by a flock of merino sheep managed to produce wool and meat. At this stage of development of the model, seasonal variation is ignored by giving all seasons the same, average weather and financial conditions.

First I describe the relevant parts of the farming system and the structure of a model to simulate it, and then present the results of runs with it to derive estimates, for four contrasted management styles, of the financial value of the competition effects of the pasture legume, and to compare these with estimates for its nitrogen- (N_2-) fixation. The model is then used to estimate the likely effect of positive interaction between the two pasture species when growing in the pasture, and to explore the possible effects of using a subterranean clover with properties varied from the standard values.

© CAB *International* 2001. *Competition and Succession in Pastures*
(eds P.G. Tow and A. Lazenby)

The Farming System

The farming system simulated is a greatly simplified version of reality. Its components have been
reduced to subterranean clover, annual ryegrass,
sheep, wheat, soil organisms and humans, with
their machines and agrochemicals. The succession
of operations in the 2-year cycle of the assumed
pasture–wheat rotation is summarized in Fig. 14.1.
Although many annual species occur in the real system as weeds of both phases, here clover and ryegrass represent the weeds of wheat. Ryegrass is in

some ways a weed of pasture, but, due to its strong
response to soil N, it is often an important component of pasture, especially early in the growing season. The pasture phase is considered first.

In autumn, immediately after the opening rains
of a new season, the sheep are moved into yards,
while, on the land that carried wheat the previous
year, the pasture phase begins, as residual seeds of
ryegrass and clover regenerate a sward within the
decaying stubble of the previous crop. After 30 days
of being fed in yards, the farm's sheep are brought
out on to the new pasture. The set stocking rate for

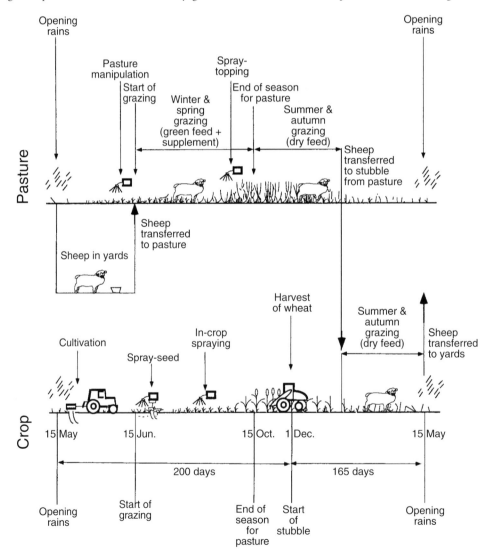

Fig. 14.1. Diagram showing main activities and time points in the 2-year cycle of the pasture–wheat
rotation assumed for the farm in the model. A season starts with the arrival of opening rains in autumn.

the year is an 'optimal' one, chosen to be the largest that can be safely fed.

As rain continues to re-wet the soil from time to time, the established pasture grows at a rate decided by stocking rate and rainfall. To remove excess grass from the sward, the farmer may intervene with herbicidal sprays at two points in the pasture growing season. Near the start, a selective spray can reduce grass seedling densities and allow clover to grow with less competition. This early spray operation is 'pasture manipulation'. The second possible point is towards the end of the growing season, when the grass seeds are starting to ripen. This late spray is 'spray-topping'. Although a non-selective spray is used for spray-topping, the timing of its application in combination with careful grazing management leads to a selective effect.

Until the pasture has accumulated sufficient biomass, the feed intake of the sheep is supplemented with oat grain. When sufficient herbage is available, supplementation is stopped. After a phase of fast growth in spring, rains cease, seeds are set, soil water becomes exhausted and the pasture stops growing and dries off. This is the 'end of season' for pasture. The dried end-of-season biomass then remains as a source of low-quality feed for grazing through the nearly rainless months of summer and autumn.

After the wheat has been harvested and little feed remains on the pasture, all the sheep are transferred to the stubble. At the stocking rate chosen, sheep can find there enough feed of sufficient quality until the opening rains of the next season.

With the arrival of the rains of a new season, land entering the cropping phase is given a shallow cultivation. This covers many of the unburied seeds of pasture species and encourages them to germinate before the wheat is sown. Sowing of the crop is accompanied by a spray of a non-selective herbicide, Spray.seed®, which kills the emerged seedlings of pasture species.

Later-emerging weed seedlings appear in the emerged crop and, provided the populations are still susceptible, can be controlled using in-crop selective herbicides. These are Hoegrass® for ryegrass and Ally® for the clover. Any plants that germinate in the stubble after summer rains die of drought without reproducing.

Outline of the Model

Because the wheat and sheep pasture (WASP) model has already been described in detail (Trenbath and Stern, 1995),[1] only some background material and the features critical to the understanding of this study are given here.

The biological basis of the system as modelled depends on three main state variables:

- Total soil N.
- The seed pool of clover.
- The seed pool of ryegrass.

In spite of the simplifications, the interactions among these components that needed to be modelled are numerous. The processes, activities and interactions included in the model are outlined in a relational diagram (Fig. 14.2). The three main state variables are highlighted in rectangular boxes and optional management activities are shown in diamonds. Most of the other items appearing in the diagram act as intermediate variables. For the sake of clarity, the diagram omits some important aspects treated in the model (e.g. the distribution of seeds between three soil layers and the effects of cultivation on this). To represent the relationships between the variables shown in Fig. 14.2, the WASP model contains 160 equations involving 173 parameters.

The model in its present form refers specifically to one location (Wongan Hills, Western Australia). The averaged seasonal climate for this location is specified by three invariant cardinal dates: expected dates of opening rains and of pasture and wheat maturities. Productivities are estimated for an average year with 375 mm of rainfall. Costs and prices in the model refer to 1994 Australian dollars (A\$).

Early runs with the model showed that a key component of the model in determining the profitability of the farm system is the size of the ryegrass seed bank at the start of each wheat phase. Although ryegrass is a useful species in pasture because with high soil N it can produce more biomass than clover, its prolific seed production makes it potentially a serious competitor with the wheat in the crop phase. Since the management of ryegrass is so critical in real farms, the model makes allowance for the use of up to 13 different methods for controlling it at different points in the cycle of pasture and crop (Fig. 14.3). While this may seem a formidable array of methods, on many farms the recent appearance of herbicide resistance has started to rule out options involving selective herbicides in both crop and pasture; furthermore, the risk of erosion has made farmers less inclined to burn stubble or pasture just before the start of the new season.

The most important of the 13 methods are considered below, in the next two sections.

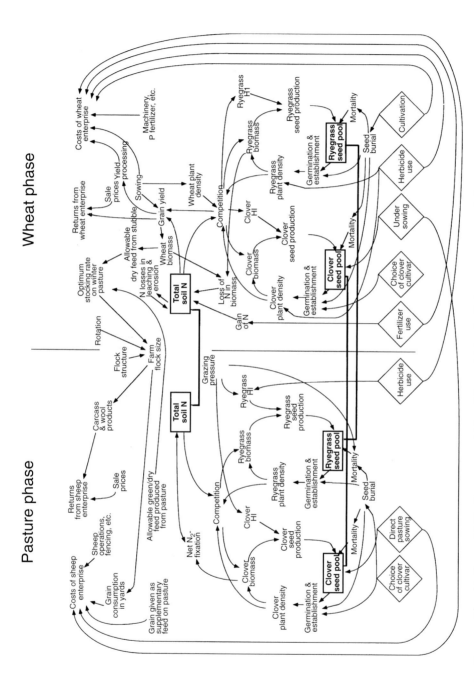

Fig. 14.2. Interactions considered in the model. The three state variables of the model (total soil nitrogen, ryegrass and clover seed pools) are shown in rectangles and appear both on the left side (pasture phase) and on the right side (wheat phase) of the diagram. Options for intervention by the farmer are shown in diamonds along the bottom. HI, harvest index = weight of seeds/total above-ground biomass at the end of the pasture growing season.

The Modelling of Plant Competition

Competition with ryegrass can greatly reduce the fixation of N in pasture and depress the yield of wheat in the crop phase. The viability of the system relies largely on how well the ryegrass population is controlled. Two methods for controlling ryegrass depend on plant competition.

A first method for controlling ryegrass is encouraging the growth of clover (or other species) in the pasture. This exerts competitive pressure on the ryegrass and lowers its biomass and seed production. In the model, the presence of clover

reduces ryegrass seed production to an extent depending on clover's plant density and competitive ability, the latter depending on the soil N level.

To mimic the interaction between the species in pasture, de Wit's (1960) competition model is used. In an ungrazed mixture of two plant species i and j, the biomass yield y_{ij} of species i is given by the equation

$$y_{ij} = \frac{y_{ii} \cdot p_i k_{ij}}{p_i k_{ij} + p_j} \tag{14.1}$$

where k_{ij} is the crowding coefficient of species i in respect of species j, y_{ii} is the ungrazed biomass yield

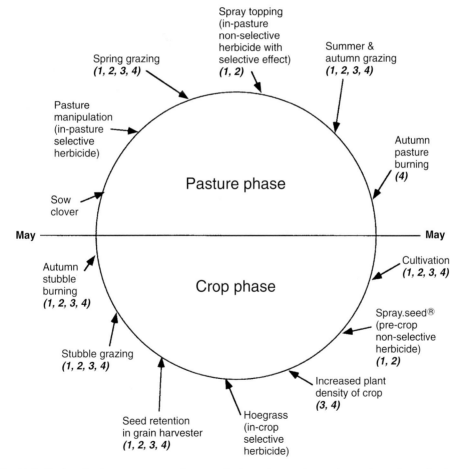

Fig. 14.3. Options for the management of ryegrass in the pasture and crop phases of a pasture–wheat rotation. The position of each arrow indicates roughly the point in the 12-month clockwise cycle from May to May at which a control measure is applied. Numbers in parentheses show which of the four simulated farms use the method. Burning of stubble or pasture is used in simulations of farm 4, but only as extra variants mentioned in the text.

of a pure stand of species i, and p_i is the proportion of plants of species i in the mixture ($p_j = 1 - p_i$). In the experimental systems to which this approach has been applied, all mixtures and pure stands are at a standard plant density, so that the range of possible mixtures represents a 'replacement series'. When $k_{ij} \cdot k_{ji} = 1$, the two species are considered to be competing for the same 'space' and there is no yield advantage from growing a mixture of species, but when $k_{ij} \cdot k_{ji} > 1$, then some kind of positive effect or synergy is present in the interaction of the two species. Under some conditions, this synergy can cause the mixture to outyield a pure stand of the more productive species (Trenbath, 1976). Even when the mixture does not outyield the pure stands, synergy will still raise the yield of the mixture. In legume–grass mixtures, synergy is commonly found, because the species differ in their sources of soil N (de Wit and van den Bergh, 1965).

According to Equation 14.1, a crowding coefficient bears no necessary relationship to the productivities of the species in their pure stands. It is usually independent of the proportions of the two species in the mixture (de Wit, 1960). In the WASP model, lacking data showing otherwise, the crowding coefficients of clover in respect of ryegrass (k_{cr}) and of ryegrass in respect of clover (k_{rc}) are assumed to be also independent of the total plant density.

In most of the simulations to be reported, $k_{cr} \cdot k_{rc} = 1$, and the k_{rc} is easily derived from a calculated k_{cr} as $k_{rc} = 1/k_{cr}$. However, to mimic an interaction between clover and ryegrass where synergy is present, a situation where $k_{ij} \cdot k_{ji} > 1$ is required. This is arrived at by simply adding a constant Δ to each coefficient calculated as above (Trenbath, 1983).

The modelling of the competitive ability of subterranean clover in respect of annual ryegrass is made difficult by the lack of published experimental data on replacement-series mixtures of them grown in the field, with grazing or cutting, over a range of plant densities and soil N levels. Accordingly, judging from the results of an undefoliated pot experiment with these species in additive mixtures by Trumble and Shapter (1937), a falling asymptotic response of log k_{cr} to increasing soil N seems appropriate (Fig. 14.4a). This response matches common experience, in which clover outcompetes grass at low N levels but is suppressed by it at high levels (Willoughby, 1954; Stern and Donald, 1962). Tuning a negative exponential curve so that k_{cr} values derived from it could be used in Equation 14.1

to fit the total mixture yields from the three treatments of an undefoliated field experiment in Albany, Western Australia (Moore, 1989), gave the heavy line in Fig. 14.4a. This curve was finally adjusted upwards slightly to take account of the likely higher competitive ability of clover under grazing (Greenwood et al., 1967). Using this final curve, the model generates results (Fig. 14.5a and b) similar to the unpublished results of the individual components from the two high-density treatments of Moore's Albany experiment (Fig. 14.5c and d; J.H. Moore, Albany, 1996, personal communication). Setting the parameter $\Delta = 0.6$ modified these modelled results (Fig. 14.5e and f) to imitate quite well the degree by which some mixtures outyielded the pure stands in two treatments of a defoliated replacement-series experiment with barrel medic and oats carried out in the field near Adelaide (Fig. 14.5g and h; E.V. Naidu and P.G. Tow, Adelaide, 1998, personal communication). The similarity of the observations and the simulations can be assessed using a measure of the degree of synergy in replacement-series mixtures, relative yield total (RYT) (de Wit and van den Bergh, 1965); in the top row of graphs in Fig. 14.5, the average RYTs are 1.00 (simulated) and 1.13 (observed), and in the lower row in Fig. 14.5, maximum RYTs are 1.26 (simulated) and 1.30 (observed).

A second method of controlling ryegrass by competition can be used in the crop phase. The competitive effect of ryegrass on wheat yield is shown in Fig. 14.4(b and c). The similarity of the responses of wheat yield to added ryegrass and clover plants (Fig. 14.4d) allowed the model to treat plant populations of the two species as competitively equivalent in respect of wheat yield depression. However, in the crop, an increased seed rate of the wheat raises the crop's competitive ability, so that weeds are more strongly suppressed. Although the use of increased seed rates is apparently more popular among wheat farmers in the eastern states of Australia (S.B. Powles, Perth, 1998, personal communication), in the Wongan Hills area of Western Australia it seems equally feasible; in an average season, a threefold increase of seed rate over the conventional level will increase weed suppression without jeopardizing yields (W.K. Anderson, Perth, 1998, personal communication). For the cost of the additional wheat seed, this increased weed suppression allows the wheat yield to rise towards a weed-free level (Fig. 14.4e; M.L. Poole, Perth, 1978, personal communication).

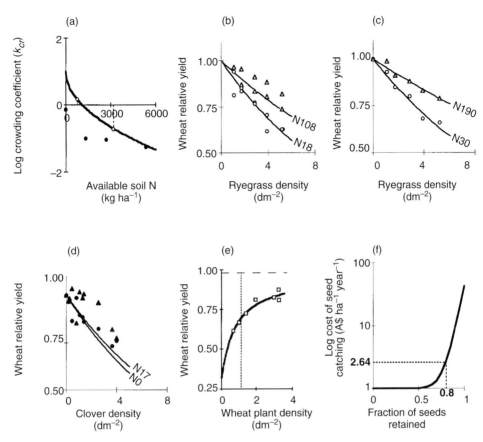

Fig. 14.4. (a) Effect of soil N level on the interaction between ryegrass and clover. Estimates of the crowding coefficient k_{cr} (filled circles) at different levels of available soil nitrogen, based on results of an undefoliated pot experiment using ryegrass and clover grown in mixtures and pure stands (data of Trumble and Shapter, 1937). An exponential curve (heavy line) is drawn to follow the same trend as the experimental points and to pass through two points (open circles) which allow the model to mimic Moore's (1989) results. The levels of 'available' soil N (783 and 3159 kg N ha^{-1}) that had to be assumed to fit Moore's experiment are shown as vertical broken lines. These levels are expressed in terms equivalent to total soil N at Wongan Hills. (b) to (d) Response of wheat grain yield at various sites in Western Australia to density of seedlings of weeds: ryegrass, open symbols; clover, filled symbols (for the unpublished data sources, see Trenbath and Stern, 1995). Wheat yield is presented as relative to its weed-free value. Symbols indicate the N treatment as either 'high' (triangles) or 'low' (circles). Lines are modelled responses based on levels of applied N fertilizer (kg N ha^{-1}) chosen to match those applied to the experiments. (b) Ryegrass at Perth, (c) ryegrass at Wongan Hills, (d) clover at Wongan Hills. (e) Response of relative wheat yield to increasing wheat plant density with ryegrass plant density, constant at 4.5 dm^{-2}, showing fitted curve used in the model (M.L. Poole, Perth, 1978, personal communication). Yield depression decreases with increasing crop density. The asymptote of the curve is indicated by the dashed horizontal line, and the local normal wheat density of 1.1 dm^{-2} by the dashed vertical line. (f) Curve assumed in the model to relate the cost of a seed-catching modification of the harvester to its effectiveness. For an effectiveness of 80%, the cost shown is equivalent to an estimate by J.M. Matthew (Perth, 1998, personal communication) (see text).

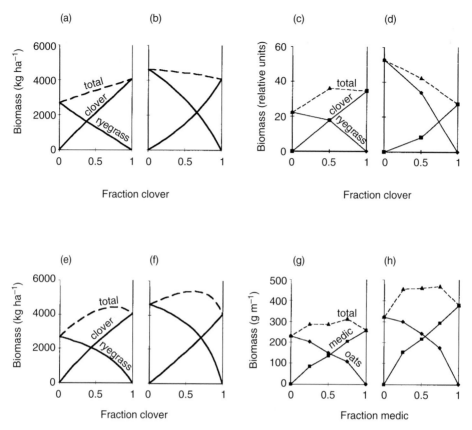

Fig. 14.5. Modelled replacement series results of grass and legume compared with experimental data, without and with synergism. Composition of mixtures is given as fraction of legume. (a) Modelled results with relative yield total = 1 and assuming available soil nitrogen (ASN, kg ha^{-1}) = 350; (b) as (a) but ASN = 750; (c) low-N treatment of field experiment using ryegrass and clover, Albany, Western Australia (Moore, 1989); (d) as (c) but high-N treatment; (e) modelled results with ASN = 350 and Δ = 0.6, producing relative yield totals up to 1.28; (f) as (e) but with ASN = 750 and relative yield totals up to 1.24; (g) frequent-defoliation treatment of field experiment using oats and Jemalong medic (*Medicago truncatula* Gaertn.), Roseworthy, South Australia (E.V. Naidu and P.G. Tow, Adelaide, 1998, personal communication); (h) as (g) but infrequent defoliation.

Modelling of Other Processes Affecting Populations of Ryegrass and Clover

Besides approaches involving plant competition, other commonly used methods for managing population levels of ryegrass and clover are cultivation, spraying herbicides, using a modification to the harvester that catches ryegrass seeds during the wheat harvesting, grazing by sheep and burning pasture and stubble. These will be considered in turn.

Cultivation

In the systems to be simulated, it is assumed that, within 2 days of the rains that open the season, the soil receives a shallow cultivation. The intention is to bury surface seeds of ryegrass to maximize the number that germinate in the following 2 weeks (Pearce and Quinlivan, 1971). At this time, many of the seeds of both pasture species, ryegrass and clover, germinate and emerge. The seeding of the wheat is accompanied by the application of non-selective Spray.seed® to the soil surface to kill weed seedlings, and by applications of superphosphate and any nitrogenous fertilizer required. Unless stated otherwise, all costs are according to Department of Agriculture Western Australia (1993).

The depth of burial of a seed of ryegrass or clover largely determines its later fate and so in the model the soil profile is divided into three layers, each of 3 cm thickness. To mimic the effect of the two soil disturbances before and after the seeding of wheat, probabilities of redistribution of seeds between layers and of germination and seedling establishment were calculated for the model from data of Gramshaw (1974) and Saoub (1994). Corresponding probabilities for the pasture phase were found from the same sources. In the model, the outcome of these processes strongly affects the plant population densities in the crop and in pasture.

Herbicidal sprays

In the simulations to be reported, it is assumed that herbicide resistance is already sufficiently common in the ryegrass for the use of grass-selective sprays, such as Hoegrass®, to have been abandoned. However, spraying of the seed heads of ryegrass with the non-selective glyphosate ('spray-topping') does still, if well timed, produce a strongly selective reduction in ryegrass seed production. The standard dose of Glyphosate CT® used in the model produces a 91% non-development of ryegrass seeds (a corresponding 18% of clover seeds are also prevented from developing).

The selective action of Ally® against clover in wheat is still exploitable and, if it is to be used in a run, a standard rate is applied when clover plant density in the crop exceeds a threshold value decided by the farmer; this rate kills 90% of the clover seedlings growing as weeds.

If it is to be used, the non-selective spray, Spray.seed®, is applied at the recommended rate when the crop is seeded. At present, it is still effective against both clover and ryegrass, killing 95% of all seedlings.

Retention of ryegrass seeds in the harvester

As herbicide resistance narrows their range of options, farmers are starting to experiment with modifications to their grain harvesters to catch and retain ryegrass seeds present in the stand of wheat.

One modification, the discontinued Ryetec unit (Matthew et al., 1996), cost A$9000 at the time (1994) when the prices in the model were estimated, and this caught and retained usually about 80% of the ryegrass seeds growing in wheat (J.M. Matthew, Adelaide, 1998, personal communication). Lying on the surface or buried, clover seeds are not caught. Further research is under way in South Australia, suggesting that new modifications will soon be available. An assumed relationship between the logarithm of the cost of using these future machines and their efficiency is given in Fig. 14.4(f). This curve passes through the log cost of using the Ryetec unit with its 80% efficiency, assuming a 10% interest rate on a mortgage to cover the purchase and a 5-year life of the machinery. Starting at a low level representing a minimum cost of any modification (about A$3000), the curve climbs steeply to reflect the rapid increase in sophistication required in the machinery as it approaches perfection of performance.

Grazing by sheep

Sheep grazing pressure on stubble and pasture is governed by a general assumption that, on the farm, grazing is always 'prudent'. The 'optimal' flock size selected by the model is the largest one possible, subject to two conditions ensuring: (i) a humane use of sheep; and (ii) an environmentally benign use of the land. Specifically, the conditions require that the sheep take no more than half of the end-of-season pasture biomass and no more than two-thirds of the initial stubble biomass, and that neither pasture nor stubble should ever carry less than 1500 kg ha^{-1} of biomass when grazing ends. The proportional thresholds ensure that sheep can find sufficient feed of reasonable quality to prevent them from consuming clover seeds, and the 1500 kg ha^{-1} threshold ensures that there is enough biomass to prevent wind erosion. Although the model moves with a yearly time step, functions are included that have been based on simulations of grazing using a daily time step to make sure that the conditions are not violated at critical times. These critical times are just before the stubble becomes available and just before the opening rains of the next season.

To simplify the calculation of grazing pressures of sheep on pasture and stubble, it is assumed in the model that grazing of the mature pasture proceeds until one of the above conditions is breached, at which time all the sheep are transferred from pasture to the stubble. Simulations show that under most management styles the sheep finish the grazing year on stubble. When this occurs in the model, all the nutritious and very visible ryegrass head seed (an assumed 5% of the total production in the crop) is presumed grazed and destroyed, whereas shed ryegrass seed (like surface and buried clover seed) is not eaten.

Loss of seed production due to grazing of the pasture varies with stocking rate. In the model, before the transfer to stubble, the pasture carries the whole farm flock so that the stocking rate is the flock size divided by the area of pasture. The weights of seed produced by the two pasture species are found by multiplying the end-of-season grazed biomass of each species by its calculated 'harvest index'. The end-of-season total biomass under this stocking rate is found from previous simulations, depending on initial seedling densities, stocking rate and soil N. To include the effect of competition on seed production, this total biomass is then partitioned between the two species according to the ratio of the individual ungrazed biomasses, calculated using the competition model of Equation 14.1. The appropriate harvest index is found from estimated relationships between harvest index and grazing pressure, expressed as stocking rate per unit of end-of-season biomass. Three grazing trials in southern Australia agree that ryegrass harvest index declines with grazing pressure and two trials agree that of clover ultimately declines but only after an initial rise. For the model, these results are translated, rather speculatively, into the heavy lines in Fig. 14.6a and b.

At the end of the pasture's growing season, an assumed 30% of the freshly matured seeds of ryegrass remain in the seed heads and suffer loss by grazing. These head seeds suffer a proportional loss that depends on the intensity and duration of the stocking (Fig. 14.6c; Gramshaw, 1974). As in stubble, shed seed of ryegrass and all seeds of clover are assumed to be safe from grazing by sheep.

The costs of imposing the grazing are absorbed into the costs of the sheep enterprise, estimated mainly from Pannell and Bathgate (1994).

Burning

As a means of selectively killing ryegrass seeds, the burning of old pasture or stubble can be effective, but is now generally not favoured because it leaves the soil unprotected against erosion. For the purpose of the model, its effect on ryegrass seeds is estimated from an experiment (Davidson, 1994) where paddocks carrying varying amounts of stubble were burned and seedling counts were made in the following season. Unfortunately, the experimental design did not include a control, and so a curve fitted to the three experimental points was extrapolated backwards to give a plausible result for an unburned treatment (Fig. 14.6d). This curve was then standardized by division by the estimated control value to give a relationship between survival through burning and biomass burned.

'Experimental Design' for the Simulations of Four Contrasting Farms

In preparation for a comparison of whole-farm simulations, in which species either have their properties changed or in which species are deleted, standard forms of four farm types are first defined. These farm types show a range of possible management styles within the same rotation. They can be viewed as two pairs of systems, each pair with N fertilizer either used or not, but with less use of agrochemicals in the second pair:

1. Conventional + N:
   (a) normal wheat seed rate;
   (b) Ally® sprayed against clover in wheat;
   (c) the ryegrass is resistant to selective herbicides and is controlled by catching 80% of the seeds in the harvester and by spray-topping with glyphosate in the pasture phase;
   (d) fertilizer N is applied each season at the current economically optimal rate;
   (e) superphosphate (100 kg ha^{-1}) is applied in each wheat phase.
2. Conventional – N: as in farm 1, but no fertilizer N used.
3. Reduced-input: fertilizer N and superphosphate applications as in farm 1, but:
   (a) no sprays used except at seeding (when Spray.seed® is used);
   (b) triple the standard wheat seed rate;
   (c) catching 90% of the ryegrass seed in the harvester.
4. Low-input (near-organic): As in farm 3, but no N fertilizer is applied, and:

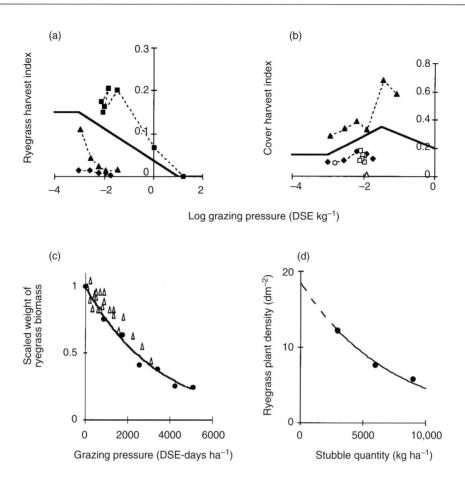

Fig. 14.6. (a) and (b) The response of harvest indexes of ryegrass and clover to intensity of grazing. Grazing pressure is measured on a log scale as stocking rate (dry sheep equivalent (DSE) ha^{-1}) divided by the biomass (kg ha^{-1}) of the pasture at the end of its growing season. The model uses the heavy lines interpolated between the data points (for the unpublished data sources, see Trenbath and Stern, 1995). (a) Ryegrass. Experiments at Adelaide, South Australia (filled squares) and Albany, Western Australia (WA) (filled diamonds and triangles). (b) Clover. Experiments at Albany (filled diamonds and triangles, and open squares) and Katanning, WA (open triangle and open circle). (c) Response of dried ryegrass biomass (and seed) to grazing pressure during the summer–autumn period (data of Gramshaw, 1974). Experiments at Perth, WA (triangles) and Bakers Hill, WA (circles). (d) Effect of stubble quantity on survival of ryegrass seeds through autumn burning (data of Davidson, 1994). The seeds were produced in the past season and were situated on or just above the soil surface. Survival was judged by ryegrass plant density in the succeeding wheat crop. An exponential curve was fitted to the three data points and extrapolated to estimate a zero-stubble 'unburned' value.

(a) no Spray.seed® used;
(b) percentage of ryegrass seeds caught reduced to 80%, but with the cost of a machine originally catching 90%;
(c) disc ploughing and extra cultivations used to compensate for the lower percentage of ryegrass seeds caught.

The order 1 to 4 can also be seen as representing a possible evolutionary sequence through time, in which the farmer is adapting to a more and more difficult situation: the farmer is devising ways of protecting the financially critical wheat crop from competition from ryegrass, which is becoming progressively harder to kill, and at the same time reacting to public pressure to reduce environmental pollution. Farm 1 already has ryegrass resistant to selective herbicides, but the farmer maintains a control by spray-topping the pasture using a non-selective chemical, together with the new technology of seed catching. Farm 2 is managed in a 'greener' fashion, because the farmer has been persuaded to discontinue N fertilizer application in order to minimize leaching of N to the groundwater. The wheat therefore depends for N on fixation by clover; the generally lower soil N levels reduce the competitiveness of ryegrass. In farm 3, to avoid dependence on selective sprays altogether and to extend the effectiveness of non-selective sprays as long as possible on the property, the farmer has stopped using all sprays except the non-selective one, at seeding. To retain control of ryegrass, he or she uses a more sophisticated harvester modification. Nitrogen fertilizer is still used.

Farm 4 with its near-zero input of agrochemicals, might represent the gloomy scenario in which sprays are no longer used because herbicide resistance has become widespread and N fertilizer is too heavily taxed ever to be economic to use in extensive farming. It is further supposed that use of the harvester modification has selected ryegrass populations that set seed lower in the crop, so rendering it less effective. In an effort to compensate for this, very thorough cultivation is performed, starting immediately after the opening rains. The soil disturbances occur, as in the other farms, in two phases, with a fortnight in between to allow weed germination. The first phase is disc ploughing and a cultivation and the second is another cultivation and then seeding (Taylor, 1985). This is taken to give the same 95% kill of weeds as in a successful Spray.seed® operation (T.J. Piper, Perth, 1997, personal communication). Although the cost of Spray.seed® application is saved, the extra passes of plough and cultivator cost much more than the single cultivation assumed in Farms 1 to 3 (A.D. Bathgate, Perth, 1998, personal communication). The thorough cultivation leads to a more even redistribution of seeds down the soil profile.

To illustrate the basic form of output from the model, the results of a simulation of the Conventional + N farm 1 are shown in Fig. 14.7. Simulated time courses are shown for 30 cycles of the 2-year rotation – that is, for 60 years. The 'pasture' graphs in the top row give the plant densities in pasture, the stocking rate on the winter pasture and the clover crowding coefficient (k_{cr}), the grazed

pasture biomasses at the end of season and the numbers of seeds produced. The 'crop' graphs in the second row indicate the plant densities of the pasture species growing in the crop after any spraying, show whether the spray Ally® has been used to control clover, give the wheat crop's potential yield at its level of soil N together with the actual yield as reduced by competition from weeds, and record the cycle's initial total soil N level and fertilizer N application to the wheat. The 'economics' graphs in the bottom row show the incomes and costs for the wheat and the sheep enterprises, the two enterprise net incomes and the total farm net income ('operating surplus'). Questions of farm ownership, debts and taxation are not considered.

While the standard form of farm 1, as defined above, leads to an apparently sustainable system with a high steady net income of about A$150,000 year^{-1}, the simulation results shown in Fig. 14.7 are not of this standard form, but of a slight variant. It has been chosen to illustrate the narrowness of the margins of error that the farmer faces, and also some of the dynamic behaviour that the system can display. Instead of the usual seed catcher of farm 1, which catches 80% of the ryegrass seeds at harvest, the farmer has for this simulation chosen a cheaper and less efficient kind, which only catches 70% of the seeds. In this variant, the ryegrass population has not been controlled well and the wheat yield varies cyclically, due to variation in the level of infestation of clover in the crop, and the consequent periodic spraying with Ally®. Severe losses, amounting to half the potential yield occur about every 6 years, causing farm income to oscillate between A$55,000 and A$110,000 year^{-1}.

Because of the difficulty of comparing variable income streams lasting 60 years, in the simulations reported below the net present value (NPV) in Australian dollars of the streams of operating surplus is used as an indicator of farm profitability. The NPV of a future stream of operating surplus is the sum of money that would need to be invested now in an account bearing interest at a fixed rate to allow the stream to be exactly paid from it, year by year, without any final residue. For this calculation, a conventional 10% interest rate is assumed. As an example of its use, it can be applied to the previous comparison of the standard farm 1 simulation and the variant with the less efficient seed catcher shown in Fig. 14.7. With the better 80% catcher, the oscillations in operating surplus disappear and the NPV, with a 10% discount rate, of this same

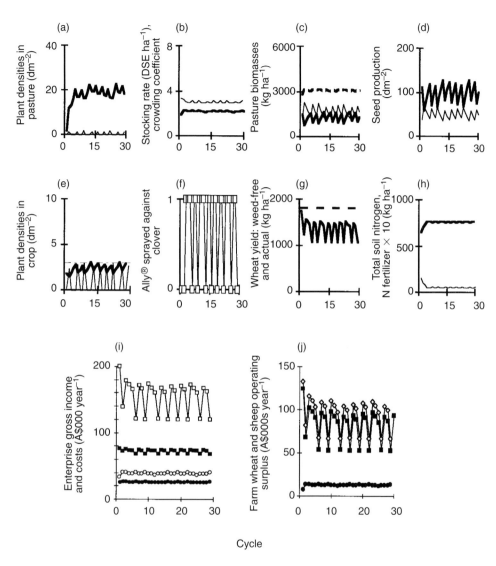

Cycle

Fig. 14.7. Simulation of the conventional + N farm 1, in which ryegrass is not well controlled by the combination of light cultivation, Spray.seed®, spray-topping and grazing, but with only 60% seed retention in the harvester. Types of graph: top row, pasture phase (various units); middle row, crop phase (various units); bottom row, financial indicators (thousands of A$ year^{-1}). Where the two pasture species are plotted together, ryegrass is shown by thick and clover by thin lines. (a) Plant density (dm^{-2}) of ryegrass and clover; (b) stocking rate of pasture in winter (dry sheep equivalent (DSE) ha^{-1}) (thin line) and crowding coefficient of clover in respect of ryegrass (thick line); (c) grazed pasture biomass at end of its growing season (kg ha^{-1}): ryegrass, clover and total (broken line); (d) number of seeds produced (dm^{-2}); (e) plant density (dm^{-2}) of ryegrass and clover. The horizontal dashed line is the threshold plant density of clover for the spraying of Ally®; (f) use of Ally®: yes (1), no (0); (g) weed-free wheat yield (kg ha^{-1}) (broken line), actual yield (kg h^{-1}) (solid line); (h) total soil nitrogen (kg ha^{-1}) (thick line) and N fertilizer applied (kg N × 10 ha^{-1}) (thin line); (i) gross income (open symbols) and costs (filled symbols) of two farm enterprises, wheat (squares) and sheep (circles); and (j) operating surpluses of wheat (squares) and sheep (circles) enterprises and the total farm operating surplus (diamonds).

farm's future operating surpluses is A$1.4 million; the less efficient catcher causes the NPV to be only A$1 million.

The first group of comparisons made with the model simulated, for each farm, a 30-cycle run of seasons with the following features of the plants:

(a) standard form, with Δ = 0 to always give pasture RYT = 1;
(b) as (a) but with Δ = 0.6 to give pasture RYT values of up to about 1.3;
(c) as (a) but without N_2-fixation;
(d) as (a) but without clover;
(e) as (a) but without ryegrass.

The second group of runs compares farms that have clover differing in: (i) potential end-of-season biomass; and (ii) crowding coefficient.

Results and Discussion

Results of simulations

In most runs, the systems were close to equilibrium soon after ten cycles; short-period oscillations, as seen in Fig. 14.7, were usually due to the rising population of clover periodically reaching the threshold at which the crop was sprayed with Ally®. Long-period oscillations, as in Fig. 14.9(a) below, were uncommon.

Comparing the standard forms of the four farms in Fig. 14.8a, the NPVs decline in the assumed time sequence 1 to 4, as constraints due to herbicide resistance or environmental considerations accumulate. The trend is due to the increase in use of originally less effective and/or more costly means to control the ryegrass. From an environmental viewpoint, the use of intensive cultivation in farm 4 to match the effectiveness of the abandoned Spray.seed® can be seen as a retrogressive measure bred of desperation. Although effects of tillage on soil structure are not modelled, intensive cultivation degrades the soil structure and thus represents a return to the aggressive approach to farming used earlier on Australian wheat farms. Indeed, other backward steps might be hard to avoid as farmers saw income dwindling: according to the model, the standard form of farm 4 could increase its NPV by respectively 33%, 36% or 100% if its stubble or pasture or both were burned just before each new season began. Flame or steam treatments

to kill weed seedlings are environmentally benign alternatives but are costly.

Figure 14.8(a) shows that in all farms the variant with enhanced RYTs and therefore greater pasture production has either the same or lower profitability. This result is so unexpected that it will be analysed further below.

When N_2-fixation by clover is suppressed in the model, the profitabilities of farms 1 and 3 are little affected, but those of farms 2 and 4 collapse. The low-input farmer in 4 even makes a loss. The pattern of results is similar when clover is removed from the system, with farms 2 and 4 suffering greatly. Clearly, the provision of fertilizer N protects the system when biological N is no longer supplied by the clover.

The removal of ryegrass favours the profitability of all farms, but, in the case of the low-input farm, the improvement is dramatic. The reason for this exceptional behaviour is that, in this farm, the wheat crops are much more strongly infested by ryegrass than elsewhere. Yields are only half of the potential level. Partial removal of this limitation on wheat yield allows it to rise to 70% of its potential and so net income can more than double.

The results given in Fig. 14.8(a) suggest a way to quantify the financial benefit derived by farmers from the use of clover in their rotations. Following a study by Ewing et al. (1987) using the mathematical modelling program MIDAS to estimate the benefit from inclusion of lupin in a lupin–wheat rotation, I use the same principle of running the WASP simulations with and without clover present and judging the benefit by subtracting the farm profit without legume from that run with the legume.

Using a similar principle borrowed from another MIDAS study (Pannell and Falconer, 1987), we partition this benefit into two additive parts. Thus, for each farm, after finding the 'clover effect' (third column in Fig. 14.8b) as the profit from a with-clover system minus that of the no-clover system, we see that the clover effect can be viewed as the sum of a 'N_2-fixation effect' and a residual effect. The fixation effect is obtained as the profit from a standard system minus the profit from one with a non-fixing clover, while the residual effect is equal to the profit from the non-fixing clover system minus that from a no-clover one. Since there are no special sheep-nutritional/pasture quality effects of clover biomass as opposed to ryegrass included in the model and disease has been ignored, the residual

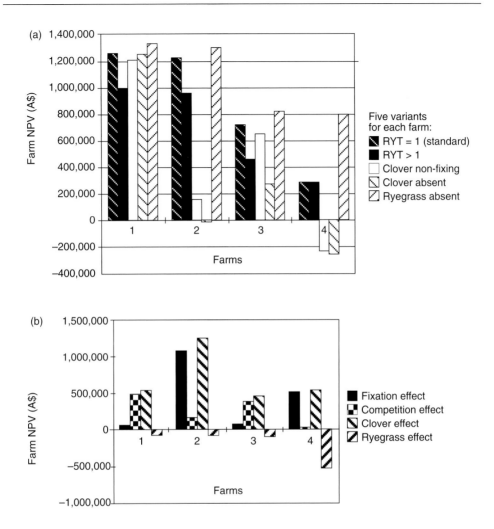

Fig. 14.8. (a) Results of varying the properties of the plant species or removing some species on net present values (NPVs in A$) of the simulated streams of operating surplus of four farms. The four farms are: 1, conventional + N; 2, conventional − N; 3, reduced-input; and 4, low-input. For each farm, column 1: the standard form of the farm with no synergy between ryegrass and clover, i.e. relative yield total (RYT) = 1; column 2: standard form but with synergy, i.e. RYT up to 1.3; column 3, standard form but no N_2-fixation by clover; column 4, standard form but clover absent; and column 5, standard form but ryegrass absent. (b) Numerical effects on farm NPVs of: N_2-fixation, plant competition, presence of clover and presence of ryegrass, in columns 1–4, respectively.

effect seems identifiable as the 'competition effect' of the clover, an effect that it would have in the system if it were present as a non-fixing species (e.g. capeweed, *Arctotheca calendula*), but with its competitive relationships unchanged.

These two effects are given in Fig. 14.8(b), where the first and second columns for each farm give, respectively, the fixation and competition

effects. It is seen that, as expected, the monetary value of the fixation effect is small in farms 1 and 3. Here, the fixation ability of the clover has been made redundant by the application of fertilizer N. In the other two farms, the importance of the two effects is reversed. At its most significant in the highly productive farm 2, the fixation effect is worth A$1 million of NPV. Consistent with the lower productivity of the low-input farm 4, fixation is there worth half of this.

The value of clover's competition effect is considerable in farms 1 and 3, although much smaller than the large fixation effect in farm 2. The values are positive but this must be because the beneficial effect of controlling ryegrass seed production by competition with it in the pasture and in the wheat crop is greater than the negative effects due to clover's part in depressing the wheat yield and lowering pasture production due to its relatively low potential end-of-season biomass. In farm 1, clover in the crop is well controlled by spraying Ally® and so in that farm the competition effect is due only to a reduction of ryegrass seed production in the pasture, offset by some loss of pasture production. Without clover in the pasture to compete with ryegrass for resources and thus reduce its growth, these farms would be much less profitable. Their NPVs would be between one-third and two-thirds lower. The abundant capeweed in Western Australian pastures must be providing a similar service.

While a two-way partitioning of the clover effect seemed straightforward, it is less easy to partition the ryegrass effect. Being negative in all farms, ryegrass clearly imposes a net cost. However, since the modelled sheep respond in the same way to biomass of both pasture species, the presence of ryegrass, with its higher potential (end-of-season) biomass, might be expected to lead to a positive production effect on the sheep, as well as negative effects of competition on wheat yield and on N_2-fixation by clover. To test for a positive effect of ryegrass on the sheep enterprise, in all four standard farms the ryegrass potential production was lowered by about 20% to equal that of clover. There were fleeting decreases in stocking rate, but these persisted for no longer than one cycle. Although the stocking rate and profit from sheep respond sensitively to end-of-season pasture production (see Fig. 14.9 below), it seems that, in the farm systems tested, positive effects are negligible. As with clover, no attempt is made to separate the effects of competition in the crop from that in the pasture.

It was noted above (Fig.14.8a) that the introduction of a synergistic interaction between the two pasture species growing in pasture led to no increases in farm NPVs. The modelled increase of RYT values up to 1.3 caused pasture production from a mixed stand to be up to 30% higher than without the synergy, and yet farm profits fell. This result is important in that it contradicts the common assumption that greater agricultural production implies greater profits for the farmer. Where pasture production varies, the modelled optimal stocking rate certainly varies in parallel (Fig. 14.9a). This is well in line with results from many grazing trials (e.g. Williams, 1978). Increased model flock size then leads to increased profits in the sheep enterprise (Fig. 14.9b). However, the wider view taken by a comprehensive model encompasses side-effects that a field-level analysis may miss. In the present case, the overall effect becomes negative at the farm level, because part of the increased production in pastures with enhanced RYT is of ryegrass. The extra seed production from the enhanced ryegrass growth is so strongly detrimental to later wheat yields that the negative effect on wheat income overwhelms or just balances the positive effect on sheep income. The small effect of enhanced RYT in farm 4 is due to the asymptotic nature of the exponential response curves used in the model (see Fig. 14.3a–c). In farm 4, the density of weeds is so huge that further ryegrass plants in the crop cause only slight further reductions in yield, which are balanced by a higher stocking rate.

Among the further questions relating to competition that the WASP model can address are several concerning the importance of competitive ability in pasture legumes in a rotational system where a pasture grass is also a weed of crops (e.g. Latta and Carter, 1998). Where herbicide resistance in ryegrass is restricting spray options and where N inputs are mostly from fertilizer (farms 1 and 3), Fig. 14.8b has already shown the high importance of competition from clover. However, de Wit's competition model used in WASP predicts the mixture yield of each pasture species and hence its survival capacity on the basis of two components: potential pure-stand end-of-season biomass and crowding coefficient. This led to a new series of simulations to compare these two components' relative importance by varying each independent of the other. When both were set at standard values from which one at a time was increased by 50%, the changes had some remarkable effects on the relative contributions of the wheat and the sheep to farm income. The results depended strongly on the farm considered.

The effects of changes in clover's potential end-of-season biomass are shown in Fig. 14.10. They are presented as the final values of the annual operating surpluses of the wheat and sheep enterprises,

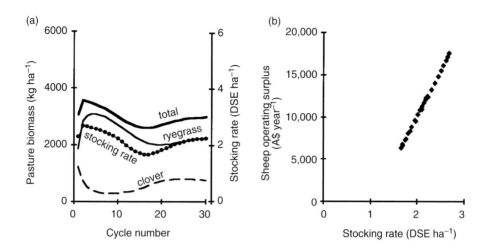

Fig. 14.9. Dependence of simulated stocking rate (dry sheep equivalent (DSE) ha^{-1}) and net profit from the farm's sheep enterprise (A\$ year^{-1}) on variation in pasture yield and composition. These pasture biomasses are of grazed swards at the end of their growing season. (a) Time courses of pasture in the standard form of farm 4, with clover crowding coefficient set at half the standard value, resulting in poor control of ryegrass and damped oscillations of pasture composition. (b) Scatter graph showing stocking rates in (a) plotted against the operating surpluses of the farm's sheep enterprise to show sensitivity of the latter to variation in stocking rate over the range commonly found in runs of the model.

in, first, the standard forms of the farms and, secondly, the same farms but with clover potential biomass 50% higher. In farms 1 and 2, the increased pasture production enhances the sheep profits, but has no effect on profit from wheat. In farm 3, increased pasture growth helps the sheep enterprise, but the consequent extra production of clover seeds strengthens the already strong suppressive effect on wheat of clover in the crop. Thus, the introduction of a higher-yielding clover causes a change in the farm from one where farm finances are strongly dominated by the wheat enterprise to one where the enterprises make near-equal contributions. The effect on the low-input farm is even more dramatic and unexpected. The use of higher-yielding clover here causes the reverse: a farm with nearly all its income from sheep becomes a farm in which, quite paradoxically, the wheat contribution now dominates. The explanation of the difference of response is that, in farm 3, the clover problem in the wheat is made worse, while, in farm 4, the serious problem is ryegrass in the crop; the enhanced growth of the clover not only allows a higher stocking rate but also suppresses the ryegrass seed production, so that the wheat profit is hugely increased.

The results of changing the crowding coefficient of clover in respect of ryegrass are not shown because they are simpler. When the crowding coefficient is increased by the same 50%, there is no effect on profit in farms 1 to 3, where the pasture still rapidly becomes 100% clover. The single but very large effect is in farm 4, where wheat profit is increased 30-fold. In this case only, the enhanced competitive ability is extremely valuable, more than tripling the farm's operating surplus. In respect of effects on profits, there is therefore some limited interchangeability between changes in crowding coefficient and in potential biomass production, but only in farm 4.

If new pasture legumes are being selected to give markedly greater end-of-season biomass through improved water-use efficiency (Revell *et al.*, 1998), it will be desirable to have some foreknowledge of the likely magnitude and direction of effects when they are introduced. A WASP-type model extended to simulate a wider range of rotations may be able to assist in this. Clearly, the results in Fig. 14.10 suggest that introduction of higher-yielding legumes at different stages of the farmers' hypothesized sequence of responses to increasing herbicide resistance will have very different effects. If real farms come to resemble in any way the four theoretical ones postulated here, the largest financial gains from greater legume biomass production will be felt in those like the worst affected, least profitable farm 4.

Role of simulation models

Even the ample research budgets of earlier times could probably never have funded a real performance of the rotational experiments reported here. But, through simulation modelling to conduct them theoretically in a computer, we gain the impression that we have real answers. Similarly, although some of the treatments used here, such as switching N_2-fixation in a legume either on or off, have been achieved in pot experiments (de Wit *et al.*, 1966), there are still no technical means to do it in an established species on a farm scale. Such a theoretical switching puts this work truly in the realm of 'thought experiments'. As modelling extrapolates ever further from actual experimentation, the validity of the results becomes less certain. The results reported here have been checked for internal consistency and apparent good sense but are quite unvalidated. They can be viewed only as 'indicative'.

Since some surprising but explicable results have been encountered in the simulations, it is interesting to note that similar surprises are found in other modelling studies of whole farms. The extensively used mathematical programming (MP) approach calculates allocation of the various farm resources to farm activities in a way that maximizes profit. Applying the MP model MUDAS to the management of similar wheat-belt farms, Kingwell (1998) and Kingwell *et al.* (1992) found that an improvement of the pasture on a very poor soil led unexpectedly to the optimal farm plan including a smaller total area of pasture. Once obtained, this result was readily interpreted, but the appearance of apparently counter-intuitive results from the use of both kinds of approach underlines the value of modelling as a way of deepening understanding of wheat-belt, as well as other, farming systems.

The MUDAS model used by Kingwell (1998) is a recent extension of the older model (MIDAS: Morrison *et al.*, 1986), in which a highly detailed consideration of the many aspects of real farms lends great plausibility to the results. The MUDAS extension now considers seasonal and price variability, which further improves acceptability. However, the approach does still assume an underlying steady state in the system, whereas many investigations of biological systems suggest that arbitrarily created system states are seldom near equilibrium. Illustrating this, in most runs of the WASP model the state of the system departs immediately from the initial condition (e.g. Fig. 14.7). The time

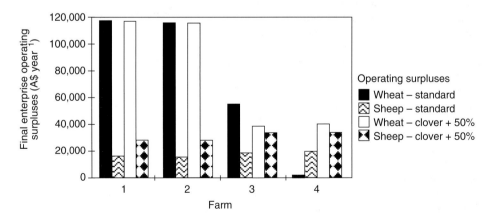

Fig. 14.10. Effects of increased potential biomass production by the clover on the profitability of the wheat and sheep enterprises. Farm operating surpluses (A\$ year^{-1}) for the 30th cycle on each of the four farms are given for the wheat enterprise (columns 1 and 3) and sheep enterprise (columns 2 and 4) for the standard form of the farm (columns 1 and 2) and for the same farm but with a 50% increase in clover potential biomass (columns 3 and 4). By the 30th cycle, all simulations are effectively at equilibrium. The four farms are: 1, conventional + N; 2, conventional – N; 3, reduced-input; and 4, low-input.

course of pasture production in Fig. 14.9(a) is actually quite unusual in that it settles near to its initial state, but it only does so after a long excursion away from it. In spite of the elaboration of the MIDAS method to deal with a greater range of farm types and environments (e.g. Pannell and Bathgate, 1994; Robinson *et al.*, 1996), the assumption of steady state seems to weaken the claim of MP methods to realistically address farm problems, because, first, MIDAS/MUDAS optimal farm plans apply only to the farm in its initial or proposed state and, secondly, the direction of system change is not known, so that the costs of maintaining the assumed steady state may well have been omitted. On the other hand, in a mechanistic model like WASP, if some given steady state is to be maintained, control methods can be specified and their costs quantified; also, limited optimization capabilities are already available. But simulation methods lack the ability of MP to easily address detail. The best modelling method is still to be found, but will ideally combine the flexibility of the MP methods with a full acceptance of dynamic behaviour.

General conclusions

The reported simulations backed up by costly experience on real farms has shown that, with present price structures, uncontrolled competition between ryegrass and other components of the Western Australian pasture–wheat system can make the system non-viable. This is because increasing competition of ryegrass with the farm's most valuable output, the wheat, can cause its enterprise costs that exceed profit. Farmers have relied in the past on selective sprays to reduce ryegrass populations, but the widespread appearance of populations resistant to these threatens this practice and suggests that resistance to even broad-spectrum herbicides, such as paraquat, will shortly follow. Assuming this and also pressure to reduce applications of nitrogenous fertilizer, four scenarios (farms 1–4) have been proposed to represent a likely sequence in time as farmers adapt to such developments. Simulations suggest that, with present prices, the

appearance of broadening resistance in ryegrass will seriously decrease profitability of farming. This could entice farmers to return to thorough cultivation and burning of stubble and pasture in order to control ryegrass. Effects of these on soil N dynamics and soil erosion have not been modelled, but are likely to threaten sustainability of the real system.

The four scenarios have provided a range of farming environments in which to test some system variants. The introduction of synergy between clover and grass in the productivity of mixed pasture led unexpectedly to predicted falls in income from all farms, due to the negative effect of exacerbation of weed problems in the wheat crop exceeding the positive effect of an increase in stocking rate. This strongly suggests the need to quantify experimentally the actual level of synergy between pasture legumes and grasses. The elimination of clover from the farms or suppression of its N_2-fixation caused collapse in the farms not using N fertilizer. The presence of a theoretical non-fixing clover with an unchanged competitive ability was valuable in simulated farms using N fertilizer, because competition between it and the ryegrass in pasture greatly reduced the ryegrass problem in wheat. This suggested an important role for competitive nonlegume species in pasture in reducing ryegrass seed production. Elimination of ryegrass from the farms led to increases in profitability, especially in the farm with the worst weed problem.

Acknowledgements

I am grateful to many colleagues in Agriculture Western Australia and elsewhere who have guided me to the unpublished reports of their research and helped me with interpretation. I also thank Bill Bowden and Ross Kingwell for their comments on the manuscript.

Note

1. The model can be obtained as an Excel® file by sending an unused disc to the author.

References

Davidson, R.M. (1994) Biology and control of herbicide resistant *Lolium rigidum*. MSc thesis, La Trobe University, Melbourne, Australia.

Department of Agriculture Western Australia (1993) *Elders Weekly Farm Budget Guide*. Department of Agriculture Western Australia, Perth, Australia.

de Wit, C.T. (1960) On competition. *Verslag Landbouwerk Onderzoek* 66(8), 1–82.

de Wit, C.T. and van den Bergh, J.P. (1965) Competition between herbage plants. *Netherlands Journal of Agricultural Science* 13, 212–221.

de Wit, C.T., Tow, P.G. and Ennik, G.C. (1966) *Competition Between Legumes and Grasses*. Institute for Biological and Chemical Research on Field Crops and Herbage Agricultural Research Reports 687, Pudoc, Wageningen, The Netherlands.

Donald, C.M. (1963) Competition among crop and pasture plants. *Advances in Agronomy* 15, 1–118.

Ewing, M.A., Pannell, D.J. and James, P.K. (1987) The profitability of lupin : cereal rotations. In: Kingwell, R.S. and Pannell, D.J. (eds) *MIDAS, a Bioeconomic Model of a Dryland Farm System*. Pudoc, Wageningen, The Netherlands, pp. 82–90.

Gramshaw, D. (1974) Survival of annual ryegrass (*Lolium rigidum* Gaud.) seed in the pasture–crop rotation. PhD thesis, University of Western Australia, Perth, Australia.

Greenwood, E.A.N., Lloyd Davies, H. and Watson, E.R. (1967) Growth of an annual pasture on virgin land in south-western Australia including effects of stocking rate and nitrogen fertilizer. *Australian Journal of Agricultural Research* 18, 447–459.

Harper, J.L. (1977) *Population Biology of Plants*. Academic Press, London, UK.

Kingwell, R.S. (1998) Dryland pasture improvement given climatic risk. *Agricultural Systems* 45, 175–190.

Kingwell, R.S., Morrison, D.A. and Bathgate, D.A. (1992) The effect of climatic risk on dryland farm management. *Agricultural Systems* 39, 153–175.

Latta, R.A. and Carter, E.D. (1998) Increasing production of an annual medic–wheat rotation by grazing and grass removal with herbicides in the Victorian Mallee. *Australian Journal of Experimental Agriculture* 38, 211–217.

McCown, R.L., Cogle, A.L., Ockwell, A.P. and Reeves, T.G. (1987) Nitrogen supply to cereals in legume ley systems under pressure. In: Wilson, J.R. (ed.) *Advances in Nitrogen Cycling in Agricultural Ecosystems*. CAB International, Wallingford, UK, pp. 292–314.

Matthew, J., Llewellyn, R., Jaeschke, R. and Powles, S. (1996) Catching weed seeds at harvest: a method to reduce annual weed populations. In: *Proceedings of the 8th Australian Agronomy Conference*. Australian Society of Agronomy, Toowoomba, Australia, p. 684.

Moore, J. (1989) Pasture manipulation. In: Maling, I. and Madin, R. (eds) *Proceedings of a Workshop on Grass Control in Pastures for the Wheat/Sheep Zone, 6–7 April 1989, Forrestfield (WA)*. Miscellaneous Publication 41/90, Western Australian Department of Agriculture, Perth, Australia, pp. 186–187.

Morrison, D.A., Kingwell, R.S., Pannell, D.J. and Ewing, M.A. (1986) A mathematical programming model of a crop–livestock farm. *Agricultural Systems* 20, 243–268.

Pannell, D.J. and Falconer, D.A. (1987) The value of nitrogen in a crop–livestock farm system: a bioeconomic modelling approach. In: Bacon, P.E., Evans, J., Storrier, R.R. and Taylor, A.C. (eds) *Nitrogen Cycling in Agricultural Systems of Temperate Australia*. The Australian Society of Soil Science, Riverina Branch, Wagga Wagga, Australia, pp. 449–466.

Pannell, D.J. and Bathgate, A. (1994) *Model of an Integrated Dryland Agricultural System: Manual and Documentation for the Eastern Wheatbelt Model Version EWM 94–1*. Economics Analysis Branch, Department of Agriculture Western Australia, Perth, Australia.

Pearce, G.A. and Quinlivan, B.J. (1971) The control of annual ('Wimmera') ryegrass in cereal crops. *Journal of Agriculture Western Australia (4th Series)* 12, 58–62.

Revell, C., Nutt, B. and Ewing, M. (1998) Success with Seradella in the wheatbelt. *Journal of Agriculture Western Australia (4th Series)* 39, 24–29.

Robinson, J.B., Kearns, B.F., Armstrong, E.L. and Butler, G.J. (1996) PRISM: applying bioeconomic models to farm management issues. In: *Proceedings of the 8th Australian Agronomy Conference*. Australian Society of Agronomy, Toowoomba, Australia, p. 711.

Saoub, H.M. (1994) Persistence of two annual pasture legumes under different rotation and cultivation systems, in a Mediterranean-type environment. PhD thesis, University of Western Australia, Perth, Australia.

Stern, W.R. and Donald, C.M. (1962) Light relationships in grass clover swards. *Australian Journal of Agricultural Research* 13, 599–614.

Taylor, G.B. (1985) Effect of tillage practices on the fate of hard seeds of subterranean clover in a ley farming system. *Australian Journal of Experimental Agriculture* 25, 568–573.

Trenbath, B.R. (1976) Plant interactions in mixed crop communities. In: Papendick, R.I., Sanchez, P.A., Triplett, G.B. and Bronson, R.D. (eds) *Multiple Cropping*. American Society of Agronomy Special Publication, Madison, USA, pp. 129–169.

Trenbath, B.R. (1983) The dynamic properties of mixed crops. In: Roy, S.K. (ed.) *Proceedings of the Indian Statistical Institute Golden Jubilee International Conference on Frontiers of Research in Agriculture*. Indian Statistical Institute, Calcutta, India, pp. 265–286.

Trenbath, B.R. and Stern, W.R. (1995) *WASP (Wheat and Sheep Pasture): a Virtual Farm*. Miscellaneous Publication 13/95, Department of Agriculture Western Australia, Perth, Australia.

Trumble, H.C. and Shapter, R.E. (1937) The influence of nitrogen and phosphorus treatment on the yield and chemical composition of Wimmera ryegrass and subterranean clover, grown separately and in association. In: *Bulletin 105*, paper 2. Commonwealth of Australia Council for Scientific and Industrial Research, Melbourne, Australia, pp. 25–36.

Watson, E.R. and Lapins, P. (1964) The influence of subterranean clover pastures on soil fertility. II. The effects of certain management systems. *Australian Journal of Agricultural Research* 15, 885–894.

Williams, C.M.J. (1978) Studies of herbage availability and plant density in relation to animal performance. PhD thesis, University of Adelaide, Adelaide, Australia.

Willoughby, W.M. (1954) Some factors affecting grass–clover relationships. *Australian Journal of Agricultural Research* 5, 157–180.

15 Some Concluding Comments

Alec Lazenby[1] and Philip G. Tow[2]

[1]63 Kitchener Street, Hughes, Australia; [2]Department of Agronomy and Farming Systems, University of Adelaide, Roseworthy Campus, Roseworthy, Australia

The importance and complexity of plant competition and succession in pastures is reflected in the extent and diversity of the relevant information and research reviewed in this book. There can be no doubt that the considerable work undertaken has both enhanced our knowledge of such competition and succession and improved the reliability of predicting the effects of a number of factors on the botanical composition and performance of natural and artificial grasslands. It is equally clear that much remains to be done if we are to understand the principles and processes involved in: (i) determining which species occur in these grasslands; and (ii) managing them properly.

Terminology and Experimentation

Inconsistency in the use of terms by those involved in studying competition has sometimes confounded interpretation of their findings. Further, the varying rigour with which some of the investigations have been conducted, reflected in the questions asked, experimental designs, analysis of data and interpretation of results, has added to the difficulty in formulating general principles (see Sackville Hamilton, Chapter 2, this volume). It has also contributed to the sometimes dogmatic and emotionally charged debate, a not uncharacteristic feature in the development of important scientific topics. The comprehensive description of the types of experiment available for studying various aspects of competition, together with the critical analysis of their advantages and disadvantages in seeking to answer specific questions

(Sackville Hamilton, Chapter 2, this volume), represents a significant contribution to the clarity of thinking on and thus the overall study of competition.

Many of the earlier studies on competition were undertaken on annual crops grown in monoculture or simple two-species mixtures, while those involving pasture plants usually consisted of measurements from short-term experiments in controlled environments or small plots, many of them cut rather than grazed. Such measurements of competitive interactions between species or genotypes, usually made in terms of yields per unit area, can provide valuable agricultural information. However, they have been collected from competitive environments that are considerably less complex than long-lived grasslands, which almost always contain a range of plant species. Thus the information has limited use for predicting long-term ecological outcomes in the botanical composition of grassland communities, which are important for understanding the persistence and stability of pastures. Further, it has contributed little to our knowledge of the processes involved in competition between pasture plants. Increased understanding of competitive mechanisms presents the researcher with one of the biggest challenges in the future study of competition.

In Chapter 5 of this volume, Kemp and King highlight the changing focus of studies on competition which has occurred over time. Many early investigations sought to identify the resource(s) for which plants competed, often seeking 'to isolate responses to single variables'. As Kemp and King conclude, because the resource(s) limiting growth almost invariably change(s) over time, the value of any information collected from such experiments is likely to be marginal for predicting the outcome of competition. More recent work on competition has shown an increasing focus on this outcome. Often, a comparison of the relative growth rates or bio- mass of competitors has been used as a measure of 'the net outcome of resource capture', while less attention has been paid to the specific resource(s) for which the plants are competing. There have been more investigations on competition for light, between plants associated with high-yielding condi- tions, than for soil resources, which almost invari- ably limit growth in resource-poor environments.

Plant Characteristics and Competitiveness

Contributors to this publication show that there is no shortage of ideas on the features and strategies of plants that confer competitiveness (see Nurjaya and Tow, Chapter 3; Kemp and King, Chapter 5; Wolfe and Dear, Chapter 7; Skarpe, Chapter 9; Peltzer and Wilson, Chapter 10, this volume). Good progress has been made in identifying the morphological and physiological traits that give plants a competitive advantage in particular grow- ing conditions. Whilst plants establishing and occupying the ground quickly have an obvious competitive advantage, at least initially, there is considerable evidence that characteristics such as plant height, leaf area and root mass are all associ- ated with competitiveness; so also are traits such as rapid growth rate of shoots and roots and leaf area ratio. These features all indicate an association of competitiveness with the ability of a plant to pre- empt light, nutrient and water resources (see Nurjaya and Tow, Chapter 3, this volume).

Many plant traits are highly heritable, though with the extent of their expression being modified not only by competition, but also by environmental conditions, whether abiotic, such as temperature and rainfall, or the direct result of human activities, e.g. management decisions. Morphological plastic- ity is one obvious expression of a plant's ability to adapt to different growing conditions; for example, the same genotype of a grass such as *Lolium perenne* may thrive as a plant with either a handful of or more than a thousand tillers, according to whether grown in a dense sward or as a widely spaced plant.

Plants vary in their ability to adapt to and be competitive in different growing conditions.

Characteristics conferring a competitive edge in specific environments can sometimes be identified. Amongst these, an obvious example is the advan- tage of plants with a C_4 pathway of carbon fixation over C_3 plants in high-temperature conditions. It seems clear that plants are most competitive in environments to which they are best adapted. For instance, the competitiveness of the slow-growing *Festuca ovina* in some resource-poor conditions can be attributed to its long tissue retention and perma- nently functioning root system, which enable it to take advantage of nutrient pulses lasting only a few hours or even minutes. In contrast, grasses such as *Arrhenatherum elatius*, with their much shorter tis- sue retention, need to grow new roots before they can take advantage of any increase in nutrient lev- els; they are thus less competitive than *Festuca* in such poor growing conditions (see Peltzer and Wilson, Chapter 10, this volume).

A further example not only identifies plant traits that confer competitiveness, but also indicates both the complexity of competition and how finely bal- anced the outcome can be (see Nurjaya and Tow, Chapter 3, this volume). Experiments involving competition between *Molinia caerulea* and *Calluna vulgaris* indicate that the extensive root system of the faster-growing, lower nutrient-retentive *Molinia*, coupled with its ability to intercept light through leaf elongation, is responsible for its com- petitive edge under good growing conditions. In poor conditions, however, such features were unable to outweigh the high nutrient retention ability of *Calluna*. Yet high-density stands of *Calluna* can compete successfully with *Molinia* in better growing conditions by cutting out light in the early growth stages of the grass.

The concept of slow-growing species with a low nutrient loss being competitively superior to faster- growing plants with greater nutrient loss in poor, but not in good, growing conditions has been used in a novel way by Berendse (see Nurjaya and Tow, Chapter 3, this volume). He developed a model designed to: (i) provide a tool simple enough to analyse perennial species in different competitive situations; and (ii) enable the qualitative prediction of the effect of changes in nutrient supply on the outcome of competition between the species.

Two of the most influential researchers in influ- encing our thinking on competitiveness – J.P. Grime and D. Tilman – have both used autecologi- cal information to develop theories of plant strate- gies that enable the prediction of the outcomes of competition (see Kemp and King, Chapter 5, this volume). Grime argues that a plant's ability to dominate a community is determined by a combination of features, including maximum relative growth rate (RGR_{max}), net assimilation rate (NAR) and leaf area ratio (LAR). He divides successful competitors into three main categories namely: (i) highly competitive plants (C), which can exploit good growing conditions by rapidly absorbing nutrients and growing quickly, thereby dominating the vegetation; (ii) stress-tolerant species (S), which are adapted to, or at least tolerant of, resource-poor conditions; and (iii) ruderal species (R) with the ability to invade and grow in disturbed conditions. CSR scores have been determined for a number of British plants, while Wolfe and Dear (see Chapter 7, this volume) have applied Grime's three categories to a range of common temperate grassland plants and weeds; their tolerance to competition, stress and disturbance enables broad prediction of their performance under a range of conditions.

Tilman's theory – that a good competitor is able to perform well despite shortage of resources – was based on his belief that the ability of a plant to extract nutrients down to very low concentrations gave it a competitive edge. This ability (designated R^*) has been determined experimentally for the extraction of soil N by a number of plants; the R^* values of such plants were shown to predict the outcome of competition when grown in pairwise mixtures (see Kemp and King, Chapter 5, this volume). There is also experimental evidence from natural grasslands to support Tilman's claim that plants with such ability will ultimately dominate the vegetation.

At first sight, the theories of Grime and Tilman might appear incompatible. Certainly there is some difference in both their basis and interpretation. For instance, Grime believes that there is a trade-off between stress tolerance and competitive ability, whereas Tilman argues that this concept cannot always apply, as competitive ability and stress tolerance may both be conferred by the same traits (see later). Yet, as a number of researchers have concluded, the theories of Grime and Tilman are not necessarily mutually exclusive. Whereas Grime's proposition appears more applicable to predicting competitive performance in good growing conditions, that of Tilman seems more relevant to low-fertility situations, where it both enables the prediction of competitive outcomes in resource-poor environments and makes a contribution to our understanding of competitive systems in such conditions. The

fact that plants with the highest competitive score (based on Grime's theory) and the lowest R^* value (from Tilman) appear likely to dominate communities in the absence of the grazing animal provides the basis of an important ecological principle.

Peltzer and Wilson (see Chapter 10, this volume) critically evaluate the two broad types of competitive ability, namely, good response plants, able to resist suppression by others, and good effect competitors, which can reduce the performance of neighbouring species. Some plant traits, including high root : shoot ratios, low growth rates, small size, nutrient-conserving mechanisms, carbon-based defences and storage organs, which are shown to confer competitive response ability in unproductive environments, are almost identical with those associated with stress tolerance. This correlation is not only interesting but is also a valuable tool for increasing our understanding of the basis of competition. However, as Peltzer and Wilson conclude, it is unclear whether the traits identify good stress tolerance or good competitors in a stressful environment. Evidence on the relative importance of competition and stress tolerance in determining the botanical composition of communities growing in resource-poor environments is mixed. Further work is thus needed to clarify and better understand this relationship.

Competition – Some Agronomic Implications

In seeking to analyse the effects of competition on yields per unit area, it seems logical to postulate that the more coincident the requirements of plants for resources, the greater the competition that can be expected and hence the more similar the dry-matter production of monocultures and mixtures. Thus, a continuum can be envisaged, with competition being greatest between identical genotypes (say, plants of a self-pollinating legume or cereal crop) and reducing in stages from growing together: (i) genotypes of one species (e.g. mixing cultivars of a clover); through (ii) plants of similar growth habit and phenology (such as the grasses perennial ryegrass and tall fescue); (iii) plants of different growth habit but of broadly similar growth rhythm (e.g. temperate grasses and clovers); to (iv) plants with distinctly different growth rhythms (typified by perennial ryegrass and paspalum, respectively C_3 and C_4 species).

Results from relevant experiments have been

mixed, with some data supporting the above hypothesis and others not showing the effects of competition that might have been expected. In a number of investigations, no increase was detected in biomass production per unit area – used as an indicator of competitive effect – of pairwise mixtures compared with monocultures. These studies involved yield comparisons of monocultures and mixtures not only of genotypes of the same species and of plants of different species with similar growth form and phenology, but also of plants with widely different growth rhythms.

It is not difficult to envisage that plants similar in form and phenology, whether from the same or different species, would be competing for essentially the same resources at the same time; hence, a similar biomass production per unit area could be anticipated, whether from monocultures or mixtures of such plants. However, over a period of a year, greater production might be expected from a mixture of plants with a similar growth habit, but widely differing phenology. Yet, if one or more resources become(s) limiting during their growing season plants will almost certainly fail to reach their potential. For instance, on the northern tablelands of New South Wales (NSW), mixtures of paspalum and perennial ryegrass produced significantly more dry matter during the year than the monocultures only when soil moisture was available throughout the growing period (Harris and Lazenby, 1974).

Because competition is site-specific, the prevailing growing conditions are also highly significant in determining competitive outcomes and thus, as conditions change, the plants dominating the community may also change. For example, improved pastures on the tablelands of NSW contain both C_3 and C_4 grasses. The latter are better adapted not only to high summer temperatures than the C_3 plants but also to low-fertility conditions. It is not surprising, therefore, that reduced use of superphosphate since the early 1970s has been associated with a change in the botanical composition of many pastures on the northern tablelands of NSW, with grasses such as *Bothriochloa macra* becoming increasingly common at the expense of the higher-fertility-demanding species with lower tolerance of high temperature, such as *L. perenne* (Cook *et al.*, 1978).

A significant proposal for studying changes in the botanical composition of pastures is described by Kemp and King (see Chapter 5, this volume). It provides a framework whereby the many species found in a typical long-lived pasture can be divided into a small number of categories based on plants with similar characteristics, e.g. valuable perennial grasses, legumes, undesirable annual grasses and weeds. The effects of a number of variables, such as available moisture or grazing intensity, on the population dynamics, growth rates, fecundity, seed set and seedling establishment of the various categories can then be measured over time. These data cannot be used to distinguish the relative influence of competitive and non-competitive factors on composition and succession in pastures, for which objective few experimental data have been collected and further work is obviously necessary. However, not only does the information collected on population dynamics have the advantage that it sheds light on the effects of the variables on pasture plant composition but it is clearly valuable in making management decisions and advisory recommendations.

Heterogeneity, Diversity and Stability

Between-site differences in botanical composition indicate variation in growing conditions in pastures. The significance of such heterogeneity, common to a greater or lesser extent in all grasslands, is only now being appreciated. It is highly relevant for niche occupation and thus botanical diversity in pastures. The value of some diversity in pastures is argued by Clark and Harris (see Chapters 6 and 8, this volume), while its likely role as a buffer to long-term climatic change is advocated by Campbell and Hunt (see Chapter 12, this volume).

Many plants that invade grasslands are weeds, but there is evidence that, in some situations, botanical diversity can lead to increased pasture output (see Clark, Chapter 6, this volume). Further, provided the right species are present – appropriate perennial grasses appear to be especially important – diversity increases the stability of grassland (see Clark, Chapter 6, and Garden and Bolger, Chapter 11, this volume). This contrasts with annual species, which contribute to pasture instability and environmental degradation. Garden and Bolger present a good case for a correlation between loss of perennial grasses in pastures and a decline in 'ecosystem function' and subsequently in the clean water and air expected by society. They argue for the positive effects of perennial grasses on the hydrological cycle, including reduction of deep drainage, leaching of N and soil erosion; this is in addition to the obvious agricultural value of such grasses in providing perennial forage for grazing by ruminants. Economic advantages of stability in pastures include a reduced need for frequent reseeding.

Diversity is also seen as an important factor in preserving flora and fauna, enhancing the aesthetic value of the landscape and increasing its recreational and amenity value, with one contributor, Chapman (see Chapter 13, this volume), arguing a case for the re-establishment of more botanically diverse grasslands. Re-creating species-rich pastures is neither easy nor short-term, whether from high-fertility grassland, which would require a considerable reduction in available soil nutrients, or from sites with soil disturbed by mining or engineering operations, which would need both more available minerals and an improved soil structure. Further, according to Chapman, the aim is to provide competitive conditions enabling the growth and coexistence of competitive, stress-tolerant and ruderal species. This difficult objective requires both the right plants to be present and appropriate management, including a suitable grazing regime.

Changing the botanical composition of grassland may be limited by the inability to recruit suitable species; recruitment limitation has been accepted only recently as an important determinant of community structure, complementary to competition (see Garden and Bolger, Chapter 11, this volume). If a suitable seed mixture is not sown to hasten the process, attaining a desirable botanical composition may take a considerable time, particularly if the pasture is separated by some distance from a source of appropriate plants to colonize the site.

Legumes and Grasses – Competition and Coexistence

Legumes and grasses have vital and complementary roles in pastures. The ability of legumes to fix N, not only for their own growth but also to both enhance the dry-matter production of grasses and improve soil fertility, makes them the key to grassland production in many parts of the world; for example, they are integral to high output from improved temperate pastures in Australia and New Zealand. Even where legumes are less important in influencing pasture production, e.g. in some intensively managed grasslands in the UK, The Netherlands and North America, where high output depends on applying nitrogen (N) as a fertil-

izer, pressure is increasing for their wider use. Whilst such greater use of legumes in improved pastures can be expected, it would be hastened if their reliability could be improved. The value of grasses in pastures lies not only in their greater herbage yield potential than legumes in many conditions, but also in: (i) providing a more complete ground cover, thus reducing weed invasion; and (ii) contributing to the stability of long-lived pastures.

In order to combine the advantages of grasses and legumes in pastures, they need to coexist in proportions where both can make a significant contribution to grassland output. Attaining this objective can be difficult, especially where the plants best adapted to prevailing growing conditions differ in characteristics, such as growth habit or growth rhythm. However, provided neither the legume nor the grass remains at a competitive disadvantage over a lengthy period – legumes are generally more susceptible than grasses to adverse growing conditions – it is often possible to maintain both at a level sufficient to enable each to make a major and continuing contribution to pasture production. Further, the indications are that, in some situations at least, a dynamic equilibrium may develop between such grasses and legumes (see Tow and Lazenby, Chapter 1, and Davies, Chapter 4, this volume).

The importance of grasses and legumes in our grasslands is the reason for much of the work undertaken to improve understanding of their competitive relationship and make better use of their ability to coexist – both major objectives of this book. The grass : legume relationship has been studied most among temperate pasture plants, particularly perennial grasses and clovers. Results from a number of experiments, such as those designed to better understand the morphological expression and physiological basis of growth, both seasonally and under different conditions, have added to knowledge (see Davies, Chapter 4, this volume). These include the findings that the poor performance of white clover during periods of moisture stress was a result of the inability of the leaf to adapt to a hydration deficit and, more specifically, to close its stomata quickly under such conditions. Some of the other conclusions from studies which have increased our understanding may seem more surprising. Amongst these are: (i) the effect of reducing day length at the end of the growing season on petiole length of white clover; and (ii) the finding that the reason why white clover growth in early spring is slower than that of perennial ryegrass is a result of differences in the rate of leaf expansion of the plants at low temperatures, not in their rates of photosynthesis.

In seeking a favourable grass : legume balance, different challenges are often presented for the range of growing conditions prevailing in the world's grasslands. For example, tropical legumes, generally upright and often twining plants, require more infrequent defoliation than tropical grasses to produce high dry-matter yields. Lucerne, a normally upright plant adaptable to a wide range of temperature conditions, is also susceptible to frequent defoliation. Variation in its competitive performance, detected in a study involving drills sown alternately with lucerne and the grass digitaria, was clearly relatable to the growing conditions that obtained in the different treatments. Fourth-order interactions showed that the competition between the two plants was sometimes finely balanced and could be affected significantly by temperature, available moisture and N regime (Tow and Lazenby, Chapter 1, this volume). The data provided clear pointers to the conditions in which lucerne can be expected to: (i) perform well (moderate temperatures, available soil moisture and low soil N); and (ii) be less competitive and lower-yielding (a combination of high temperatures and flooding having particularly adverse effects).

To achieve a major and continuing contribution to grassland production in many temperate pastures, the management needs to accommodate both the essentially prostrate clover with its horizontal leaves and the more upright and vertical-leaved grasses. Temperate grasses and clovers both have a high proportion of their growing points near ground level and thus are adapted to a grazing regime, provided it is not too severe and prolonged. Investigations on the effect of a number of grazing systems on grass : clover balance in pastures (see Davies, Chapter 4, this volume) include some results that are already incorporated, to a greater or lesser extent, in good management practice. These results include the detrimental effects of severe grazing, especially by sheep, which are often apparent first in moisture-stressed pastures, where white clover is normally affected before the grasses. Other interesting and valuable data from some recent agronomic experiments show the importance of stolon development and decline in white clover in helping to both indicate early signs of a clover failure and develop management systems to prevent such an occurrence.

Plant breeders have made considerable efforts to improve legume performance. New cultivars have been selected, e.g. with enhanced levels of N_2-fixation, better competitive performance with grasses, increased dry-matter production and improved resistance to pests and diseases. There is no doubt that some progress has been made. However, breeders have to accept that improvements shown by cultivars in plots used for their selection and evaluation are not always realized in the more complex grassland ecosystems. For instance: (i) rhizobia found naturally in the soil not only normally fix less N, but are also usually more competitive, than strains selected for their high N_2-fixation when grown with compatible host plants; selected strains may thus fail to persist; (ii) cultivars better able to compete with grasses under a lenient defoliation regime are likely to be more sensitive to severe grazing; and (iii) when grown in mixtures with grasses, selections made for increased monoculture yield often produce similar yields per unit area as mixtures with a lower-yielding legume.

Modelling

Models have been used to throw light on a number of processes and factors influencing the grass : legume balance in pastures including: (i) simulating the distribution of white clover at different stages in the grass/clover cycle, which is associated with levels of soil N; and (ii) modelling the relationship between the extension rate of white clover stolons and the density of neighbouring stolons, and relating the predictions to the expected persistence of clover patches (see Davies, Chapter 4, this volume). Whilst the conclusions from modelling do not always coincide with results in the field – as a result, for example, of a situation being oversimplified and thus too few variables being included in a model – they have been valuable in indicating some outcomes of competition in pastures and how a number of management decisions influence the grass : legume balance.

The value of modelling the effects of competition at a whole-farm level is clearly shown in the interesting contribution of Trenbath (see Chapter 14, this volume). Both the biological and the economic outcomes of a system based on pasture and wheat are shown to depend fundamentally on the level of competition between annual ryegrass and wheat, which, if not controlled, can lead to the system becoming unviable. The advantages of modelling a whole-farm system include the opportunity to predict the results in a wider context than is possible from traditional, more analytical-type, experiments. Such simulation thus provides both a chance to better understand the system as a whole and more possibility of explaining any unexpected results. For instance, in Trenbath's model, the surprising information that high pasture output had a strong negative effect on farm profits can be explained by the negative effect of weed competition, particularly from annual ryegrass, on wheat (the most profitable part of the system); this was greater than the positive effect of increased production from the pasture.

There can be no doubt that simulation will become increasingly important in future studies of competition in grassland. Not only can more modelling be predicted confidently in the types of work discussed above, but such simulation is highly likely to be extended into other relevant fields. For example, Wolfe and Dear (see Chapter 7, this volume) argue for increased modelling to 'provide a more theoretical approach to grassland dynamics', while Campbell and Hunt (see Chapter 12, this volume) conclude that 'modelling is critical for examining long-term effects of global change on resource availability, competition and succession'.

Climate Change

Possible effects of climate change on competition and succession in pastures are considered by Campbell and Hunt in an intriguing contribution (see Chapter 12, this volume). They postulate that such change is likely to have a long-term and indirect effect on the botanical composition of pastures, specific to sites and environments, and acting through processes and events that affect the morphology and physiology of plants. The effect of components of climate, such as temperature and rainfall, on the growth and development of plants is better understood than the longer-term effects of changes in the global climate on soil resources and agents of disturbance, said to be of much greater significance for the structure and function of ecosystems.

Limited evidence exists to indicate the accuracy of predicting the effects of climate change – namely, increased atmospheric CO_2, decreased stratospheric ozone and changes in temperature and in rainfall distribution and intensity. Predictions supported by experimental data and field observations include: (i) the improved competitive performance of white clover grown with perennial ryegrass at higher CO_2 levels; and (ii) the intrusion of C_4 plants into vegetation dominated by C_3 species following a rise in temperature in the North Island of New Zealand, where temperate and tropical plants converge.

However, other probable global climate changes – changes in rainfall patterns are a good example – remain largely unknown at present, even though their effect on competition, succession and productivity in pastures could be overriding. Campbell and Hunt believe human intervention to be the key to maintaining or increasing grassland output in a changed climate; they see input of fertilizers, plant type (a major role is envisaged for selecting cultivars and more complex mixtures, both better adapted to the changed growing conditions of a globally modified climate) and timing management being especially important.

Grazing – Some Effects on Botanical Composition and Succession

The key role of grazing in influencing the competitive environment and thus the botanical composition of and plant succession in both natural and artificial grasslands is discussed by a number of contributors (see Kemp and King, Chapter 5; Wolfe and Dear, Chapter 7; Skarpe, Chapter 9; Garden and Bolger, Chapter 11, this volume). Species that dominate the vegetation in native grassland communities are well adapted to the prevailing conditions, including the characteristically infrequent and lenient defoliation, e.g. of marsupials on the tablelands of south-east Australia or nomadic large mammals on the savannahs of Africa. The adaptation strategies used by these dominant species differ from one native grassland type to another. For instance, the most common plants in the native grasslands of the tablelands were tufted grasses, such as species of *Themeda*, *Stipa* and *Poa*, all relying on seed regeneration for their continuing presence. The survival strategies of characteristic savannah species range from those plants found in good growing conditions and able to produce new regrowth rapidly after defoliation to those with a high proportion of their resources underground and thus unavailable to browsing or grazing animals or with well-developed physical or chemical defences (see Skarpe, Chapter 9, this volume).

The effect of domestic animals has been broadly similar on both these native grasslands. There seems no doubt that the increased intensity of grazing, following the introduction of cattle and sheep on to many natural grasslands of temperate Australia, was the main catalyst for the changes in the competitive relationships and succession in their botanical composition which occurred in the 19th and 20th centuries. The original grasses were unable to withstand the more intensive grazing of domestic ruminants. The perennial exotic grasses and white clover, which replaced the native plants in improved pastures, were generally poor seed regenerators, which performed well provided soil fertility remained high and moisture reasonably plentiful. However, they are much more susceptible to drought than the natives. White clover, particularly sensitive to dry conditions, was usually the first species to die out, 'drying up' the supply of N, which is pivotal to high pasture production. Such poor growing conditions meant less available herbage and thus increased grazing pressure, further contributing to the dying out of the introduced plants (see Wolfe and Dear, Chapter 7, and Garden and Bolger, Chapter 11, this volume). Similarly, in savannah grasslands, the introduction of domestic animals has been shown to reduce the regeneration of desirable species, reduce species diversity and, in extreme cases, result in vegetation dominated by annuals (see Skarpe, Chapter 9, this volume).

A few of the native perennial grasses tolerant of the dry, lower-fertility conditions of the tablelands, notably members of the genera *Microlaena* and *Danthonia*, have agronomically desirable features – including a long growing season and better quality than the C_4 species that dominated the natural grasslands. The potential value for Australian pastures of grasses such as *Microlaena* is only now becoming widely appreciated.

As long ago as the early 1930s, in a series of seminal papers, Martin Jones (1933a, b, c, d) demonstrated the considerable influence which the timing and intensity of grazing could have on the botanical composition of differing UK pastures. By resting valuable species during their most vulnerable growth stages (early season and new regrowth for perennials) and intensively defoliating undesirable plants at such times (e.g. at heading for annual weeds), he showed that: (i) the competitive balance between such plants in the pasture could be changed drastically; and thus (ii) it was possible to develop systems

of management to achieve a desirable botanical composition. These findings are just as significant today for developing good systems of pasture management as they were almost 70 years ago. Yet any system of good management, needed to achieve the most desirable botanical composition of pastures, requires some understanding of the competitive interaction between grassland species and an appreciation of any differences in their growth cycles.

Seed Mixtures and Cultivar Evaluation

Although it may be argued that there is too much emphasis on the reseeding of long-lived grassland and too little on appropriate management to maintain production (see Clark, Chapter 6, this volume), seed mixtures remain a very important consideration. In addition to their use for pastures, they are increasingly required for environmental and amenity purposes. The selection of plants for inclusion in seed mixtures can be considered in the context of both intra- and interspecific competition. The objective of sowing a seed mixture for grassland is to obtain a predetermined botanical composition of a sward as the basis for high output. However, not only do unwanted plants invade the pasture, but there is evidence that other rapid changes may occur, in both the species (see Clark, Chapter 6, this volume) and the genotypes (see Harris, Chapter 8, this volume) which may be present.

In considering simple and complex seed mixtures, Harris argues for the need to select seed of plants that best fit the growing conditions, cover the ground quickly to prevent weed ingress and provide adaptation to the inevitable soil heterogeneity. Some complexity in the mixture is said to increase botanical diversity and improve the opportunity for sustaining grassland output. It is certainly a fact that those permanent pastures in England that are prized for producing fat cattle contain a fairly large number of species. Few critical studies have been undertaken to quantify the comparative effects of simple and complex mixtures on the botanical composition, output and stability of pastures. Further investigations – involving a range of mixtures and sites, monitored over a number of years – are needed to improve our knowledge in this field.

In recent years, there has been a major increase in the number of pasture plant cultivars available commercially, a development attributable, in part at least, to plant breeders' rights (PBR) legislation. Payment of royalties for cultivars accepted as being 'distinct, uniform and stable' (DUS) has resulted in more emphasis being put on breeding cultivars that can pass the DUS test and less on selecting plants with distinct agronomic advantages. The genetic variation of proprietary cultivars has been reduced and, with it, the range of their adaptation, when compared with the ecotypes and cultivars available before the PBR legislation. Harris points to the dangers of the indiscriminate use of proprietary cultivars in seed mixtures. Increasing pressure from those with vested interests to include a number of such cultivars of one species in seed mixtures can result in sowing plants with almost identical growth requirements; competition is thereby increased, often to the detriment of the ecological or agronomic advantages flowing from the selected use of more broadly based cultivars.

The evaluation of pasture plant cultivars, as currently practised, is far from perfect. The methods most commonly used – basically the prediction of field performance of such cultivars from small-plot data – are justifiably criticized by a number of contributors (see Kemp and King, Chapter 5; Clark, Chapter 6; Wolfe and Dear, Chapter 7; Harris, Chapter 8, this volume). Deficiencies of current testing methods include differences between small plots and grassland communities in their competitive environment, and an undue emphasis on herbage dry-matter production in evaluating performance. Further, citing evidence of a poor correlation between herbage yield and animal production, in both cultivars and species, Clark (see Chapter 6, this volume) argues the need to develop field-scale or landscape-based protocols for species and mixtures to measure performance in a meaningful way.

This proposal has merit. There can be no doubt that cultivar evaluation must be determined ultimately on measurements that are appropriate; animal production data are normally the most relevant measure in an agricultural context. However, there are dangers in accepting Clark's proposition uncritically. Not only would it be very expensive to use as a routine testing procedure, but the interpretation of the results could be difficult, with differences, for example, in grazing management or fertility level, perhaps overriding relative cultivar performance. There remains an important role for laboratory and small-plot testing to determine differences between pasture plant cultivars in, for example, quality, the presence of toxic substances, compatibility with other species and response to grazing.

The Last Word

All successful researchers can argue, with conviction, the need for further work in their field of interest. This principle is clearly evident in the contributions to this publication, which include a raft of recommendations for more studies on topics with relevance to competition and succession in pastures. It would be presumptuous to attempt to pass judgement on the relative importance of such proposals to increase our knowledge of competition and succession in grassland. However, improved understanding is fundamental for achieving and maintaining the optimal botanical composition of many of the world's grasslands, as well as developing systems for managing them properly.

There are some topics that obviously require further work. These include the need to investigate the interrelationship between abiotic factors, biotic factors and competition and subsequently succession in determining the botanical composition of communities. In relatively undisturbed areas, climate, particularly temperature and rainfall, and the inherent physical and chemical conditions of soils provide the main background for competition. Investigating the relative importance of such abiotic factors and competition in natural pasture communities could help resolve the question of just how important the latter is in determining the botanical composition in these generally resource-poor conditions.

In contrast, in 'improved' pasture, found typically in more productive environments, it is the management decisions taken by humans – fertilizers applied, plants introduced and the grazing regime practised – which influence the competitive environment and thus the plants that are able to survive and perform well under the changed growing conditions. The precise effects and interrelationships of such management factors warrant further study. In addition, invertebrate herbivores deserve more attention. Although they contribute so much to the biomass of improved pasture ecosystems especially, their effect on competition in and production from such grasslands is at present little understood.

It goes without saying that, whether seeking to improve understanding or attempting to solve a more practical problem, the researcher needs to formulate a precise question and select the methodology most suited for its investigation. While a great deal of work has been put into investigating various aspects of competition and succession, few long-term studies have been undertaken. Such experiments are vital to provide worthwhile data on many important questions and issues. These include the study of changes in the botanical composition of native pastures, and providing reliable information on any possible association between plant type and ecosystem function and on the conflicting claims of a decline in the legume content and productivity of improved pastures in Australia. Studies extending over many years are also needed to help unravel the complexity of any effects on competition and succession in pastures associated with global change. Unfortunately, investigations of lengthy duration do not fit well with the present research funding, which, in Australasia at least, is weighted heavily in favour of short-term studies.

There can be no doubt that improved understanding of competition, involving a major effort to investigate the processes involved, should be a priority research objective. Work so far includes investigations on: (i) the differential effects of a number of environmental factors on some physiological processes of competing plants; and (ii) the role of these processes in giving some plants a competitive edge in specific growing conditions. Nevertheless, this field, which is so important to the ecologist and agronomist alike, still contains many important challenges for the researcher.

Other topics that are highly relevant to competition and succession in pastures and deserving of considerably more work include the interrelated fields of soil heterogeneity, niche occupation, botanical diversity and stability. Further investigation of these fields should provide data of considerable value to the ecologist and agronomist, for both the individual pasture and the wider ecosystem. In addition, such data could well shed more light on the selection of seed mixtures and cultivar evaluation. More information is also needed on the extent

to which, and under what conditions, competition and succession may be limited by an inability to recruit suitable species.

A number of agricultural and environmental goals – including stability and productivity – need to be incorporated in any management systems developed for long-lived pastures. It follows that agricultural and environmental goals should become more coincident, and there should be increasing collaboration between agronomists and environmentalists in research and development (Fig. 15.1). Prevention of further land deterioration from overgrazing, salinization and erosion and reversal of such effects of bad management are all objectives in both agriculture and conservation; the conservationist should also accept the need, in land use systems, for grasslands to be managed at a level that provides a viable living for the grazier.

Fig. 15.1. Grassland in the Kingdom of Jordan, comprising a mixture of grasses and native flowers, including the striking black iris (*Iris nigricans*). This picture symbolizes the need for a combined conservation/agricultural approach to grassland management in order to both use and protect these resources effectively.

References

Cook, S.J., Lazenby, A. and Blair, G.J. (1978) Pasture degeneration. 1. Effect on total and seasonal pasture production. *Australian Journal of Agricultural Research* 29, 9–18.

Harris, W. and Lazenby, A. (1974) Competitive interaction of grasses with contrasting temperature responses and water stress tolerances. *Australian Journal of Agricultural Research* 25, 227–246.

Jones, Martin G. (1933a) Grassland management and its influence on the sward. I Factors influencing the growth of pasture plants. *The Empire Journal of Experimental Agriculture* 1, 43–57.

Jones, Martin G. (1933b) Grassland management and its influence on the sward. II The management of a clovery sward and its effects. *The Empire Journal of Experimental Agriculture* 1, 122–128.

Jones, Martin G. (1933c) Grassland management and its influence on the sward. III The management of a grassy sward and its effects. *The Empire Journal of Experimental Agriculture* 1, 223–234.

Jones, Martin G. (1933d) Grassland management and its influence on the sward. IV The management of poor pastures. V Edaphic and biotic influences on pastures. *The Empire Journal of Experimental Agriculture* 1, 361–367.

Index

_____

Bold refers to the five sections in this index: botanical composition; competition; grass–legume relationships; plant species; and succession.

Botanical Composition
Annual cycles *see* **Succession**

Biodiversity 91
Botanically diverse grassland communities 109, 110, 111

Co-existence 51, 204, 234
Complementary exploitation of habitat 227, 314
Complexity of mixtures 104, 106, 107, 108, 109, 110
    see also Diversity

Definition 2
Diversity
    genetic 51, 113, 168
    landscape level vs. small plots 109, 313
    species 11, 103, 112, 113, 142, 177, 185, 203, 227, 238, 263, 264
Dominance 214, 216, 227, 245, 266

Ecological sustainability 85, 111, 226–228
Environmental heterogeneity 109, 110, 111, 165

Grazing-tolerant species *see* **Succession** (Factors affecting succession, biotic)

Hard-seededness 136

Instability 222, 228
    see also Stability

Legume decline *see* **Succession**

Manipulation by seeds mixtures 312, 313
    for early ground cover, weed suppression 151, 153
    for environmental matching 154, 155
    genetic base 153, 168
    multipurpose, versatile 151, 152, 156
    for recreating botanically diverse grasslands 275, 278
    seed rates 156, 157
Mass ratio hypothesis 226

Natural selection 168, 170
'Naturalized' state 85
Niche(s) 8, 109, 110, 204, 275
    mosaic, diversity 110
    partitioning 161, 204

Pasture decline, degeneration 96, 120
    see also **Succession**
Pasture species composition matrix *see*
    Competition (Methodologies for study of); **Succession** (Models)
Patchiness, heterogeneity
    of soil conditions, resources 110, 111, 197, 204, 227, 234, 264
    of species 8, 227
Perennial grasses
    effect on hydrological balance 130, 131
    persistence 132, 133
Plant recruitment *see* **Succession** (Recruitment of species)

Resilience 10, 132, 133
    see also **Succession**
Ruderal species 135
    see also **Succession**

Species decline 110, 139
    see also Legume decline; Pasure decline
Stability 10, 103, 104, 107, 132, 133, 222, 225,
    309, 314
    see also **Succession** (Stability of ecosystem)
Stress tolerating species 5, 133, 134, 136
    see also **Succession** (Stress tolerators)

Target grass composition 96

Weed invasion 97, 115
    see also **Succession**

Competition
Agronomic implications 307–308
Apparent 9
Asymmetric 5, 6, 88

Competitive
    ability 5, 23, 194, 202
    balance 9
    effects, effects ability 4, 21, 193, 197,
        202
    equilibrium 6
        see also **Grass–Legume Relationships**
        (Dynamic equilibrium)
    exclusion 51, 154 177, 185, 249
    hierarchy 177, 181, 182, 184, 186, 194
    intensity 6, 162, 170, 239, 245, 249, 250,
        270
    outcomes 6, 21, 89, 90, 220, 233
        coexistence 8, 51
        dominance 6, 8, 9, 43, 139, 159, 184,
            214, 245
        equilibrium (stable, final) 5, 16
        population density effects 138, 139
        suppression 157, 159, 248, 277
    response, response ability 5, 193, 194, 195,
        196
        to resource heterogeneity 196, 234
        to soil nutrient patchiness 196
Competitiveness see Competitive, ability

Definitions, concepts 3, 15, 16, 159, 160, 161,
    162, 177, 178, 193, 194
Density dependence 22, 238

Ecological implications 248–251, 308–309

Factors influencing competition
    allelopathy 140
    canopy development 48, 65, 66, 67, 68, 69,
        77
    climate change 311
        CO_2 increase 235–239
        O_3 decrease 239–241
        rainfall 245–248
        temperature 238, 241–245
    compensatory growth 178, 179, 180,
        181
    defoliation 47, 48, 53
        see also grazing management
    disturbance agents 234
    environmental adaptation 52, 156
        moisture 50, 64, 65, 245, 246, 248
        soil factors, soil conditions 53–58
        temperature 53, 161
    environmental stress 64, 65, 197
    filter effect 226
    genotype × environmental interaction 8,
        50–52, 156, 163, 164, 200, 244
    grazing management 155, 167
    hard-seededness 136
        see also **Botanical Composition**; seed
        bank
    herbivory, herbivore pressure 179, 180, 181,
        182, 183, 184
        defences against 180, 181, 202
    nutrient availability 179
    pH 55, 133, 134
    physiological adaptation
        C_3 vs. C_4 plants 161, 244, 245
            see also plant traits, physiological
    plant characters 87
        see also plant traits
    plant traits 9, 186, 202, 226, 306
        agronomic 157, 158, 164
        morphological 9, 44–50, 88, 195
        physiological 9, 44–50, 161, 195
    pulses of nutrients 196, 276
    seed bank 270, 285, 286
    standing crop 193, 198, 200

stolon development *see* **Grass–Legume Relationships** (Stolons, stolon networks)
stress gradients 193, 199
trade-offs 225
treading 71, 73, 77
Financial implications 283
Frequency dependence 22, 39

Indices of competition 21
   competitive ability 25, 27, 87
   competitive effect, relative competitive effect 24, 25, 29
   competitive intensity, relative competitive intensity 6, 16, 23, 24, 198, 199, 200, 234
   crowding coefficient (relative crowding coefficient) 7, 56, 57, 287, 289
   outcome of competition 27
   relative replacement rate, relative yield ratio 7, 8, 54
   relative reproductive rate 28, 29
   relative yield, relative yield total 7, 26, 288, 289, 290
   resource complementarity 25, 26, 29
Interspecific 16, 26
Intraspecific 16, 26
Invasion of gaps or established swards *see* **Succession**

Light, competition for 2, 67, 176

Measurement of *see* Indices of competition; methodologies for study of
Methodologies for study of 29, 30, 305, 307, 313
   experimental design
      additive design 34, 36, 37
      community density series 201
      cross sectional design 34
      pairwise mixtures 37
      removal experiments 31
      replacement series 7, 34, 36
      response surface 32, 33
      stress-gradients 199
      target–neighbour approach 31, 37
      transplant experiments 31, 199
   field studies *see* removal experiments and transplant experiments
   grass–legume competition 37–39

see also **Grass–Legume Relationships** (Competition and Coexistence)
   mechanistic framework 4
multivariate statistical techniques 89
pasture species composition matrix 89, 90, 92, 94, 95, 96, 97
   see also **Botanical Composition; Succession** (Models)
Models, concepts of 310, 311
   analytical model 46
   CSR theory 87
   de Wit competition model 19, 20, 287, 288
   framework for predicting effect of climate change 233, 234
   general yield loss function 88
   logistic curve of population growth 16
   Lotka–Volterra equations 16, 18, 19, 46
      stable coexistence 18
      yield–density relationships 19, 20
      zero isocline 17, 27, 28
   mechanistic models 88
   resource reduction model, resource ratio hypothesis, R* 5, 88, 194, 225
   resource space 159
   simulation models
      comparison with mathematical models 300–301
      competition in mixed farm situations 283, 293–300

Nutrient deficient soils, competition in 275, 276
Nutrients and water, competition for 176

Responses to competition, phenotypic plasticity 43, 234
   see also Competitive, response
Roots, competition between 5, 201

Yield–density relationships 19

Grass–Legume Relationships
Clover crashes 71–77
Clover patchiness 73
Competition and Coexistence 309–310
Cyclical fluctuations in clover content 72, 74

Dynamic equilibrium 8, 39, 238

Factors influencing
    canopy development, leaf area distribution
        65–68, 69, 77
    defoliation, defoliation height 71, 77, 155
    grazing methods, patterns, rates 69, 70, 71,
        72, 73, 77, 78
    moisture 64, 65
    N addition, transfer 39, 66, 78
    N cycling 8, 74, 78
    N₂-fixation 75, 76
    pests and diseases 76, 77
    petiole length 48, 66, 72, 137, 138, 140
    red, far-red light 66
    selective grazing 73, 74
    silage cuts 71, 72, 77
    temperature 64, 65
        see also **Competition** (Factors influencing
            competition, climate change)
    treading 71, 73, 77
    urine, urine scorching 71, 73, 77
Feed budgeting system 77

Genetic shifts *see* **Succession**

Legume decline 121, 125, 140

Models 68, 74

Space, resource space 159, 160, 161
Stolon dieback 70, 78
Stolons, stolon networks 69, 70, 71, 77
Sward patchiness, heterogeneity 73
Synergy between grass and legume
    relative yield total 7, 26, 288, 289, 290
    yield increase 288, 290

Tillers 70, 78

Plant Species
Grasses
Agropyron (wheat grasses)
    cristatum (crested wheat grass) 200
    desertorum (crested wheat grass) 46, 111

repens (couch grass, quack grass) 111
smithii 204
spicatum 111
Agrostis (bent grasses) 104
    capillaris (browntop, common bent) 158
    tenuis (browntop) 55, 110
Amphibromus scabrivalis 196
Anthoxanthum odoraturn (sweet vernal) 201
Aristida ramosa (three awned spear grass) 122,
    215, 220
Arrhenatherum elatius (tall oat grass) 110, 196,
    203, 266, 306
Austrodanthonia (wallaby grasses) 215, 216, 217,
    218, 219, 220, 221, 222, 223, 225, 228
    bipartita (syn. *Danthonia linkii*) 223
    racemosa 220, 221
    see also Danthonia
Austrostipa (spear grasses) 214, 215, 216, 220
    see also Stipa
Axonopus compressus (carpet grass) 182

Bothriochloa
    ambigua (red grass) 122, 123, 244
    macra (red grass) 53, 215, 220, 221, 224,
        308
    saccharides 203
Bouteloua gracilis (blue grama grass) 200, 204
Brachypodium pinnatum 266
Bromus 138, 214, 224
    hordeaceus 92
    mollis (soft brome) 64, 264
    rigidus 64
    tectorum 46, 204

Cenchrus biflorus 57
Chloris truncata (windmill grass) 123, 217
Cynodon plectostachyus 181

Dactylis glomerata (cocksfoot, orchard grass) 3, 44,
    48, 55, 91, 94, 112, 113, 132, 134, 214,
    244, 263, 266
Danthonia (wallaby grasses) 122, 123, 152
    see also Austrodanthonia
Dichanthium sericeum (Queensland bluegrass) 4,
    220
Digitaria eriantha (digit grass, finger grass) 52, 53,
    162, 179, 182, 244

Elytrigia repens (couch grass) 55
Enneapogon nigricans 217
Eragrostis curvula (African lovegrass) 48

Festuca 203
  arindinacea (tall fescue) 44, 48, 64, 132, 154, 214, 235, 244
  nova-zelandii (fescue tussock) 160
  ovina (sheep's fescue) 110, 196, 263
  rubra (red fescue, chewings fescue) 106, 152

Glyceria fluitans 266

Hemarthria altissima (limpo grass) 161
Holcus lanatus (Yorkshire fog) 111, 112
Hordeum leporinum (barley grass) 44, 64, 123, 124, 129, 138, 224
Hyparrhenia rufa 182

Joycea pallida (syn. *Chionochloa pallida*) 215

Lolium
  multiflorum (Italian ryegass) 164, 167, 168
  perenne (perennial ryegrass) 49, 50, 51, 53, 55, 63–78, 87, 91, 104, 105, 107, 110, 111, 112, 129, 132, 134, 153, 154, 156, 158, 161, 165, 166, 167, 196, 214, 235, 237, 238, 244, 245, 263, 266, 276, 284, 308
  perenne × multiflorum (H1, short rotation ryegrass) 152, 155, 158, 165, 166
  rigidum (annual ryegrass, Wimmera ryegrass) 44, 63, 94, 95, 123, 129, 130, 138, 214

Microlaena stipoides (weeping meadow grass, weeping rice-grass) 124, 215, 216, 217, 218, 220, 221, 222, 223, 228
Molinia caerulea 46, 47, 263, 306

Nasella trichotoma (senated tussock) 127, 224

Panicum
  effusum 123, 217
  maximum (guinea grass) 182
  maximum var. *trichoglume* (green panic) 58

Paspalum
  dilatatum (paspalum) 50, 53, 152, 161, 244, 246
  virgatum (seaside paspalum) 182
Pennisetum clandestinum (kikuyu grass) 161
Phalaris
  aquatica (phalaris, harding grass) 3, 55, 64, 91, 96, 97, 98, 121, 129, 132–134, 156, 214, 221
  arundinacea (reed canary grass) 44
  coerulescens 44
Phleum pratense (timothy) 152
Poa 214, 215
  alpina 238
  caespitosa 123
  compressa (Canada bluegrass) 112, 199
  labillanderi (poa tussock) 122, 127, 222
  pratensis (Kentucky bluegrass, smooth stalked meadow grass) 104, 106, 110, 113, 134, 199
  sieberiana 220
  trivialis (rough stalked meadow grass) 160

Schizachryum scoparium (little bluestem) 195, 198, 203
Setaria anceps (setaria, golden timothy) 56
Stipa (spear grasses) 122
  aristiglumis (plains grass) 122
  falcata (spear grass) 123
  leucotricha (Texas winter grass) 203
  see also Austrostipa

Themeda triandra (kangaroo grass) 122, 123, 179, 214, 215, 216, 220, 221, 222, 224, 225, 228

Vulpia 6, 96, 123, 124, 129, 135, 138, 139, 214, 224
  bromoides (silvergrass) 92, 95, 97
  fasciculata (silvergrass) 6
  myuros (silvergrass) 64, 95, 97

Legumes
Alysicarpus ovalifolius 57

Desmodium intortum (greenleaf desmodium) 56

Lotus
    corniculatus (bird's-foot trefoil) 105
    pendunculatus (greater lotus) 55
    uliginosus 152

Medicago 125, 129, 141, 142
    lupulina (black medic) 264
    minima (woolly burr medic) 123
    murex (murex medic) 142
    polymorpha (burr medic) 123
    sativa (lucerne/alfalfa) 3, 6, 50, 52, 53, 87,
        129, 134, 135, 140, 162, 244
    truncatula (barrel medic) 5, 11, 121, 288

Neonotonia wightii (glycine) 58

Ornithopus compressus (serradella) 129

Trifolium
    glomeratum (cluster clover) 123, 214
    hirtum (rose clover) 9
    michelianum (balansa clover) 129, 142
    pratense (red clover) 152, 155
    repens (whiteclover) 3, 49, 51, 55, 63–78, 94,
        104, 105, 107, 111, 113, 125, 154, 155,
        156, 158, 161, 163, 168, 214, 237, 238,
        248, 276
    resupinatum (Persian clover) 129
    subterraneum (subterranean clover) 3, 48, 63,
        64, 65, 92, 94, 96, 120, 122, 125, 129,
        135–140, 142, 152, 214, 284

Other species
Acacia
    karoo 180
    tortilis 181
Amaranthus retroflexus 45, 49, 50
Arctotheca calendula (capeweed) 64, 123, 124, 129,
    138, 214
Artemesia tridentata (sagebrush) 111

Calluna vulgaris (ling heather) 46, 306
Caltha palustris 266
Carthamus lanatus (saffron thistle) 129
Chenopodium album (fat hen) 49, 50
Chicorium intybus (chicory) 3, 94, 95, 244

Chondrilla juncea (skeleton weed) 129
Cirsium vulgare (spear thistle) 94

Echium plantagineum (Salvation Jane, Paterson's
    curse) 92, 94, 95, 123, 129
Erica tetralix (cross-leaved heath, bog heather) 46,
    47
Erodium botrys (erodium) 64, 123, 138

Galium
    hercynicum (syn. *Galium saxatile*) (heath
        bedstraw) 54
    pumilum (syn. *Galium silvestre*) (slender
        bedstraw) 54

Hieraceum floribundum (hawkweed) 199, 200
Hieraceum spp. (hawkweeds) 160

Iris nigricans (black iris) 314

Oxalis pes-caprae (soursob) 124

Polygonum aviculare (knotgrass, knotweed) 123
Primula
    elatior (oxlip) 54
    officinalis 54

Rumex 214
    acetosella (sheep's sorrel) 161
    crispus (curled dock) 276

Silybum marianum (variegated thistle) 124, 214

Taraxacum officinale (common dandelion) 111,
    113, 214

Succession
Annual cycles 91
Annualization 221, 223, 224
Arrested successional development 93
Botanical change, progression 85, 123, 124, 126,
    185, 215, 216, 217, 222, 250, 251

Carbon-nutrient hypothesis 181
Chemical defences 180
Climax vegetation 9, 10, 122, 124

Defence mechanisms 180, 181, 202
Definitions 9, 103, 127
Disclimax communities 122
Dominance
    factors affecting 161, 162, 183
    by legumes 124
    by non-legumes 124, 245
    seasonal shifts 161, 162
    shifts due to episodic events 183
    see also **Botanical Composition**; **Competition**
        (Competitive, outcomes)

Equilibrium, dynamic equilibrium 10, 103
    see also **Competition** (Competitive,
        outcomes)

Factors affecting succession
    abiotic
        episodic climatic events 127, 219, 220
        seasonal climatic conditions 94, 219
        soil conditions (fertility) 123, 225, 272,
            273
        soil nutrient patchiness, resource
            patchiness, heterogeneity 110,
            197
    biotic
        grazing tolerance, intolerance 86, 123,
            178
        herbivory 23, 183, 184, 185, 186, 218,
            225
        see also **Competition**
        nutrient cycling 184
        persistence 93, 130, 132, 133, 167
        resilience 124, 132, 133
        seed bank 270, 271
    management 169
        cultivation (renovation) 98, 223
        fertilizers 124, 221 222, 225
        grazing 123, 225, 264, 271, 311–312
            continuous vs. rotational 69, 70, 77
            crash 93
            deferment 95
            frequency 95

pressure, intensity 127, 185, 219
stocking rate 10
timing 93
herbicides 96, 97, 98
Following fire, forestburns 98, 152, 222–223

Genetic shifts 51, 164, 165
Genotype × environment interaction 163, 164

Invasion of gaps or established swards 2, 97, 115,
    155, 223, 263

Legume decline 121, 125, 140
Long-term outcomes 314
    methodologies 104, 105, 106
    multivariate statistical techniques 10, 89

Models 121, 122
    Clementsian model, range succession model
        10, 122
    CSD model, CSR theory 122, 127, 128,
        263
    pasture species composition matrix 89, 90,
        92, 94, 95, 96, 97
    state and transition model 10, 11, 90, 122,
        124, 126
Multidimensional scarcity-accessibility hypothesis
    181

Natural succession processes vs. assisted/managed
    processes 166, 274, 275

Pasture decline, degeneration 96, 120, 127, 216,
    218, 221, 222
Perennial : annual ratio 218, 225
Plagioclimax 261
Plant apparency hypothesis 180

Recruitment limitation 224, 226
Recruitment of species 221, 225, 234, 266
Resilience 10, 124, 132, 133
Resource availability hypothesis 180
Ruderal species (opportunists) 6, 135, 264

Stability of ecosystem 91, 142
States/phases of vegetation
 affecting invasions, recruitment 270, 271,
  272
 boundaries 90
 communities, community structure 201, 202,
  224
 desirable boundary limits 98
 reversion, regression 177, 226
 stable pasture phase 222, 225
 thresholds 10, 185, 226
Stress tolerators 128
Successional trajectory 278
Switches, transitions between plant community or
 vegetation types 11, 161, 162, 184

Vegetation management
 for creation, restoration of botanically diverse
  grasslands 261, 265, 270, 275
  cut and clear 267
  direct sowing of species 274, 275, 278
  fertilizer applications 265
  grazing 277, 278
  top soil removal 268
  watertable manipulation, hydrological
   balance 130, 131, 266, 269
 for ecosystem function 226, 227, 228
 for multiple use 227, 314
 for sustainability 226, 227, 228

Weed invasion 97, 115

Browse Read and Buy

www.cabi.org/bookshop

ANIMAL & VETERINARY SCIENCES
BIODIVERSITY CROP PROTECTION
HUMAN HEALTH NATURAL RESOURCES
ENVIRONMENT PLANT SCIENCES
SOCIAL SCIENCES

 CABI *Publishing*
A division of CAB International

 Online BOOK SHOP

Subjects

Search

Reading Room

Bargains

New Titles

Forthcoming

Order & Pay Online!

 MasterCard

VISA

AMERICAN EXPRESS

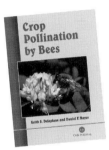 Crop Pollination by Bees
Keith S. Delaplane and Daniel F Mayer

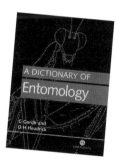 A DICTIONARY OF Entomology
G Gordh and D H Headrick

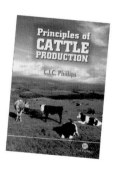 Principles of CATTLE PRODUCTION
C.J.C. Phillips

 Seeds THE ECOLOGY OF REGENERATION IN PLANT COMMUNITIES 2ND EDITION
Edited by Michael Fenner

 FULL DESCRIPTION   BUY THIS BOOK   BOOK OF THE MONTH

Tel: +44 (0)1491 832111 Fax: +44 (0)1491 829292